Contemporary Issues in Exploratory Data Mining in the Behavioral Sciences

This book reviews the latest techniques in exploratory data mining (EDM) for the analysis of data in the social and behavioral sciences to help researchers assess the predictive value of different combinations of variables in large data sets. Methodological findings and conceptual models that explain reliable EDM techniques for predicting and understanding various risk mechanisms are integrated throughout. Numerous examples illustrate the use of these techniques in practice. Contributors provide insight through hands-on experiences with their own use of EDM techniques in various settings. Readers are also introduced to the most popular EDM software programs. The website http://mephisto.unige.ch/pub/edm-book-supplement/ contains color versions of the book's figures, a supplementary paper to Chapter 3, and R commands for some chapters.

The results of EDM analyses can be perilous—they are often taken as predictions with little regard for cross-validating the results. This carelessness can be catastrophic in terms of money lost or patients misdiagnosed. This book addresses these concerns and advocates the development of checks and balances for EDM analyses. Both the promises and the perils of EDM are addressed.

Editors McArdle and Ritschard taught the "Exploratory Data Mining" Advanced Training Institute class for the American Psychological Association (APA). All contributors are top researchers from the USA and Europe. Organized into two parts—methodology and applications—the techniques covered include decision, regression, and SEM tree models, growth mixture modeling, and time-based categorical sequential analysis. Some of the applications of EDM (and the corresponding data) explored include:

- selection to college based on risky prior academic profiles;
- the decline of cognitive abilities in older persons;
- global perceptions of stress in adulthood;
- predicting mortality from demographics and cognitive abilities;
- risk factors during pregnancy and the impact on neonatal development.

The book is intended as a reference for researchers, methodologists, and advanced students in the social and behavioral sciences including psychology, sociology, business, econometrics, and medicine, interested in learning to apply the latest exploratory data mining techniques to their work. Prerequisites include a basic class in statistics.

John J. McArdle is Senior Professor of Psychology at the University of Southern California where he heads the Quantitative Methods training program.

Gilbert Ritschard is Professor of Statistics at the University of Geneva in Switzerland and project leader at the Swiss National Center of Competence in Research LIVES.

QUANTITATIVE METHODOLOGY SERIES
George A. Marcoulides, Series Editor

This series presents methodological techniques to investigators and students. The goal is to provide an understanding and working knowledge of each method with a minimum of mathematical derivations. Each volume focuses on a specific method (e.g. Factor Analysis, Multilevel Analysis, Structural Equation Modeling).

Proposals are invited from interested authors. Each proposal should consist of: a brief description of the volume's focus and intended market; a table of contents with an outline of each chapter; and a curriculum vitae. Materials may be sent to Dr. George A. Marcoulides, University of California—Riverside, george.marcoulides@ucr.edu.

Marcoulides • Modern Methods for Business Research

Marcoulides/Moustaki • Latent Variable and Latent Structure Models

Heck • Studying Educational and Social Policy: Theoretical Concepts and Research Methods

Van der Ark/Croon/Sijtsma • New Developments in Categorical Data Analysis for the Social and Behavioral Sciences

Duncan/Duncan/Strycker • An Introduction to Latent Variable Growth Curve Modeling: Concepts, Issues, and Applications, Second Edition

Heck/Thomas • An Introduction to Multilevel Modeling Techniques, Second Edition

Cardinet/Johnson/Pini • Applying Generalizability Theory Using EduG

Creemers/Kyriakides/Sammons • Methodological Advances in Educational Effectiveness Research

Hox • Multilevel Analysis: Techniques and Applications, Second Edition

Heck/Thomas/Tabata • Multilevel Modeling of Categorical Outcomes Using IBM SPSS

Heck/Thomas/Tabata • Multilevel and Longitudinal Modeling with IBM SPSS, Second Edition

McArdle/Ritschard • Contemporary Issues in Exploratory Data Mining in the Behavioral Sciences

Contemporary Issues in Exploratory Data Mining in the Behavioral Sciences

Edited by John J. McArdle and Gilbert Ritschard

Routledge
Taylor & Francis Group

NEW YORK AND LONDON

First published 2014
By Routledge
711 Third Avenue, New York, NY 10017

Simultaneously published in the UK
By Routledge
27 Church Road, Hove, East Sussex BN3 2FA

Routledge is an imprint of the Taylor & Francis Group, an informa business

Library of Congress Cataloging in Publication Data
Contemporary issues in exploratory data mining in the behavioral
sciences / edited by John J. McArdle, University of Southern
California and Gilbert Ritschard, University of Geneva.
    pages cm. – (Quantitative methodology series)
1. Social sciences–Statistical methods. 2. Social
sciences–Research–Data processing. 3. Data mining–Social aspects.
I. McArdle, John J. II. Ritschard, Gilbert, 1950–
HA29.C746 2014
006.3'12–dc23   2013006082

ISBN: 978-0-415-81706-6 (hbk)
ISBN: 978-0-415-81709-7 (pbk)
ISBN: 978-0-203-40302-0 (ebk)

Typeset in Bembo
by Cenveo Publisher Services

Contents

About the Editors

John J. McArdle, Ph.D., is Senior Professor of Psychology at the University of Southern California U.S.A. where he heads the Quantitative Methods training program. For the past 30 years he has led graduate classes in topics in psychometrics, multivariate analysis, longitudinal data analysis, exploratory data mining, and structural equation modeling. His research has focused on age-sensitive methods for psychological and educational measurement and longitudinal data analysis. He has published on methodological developments and applications using factor analysis, growth curve analysis, and dynamic modeling of adult cognitive abilities. Dr. McArdle was recently awarded an NIH-MERIT grant from the NIH-NIA for his work on "Longitudinal and Adaptive Testing of Adult Cognition." He has also worked with the APA Advanced Training Institute on Longitudinal Modeling (2000–2011).

Gilbert Ritschard, Ph.D., is Professor of Statistics for the Social Sciences at the Department of Economics and Vice Dean of the Faculty of Economics and Social Sciences of the University of Geneva, Switzerland. For the past 25 years, Dr. Ritschard has carried out teaching and research in data analysis and statistical modeling for social sciences, including longitudinal data analysis, Event history analysis, and sequence analysis. He has also contributed for more than 12 years in the domain of Knowledge Discovery from Data (KDD), data mining, and text mining. His recent applied research involves data mining of event histories, especially the exploration of individual life trajectories. He has produced new insights into personal and occupational life courses in Switzerland. He has also published research on life courses in 19th century Geneva and on social dialogue regimes. He currently leads a methodological project within the Swiss NCCR "LIVES: Overcoming vulnerability, a life course perspective," and developed with his team the TraMineR toolbox, now used worldwide for the exploration of state and event sequence data.

About the Contributors

Thomas Augustin is Professor at the Department of Statistics at the Ludwig-Maximilians-Universität (LMU) Munich, Germany, and head of the Group "Foundations of Statistics and their Applications" there. His main area of research is statistical inference and decision making under complex uncertainty and non-idealized data, including measurement error modeling and imprecise probabilities.

Cindy S. Bergeman, Ph.D., is Professor of Psychology at the University of Notre Dame, Indiana, U.S.A. She is a lifespan developmental psychologist with research interests in resiliency and aging, behavioral genetics, and the theory–method interface.

Andreas M. Brandmaier, has a Ph.D. in computer sciences, and is now a research scientist at the Center for Lifespan Psychology in the Max Planck Institute for Human Development, Berlin, Germany. He specializes in alternative ways of doing Structural Equation Models (SEM) and alternative ways of doing machine learning. He has released several software packages for modeling and exploring data.

Reto Bürgin, M.Sc. in statistics, is a LIVES Ph.D. student at the Institute for Demographic and Life Course Studies of Geneva University, Switzerland. He is developing new tools for ordinal sequence data analysis mixing data-mining-based and statistical approaches.

Cees H. Elzinga, Ph.D., is Professor in Pattern Recognition in Discrete Data Structures, and Head of the Department of Sociology at the Free University of Amsterdam, the Netherlands. Dr. Elzinga studies kernel methods and probability mixture models for categorical time series as they arise in, e.g., life course research and labor market careers. He has published on both the theory and the application of such techniques.

Paolo Ghisletta, Ph.D., is Associate Professor of Psychology at the University of Geneva, Switzerland. His current work includes studying latent variable models and both linear and nonlinear mixed effects models applied to repeated-measures (longitudinal) short-term and long-term data. Application include mathematical modeling of neuro-logical risk and terminal decline in various samples of elderly individuals.

Kevin J. Grimm, Ph.D., Associate Professor, Department of Psychology, University of California at Davis, U.S.A., has studied all classes of multivariate statistical models, including latent change score models and latent mixture models. His substantive work has bridged early education in math and sciences and complex learning issues in adult development. His current statistical work includes analyses of individual behaviors over time, and the statistical separation of change trajectories.

Kelly M. Kadlec has a Ph.D. in psychology from the University of North Carolina at Greensboro, U.S.A.; she spent several post-doctoral years at the University of Southern California as an Assistant Research Professor, and is now doing a great deal of work from home in Irvine, Southern California.

Julia Kopf has a Diploma in sociology and a Master's in statistics. She is a Ph.D. student at the Department of Statistics at the Ludwig-Maximilians-Universität (LMU) Munich, Germany, interested in statistical method-ology for the social sciences. Her thesis research is concerned with differential item functioning in item response models.

Walter Leite is an Associate Professor in the Research and Evaluation Methodology Program, within the School of Human Development and Organizational Studies in Education Psychology, College of Education, at the University of Florida, U.S.A. His primary research interest is statistical modeling of data from longitudinal studies, with a particular interest in large longitudinal surveys. The overarching goal of his research is to advance the applicability of structural equation modeling (SEM) and multilevel modeling to longitudinal datasets.

Ulman Lindenberger, Ph.D., is Director, Max Planck Institute of Human Development, Berlin, Germany. Dr. Lindenberger has published research on the heterogeneity of cognitive aging, how risks accumulate over the full lifespan, and how these events lead to critical risks of mental, physical, and social deterioriation in later ages.

Lawrence L. Lo, M.A., is a graduate student at Department of Human Development and Family Studies, the Pennsylvania State University, U.S.A. Mr. Lo's primary interest is in dynamic systems models for developmental phenomena. He implements both structural equation modeling (covariance structure analysis) and Kalman filtering (recursive analysis) methods to address different kinds of scientific questions. Currently, Mr. Lo's focus is on developing and validating exploratory data analysis programs that automatically search out suitable models for researchers.

Fabio B. Losa holds a Ph.D. in economics from the University of Fribourg, Switzerland. He is currently Monitoring and Evaluation Specialist for the water sector at the African Development Bank in Tunis. He is an associate researcher at the Centre for Research in Economics and Management (University of Rennes 1, France). His main research interests include development, labor economics, and complex decisions.

George A. Marcoulides, Ph.D., is Distinguished Professor of Education, University of California at Riverside, U.S.A. Dr. Marcoulides is a leading contributor to machine learning and optimization analysis in behavioral sciences, include the Tabu and Ant Colony procedures. He is the editor of the leading journals, *Structural Equation Modeling* and *Educational and Psychological Measurement*.

Pau Origoni graduated in sociology at the University of Lausanne, Switzerland. He has mainly worked in public statistics, specializing in social economy and education. He is currently head of the Demography Unit at the Statistical Office of Canton Ticino, Switzerland.

Thomas S. Paskus, Ph.D., is Principal Research Scientist, National Collegiate Athletic Association, Indianapolis, U.S.A. For the past decade, Dr. Paskus has studied the risk of academic failures of college student athletes using national samples from the U.S.A.

Raffaella Piccarreta, has a Ph.D. in Statistics from the University of Trento and is currently research scientist at the Bocconi University in Milano, Italia. Her research areas are multivariate data analysis including classification and regression trees and she has produced several articles on sequence analysis.

Lindsay Pitzer holds a Ph.D. in human development and family studies from Purdue University, West Lafayette, Indiana, U.S.A. Dr. Pitzer is interested in aging, stress, and early childhood traumatic experiences, as well as statistical methodology for analyzing longitudinal data. She is currently a research analyst at the Western Psychiatric Clinic and Institute at the University of Pittsburgh Medical Center.

Carol A. Prescott, Ph.D., is Professor of Psychology, University of Southern California, Los Angeles, U.S.A. Dr. Prescott studies risk factors for substance abuse and related psychopathology from a behavioral genetic and molecular genetic perspective.

Nilam Ram, Ph.D., is Associate Professor, Department of Human Development and Family Studies, the Pennsylvania State University, U.S.A. Dr. Ram's research interests have grown out of a history of studying change. Dr. Ram received his undergraduate degree in economics, worked as a currency trader in New York, and moved on to the study of human movement, kinesiology, and eventually psychological processes—with a specialization in longitudinal research methodology. Generally Dr. Ram studies how short-term changes (e.g., processes such as learning, information processing, etc.) develop over the course of the lifespan and how intra-individual change and variability study designs (e.g., measurement bursts) might contribute to our knowledge base.

Stacey B. Scott, Ph.D., is a research scientist in the Center for Healthy Aging, at the Pennsylvania State University, U.S.A. Dr. Scott has been involved in exploratory data mining for several years, including in her graduate research at the University of Notre Dame, her continuing postgraduate work at the Georgia Institute of Technology, and now Penn State. Her interests are in understanding stress and well-being across the adult lifespan, with a focus on the contexts (e.g., developmental, historical, social-environmental) in which stressors occur.

Mariya P. Shiyko, Ph.D., is an Assistant Professor, Department of Counseling and Applied Educational Psychology, Northeastern University, U.S.A. Dr. Shiyko's primary research interests are in the area of intensive longitudinal data (ILD) design and analysis. Dr. Shiyko's research covers questions related to design of studies employing ecological momentary assessments, including sampling frequency and questionnaire development. In addition, she is interested in developing, validating, and applying statistical models that allow for a full exploration of ILD.

Carolin Strobl, Ph.D., is a Professor in the Department of Psychology of the University of Zurich, Switzerland. Dr. Strobl's research covers methods for unbiased variable selection and variable importance in recursive partitioning and new approaches for detecting violations of measurement invariance in educational and psychological testing.

Matthias Studer has a Ph.D. in socioeconomics from the University of Geneva, Switzerland. He is one of the authors of the TraMineR toolbox for mining sequence data. He is currently studying the life courses of socially assisted people.

Timo von Oertzen, has a Ph.D. in computer sciences, and is an Assistant Professor of Quantitative Methods in Psychology at the University of Virginia, U.S.A., and a research scientist at the Max-Planck-Institut für Bildungsforschung in Berlin, Germany. He is an expert in all kinds of statistical principles, including simplifying the many ways to create model identification in SEMs, and the use of Monte Carlo methods of simulation.

Brenda R. Whitehead has a Bachelor's degree in psychology and a Master's degree in developmental psychology. She is currently a Ph.D. candidate in the Developmental Psychology program at the University of Notre Dame, Indiana, U.S.A., where she is studying factors associated with health and well-being in midlife and older adults. Her dissertation investigates the biasing effects of trait and state affect on self-reported health assessments.

Keith F. Widaman, Ph.D., is Professor and Head of Psychology, University of California at Davis, U.S.A. Dr. Widaman studies risk factors during pregnancy and the neonatal period associated with the development of mental retardation using data collected throughout the U.S.A. He is the new editor of the leading journal, *Multivariate Behavioral Research*.

Yan Zhou has a Ph.D. in psychology from the University of Southern California, U.S.A., and is now Data Manager at the Alzheimer's Disease Archive databank at the University of California at Los Angeles.

Preface

In the summer of 1981, a Post-Doctoral fellow in Psychology from the University of Denver (John McArdle) was given the opportunity to work at the Département d'Econométrie at the University of Geneva with the well-known scientist and Emeritus Professor Dr. Herman O. Wold, who was then creating what has become the popular *Partial Least Squares* (PLS) approach to statistical modeling. The Post-Doc had never been to Europe and was eager to learn anything, but he soon learned that the language barrier between French and English was only surpassed by the language barrier between Psychology and Economics. However, one young Econometric Faculty member from Geneva (Gilbert Ritschard) decided to break down these barriers; he had intimate knowledge of both French and Econometrics, and he guessed that English and Psychology could prove to be an interesting adventure, so he taught the Post-Doc some finer aspects of the Genevian culture (e.g., how to correctly pronounce, "bonjour") and they did some work together.

In the summer of 2002, again in Geneva, the Psychology Post-Doc (now a Senior Faculty member) again met up with the young Econometrics Faculty member (now a Senior Faculty member) and asked what he was doing—Ritschard described his work as "longitudinal exploratory data mining of categorical data" and the Post-Doc found this both foreign and fascinating. But he was willing to learn this approach (which is documented herein) because it seemed it could provide a solution to several problems of longitudinal data that the Post-Doc had been considering. To tell the truth, the Post-Doc had used this kind of approach in his early computer work, and he wondered where it went astray (or so he assumed). This sequence of events led to this continued collaboration and the creation of this book.

The APA Workshop

But first there was a workshop. The methods we term *Exploratory Data Mining* (EDM), are a collection of procedures which are largely new algorithmic approaches to data analysis, with a bit of effort put into the statistical basis. Nevertheless, and sometimes even without a firm statistical foundation, EDM is now widely used in practical applications, such as in business research and marketing, as well as in medicine and genetics. In fact, it could be said that EDM dominates the statistical analysis used in these disciplines. A few years ago, we recognized that EDM was not taught in much detail in academic circles, or in polite company of any kind. This led us to propose its use to the *American Psychological Association* (APA) for their *Advanced Training Institute* (ATI) series. The first chance we had to illustrate these methods was at the "Exploratory Data Mining" sessions of the APA–ATI in 2009 and 2010 at the University of Southern California, and again at the University of California at Davis in 2011 and 2012. The EDM topics provided such a reasonable approach and such surprising success that we intend to host these kinds of meetings many times more.

The International Colloquium

In order to explore and learn about more complex EDM topics a small International Colloquium was held at Le Château de la Bretesche (Missillac, France). This meeting was completely supported with funds provided by the Albert and Elaine Borchard Foundation (through the University of Southern California). The colloquium was intentionally small, and included only 13 participants (the co-editors and most of the authors of the chapters listed below). The meeting was co-directed by the editors, John McArdle (University of Southern California) and Gilbert Ritschard (University of Geneva, Switzerland). The present book is the follow-up to this Colloquium.

The meeting provided an open forum for state-of-the-art discussion of recent methodological advances in EDM for the analysis of data from behavioral sciences, including psychology, sociology, and econometrics. In contrast to traditional hypothesis-driven approaches to analysis, EDM enables investigators to assess the predictive value of all possible combinations of variables in a data set. Among many popular slogans we heard about EDM, the term "knowledge generation" seemed to describe the process best. Indeed, EDM has emerged in recent years as a major area of statistical research and practice and is increasingly employed by behavioral scientists. A wide range of behavioral problems have recently been studied using EDM, including: (a) the genetic analysis of risky behaviors, (b) selection to college based on risky prior academic profiles, (c) the behavioral precursors of Alzheimer's disease, and (d) cultural/educational differences in women's labor participation. Among many other issues, these topics are presented in further detail here.

The Book

There is no doubt in our minds that EDM techniques are particularly promising in behavioral science research. EDM is actively being used for the analysis of individual risk assessment from very large data sets collected in clinical survey, psychometric, economic, genomic, sociology, demography, and life course research. These modern techniques are often used to follow-up classical multivariate analyses in cases in which investigators have: (1) obtained significant results and seek to know whether there are other important patterns in the data; (2) obtained no notable results and wonder whether there are any important patterns to be found; or (3) developed questions that are far too general or imprecisely formulated to be addressed through hypothesis testing.

On the other hand, there is also no doubt in our minds that the results of EDM analyses are often filled with perils. EDM results are quickly taken as serious prediction systems with little regard for the need for cross-validation or explicit checks on the forecasting of future events. The problems that are created can be catastrophic, e.g., in terms of money lost, students misclassified, or patients misdiagnosed with or without disease. It follows that some rudimentary checks and balances need to be a key feature of EDM analyses, and a set of more rigorous methodological guidelines may be required. We offer a few suggestions herein.

This book covers the conceptual bases and strategies of the newest techniques in EDM, including a review of leading techniques and software, such as those based on canonical regression models and on pattern searches with recursive partitioning (i.e., *Classification and Regression Trees*). The book covers Technical Features, Computer Programs, and Substantive Applications. The application chapters report contributors' experiences with their own data and research questions, while some of the methodological chapters introduce new original EDM techniques.

The contributors are scientists from the USA and Europe who are active and interested in the investigation of various aspects of the promises and perils of EDM. These scientists have been conducting cross-sectional and longitudinal studies examining a wide range of different kinds of risk factors that can be examined using EDM techniques. The topics range from impact of economic stimuli, the academic risk of selection of student athletes for college, or the biological risk for later life health problems. The overall goal of the book is to integrate methodological findings and conceptual models that may explain and guide more reliable techniques for predicting and understanding all kinds of risk mechanisms underlying many different behavioral outcomes. Both the promises and the perils of EDM are addressed in detail, with the hope that this will lead to directions for future research.

Intended Audience

This book is intended for a broad array of behavioral scientists, at the least from psychologists to economists, and possibly from medical students to marketing students. The book is divided into two Parts: on methodological aspects and substantive applications. Part I should help users to understand the aims and workings of the computer programs while Part II demonstrates how outcomes can be interpreted and can improve knowledge on research questions. These subject areas of EDM are relatively new, but we know EDM is being used in a wide variety of academic research when multivariate problems are posed but no standard solution is found to be acceptable—of course, this includes business, medicine, and genetics. Persons who read this book would necessarily have quantitative interests, but the substantive topics are so diverse, we expect EDM could be of wide interest.

Our book is intentionally different from others in the area in two ways: (1) we emphasize practical examples and relevant applications from behavioral sciences, and (2) we show some skepticism about the accuracy of the EDM techniques.

Structure of the Book (in Order of Presentation)

All presenters from the above colloquium were encouraged to write up their presentations. Two additional chapters were solicited by the co-editors. As stated, the current volume is organized in two parts, one on methodological aspects and a second one focusing on applications. The first part is mostly technical and includes nine chapters.

The first contribution by co-editor John McArdle serves as an introduction to EDM by using decision trees. This chapter is a general introduction and attempts to motivate exploratory data mining by demonstrating how EDM can be used in conjunction with classical confirmatory analysis. The chapter also gives a broad overview of how classification and regression trees work and stresses their scope for behavioral studies through a series of four application examples.

The second chapter by co-editor Gilbert Ritschard is titled "CHAID and Earlier Supervised Tree Methods." In this chapter, Ritschard reviews the development of supervised tree methods from the first attempts in the 1950s until the emergence of CHAID, one of the most popular p-value-based tree growing methods. The chapter describes the differences in different versions of CHAID in detail, and illustrates these principles with data on the first year success of students from the University of Geneva.

The next two chapters are about extensions of regression trees. First, Julia Kopf, Thomas Augustin, and Carolin Strobl provide a broad technical chapter titled, "The Potential of Model-based Recursive Partitioning in the Social Sciences—Revisiting Ockham's Razor." Here they give an introduction to "model trees" which are designed to split a data set into sub-groups with significant differences in some fitted regression model, i.e., trees with regression models in the nodes. As an illustration, the authors use the German Ecomomic Panel (SOEP) data from which they identify five groups in terms of additional covariates with clear different rates of return from education.

The contribution of Andreas Brandmaier, Timo von Oertzen, John McArdle, and Ulman Lindenberger is titled, "Exploratory Data Mining with Structural Equation Model Trees." This work proposes and illustrates a new extension of standard Structural Equation Models (SEMs) by including some uses of SEM with decision trees (termed SEMTrees). This, of course, is related to the longitudinal SEM approach described by McArdle in the first chapter, but it is a much more general statement. In this chapter SEMTrees are introduced and here the nodes contain whole SEMs for different measured groups. The scope of the method is demonstrated through a series of illustrative examples ranging from univariate SEMTrees to longitudinal curve models, confirmatory factor models, and even hybrid SEMTrees where the model specification may change across the nodes.

The fifth contribution by Gilbert Ritschard, Fabio Losa, and Pau Origoni titled "Validating Tree Descriptions of Women's Labor Participation with Deviance-based Criteria" presents both an application and technical aspects. The chapter reports a full scaled application of induction trees for non-classificatory purposes in socio-economics. This is an informative application on female labor market data from the *Swiss 2000 Population Census* (SPC). The technical aspects concern the validation of decision trees. The authors evaluate the goodness-of-fit of the decision tree models with "deviance" measures between the original data distribution and the expectations created from the decision tree model. They show how this can be accomplished, and how difficult this can be. They also suggest alternative deviance-based criteria for comparing decision tree models.

The next chapter, by George Marcoulides and Walter Leite, is titled "Exploratory Data Mining Algorithms for Conducting Searches in Structural Equation Modeling: A Comparison of Some Fit Criteria." This chapter does not use the tree approach but instead focuses on other important aspects of EDM. This contribution uses the framework of SEM but addresses the question of model specification by means of selection algorithms. More specifically, the chapter compares two important model selection methods, namely "Ant Colony" and "Tabu Search." The scope of the methods is illustrated by means of synthetic data which permit the authors to completely control the expected outcome and demonstrate the accuracy of the techniques.

Kevin Grimm, along with his co-authors, Nilam Ram, Mariya Shiyko, and Lawrence Lo, has contributed a new technical chapter titled, "A Simulation Study of the Ability of Growth Mixture Models to Uncover Growth Heterogeneity." Here the authors provide an excellent example of how we should evaluate our exploratory tools before using them extensively. These researchers examine the very popular SEM technique of *Growth Mixture Modeling* (GMM), a largely exploratory device for separating different classes of individuals who have been measured on multiple occasions. They simulate data for common situations, and they show just how difficult it is to use current programs to recover this known reality. In essence, they demonstrate conditions when GMM will not work as hoped, and it seems that this is a common problem that requires special attention.

The next two chapters of our methodological section are about dealing with time-based categorical sequential data.

The chapter by Raffaella Piccarreta and Cees Elzinga is a technical piece titled "Mining for Associations Between Life Course Domains." The authors explore different and original ways of measuring the association between two channels of state sequences, describing, for example, linked lives or the trajectories of parents and those of their children. The discussion is nicely illustrated by a study of the link between family and work careers based on data from the Panel Study on Social Integration in the Netherlands.

The first part ends with a chapter by Gilbert Ritschard, Reto Bürgin, and Matthias Studer titled, "Exploratory Mining of Life Event Histories." This work is different from the previous chapter because it considers the sequencing and timing of event sequences rather than of state sequences. It demonstrates how methods inspired by those developed by the data mining community for learning buyers' behaviors, or for physical device controls, can serve to discover interesting patterns in life trajectories. Data from a biographical survey carried out in 2002 by the Swiss Household Panel serves as illustration.

The second part of the book is based on EDM applications and it includes eight new chapters, ordered according to complexity of analyses. The first five chapters exploit basic classification or regression trees, and in addition, the last few of these chapters experiment with random forests as well. The subsequent two applications highlight the scope of survival trees and random forests. The final chapter demonstrates the use of other aspects of EDM, namely *Multivariate Adaptive Regression Splines* (MARS).

The first chapter in this section is by Carol Prescott and is titled, "Clinical versus Statistical Prediction of Zygosity in Adult Twin Pairs: An Application of Classification Trees." Here Prescott exploits classification trees to classify the unknown zygosity in adult twin pairs (i.e., to predict whether twins result from the splitting of a single zygote—egg—or from two eggs). The author uses classification tree methods with predictors derived from self-reported information, and she shows how such exploratory methods can improve zygosity classification accuracy over the usual clinical and statistical methods employed.

Next, John McArdle contributes a chapter titled, "Dealing with Longitudinal Attrition Using Logistic Regression and Decision Trees." Here he describes and demonstrates the potential problems of longitudinal dropout in his *Cognition in the USA* (CogUSA) study. Using standard Logistic Regression techniques, he tries to understand and correct for the classical problems of selection bias due to person dropout. McArdle then shows how the same problems may be better understood and more clearly corrected with a classification-based decision tree analysis. This application illustrates how decision tree analyses can be practically useful in situations where there is little prior knowledge, but there is a need for a highly accurate predictions (to eliminate sampling biases), and where other statistical analyses are typically used.

In another chapter using the same CogUSA data, John McArdle introduces regression trees with large numbers of predictors to provide an application of "Adaptive Testing of the Number Series Test Using Standard Approaches and a New Decision Tree Analysis Approach." We have included this chapter because it seems vastly different from anything we have seen before. Here McArdle demonstrates some advantages of using a regression tree approach to identify an optimally shortened version of longer psychometric tests. Such tests typically consist of a series of about 50 items of increasing difficulty. Identifying efficient shortened versions of them, based on six to seven items, is important for administration costs and for getting people to take the tests at all. The author shows that regression trees clearly outperform other strategies classically used to determine shortened versions.

The presentation by Thomas Paskus comes to us in the form of an applications chapter titled, "Using Exploratory Data Mining to Identify Academic Risk Among College Student-Athletes in the United States." This is a study of the academic risk of college students, and the author creates and compares prediction models successively obtained with classification trees, regression trees, and random forests. This chapter by Paskus (now the primary researcher at the National Collegiate Athletic Association (NCAA-USA)), presents an important analytic model which shows exactly how the nonlinear and interactive approach can be useful. This work is designed to create a prediction model for academic risk of any college student, not just college student-athletes. Of course, it is interesting that the NCAA-USA has emerged as a leader in this field, but this is partially due to the necessity of making accurate predictions about college students academic behaviors.

The chapter by Stacey Scott, Brenda Whitehead, Cindy Bergeman, and Lindsay Pitzer is an application titled, "Understanding Global Perceptions of Stress in Adulthood through Tree-Based Exploratory Data Mining." The chapter describes the utility of applying recursive partitioning to the analysis of the global perception of stress. The contribution motivates the combined usage of regression trees and random forests as the means for an idiographic and a nomothetic analysis. It relates concepts of developmental psychology such as pathways, equifinality, and multifinality to trees, and gives results for two empirical datasets of stress perception in middle and old age.

The contribution by Paolo Ghisletta illustrates the potential of survival tree and random forest approaches within a study of the decline in cognitive abilities of older persons. This applications chapter is titled, "Recursive Partitioning to Study Terminal Decline in the Berlin Aging Study." It uses data from what is known as the BAS-I and provides empirical evidence on the final decline hypothesis. Here Ghisletta demonstrates the extensive uses of a new form of survival trees.

A new presentation is contributed by Yan Zhou, Kelly Kadlec and John McArdle. This is an applications chapter titled, "Predicting Mortality from Demographics and Specific Cognitive Abilities in the Hawaii Family Study of Cognition." Here the authors use classic cognitive data on the Hawaii Family Study of Cognition to examine the relations between multiple cognitive abilities and subsequent mortality using classical hazard regression models and survival trees and forests. The results of the survival trees suggest a pattern of prediction from earlier cognitive abilities that differs over gender, and this could be important for future uses of the test battery. The chapter provides interesting insights on how those classical models and survival trees can complement each other.

The final chapter by Keith Widaman and Kevin Grimm is an applications chapter titled, "Exploratory Analysis of Effects of Prenatal Risk Factors on Intelligence in Children of Mothers with Phenylketonuria." The authors use techniques of *Multivariate Adaptive Regression Splines* (MARS) to explore risk factors during the pregnancies of mothers with phenylketonuria (PKU), an in-born error of metabolism of phenylalanine. Essentially the approach consists in finding out critical levels (knot points) after which there are significant changes in the slope of the child's IQ decline with phenylalanine. Identifying such critical levels is of primary importance for prescribing suitable phenylalanine diets.

Reviewing process

The chapters in this volume have been cross-reviewed, meaning that each of them received comments and improvement suggestions from contributors of other chapters.

Accompanying website

We also include supplementary materials (including color figures from the book), a supplementary paper to chapter 3, and R commands for some chapters at http://mephisto.unige.ch/pub/edm-book-supplement/.

Please visit the above site if you see this symbol:

If you have any problems with this link please write to one of the co-editors.

Acknowledgements

Many thanks must go to the many people who created the colloquium from which the book originated, to those who participated in our interesting event, and to those who followed through our continual requests for small alterations, providing us with high quality chapters and internal reviews.

There were also several other people in attendance who helped make our colloquium into a special event. First, we want to thank Max Baumgrass and Andreas Brandmaier (Max Planck Institute) for filming and editing the film of our entire three-day event. Interested readers can find the final cuts of the streaming video on the web (see www.mpib-berlin.mpg.de/edm). Secondly, we want to thank several other people who were visitors at the colloquium, including Laura Marcoulides, Lucy Paskus, Christiane Ritschard, and Rachael Widaman. This cheerful group helped make our time at the Chateau into a very special event. And finally, we want to thank our friendly hostess and knowledgeable advisor, Dr. Janna B. Beling, for her generous help and support at the Chateau. We would not have been able to hold the event, or produce this book, without her.

We would also like to thank the individuals who reviewed the initial book proposal, Huimin Zhao, University of Wisconsin–Milwaukee, Emilio Ferrer, University of California, Davis, Riyaz Sikora, University of Texas at Arlington, and one anonymous reviewer.

As co-editors and authors, we hope this book contributes to the literature on EDM and that readers learn as much as we did in creating it.

John J. McArdle, *University of Southern California,*
Los Angeles, California, USA.
Gilbert Ritschard, *University of Geneva,*
Geneva, Switzerland

Part I
Methodological Aspects

1 Exploratory Data Mining Using Decision Trees in the Behavioral Sciences

John J. McArdle

Introduction

This first chapter starts off with a discussion of confirmatory versus exploratory analyses in behavioral research, and exploratory approaches are considered most useful. *Decision Tree Analysis* (DTA) is defined in historical and technical detail. Four real-life examples are presented to give a flavor of what is now possible with DTA: (1) Predicting Coronary Heart Disease from Age; (2) Some New Approaches to the Classification of Alzheimer's Disease; (3) Exploring Predictors of College Academic Performances from High School; and (4) Exploring Patterns of Changes in Longitudinal WISC Data. In each case, current questions regarding DTA are raised. The discussion that follows considers the benefits and limitations of this exploratory approach, and the author concludes that *confirmatory* analyses should be always be done first, but this should at all times be followed by *exploratory* analyses.

The term "exploratory" is considered by many as less than an approach to data analysis and more a confession of guilt—a dishonest act has been performed with one's data. This becomes obvious when we reflexively recoil at the thought of exploratory methods, or when immediate rejections occur when one proposes research exploration in a research grant application, or when one tries to publish new results found by exploration. We need to face up to the fact that we now have a clear preference for confirmatory and *a priori* testing of well-formulated research hypotheses in psychological research. One radical interpretation of this explicit preference is that we simply do not yet trust one another.

Unfortunately, as many researchers know, quite the opposite is actually the truth. That is, it can be said that exploratory analyses predominate in our actual research activities. To be more extreme, we can assert there is actually no such thing as a true confirmatory analysis of data, nor should there be. Either way, we can try to be clearer about this problem. We need better responses when well-meaning students and colleagues ask, "Is it OK to do procedure X?" I assume they are asking, "Is there a well-known probability basis for procedure X, and will I be able to publish it?" Fear of rejection is strong among many good researchers, and one side effect is that rejection leaves scientific creativity only to the bold. As I will imply several times here, the only real requirement for a useful data analysis is that we remain honest (see McArdle, 2010).

When I was searching around for materials on this topic I stumbled upon the informative work by Berk (2009) where he starts out by saying:

> As I was writing my recent book on regression analysis (Berk, 2003), I was struck by how few alternatives to conventional regression there were. In the social sciences, for example, one either did casual modeling econometric style, or largely gave up quantitative work ... The life sciences did not seem quite as driven by causal modeling, but causal modeling was a popular tool. As I argued at length in my book, causal modeling as commonly undertaken is a loser.
>
> There also seemed to be a more general problem. Across a range of scientific disciplines there was often too little interest in statistical tools emphasizing induction and description. With the primary goal of getting the "right" model and its associated *p*-values, the older and more interesting tradition of exploratory data analysis had largely become an under-the-table activity: the approach was in fact commonly used, but rarely discussed in polite company. How could one be a real scientist, guided by "theory" and engaged in deductive model testing, while at the same time snooping around in the data to determine which models to test? In the battle for prestige, model testing had won.
>
> At the same time, I became aware of some new developments in applied mathematics, computer sciences, and statistics making data exploration a virtue. And with this virtue came a variety of new ideas and concepts, coupled with the very latest in statistical computing. These new approaches, variously identified as "data mining," "statistical learning," "machine learning," and other names, were being tried in a number of natural and biomedical sciences, and the initial experience looked promising.
>
> As I started to read more deeply, however, I was stuck by how difficult it was to work across writings from such disparate disciplines. Even when the material was essentially the same, it was very difficult to tell if it was. Each discipline brought it own goals, concepts, naming conventions, and (maybe worst of all) notation to the table ... Finally, there is the matter of tone. The past several decades have seen the development of a dizzying array of new statistical procedures, sometimes introduced with the hype of a big-budget movie. Advertising from major statistical software providers has typically made things worse. Although there have been genuine and useful advances, none of the techniques have ever lived up to their original billing. Widespread misuse has further increased the gap between promised performance

and actual performance. In this book, the tone will be cautious, some might even say dark …

(p. xi)

The problems raised by Berk (2009) are pervasive and we need new ways to overcome them. In my own view, the traditional use of the simple independent groups *t*-test should have provided our first warning message that something was wrong about the standard "confirmatory" mantras. For example, we know it is fine to calculate the classic test of the mean difference between two groups and calculate the "probability of equality" or "significance of the mean difference" under the typical assumptions (i.e., random sampling of persons, random assignment to groups, equal variance within cells). But we also know it is not appropriate to achieve significance by: (a) using another variable when the first variable fails to please, (b) getting data on more people until the observed difference is significant, (c) using various transformations of the data until we achieve significance, (d) tossing out outliers until we achieve significance, (e) examining possible differences in the variance instead of the means when we do not get what we want, (f) accepting a significant difference in the opposite direction to that we originally thought. I assume all good researchers do these kinds of things all the time. In my view, the problem is not with us but with the way we are taught to revere the apparent objectivity of the *t*-test approach. It is bound to be even more complex when we use this *t*-test procedure over and over again in hopes of isolating multivariate relationships.

For similar reasons, the one-way *analysis of variance* (ANOVA) should have been our next warning sign about the overall statistical dilemma. When we have three or more groups and perform a one-way ANOVA we can consider the resulting F-ratio as an indicator of "any group difference." In practice, we can calculate the optimal contrast weights assigned to the groups to create an optimal linear combination of the three groups—credit for this clarification is due to Scheffe (1959). The F is an indicator of the t^2 value that would obtain if we had, on an *a priori* basis, used this new linear contrast. As a corrective, we typically evaluate this F-value based on the number of groups ($df = G-1$), but we often struggle with the appropriate use of planned contrasts, or with optimal post-hoc testing strategies. And when the results are not significant we start to worry about the rigid model assumptions (like equal variances) and we try transformations and outliers, typically until we find a significant result. Of course, we do not want to burden the reviewer or reader with such technical details, so our true adventure is never written up. I know I am guilty of breaking this unusual statistical law.

A lot of well-known work is based on *multiple linear regression analysis* (MLR), and these are the techniques we will focus on herein. We typically substitute MLR in place of controlled manipulations, when *random assignment to groups* (RAG; following Fisher, 1936; Box, 1978) is not possible for practical or ethical reasons. There are many reasons to think MLR is a good substitute for RAG, because all we need do is include the proper variables in the equation to estimate the independent effects of each one "controlling" for all others. There are also many good reasons to think that MLR is a poor substitute for RAG, including the lack of inclusion of the proper X variables, the unreliability of Ys and Xs, the often arbitrary choice of Y as a focal outcome. It is really surprising that so many social and economic impacts are simply indicated by B-weights from relatively extensive regression equations with little attention paid to the underlying model assumptions. Is this because they work anyway? I doubt it! Of course, I also know that if we did pay attention to every detail we would never get anywhere. Nevertheless, very little attention is paid to the obviously key requirement of result replication, possibly relying more on techniques of cross-validation.

For the more advanced researcher, we seem to go ahead blindly thinking this could not happen to us but, of course, it probably does. We think we never miss the next fairly obvious exploratory clues in our multivariate procedures. In the context of *principal components analysis* (PCA), we define the first component by a technique where the variable weights are chosen so the resulting variable has maximum variance. In *canonical correlation analysis* (CCA) we choose both the set one composite weights and the set two composite weights so the resulting correlation of the correlation is as high as possible. Of course, CCA analysis can be algebraically derived from a primary component (PC) of the PCA. In the subset of CCA known as T^2, we choose the dependent variable weights that maximize the resulting t-value. In the subset of CCA known as *multivariate ANOVA* (MANOVA), we have independent groups of people and we create the composite of the dependent variables that maximizes the Scheffe contrast among the groups. We then evaluate the probability of all this happening by chance alone—or how difficult it as to get to these solutions.

While all this optimization and statistical correction strategy sounds perfectly straightforward, the fact that the weights are not chosen on an *a priori* basis could be taken to mean this aspect of our classical techniques is actually quite an exploration. That is, we do not actually know the proposed weights in advance of the calculation. While the multivariate test statistic requires *degrees of freedom* (*df*) that attempt to take into account all possible weights, we probably should question if this is the proper strategy to begin with.

Several scientists have come to view the process of research as partly confirmatory and partly exploratory. Tukey (1962; see Mosteller & Tukey, 1977) was considered by many as a radical—one key message was "plot the data and take a good look at it." I think another more subtle message was more important—"confirmatory analysis should be done first, and when it fails, as it so often does, then do a good exploratory analysis" (these are my words). The overviews by Cattell (1966) and Box (1976) point to a similar theme. The general message is that an "inductive–deductive spiral" is an essential process in a good science (after Cattell, 1966). Cooley & Lohnes (1962, 1971, 1985) showed how multivariate techniques also generally require this approach. The problem is that our standard statistical methods seem to lag behind this interesting formulation. But, as we try to show here, they need not.

The literature on *Common Factor Analysis* (CFA; e.g., McDonald, 1985) has tried to come to grips with such problems, but confusion is still apparent. It seems that Tucker & Lewis (1973) tried to distinguish between what they termed "exploratory" and "confirmatory" CFA, and these labels seem to have stuck. Specifically, what Tucker suggested was that CFA models with enough restrictions could not be rotated any further and retain the same goodness-of-fit, while exploratory models were "just identified" and could be rotated further without change in the common factor space or goodness-of-fit (see McArdle & Cattell, 1994). This, of course, leads to a whole range of possible labels based on the kinds of identification conditions, and this "hybrid" mixture of exploration and confirmation now seems reasonable (see Jöreskog & Sörbom, 1979; McDonald, 1985). Nevertheless, it is very rare to find anyone advocating *a priori* known weights for the fixed effects in these models, the true confirmatory models, or even a large number of equality constraints (i.e., as in the Rasch model; see McDonald, 1999).

These kinds of *a priori* restrictions would be truly confirmatory. Instead, many parameters are allowed to be estimated from the data (as in MLR) and only the pattern of non-zero parameters is actually evaluated. Further confusion abounds in more recent model fitting exercises where some researchers think it is reasonable to add what they term "correlated errors" based on the model fits and, in an unusual exercise in circular reasoning, they still seem to retain *a priori* probability testing with countable *df*s (e.g., Brown, 2006). This specific terminology is considered disingenuous for almost any model (see Meredith & Horn, 2001). Furthermore, "modification indices" (in Jöreskog & Sörbom, 1979) are one-parameter at a time re-specifications which are highly exploratory, and which have already been found to lead to undesirable results (see MacCallum, 1986; MacCallum et al., 1992).

The literature on *Item Response Theory* (IRT; see McDonald, 1985, 1999) is a bit less contradictory. In the most common form of IRT applications, a highly restricted model of item functioning is chosen (i.e., a Rasch model) and items which do not meet these desirable criteria are simply eliminated from the scale. This makes it clear that the IRT development of a measure is a technical engineering task and much less a scientific "hypothesis testing" endeavor—In this form of IRT, we are not trying to find a model for the data, but we are in search of data that fit the model! While there are now many perceived benefits of this IRT approach, it might be best thought of as a partial exploration with a noble goal in mind (i.e., the creation of a good scale of measurement). Perhaps this is a reasonable solution.

A Brief History of Decision Tree Analysis

The historical view of *Decision Tree Analysis* (DTA) starts around 1959 when Belson (1959) suggested that matching can be prediction and can be created by a binary search strategy (see Fielding & O'Muircheartaigh, 1977; Ritschard, chapter 2, this volume). Earlier researchers seemed to have a lot to say about this as well (Fisher, 1936; Lazersfeld, 1955), but more efforts were needed to make it all work. In the decade 1960–1970, the stimulus and tools for most of current DTA algorithms (except their statistical basis) were created in the early Automatic Interaction Detection (AID) programs advocated and used by James Morgan and John Sonquist (1963–1970) at ISR (University of Michigan). The work reported by Sonquist (1970, pp. iii–iv), is most instructive. In his words:

> This investigation had its origin in what, in retrospect, was a rather remarkable conversation between Professor James Morgan, the author, and several others, in which the topic was whether a computer could ever replace the research analyst himself, as well as replacing many of his statistical clerks. Discarding as irrelevant whether or not the computer could "think," we explored the question whether or not it might simply be programmed to make some of the decisions ordinarily made by the scientist in the course of handling a typical analysis problem, as well as doing computations ... This required examining decision points, alternative courses of action, and the logic for choosing one rather than the other; then formalizing the decision procedure and programming it, but with the capacity to handle many variables instead of only a few.

When we want to describe the relationships between an outcome variable and several input variables, we can choose among many methods. The typical MLR prediction offers many alternative techniques (e.g., Confirmatory, Hierarchical, Stepwise, Best Subsets). These represent semi-classical ways to accomplish the basic analytic goals of dealing with multiple predictors (see Keith, 2006; Berk, 2009). One alternative method for the same problem based on the use of DTA methods is the algorithmic approach termed *Classification and Regression Trees* (CART). This is a relatively new form of data analysis that is "computer-assisted" and based on "machine learning." It has been formally developed over the past three decades (see Breiman et al., 1984; Berk, 2009). Some new DTA methods and examples will be described in further detail in this chapter.

Let me now assume the motives of most researchers are based on honest ethics and the reason we do not report everything we do is that there is a lack of available techniques or available knowledge. Next let's consider having new tools in our toolbox and pulling out the right one when needed. The most widely known of the EDM techniques is DTA—these techniques were essentially created by Sonquist & Morgan (1964), later elaborated upon by Breiman et al. (1984), and discussed in the more recent and informative papers of Breiman (2001), Strobl et al. (2009) and the excellent book by Berk (2009). Of course, DTA is not the only EDM technique, and a variety of alternative multivariate methods should be considered (e.g., LVPLS, Neural Nets, MASS, etc.).

The application of what we will call the first DTA approach can be explained using the result presented in Figure 1.1 (from Sonquist & Morgan, 1964, p. 66; also see Morgan & Sonquist, 1963; Sonquist, 1970). In this initial analysis the investigators examined a real dataset of $N = 2,980$ non-farm working individuals in a survey. In one example, they were interested in the many precursors of one binary outcome item, "Do you own a home?"—they found 54% of the people responded "yes." They realized this binary outcome could be predicted from at least nine "independent" variables in many different ways, including many forms of multiple linear regression analysis (i.e., stepwise, logistic, stepwise logistic), but this could include many possible nonlinear terms and interactions (see Tatsuoka & Tiedman, 1954). So, instead, what they did that was somewhat revolutionary—they first looked at each variable in their dataset (nine possible predictors) at every possible cut-point or split on each variable in an attempts to see which variable at which cut led to the highest prediction accuracy with the outcome criterion—in this case they formed a two by

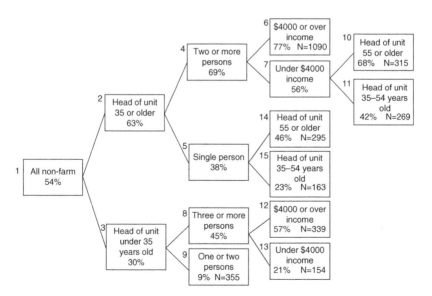

Figure 1.1 An initial decision tree using a binary outcome, from Sonquist & Morgan (1964, p. 66). Reprinted with permission from ISR.

two table with the largest chi-square value (indicating the largest correlation among the row variable and the column variable). It is fairly obvious this is an intentional exploration, and they are using a fast computer to do this work. They suggest this is exactly the technique a good scientist would be using.

In these data the first variable entering the tree was one that distinguished the home-owners was "Head of the Household Age > 35." This split the data into two sets of people: $n_1 = 2,132$ (72%) of whom 63% had a home, and $n_2 = 848$ (18%) of whom only 30% had a home. That is, this variable (Age) split at this point (Age = 35) created the maximum distance in the outcome proportion (from 54% to 63% and 30%). As an aside, Sonquist & Morgan (1964) did not suggest we evaluate the probability of the statistical test because they knew this was not an *a priori* variable selection. However, following the use of "protected test" logic, they could have required this first variable selected to meet a minimum standard of probability before going on (i.e., if this one was not significant no other one would be either).

The next radical move made by Sonquist & Morgan (1964) was to "split-up" or "partition" the data at this particular variable (Age) at this particular cut-point (35) and then reuse the same search procedure. That is, for the group under 35 years of age they repeated the process and "recursively" predicted the outcome using all nine variables again (including the possibility of another Age split). Within this upper age group they found the next best variable was "How many people are living with you?," and "Two or more persons" was at 69% whereas "Single person" was at 38%. They the used the same search procedure in the group of persons under 35 years of age and found the best split on the same variable, but at different cut-points "Three or more persons" were 45% homeowners, whereas "One or two persons" were only 9% homeowners. This partitioning was repeated (recursively) in all subgroups until it was no longer useful in making an accurate prediction (for details, see next section). This last split is interesting because they report that no other variable could be found to usefully split the 9% group, and it became the first "terminal node" (with $n_9 = 355$, or 12% of the sample). For purposes of later classification, every member of this group was assigned an expected probability of 9% homeowner.

In order to describe these optimal variables and splits Sonquist & Morgan (1964) drew the initial "classification tree," or "decision tree," copied here as Figure 1.1. This tree-like diagram (a "dendrogram") starts out with a box representing scores for all persons on one variable (later termed the "root node") followed by additional boxes for variables and cut scores that represent the optimal way to split up the data. The diagram has a set of seven final boxes (or "terminal nodes") suggesting that no further splits were useful, so all persons in these classification boxes are taken to be similar on the outcome (only later reinterpreted as an expected probability). There is no doubt that this specific pictorial representation made it easy for users to understand, and this graphic insight should not be overlooked.

This computer intensive approach to DTA was also explained in great detail by Sonquist (1970), who termed this technique *Automatic Interaction Detection* (AID), and this early treatment is worthwhile reading today. Sonquist & Morgan (1964) attribute a lot of the stimulus for this AID approach to earlier work by Riley (1964) and Goldberg (1965), and others. But there are 15 diagrams of trees in this report, so it is clearly focused on this DTA-type approach. Later, this duo used the chi-square and theta indices to define the splits, and created popular CH-AID and TH-AID programs where, again, it seems as if they were trying to "automatically mimic scientists at work." At various points they said they were searching for an "optimal split" but at no point did they claim they had created an "optimal tree." A large number of other scientists joined in these efforts (e.g., ELISEE and IDEA; see Ritschard, chapter 2, this volume) but DTA did not seem very popular with mainstream scientists.

From 1980 to 1990 the Berkeley/Stanford Statistics Department revived the AID work and invented the term CART. In fact, Leo Breiman and his colleagues (1984) developed several proofs of these earlier concepts, and Breiman also developed an original (and publicly available) CART computer program. At no point here is there any proof of an "optimal tree," because that is simply too difficult to conceive of or prove, but there are several new and important proofs of "optimal splits." For the latter reason, many contemporary researchers now attribute the entire DTA concept to Breiman et al., 1984 and call it CART (i.e., not the AID developed by Morgan & Sonquist, 1963).

From 1990 to 2000 there still seemed to be much resistance to DTA in most any form, but there were the seeds of growth in respectability among many statisticians (and many publications). However, during 2001–2009, the work of Breiman et al. (1984) caught on rapidly, and there was a massive growth of alternative CART package programs and applications. These DTAs seemed to be most popular in Business research, but also in Molecular Genetics and Medical research. In 2009 even the Psychologists held the first APA-sponsored Exploratory Data Mining conference (at the University of Southern California, sponsored by the APA Science Directorate). Since very little about these DTA techniques had changed in the 40 years since originally defined by Sonquist & Morgan (1964; see Morgan & Sonquist, 1963), something must have changed about the needs of the scientists themselves—Perhaps we are now no longer morally opposed to using computer search strategies (see Leamer, 1978; Miller, 1990; Hand et al., 2000; Hastie & Dawes, 2010).

Selected Decision Tree Applications in the Behavioral Sciences

There are many applications of the DTA techniques to formal scientific classification problems (e.g., Tatsuoka & Tiedman, 1954; Everitt, 1979; Vayssières et al., 2000; Sephton, 2001; Put et al., 2002; Miloslava et al., 2008).

Table 1.1 A chronological set of examples of Decision Tree Analysis (DTA) from the behavioral sciences

1. Sonquist & Morgan (1964) on home ownership (and see Figure 1.1)
2. Temkin et al. (1995) on problems following head trauma
3. Seroczynski et al. (1997) on different pathways to childhood depression
4. Zhang (1999) on the physical growth of infants
5. Monahan et al. (2001) on mental health and violence (also see Steadman et al., 1998, 2000)
6. Wallace et al. (2002) on well-being in later life
7. Sullivan et al. (2003) on reasons for survey fatigue in twins
8. Kerby (2003) on suicide
9. Dierker et al. (2004) on adolescent smoking
10. Rovlias & Kotsou (2004) on severe head injury
11. Weber et al. (2004) on genetics and diagnosis
12. Granger et al. (2006) on low back pain
13. Penny & Chesney (2006) on death following injury
14. Gruenewald et al. (2008) on affect in later life
15. McArdle et al. (2010) on adaptive testing, longitudinal attrition, and internet usage

Besides the original AID models reported by Sonquist (1970), most of the contemporary applications use some form of DTA or CART (see Questier, Put, Coomans, Walczak & Vander Heyden, 2005; Shmueli, Patel & Bruce, 2007). A few selected substantive examples from behavioral sciences are presented in chronological order in Table 1.1.

A few common features of these DTAs are: (1) These analyses are admittedly "explorations" of available data, and this is an attempt to "mine the data" for the most accurate relationships; (2) in most DTAs the outcomes are considered to be so critical that it does not seem to matter how we create the forecasts as long as they are "maximally accurate"; (3) some of the DTA data used have a totally unknown structure, and experimental manipulation, and subsequent "causal interpretation" is not a formal consideration; (4) DTAs are only one of many statistical tools that could have been used. A question that is asked by many researchers is, "Why do they use so many cute words to describe what they are doing?"; but the most important question seems to be, "Do you have any similar problems?"

In this chapter, several key methodological features of DTAs are next described. Then results from four examples by the author are briefly presented to illustrate specific features of DTAs applied to real problems. In each case some effort is made to show the results of alternative techniques as well. But we focus on the methodology of DTAs to highlight some special features.

Methods of Decision Tree Analysis

Some of the popularity of DTA comes from its easy to interpret dendrograms or tree structure and the related Cartesian subplots. DTA programs

Table 1.2 A few key steps in basic Decision Tree Analysis (DTA)

---

0. We define one outcome variable (or dependent variable, or left-hand side variable, usually binary or continuous) and a set (many) of predictors (or independent variables, or right-hand side variables)
1. We search for the single best predictor of the outcome variables among *all* predictor variables, considering *all* "ordered splits" of predictor variables
2. We split the data (partition) into two parts according to the optimal splitting rule developed in step 1
3. We reapply the search strategy on each partition of the data
4. We do this over and over again (recursively) until a final split is not warranted or a "stopping criterion" has been reached
5. We recognize that DTA would not be possible without modern day computations

---

are now widely available and fairly easy to use and interpret. But Figure 1.1 from Sonquist & Morgan (1964) appears to be the first application of its kind for a binary outcome, and is worth considering in some detail. Before doing so, let us state that DTAs are based on the utterly simple ideas that we can deal with a multivariate prediction/classification by (a) using only parts of the database at any time (partition), and (b) doing some procedure over and over again in subsets (i.e., recursively). These are still important statistical ideas. A few simple steps are presented in Table 1.2.

Key Technical Issues for Decision Tree Applications

There are many technical questions about computer search strategies, and not all of these will be answered here. We can start by saying that a major model tradeoff is often between "model complexity" (e.g., based on the *df*s) and "model misfit" (e.g., based on the model likelihood L^2). This tradeoff is the same as in any other issue in statistical reasoning (see Hastie et al., 2001; McDonald, 1985, 1999). But this is not really good news because it remains largely unresolved. In general, if a model is very complicated it will fit the present data very well, but it may not replicate so well. On the other hand, if the model is very simple, it will not fit the current data very well, but it is more likely to replicate. This dilemma in choosing among many statistical models is general, and many solutions have been proposed. For example, we can find this in the basic writing on *Structural Equation Modeling* (SEM) where it is written as:

> From this point of view the statistical problem is not one of testing a given hypothesis but rather one of fitting models with different numbers of parameters and of deciding when to stop fitting.
>
> (Jöreskog, 1969, p. 201, in Jöreskog & Sörbom, 1979)

So this model complexity/misfit tradeoff is a general concern of most any modeling. Nevertheless, some of the more specific questions may be essential to our understanding of any results, so a few technical issues will be described now.

One common question is "How is the best split found?" As stated earlier, for any sub-partition of the data, DTAs are designed to examine all predictors X(j) at all possible cut-points XC(k) with respect to the outcome Y. Optimal computer routines make this a feasible and relatively rapid search. The cut-point XC(k) on predictor variable X(k) which maximizes the association (positive or negative) with scores on the outcome Y (either by discrimination or similarity) is chosen as a splitting variable. With continuous outcomes, we simply find, for example, the cut-point with the highest resulting *t*-value between groups among all other splits. Interestingly, this approach is very similar to what Fisher (1936; see Box, 1978) previously suggested we do to find optimal but unknown classifications. The key difference here is that the classification we seek to mimic with other variables is measured as an outcome. With binary outcomes, the optimal cut-point is typically defined as an index of the resulting classification tables using the some index of "impurity" such as the Gini index, rather than using another index of similarity (i.e., the Phi or Tetrachoric index), but is it also possible to choose some statistical test (such as the Chi-square in CHAID). This simple ordered choice of a cut-point also makes DTAs resistant to extreme scores of outliers, as long as they are at the extremes, and this might prove to be a big benefit (see Wilcox, 2003). In the same sense, this choice can be rather arbitrary in selection of one variable over another for a specific dataset, and replication may not follow. In any case, all the data are used to define the first split, and then the same testing procedure is applied recursively (and see Efron, 1979).

There are many alternative ways to define the best splits. Given a specific data partition (D_k) we can define the "best split" in several ways. This is always based on (a) the relationship of the outcome variable ($Y|D_k$) in some subset of data, and (b) a "search" through every possible subset of the predictor variables X(j). For any cutoff or split on the k-th predictor, defined here for variable X(j) as the value XC(k), we obtain p(k) = the proportion of data that is split one way. For this simple split we can define: (1) Bayes error = $Max\{XC(k)\} = Min\{p(k), 1-p(k)\}$, (2) Gini index = $Max\{XC(k)\} = \{p(k) * 1-p(k)\}$, (3) Cross-entropy = $Max\{XC(k)\} = \{[-p(k) \ log(pk)] - [1-p(k) \ log(1-p(k))]\}$, or even (4) Chi-square/$Max\{$Chi-square$\}$. These indices all should range between 0 and 1, and the simple Gini index is typically used. However, from the perspective of the overall model built, it does not actually seem to matter what specific cut-point index is chosen (see Hastie et al., 2001).

How do we create a "Recursive Stopping Rule"? In every dataset, the splitting can go on as long as the outcomes differ within a node. But it will not be feasible to continue to create splits of the data because (a) all values are the same, or (b) the sample sizes are too small to consider, or (c) no new variable can be found. Among several choices, we can (1) define a limit on the number of people or the increment in overall accuracy found; or (2) Breiman (2001) suggested a "full" tree should be created and this should be "pruned" back to the point where these conditions are found; and (3) Loh & Shih (1997) suggested those predictor variables (i.e., X(j)) with more possible values had more possible splits, and this was supposedly corrected by Hothorn, Hornik & Zeileis (2006) when they created an innovative permutation test with Bonferroni correction for unbiased results. This was for a single split, so the desired relationship between overall "model prediction" and "model complexity" still needs to be considered (see Hastie et al., 2001; Berk, 2009).

The next question, after a tree is formed, is how can we tell the "Importance of Each Variable"? There are many ways to define the *variable importance* (VI) of each predictor variable in a specific DTA result (for details, see Ishwaran, 2007). One obvious way that is not discussed in great detail is simple and classical—remove that variable and see how much loss of fit there is—either by decrease in R^2 or increase in misclassification error E^2. Another way is to consider the numerical amount of data that has been split by a single variable. We can count how many times a specific predictor variable has been used. And then consider one variable as the most informative splitter and the VI for the rest of the variables as a simple ratio of the amount of decisions made by the first one (as done in the latest versions of CART PRO). Alternative indices of VI each summarize the key information and it may lead to model comparison and future uses of DTA. Of course, the VI may be altered quite a bit by the introduction of highly correlated variables (as in MLR fitted by OLS). Other approaches based on deviance measures are also possible (see Ritschard, chapter 2, this volume). At the same time, collinearity should not be a big problem because DTAs are designed to select the best one of a set of correlated variables.

As another note, we should not forget the "Importance of Utilities" (see Swets, Dawes & Monohan, 2000). Any formal decision split in a tree is based on a combination of *objective* data and *subjective* utilities. In the simple two choice (2 × 2) case, the exact value of the utility numbers does not matter, just their ratio across the actual outcomes—practically speaking, we can simply weight all Benefits = 0 and then the relative Costs can carry the entire relative weight (RW = (WFN)/(WFP)) merely based on the Weight applied to the False Negatives (WFN) and the Weight applied to the False Positives (WFP). Of course, the critical weights assigned to each false outcome need to be determined in a subjective fashion. The maximum cutoff could be established at the same point for all indices or it can differ due to its relative weighting scheme. Naturally, ignoring this subjective weighting gives the usual ROC, and this assumes Costs are typically one-to-one, but this is rarely intentional (see Hastie & Dawes, 2010).

From a standard psychometric view, DTA users are limited by another key fact—the key predictors and outcomes must all be measured—if key outcomes or predictors are not measured, DTA surely will be biased and unrepeatable. It follow that DTAs are sensitive to random noise in data, such as measurement error, so we cannot count on DTA to find best solutions with poorly measured data. As usual, there is no substitute for good measurement. It would also be much better if there was an overall objective function to be minimized, and then DTA would have a simple way of measuring prediction accuracy (ACC) compared to model complexity (as in the RMSEA, BIC, AIC, etc.) but, in real applications, the number of parameters actually used (and, hence, the *df*s) is often hard to define in advance of the model fitted (e.g., see Berk, 2009).

Statistical Testing Issues in Decision Trees

A reasonable criticism of exploratory DTA methods is that they take advantage of chance occurrences in a selected set of data and we are not likely to produce replicable results (e.g., Browne & Cudeck, 1989). Furthermore, as stated above, DTAs are usually created without standard statistical tests of probability, largely because it is difficult to define a full basis against which to judge the probability of any specific result. However, to better insure repeatability of a specific analysis, some statistical evaluation is useful. Thus, it is not usual for DTA users to explore one or more "cross-validation" strategies. For example, it is common to use a *Learn then Test* paradigm—in most cases about half of the data are used to develop a model (Train) and the other half are used to evaluate the model predictions (Test). Another common treatment is to use a general form of *Internal Cross-Validation*—in these cases a random sample of individuals is drawn without replacement for the Training sample, this is repeated T ($\sim$10) times, and splits that consistently obtain are reported. Finally, we can also use *Bootstrap Cross-Validation*—in these cases a random sample is drawn with replacement for the Training sample, this is repeated T ($\sim$2N) times, and splits that consistently obtain are reported (see Efron, 1979; Tibshirani & Tibshirani, 2009). Incidentally, this kind of bootstrapping can also provide empirical estimates of the standard errors of the split score cutoffs ($SE\{XC(k)\}$).

DTA performance has been shown to improve using aggregations of data and models—e.g., Bagging, Random Forests, Boosting, and similar "Ensemble" approaches (see Strobl et al., 2009; Berk, 2009). Random Forests (Breiman, 2001) can be considered as a repeated random sampling of the predictor variables, Boosting (Fruend & Schapire, 1999, as reported by Berk, 2009) can be considered as a repeated weighting of the largest residuals, and Ensembles (Strobl et al., 2009; Strobel et al., 2013) can be considered as combinations of each of these and several other approaches. We can view all of these techniques as "computer intensive" strategies that would not be possible without current computing power (for a recent discussion on cross-validation, see Tibshirani & Tibshirani, 2009).

One final issue that comes to mind in this context is the "optimality of a tree." It is not often made clear that an optimal selection can be made at each node, but this does not mean the overall tree is optimal. One way to say this is that the final set of individual predictions, based on whatever DTA or other technique is used, may not be the most accurate for any person, and these will require further evaluation and comparison. Prediction accuracy will be considered here.

Available Decision Tree Computer Programs

There are already many computer programs that can be used to carry out these calculations, and some of these are simply "smoothing techniques," including commonly used nonlinear transformations, Box–Cox powers of the dependent variables, transforms of the joint relationships (e.g., lowess transforms of the relationships), and curve fitting programs based on exhaustive searches (e.g., T2CURVE, T3CURVE, etc.). Other computer programs offer relatively advanced search techniques, including "best subsets regression," a commonly used regression technique for examining all possible (integer) combinations of predictor variables. The explicit DTA techniques can be found in many different computer packages, including those listed in Table 1.3.

A reasonable question that is often asked is, "Do these DTA programs produce different results? The answer is, "Yes, sometimes substantially." But

Table 1.3 Some available computer programs for Exploratory Data Mining (EDM)

1. *R-Code*—many state-of-the-art algorithms covering many different EDM topics which are generally and freely available (e.g., party, rpart, rattle, TraMineR, MASS). These contain many different approaches, for example, TraMineR does not offer a basic DTA method, but includes a method for growing trees from a dissimilarity matrix, useful for categorical variables (Studer et al., 2011)
2. *XLMINER*—very popular and inexpensive suite of programs built around EXCEL structures (also see XLFIT, TABU, and RTREE)
3. *SAS Enterprise, SPSS Clementine, SYSTAT Trees*—add-on EDM suites for well-known packages of programs with broad and general use. For example, SPSS allows users to choose among CART, CHAID, ExCHAID and QUEST
4. *CART-PRO* (Classification and Regression Tree)—A search approach for categorical and continuous outcomes. This program supposedly yields optimal discrimination using successive search techniques (but we need to watch out when analyses ignore utility theory). A newer version, *TREENET*, is an implementation of "Random Forests" using the CART-PRO approach
5. *C4.5*—The C5 method (from Quinlan & Cameron-Jones, 1995) is among the most popular methods used by data-mining community. (This approach is known as J48 in Weka, and is now in SPSS)
6. *MARS* (Multivariate Adaptive Splines)—A newer regression technique for examining all possible functions of predictor variables using linearity with cut-points
7. SEM based programs marketed as SEM rather than CART—*SAS NLMIXED* and *PLS, WinBUGS*, and *Latent GOLD*

this is not due to data being read in different formats, and it is not due to computational inaccuracies (as was the case with some early MLR programs; see Sawitzki, 1994). Instead, differences in resulting trees are partly due to the use of different options for split criteria, and partly due to different stopping rules (see below).

Unfortunately, from the point of view of the user, this implies that the resulting tree from different computer programs may not be unique—it may not seem to be the same for the same data even for very simple problems— the first variable (at the root node) may be different, and the cut-points may be different too, then the second variables might differ, and so on. Perhaps we should take solace in the fact that what is more likely to remain similar across different programs is the VI and the predicted outcome (Yhat) for each person. And, although not formally stated, the use of alternative DTA programs to examine the range of possible results, could be considered a benefit of "Ensemble" calculations (see Strobl et al., 2009). But this certainly means that DTA results require detailed explanations before they can be useful.

Results

In this section we discuss four examples of recent DTAs to illustrate the benefits and limitations. These examples are intended to become progressively more complex. They start with a simple one predictor binary classification, move to a multiple group classification, compare various multiple regression models, and finally build up new calculations of a longitudinal data trajectory classification.

Example 1: Predicting Coronary Heart Disease from Age

The first example is taken from Hosmer & Lemeshow (2000, pp. 2–4). In this example $N = 100$ people were examined at a hospital for the presence of *Coronary Heart Disease* (CHD). Among many variables also measured was the Age of the person (in years) at the time of the medical evaluation. The question raised by Hosmer & Lemeshow is "Is Age predictive of CHD?."

One answer to this question was provided by logistic regression, where the binary variable CHD was predicted. The relevant R outputs created here are presented in Tables 1.4a and 1.4b (the input codes and data are available on our website). In all cases, the logistic regression of CHD on Age was a significant predictor, with a prediction equation of $ln(P_n/1-P_n)$ $= -5.03 + 0.117\,\text{Age}_n$. This can be interpreted as a significantly increasing risk of 1 to 1.117 per year of age (with a confidence interval of 1.07 to 1.17). This also gives a Pseudo Explained Variance (PR^2) of $PR^2 = 0.25$ with a maximum of $PR^2 = 0.32$. By turning the LRM expectations (and CI) into expected probabilities for each value of age, we can draw the expected value plot of Figure 1.2(a). Here it is now fairly obvious that

Table 1.4a Logistic regression input/output from R-code for CHD problem

```
>
> # Simple Logistic regression
> REG2 <- glm(CHD ~ YEARSAGE, family=binomial())
> summary(REG2)

Call:
glm(formula = CHD ~ YEARSAGE, family = binomial())

Deviance Residuals:
    Min      1Q   Median      3Q      Max
-1.9718 -0.8456 -0.4576 0.8253  2.2859

Coefficients:
            Estimate Std. Error z value Pr(>|z|)
(Intercept) -5.30945    1.13365  -4.683 2.82e-06 ***
YEARSAGE     0.11092    0.02406   4.610 4.02e-06 ***
---
Signif. codes: 0 '***' 0.001 '**' 0.01 '*' 0.05 '.' 0.1
' ' 1

(Dispersion parameter for binomial family taken to be 1)
    Null deviance: 136.66 on 99 degrees of freedom
Residual deviance: 107.35 on 98 degrees of freedom
AIC: 111.35

Number of Fisher Scoring iterations: 4

> exp(REG2$coefficients)
(Intercept)     YEARSAGE
0.004944629 1.117306795

> plot(REG2, which=2)

> x <- predict(REG2)
> YHAT.REG2 <- exp(x)/(1+exp(x))
> plot(YHAT.REG2, YEARSAGE)

> PRED.REG2 <- cor(YHAT.REG2, CHD)**2
> PRED.REG2
[1] 0.2725518
```

the predicted probability of CHD for any person increases with age (in log-linear fashion).

The expected result of the alternative DTA is presented in Figure 1.2(b). Here the value of the optimal prediction is given by five linked horizontal lines over Age. Each of these lines represents the expected value of one terminal node defined as a specific age range. This is an empirical plot of the best predictions from the data, and this one is not linear. That is, up to about age 38 the CHD risk is about 15%, but increases to about 30% for

Table 1.4b PARTY input/output from R-code for CHD problem

```
> # Newer Regression Tree 2
> library(party)
> DTA2 <- ctree(CHD ~ YEARSAGE)
> plot(DTA2)

> YHAT.DTA2 <- predict(DTA2)
> table(YHAT.DTA2, CHD)
                         CHD
YHAT.DTA2               0   1
  0.142857142857143    30   5
  0.4                  21  14
  0.8                   6  24

> plot(YHAT.DTA2, CHD)

> plot(YHAT.REG2, YHAT.DTA2)

> PRED.DTA2 <- cor(YHAT.DTA2, CHD)**2
> PRED.DTA2
          [,1]
CHD 0.2865886
```

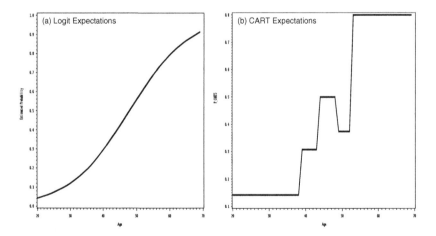

Figure 1.2 Age plots of (a) Logit Model Probabilities, and (b) CART Model
        Probabilities.

ages 40–44, then increases again to 50% for ages 45–49. Most informatively,
the risk decreases to about 40% for a specific subset of ages 50–52, before
it increases again to 80% for those 53 and older. Although the expectations
largely increase this is not always the case.

This simple one predictor illustration shows the potential for nonlinearity
within a single variable for any Y ← X relationship with DTAs. The fact

that the DTA expectations are different than logistic expectations is a well-studied question (e.g., Lemon et al., 2003). The substantive question that is commonly raised is, "Is there something special about the act of becoming 50 that makes a person enough aware of the risk to do something to prevent it?." This is clearly a hypothesis that is generated from this result, and one that may be studied further, but it is certainly not an explicit finding of the DTA.

The previous discussion did not focus on the interpretation of the tree structure, but the tree of Figure 1.2(b) is presented in Figure 1.3(a). This was initially defined by using the XLMINER program with five standard splits at ages 52.5 (splitting below 53 from 53 and above), at 58.5, at 39.5, at 33.5, and at 29.5. The built-in stopping rules suggest we stop at eight terminal nodes, and the un-weighted decision accuracy of this tree is 78%, with a calculated $PR^2 = 0.27$. This prediction accuracy is typically greater than the logistic explained variance, but only at specific cut-points, so it is not easy to compare this index of fit to the standard logistic PR^2 (see Feldesman, 2002; Lemon et al., 2003; Penny & Chesney, 2006) after the complexity of the model search (i.e., eight groups).

To examine this issue in further detail, the same data and model was refitted using the new PARTY program (in R-code, see Strobl et al., 2009; Chapter 3, this volume). With PARTY, the built-in stopping criteria suggested that only the first two splits were significant by the built in permutation rules (see Figure 1.3(b)), and the un-weighted decision accuracy of this tree is 75%, but with a $PR^2 = 0.28$.

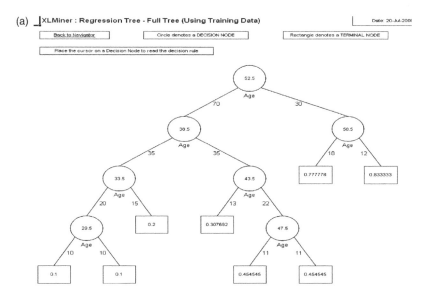

Figure 1.3(a) CART result from XLMINER.

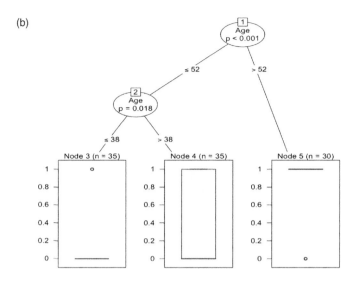

Figure 1.3(b) CART result from PARTY.

The same data and model was rerun using CART-PRO (6.0; Salford Systems, 2009), using the built-in algorithm used 10-fold cross-validation to create only one split—at age 33, and the un-weighted decision accuracy of this tree is 75%. We found we can improve the accuracy slightly by creating further parameters, and we can produce essentially the same result as the prior DTA programs using only a part of the data (see Figures 1.3(c) and 1.3(d)). Thus, using the built-in stopping rules, a model of substantially less complexity is actually needed, and we have to push this program to produce the same results with all DTA programs. Nevertheless, three different DTA programs are doing similar calculations, but they can give us slightly different results for even this simple one predictor model.

This does not end our analysis, because most DTA programs allow sampling weights or other forms of weighting, such as the inclusion of utility structures, and we can examine a few of these here. For example, suppose we add the additional restriction that the mistakes are not of equal magnitude in this classification, and the *False Negatives* (FN) are much worse than the *False Positives* (FP). For example, in cases where we do not want to miss anyone using a simple age cutoff, possibly because this is a first-stage screening, we would redo the analysis using FN:FP weights of, say, 2:1, and the result is a new cross-validated decision rule based at an Age cutoff of 32. This yields a weighted accuracy of 71%. If the opposite were true, and we did not want to say someone had CHD when they did not, we would reverse the FN:FP utility weights to be 1:2, and the resulting cross-validated CART would still have a cutoff at Age = 52. This makes it clear that any

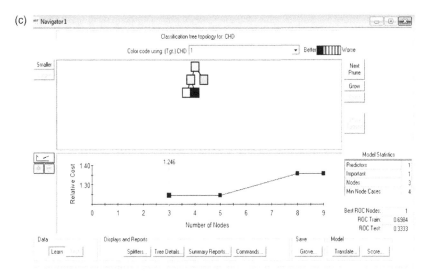

Figure 1.3(c) Initial output from CART PRO.

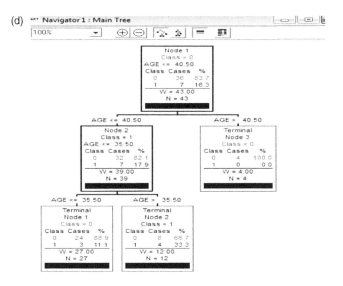

Figure 1.3(d) Selected results from CART PRO.

sampling weights, such as the utilities of the mistakes, are an integral part of any decision rule, and they can alter the point of optimal splits, or the importance of a variable. This, of course, is true of MLR models also, but utility analysis is not highlighted as much as it could be, and this is easy in CART-PRO.

Example 2: New Approaches to the Classification of Alzheimer's Disease

The DTA approach can obviously be used in a number of other situations where there are more than two groups to classify (see Fisher, 1936; Everitt, 1979; Amick & Walberg, 1976; Feldesman, 2002). As one example of this we tried to classify persons based on neuropsychological tests into subgroups of people who may have actually had Alzheimer's Disease (AD). The basic idea here was we would like to know how people are functioning in life based only on data from the tests.

In about 1995, the Massachusetts General Hospital (MGH) staff had recruited and followed a cohort of $N = 165$ individuals. When first evaluated, 123 were non-demented individuals with memory problems and 42 were controls. At baseline, all participants were administered a semi-structured interview, a neuropsychological test battery, an MRI scan, and a SPECT scan, and $N = 141$ had complete data on all variables. An annual semi-structured interview was repeated annually and made it possible to quantify "degree of functional difficulty" and this led to a clinical decision to classify each into one of three groups (for more details, see Albert et al., 2007).

In one analysis, we were also trying to classify each person into one of the three groups—(1) Normals ($n = 31$), (2) Questionables ($n = 88$), and (3) Mildly Demented ($n = 22$)—but we were using only demographic features (e.g., Educational Attainment), and the neuropsychological test battery including eight tests. The statistical idea was that members of these three groups would differ in the level, pattern, or scatter of the multivariate test scores (following Horn & Little, 1966), and by seeing this we would then be able to correctly classify them into special treatment groups at entry, or examine their differential progression over time (as in Albert et al., 2007).

There were many available techniques that could be used to do this three group classification, but we selected only three: (1) *Linear Discriminant Analysis (LDA)*—these classical multivariate techniques (see Amick & Walberg, 1976) could be used with linear assumptions to set a standard for classification; (2) *Multinomial Logit Regression (MLR)*—we could use the well-known logistic regression model with the added assumption of "proportional hazards" to give probabilities of classification for three different groups; (3) *Decision Tree Analysis (DTA)*—the current DTA approaches could be used to optimize the nonlinear selections of diagnostic groups while minimizing the incorrect classifications.

In a first set of analyses we applied the standard techniques of LDA and found we had to eliminate all persons with some incomplete data ($n = 37$). But we then found significant results for all nine variables used (Educational Attainment and the eight scales), and this led to 75.9% overall accuracy of classification (with group accuracy of 63%, 89%, and 46%). Next, using the MLR with the same data we found we could not reject the proportional hazards assumption ($p < .03$), the overall model fit pretty well ($L^2 = 70$, $df = 11$), and six of the neuropsychological tests, as well as the Educational Attainment level, met the rigid standard for significance ($p < .001$) set on an *a priori* basis. This differs a bit from the prior LDA result, but it is mainly the same. The $PR^2 = 0.47$, so the separation of at least some groups was apparent (82% when comparing G1 versus G2 plus G3).

Finally, we applied the techniques of DTA to the same problem. The way this algorithm works is to find a splitting variable which maximally separates the three groups on the means. This, of course, is very much like a MANOVA, but here the variables are selected so the groups are maximally different. In a first model classification the CART-PRO suggested we use 24 nodes, and this yielded an overall classification accuracy of about 87.6% (for three groups it was 84%, 97%, and 57%). For simplicity, this CART was rerun with a less stringent cutoff and we only obtained 10 nodes—the output from this run is displayed in Figures 1.4(a) and 1.4(b). The way this CART algorithm worked was that it started with a complete tree and then was "pruned" back to have only 10 nodes, but these 10 nodes have been repeatedly found with a 10-fold cross-validation scheme (but see Tibshirini & Tibshirani, 2009).

(a)

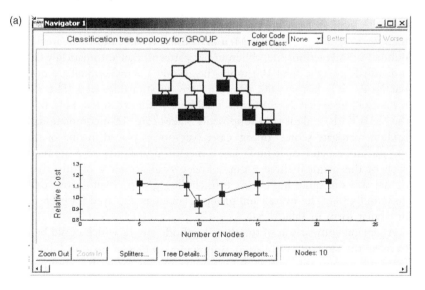

Figure 1.4(a) The initial AD classification tree (internally (V-fold) cross-validated CARTPRO).

(b)

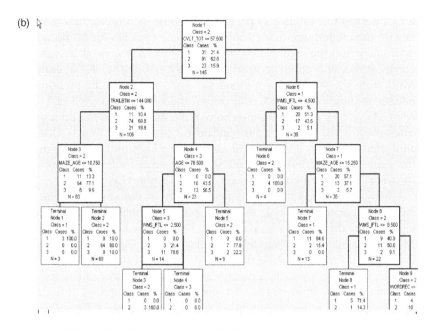

Figure 1.4(b) Details of CART2 results for three-group AD data. Note: 10 nodes with ACC = 82.8%.

As can be seen in Figure 1.4(b), the first DTA splitter is based on scores on the California Verbal Learning Test (CVLT) (if CVLT < 57 then we look at Time to complete Trails B (Trails TM); if CVLT > 57 then we look at the Wechsler Memory Scale (WMS), and there are a large number of interactions (i.e., different test patterns) that automatically classify individuals into the final 10 groups (i.e., Node 1, $n = 3$ and all are normal; but Node = 2 has $n = 80$, where 64 are questionable, etc.). The careful reader will note that $N = 145$ was used here (rather than $n = 141$), because the CART PRO algorithm supposedly "takes care" of incomplete data by finding surrogate scores. In any case, everyone is placed in one of the 10 specific nodes, and each node has a specific probability of assignment to each of the original three groups. The overall accuracy of this 10-node model was still quite high at 82.8 (including group classifications of 71%, 96%, and 48%). The overall and group accuracy is displayed for each group in tabular form in Table 1.5a.

A general comparison of the three methods, any of which could be used to make this simple classification, is presented in tabular form in Table 1.5b. Here we present the overall and group accuracies for each technique. We see that the LDA was 75.9% accurate, the MLR was 79.9% accurate, the 24-node DTA was 87.6% accurate, and the 10-node DTA was 82.8% accurate. This classification using three groups (Normal, Questionable, and Mild) was fairly good using any of the multivariable techniques. The LDA

Table 1.5a Nonlinear DTA (DTA-PRO) results with cross-validation (10 nodes)

Group	Hit Rate	Error Rate	1.Norm (N0-N4)	2.Ques (Q0-Q4)	3.Mild (Q0-M4)	Total
1.Norm	71.0	29.0	22	9	0	31
2.Ques	95.6	4.4	4	87	0	91
3.Mild	47.8	52.2	1	11	11	23
Total	82.8	17.2	31	101	13	145

Table 1.5b Comparative classification accuracy (hit %) of alternative techniques

Group	Linear Discrim	Logit1 G1-2&3	Logit2 G2-G3	DTA1 Node24	DTA2 Node10
1.Norm	61.3	60.0	---	83.9	71.0
2.Ques	88.6	86.2	84.0	96.8	95.6
3.Mild	45.5	86.2	62.5	56.6	47.8
Subtotal	---	81.6	78.2	---	---
Total	75.9	79.9		87.6	82.8

function approach was a useful baseline. The MLR yielded increased accuracy and easily permitted hypothesized nonlinear relationships to be considered. Nevertheless, the optimization using a DTA approach showed the strongest classifications, although these were certainly reduced by cross-validation. In this simple kind of comparison, the interactive nature of the DTA yielded the best overall accuracy and some easy to define classification rules. Although any diagnostic classification based on specific variables has limited accuracy, these kinds of nonlinear models provide a starting point to examine higher-order interactions in profiles of specific dementia groups (three to five groups studied). Likewise, any objective weighting of some groups instead of others could change this result.

In a more recent set of classification models we are examining a variety of alternative algorithms and profiles for direct measurements of brain functions (e.g., SPECT scores and ROI variables). Taken together with what we have already done, this is likely to require a lot more variables than could be handled effectively by LDA or MLR methods. We would need to carefully select variables in advance of these analyses, or create some new strategic plan for their inclusion together with the neuropsychological tests. But for the DTA, the prediction of the multi-group outcome using many more variables with a lack of adequate theory is not seen as a major problem. The problem of correctly classifying known groups based on measured scores is the only problem the DTA can consider. As with any technique, cross-validation of the model features is essential (see Berk, 2009). In sum, the DTA certainly seems to do this classification as well as, or better than, any other available multivariate technique.

Example 3: Exploring Predictors of College Academic Performances from High School

A completely different set of data can be used to illustrate some features when we have a continuous outcome variable. The focus here is to use High School Academic Characteristics to predict College Academic Performances. In the USA, the simple idea that High School academic characteristics are predictive of first year college outcomes provides a basis for the selection of High School students into Colleges and is used by many testing agencies to support their efforts (see Willingham et al., 1990; cf., Crouse & Trusheim, 1988).

The data used here are the same as those presented in McArdle & Hamagami (1994) on High School predictors of college graduation and McArdle (1998) on the calculation of different forms of "Test Bias." These data were collated by the National Collegiate Athletic Association (NCAA) on about $N \sim 4,000$ college students during 1984 and 1985 who were going to college on an athletic scholarship. The variables included first year academic performance outcomes, such as GPA, Credit Hours attained, and basic eligibility to continue to the second year. The predictors of these variables included High School Core GPA (in 11 similar courses), the best score obtained on any nationally standardized test of abilities (e.g., the ACT or SAT test), the student's Sex and Ethnicity, whether or not they were on a Pell Grant (an indirect indicator of low income level), and two characteristics of the size of the school, College Size and Graduation Rate (but see below; and see Paskus, chapter 13, this volume).

As one component of a larger program of research at the NCAA, we developed a prediction model for academic performance scores in the first year College GPA (*colgpa*) based on High School GPA in core courses (Z-score scaling *hsgpac*) plus the best test score on either the SAT or ACT national test (Z-score scaling of *besttest*). This was initially based on a multiple linear regression equation where the *Expected*{*colgpa*}$_n$ = 2.304 + 0.292 *hsgpac*$_n$ + 0.137 *besttest*$_n$, and this equation had an $R^2 = 0.32$ (the R-code is on the website for the book; see Fox, 1997; Keith, 2006). Although this equation was new it was similar to a previous equation where the predicted criterion of *colgpa* = 2.0 was used as a cutoff on initial eligibility (see McArdle, 1998). It is also worthwhile stating that this equation seemed a bit different from the multi-level logistic prediction model for eventual five-year college graduation (*colgrad*; see McArdle & Hamagami, 1994). In terms of graduation (*colgpa*) we found the weights for both Z-scaled variables were nearly equal and an equally weighted variable (*index*) was created for use in initial eligibility regulations.

Of course, we noted that this expectation was formed as a simple additive linear model with no interactions. We tried a few simple ones but found relatively few improvements in fit. So then we moved to a DTA, and this exploration was mainly to see what nonlinearities and interactions we might have missed up to this point in our analyses. The initial result from CART-PRO is given in Figures 1.5(a) and 1.5(b). Figure 1.5(a) shows the evaluation of the relative error versus number of nodes in the increasing size of the tree, and here we see a clear minimum of costs (complexity)/benefits (reduced prediction error) in a tree with 31 nodes and 15 terminal nodes (i.e., sub-groups). Figure 1.5(b) shows this selected tree in greater detail, with a root node of *hsgpac* > 3.10, and so on down the tree. This illustrates how complicated the direct interpretation of a tree structure can be.

A simpler way to describe the most important features of this analysis is that the scaled variable importance shows that, relative to the largest VI(*hsgpac*) = 100, then the second predictor VI(*besttest*) = 70. That is, the HS grades are more predictive than the test scores by about 1.5 times (100/70). This was not surprising, but it does not seem to be so well known (e.g., see Willingham et al., 1990). Also, the CART-PRO program lists the $R^2 = 0.32$, suggesting that considering the nonlinear and interactive nature of these two variables we did not add very much to our prediction of this outcome. That is, the simpler linear additive system based on multiple regression was almost as good as the seemingly more complex tree structure.

We also noted that other demographic variables may have interactions with these interpretations and, if so, any rules based on these simpler predictions would be biased. We initially studied both Gender and Ethnicity, and found some differences for Gender (the intercepts were higher for Females), but no additional differences were found for different Ethnic groups. However, additional variables such as individual Pell Grants and Financial Aid (indicating family income), as well as some features of the colleges themselves, such as their Size and Graduation Rate, also proved useful. In a traditional way, seven more variables were added as IVs into the multiple regression analysis. We found several significant results, for Sex, Ethnicity, Size, and Graduation Rate, yielding an $R^2 = 0.34$, representing a small but significant increase in prediction of first year college grades ($dR^2 = 0.02$).

When faced with the potential for nonlinearity and interactions among all nine variables we again turned to DTA, and the results are displayed in Figures 1.5(c) and 1.5(d). This model yields 50 terminal nodes and attains an overall $PR^2 = 0.36$. We also note that the variable importance (in Figure 1.5(d)) shows that the variables with largest variable importance remained High School GPA (VI = 100%) and HS Besttest scores (VI = 70%), with the others being relatively small (School Size, VI = 15%). It is not that these additional variables are not important, but when we take into account seven additional variables, and all possible nonlinearities and interactions, we actually gain very little in prediction accuracy ($dPR^2 = 0.02$).

(a)
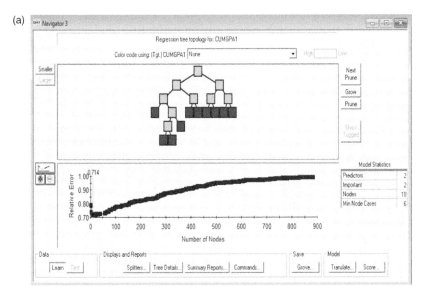

Figure 1.5(a) CART PRO output of NCAA regression IV = 2.

(b)
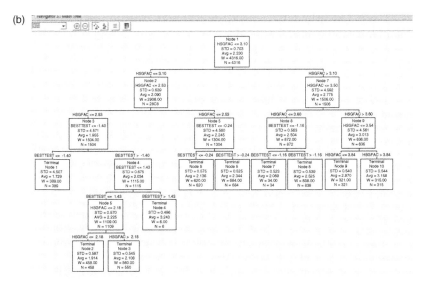

Figure 1.5(b) CART PRO output from NCAA regression IV = 2.

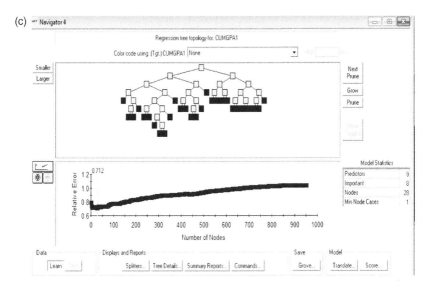

Figure 1.5(c) CART PRO output of NCAA IV = 8.

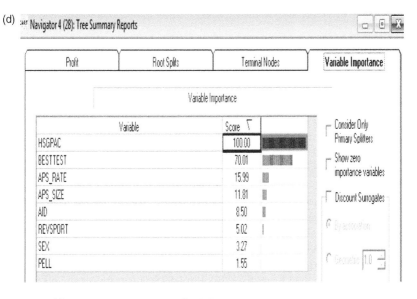

Figure 1.5(d) CART PRO output of NCAA IV = 9.

In these examples, DTA was used in this multiple prediction analysis to calculate the "optimal prediction" model. This result could be used directly, or we simply might say DTA is useful as a baseline against which to judge the success of any previous MLR analysis. The fact that we could not improve our basic regression model predictions using DTA gave us added confidence in the simpler MLR models. But we also knew that these results would be used in subsequent rank ordering of potential college admissions, so we wanted to make sure we were as accurate as possible.

In subsequent research we have studied other School-level variables, such as Type (Public, Private, HBCU), Size (Enrollment and Locations), Admissions Policies (Selective vs Open), Economic Considerations (Per/Pupil Expenditures and Pell Grants), NCAA Division (I, IA, IAA, IAAA), and Conference Affiliations. We have also used DTAs to approach the assessment of academic "misfit," defined by Year-by-Year Retention, Year-by-Year Credit Hours and GPAs, Academic Performance Rates of Individuals and Teams, Graduation Success Rates of Individuals and Teams. In all cases there will be variations between the academic outcomes of different schools, and in the distinction of the Student Body and Student Athletes. Our main results suggest different profiles of schools with different "Institutional Missions."

In general, we have learned that there is very little one can do with standard statistical models like MLR when faced with an enormous database including many potential predictors of variation which must be explored. In such cases, the DTA approach can be very useful.

Example 4: Exploring Patterns of Changes in Longitudinal WISC Data

There are many classical models for the analysis of multivariate repeated measures, and each asks a slightly different statistical question. The traditional SEMs can combine factor analysis, time-series, and MANOVA, and are widely available (see Lawley & Maxwell, 1971; McDonald, 1985). Of course, subtle differences in the choice of models define the nature of developmental process and change, i.e., the dynamic systems (see McArdle, 2001, 2009). But often the researcher does not know what to fit as an optimal structure for the longitudinal data. Also there are often many exogeneous variables that can be examined. Recursive Partitioning provides some potential assistance. In addition, longitudinal data are expensive to collect so we must try every possible analysis.

Classic longitudinal data from Osborne & Suddick (1972) on the Wechsler Intelligence Scale for Children (WISC) was reported and used by McArdle & Epstein (1987) and McArdle (1988; and also see Brandmaier et al., chapter 4, this volume). These are repeated measures on the WISC cognition scales collected on $N = 204$ children at ages 6, 7, 9, and 11 (or Grades 0, 1, 3, and 5). A trajectory plot of available WISC data is presented in Figure 1.6(a). A variety of path and factor analytic SEM have already

(a) Verbal[t]

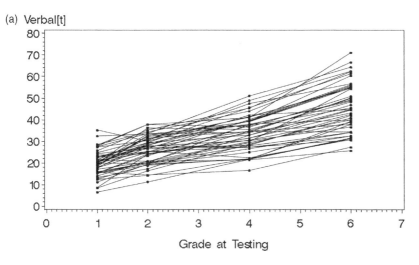

Figure 1.6(a) Individual trajectories on the WISC-verbal scales ($N = 50$).

been used to evaluate models of cognitive development for the same data (see McArdle & Aber, 1990; McArdle & Nesselroade, 1994; Ferrer et al., 2004).

Although the prior SEMs have been considered informative, there is much more to be considered when dealing with longitudinal data. For example, one key idea is that we have a longitudinal series and we are not sure if it is good to consider "heterogeneous" groups of participants (see McCall, Applebaum & Hogarty, 1973). Many recent papers deal with what is now termed "growth–mixture modeling" (e.g., Muthén, 2004; Vermunt & McCutcheon, 2004; McLachlan & Peel, 2005; Nagin, 2005; Grimm, McArdle, 2007 & Hamagami, 2007; Nylund et al., 2007; Jung & Wickrama, 2008. This approach is based on the estimation of individual latent probabilities of classes of individuals (see Grimm et al., chapter 7, this volume). However, an entirely different way of viewing this problem has been created by DTA approaches. This approach seems to have been initiated by Segal (1992), Zhang (1997, 1999, 2004; Zhang & Singer, 1999), and Abdolell et al. (2002), but these all follow the basic logic of survival-based DTA (S-DTA; Gordon & Olshen, 1985; Su & Tsai, 2005). The key idea here is that we can use multiple measured variables to split the data into sub-groups of different persons with common "patterns of longitudinal scores." It is interesting that the same form of observed-variable pattern-split model forms the basis of both S-DTA (Su & Tsai, 2005; Ture et al., 2009) and event history DTA (McVicar & Anyadike-Danes, 2002; Ritschard, chapter 2, this volume).

It is fairly clear that we can now use standard longitudinal growth models, even those of a fairly complex variety (e.g., Meredith & Horn, 2001; McArdle, 2009), to then create estimates of the common factor scores representing, say, the model based levels and slopes. Once these factor score estimates are obtained, these factor score estimates can be plotted and examined in many different ways, and the standard DTA can be used to define subgroups that are predictive of these components. This is a two-stage approach that is certainly clearer and may be favored by some.

The use of a simultaneous Longitudinal DTA (L-DTA) is relatively new, and although it is similar in several ways to other forms of DTA, it will be useful to explain some details here. The key idea is that we have a longitudinal series on a group of individuals and we want to separate them in some way or another so they appear to differ. If we fit a single mixed model to a groups of participants, then we end up with a scalar from an individual likelihood (e.g., $L_n^2\{\mu[t], \Sigma[t]\} = K + [Y[t]_n - \mu[t]] \Sigma[t]^{-1} [Y[t]_n - \mu[t]]')$ where the individual's observed scores $(Y[t])$ are compared to a population mean $(\mu[t])$ and population covariance $(\Sigma[t])$ defined by a model (and K is a constant). The total misfit for any model is the sum of these individual likelihoods (*Overall* $L^2 = \Sigma L_n^2$; see McArdle, 1991, 1994). Indeed, as we have also shown earlier (see McArdle, 1997), isolating each person's misfit can be a useful way to examine influential observations.

But in the L-DTA approach we can split the people into groups defined by a measured splitting variable (X) into any subsets a and b. We can form a new likelihood for each person as being in one of two data splits (*For* $X = s$, $L_{ns}^2\{\mu[t]_s, \Sigma[t]_s\} = K + [Y[t]_{ns} - \mu[t]_s] \Sigma_s^{-1} [Y[t]_{ns} - \mu[t]_s]')$. The total misfit for this joint model is the weighted sum of these likelihoods (*Split* $L^2 = \Sigma L_{na}^2 + \Sigma L_{nb}^2$). This means the effectiveness of the split of data can be defined by the likelihood difference (*Gain* L^2 = *Split* L^2 − *Overall* L^2). Of course, this splitting based on a measured input variable on the whole outcome vector is done automatically. There are several new R-code based programs that can be used to calculate this form of recursive partitioning on the likelihood of longitudinal data, and a few of these are listed in Table 1.6.

To illustrate this type of analysis, we will use the freely available *longRPart* program (see the book's website for input files). Since only three levels of Mother's Education (*moeducat*) are available (in this example) there are only a few possibilities. If considered ordinal, then we can only have category

Table 1.6 New R-based computer programs for Longitudinal Decision Tree Analysis (L-DTA)

1. *Masal*—Multivariate Adaptive Splines Model for the Analysis of Longitudinal Data
2. *REEMtree*—Regression Trees with Random Effects for Longitudinal Data
3. *longRPart*—Recursive Partitioning for Longitudinal Data
4. *TraMineR*—Trajectory Mining from sequence data (Gabadinho et al., 2011)
5. *SEMTrees*—Structural Equation Modeling using Trees (see Brandmaier et al., chapter 4, this volume)

"1 vs 2+3" or "1+2 vs 3." If so, we can then split up the next into "2 vs 3" or "1 vs 2." If these three categories are considered nominal, then we can have the above splits, plus "1+3 vs 2" (out of order), possibly followed by 1 vs 3.

Of the three levels of *moeducat* available in these WISC data there is a statistical difference in fit if we compare categories 1 (M <0.5 = 304 data points, 76 people, 37%) vs "2+3" (M >0.5 = 514 data points, 128 people, 63%). This is the only important split we can find in these data if we want to improve the goodness of fit. The *longRPart* result of Figure 1.6(b) shows that the group differences are largely a difference of initial levels, and not of slopes, but this was only one possibility, so this is news. Figure 1.6(c) shows the separate trajectory plots for the two groups, and this adds more detail to this outcome. Figure 1.6(d) shows the trajectories of the children split up in the way the model suggests, and then plotted using a spline smoother, and the two groups suggested by the DTA do look a lot different.

Additional levels of *moeducat* can and should be considered and, possibly more importantly, additional demographic variables can now easily be used (e.g., Gender, Income, etc.). That is, a wide variety of additional variables can now be easily included—this is quite a bit easier than tedious and repetitive runs of multi-level model alternatives, especially the kinds of growth-mixture models (see Grimm et al., chapter 7, this volume). On the other hand, this L-DTA approach requires measured variables for the splits, so there is a distinct tradeoff that must be made. Although L-DTA is only one form of the more general *SEMTrees* approach (see Brandmaier et al.,

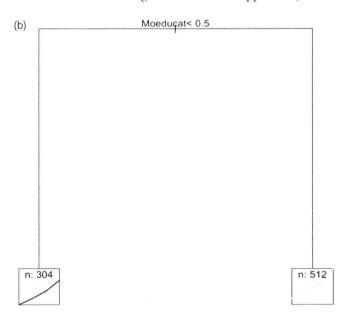

(b)

Moeducat< 0.5

n: 304

n: 512

Figure 1.6(b) Initial results from longRPart.

(c)

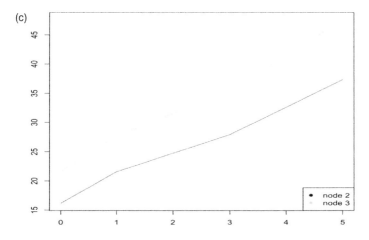

Figure 1.6(c) Resulting two groups trajectories from L-DTA.

(d)

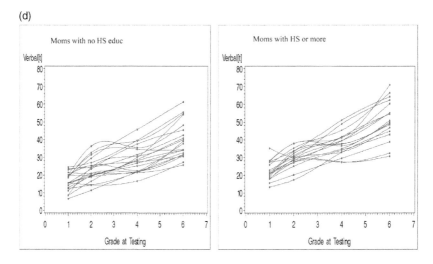

Figure 1.6(d) Spline regression estimates of individual growth curves for two groups.

chapter 4, this volume), it is an important addition to the longitudinal data analysis possibilities.

Each approach presented here can be thought to answer a slightly different statistical question. If we are using a simple linear mixed model, the least squares estimates can be calculated from algebraic statements—levels and linear slopes for each person. These simple levels and slopes carry a great deal of information about the data, and they can be used directly in standard DTA as outcome variables. The behavior observed in the final

WISC example suggested the DTA has slightly more predictive power than a comparable additive linear model prediction. In general, this kind of use of simple composites might be a useful starting point for longitudinal data analysis with DTA.

This new type of L-DTA analysis is not needed at the confirmatory/testing stage of research, but this is almost always needed nearer the end of a research enterprise when we are evaluating the single-group heterogeneity assumption. We may also need to consider the advanced possibility that the curve changes shape at some unknown time point ($B[t]$) and this "change point" itself differs over different groups (see McArdle & Wang, 2008). Of course, the same could be said for almost any substantive hypothesis, so it is always wise to compare the results of any exploratory L-DTA analysis with a more rigid multiple group trajectory hypothesis.

Conclusion

The four examples used here were designed to illustrate a broad range of DTA applications and programs. In example 1, the simplest possible relationship between CHD and Age was explored, with a direct comparison of logistic expectations and DTA classification, and several technical issues were illustrated. In example 2, DTA was used to classify individuals into three different groups based on a variety of cognitive assessments. In example 3, DTA regression was used to indicate the "predictors of college grade from high school" scores. In example 4, DTA was used to indicate the "optimal group differences" for longitudinal trajectory data. In all cases, these DTAs could be used directly, to create optimal contrasts in future analyses, or we might just say that DTA was useful as a baseline against which to judge the success of any classical analysis.

Decision Tree Analysis Successes and Failures

When do we expect DTA to be most helpful? In a lot of real applications, but especially in the typical situation when we have little *a priori* information and are not sure what to do next. It is clear that DTA is most useful when there are many predictors and we do not know which ones to consider first. But DTA is most useful when there are nonlinearities or many possible interactions and we do not really know which ones are most crucial. Of course, this happens most when the standard model assumptions are not met due to nonlinearity, or outliers, or even combinations of both.

When do we expect DTA to be least helpful? A lot of the time we can pick up on unnecessary artifacts or chance events in the data that are unlikely to replicate. This can occur when the model is linear and additive and all assumptions are met. This is also true when there are limited (specific) hypotheses we want to test, including those based on interactions—most interestingly, when the initial split is not correct, and the rest of the splits may not recover from the original mishap.

Thus, DTA is both easy and hard to use. The easy parts come because DTA is an automated stepwise procedure that considers most nonlinearity and all interactions. DTA does not need assistance with scaling and scoring and variable redundancy is allowed (we can simply put all IVs in). DTA also does a lot of numerical searching, but it is very fast. However, the hard parts come when we realize we need to know the key outcome variable, and we need to judge how many nodes should be used to fit the current data, and we still must have a repeatable model. Of course, we have other hard questions to answer, such as "Is this sample of persons an adequate representation of the population?" Furthermore, while DTA handles missing data, it only does so in a special way (see Breiman et al., 1984; Steinberg & Colla, 1999; King et al., 2001; Horton & Lipsitz, 2001), and this way of dealing with incomplete data is not always advisable.

A Final Note on Multivariate Analysis

As stated earlier, the approaches termed DTA, CART and MARS handle one outcome variable at a time, so these are not correctly termed "multivariate" techniques (see Cattell, 1996). Of course, Friedman developed an advanced version of the DTA model including linear combinations for DTA splits based on "Multivariate Adaptive Splines" (MARS; see Zhang & Singer, 1999; Dwinwell, 2000). But this form of MARS seems to be mislabeled—In a classic view (e.g., Cattell, 1966) MARS should be classified as a univariate technique because it has only one outcome variable and multiple predictors (i.e., it is "multivariable" in this sense only). Other technical issues are described in the last section here.

In any case, there is a great deal of similarity in the DTA approach and, say, the prior literature in exploratory factor analysis (see Horn 1965; Zwick & Velicer, 1985; Yates, 1987; Browne, 2001) and canonical analysis (Cliff & Krus, 1976; Liang et al., 1995). However, we have also seen several DTA-based approaches to the analysis of multivariate repeated measures (see Zhang & Singer, 1999; and see Example 4 above, and Brandmaier, chapter 4, this volume), and most of these are based on a generic likelihood splitting which could be considered for any multiple outcome model (e.g., a latent variable path model).

This brings us back to the early work by Sonquist (1970) who showed how the principles of DTA (then embodied in AID) could be useful when dealing with non-additive and nonlinear relations and, importantly, these techniques would not be misleading in situations where they were not needed. It is also fair to say that the current status of DTA has increased dramatically over the past few decades, probably because of the interest among professional statisticians (e.g., Breiman, et al., 1984; Breiman & Friedman, 1985).

But it also seems there is much left to be done to effectively deal with the simultaneous analysis of multiple inputs and multiple outcomes, and this is a typical empirical situation, so there is a need to improve multivariate versions of DTA and similar techniques (e.g., Drezner et al., 2001; Marcoulides & Drezner, 2004; Mills et al., 2005). As this work is being carried out, a DTA-based simultaneous approach to multivariate analysis (M-DTA) possibly using SEMTrees (see Brandmaier, chapter 4, this volume) can and should be compared to other popular confirmatory techniques (e.g., LISREL; Jöreskog & Sörbom, 1979; Loehlin, 2005; McArdle, 2009; Kline, 2010) and exploratory techniques (e.g., LVPLS; Jöreskog & Wold, 1992; Marcoulides et al., 2009). Virtually any M-DTA approach we use will open new avenues to understanding complex data and, given our current state of multivariate prediction, such improvements are definitely needed.

A Research Strategy Suggestion—Confirm Then Explore

It seems that if we ask for a thorough exploration of our research data then we also must find a way to speak about these analyses in public forums. Unfortunately, there are many barriers to communication in all aspects of this work in the behavioral sciences. So I would like to end by making a small suggestion about our current research strategies: *I think every researcher should first do everything they can with the wide range of available confirmatory tools before turning to any exploratory devices. This implies all researchers should Confirm first and then Explore.*

OK, this message is not really new, and it is basically how I already interpreted a key message from Tukey (1962). However, to be blunt, I do not think these two stages are now being used or encouraged by many others. But this two-stage approach will encourage most anyone (especially doctoral dissertation researchers) to start their analyses in a confirmatory fashion—by writing theoretical models which match the most appropriate theoretical ideas, collecting appropriately designed data sets, and then even using standard techniques to evaluate the probability of random behaviors in these data. Without being too pejorative, this will allow members of the academy to carefully scrutinize and evaluate the scholarship of the work. This is a big plus because most of us in the academy have already been well-trained in confirmatory statistical principles. So, and without any doubt, the confirmation stage should come first.

But we are not yet done. After all the requisite huffing and puffing of confirmatory analysis is completed, we can then get down to the really hard work of more complete plotting of and exploring the data. Remember, these were data that seemed very important to collect in the first place and we really do not want to miss anything. And all we really need to do at this stage is to declare "we are exploring," or we are "mining the data," and we can even say that no probability basis is evident or required (although we may use re-sampling techniques). Of course, we will want our exploratory findings to replicate, and not be a chance occurrence, but this is not our first goal—our first goal is to find something new! This could give new life to old data.

Why not Explore first and Confirm second? As stated above, we probably should not do exploration at the start, but only to follow the confirmation stage. Logically, if we explore first and do it well, then there should be nothing left to confirm. Also, if we do keep exploring first, we may find results faster, but we will lose all the inherent beauty of having the prior statistical knowledge, including the important probability values that helped guide our statistical decisions. We may even be stuck wanting to use confirmatory techniques to present our findings. And then we would have to wait.

Of course, I realize that exploration is now often done much sooner than suggested—witness the many researchers who suggest we should immediately plot our data at the very start of an investigation to examine relations, to suggest appropriate transformations, and to detect outliers. While I certainly agree, we need to avoid incorrect data coding, I can think of many ways to do so without sneaking a peak at the data and altering our ideas about the data generating process. In sum, I do not think we should explore the data and then use confirmatory techniques to explain our results—It seems to me that this is unethical behavior and it should be avoided at all costs (see McArdle, 2010).

For these and other reasons, then, the process of DTA exploration described here should only be considered at a second stage of research—maybe after all else fails. This set of analyses may allow us to find important results for others (or ourselves) to check out in future work. One of the major techniques for exploring data (mining data) in this way is based on DTA. When used in this sequence, after using up all the existing confirmatory tools, I cannot see how the use of DTA, or any other EDM technique for that matter, can be misleading, or in any way dangerous.

Note

I thank the National Institute on Aging (Grant # AG-07137-22) and the APA Science Directorate for funds to support this ongoing research. Portions of this research have been previously published in McArdle (2012). I also thank my many colleagues, especially Dr. Gilbert Ritschard, for pointing out where the Decision Tree Analysis approach can be useful and where it can fail.

References

Abdolell, M., LeBlanc, M, Stephens, D. & Harrison, R.V. (2002). Binary partitioning for continuous longitudinal data: Categorizing a prognostic variable. *Statistics in Medicine, 21*, 3395–3409.
Albert, M., Blacker, D., Moss, M.B., Tanzi, R., & McArdle, J.J. (2007). Longitudinal change in cognitive performance among individuals with mild cognitive impairment. *Neuropsychology, 21*(2), 158–169.

Amick, D.J. & Walberg, H.J. (1976; Eds.) *Introduction to Multivariate Analysis.* Berkeley: McCutcheon.

Belson, W.A. (1959). Matching and prediction on the principle of biological classification. *Applied Statistics*, *8*(2), 65–75.

Berk, R.A. (2003). *Regression Analysis: A Constructive Critique.* Newbury Park, CA: Sage Publications

Berk, R.A. (2009). *Statistical Learning from a Regression Perspective.* New York: Springer.

Breiman, L. (2001) Statistical modeling: The two cultures. *Statistical Science*, *16*(3), 199–231.

Breiman, L. & Friedman, J.H. (1985). Estimating optimal transformations for multiple regression correlation. *Journal of the American Statistical Association*, *80*(391), 580–598.

Breiman, L., Friedman, J., Olshen, R., & Stone, C. (1984). *Classification and Regression Trees.* Pacific Grove, CA: Wadsworth and Brooks/Cole.

Brown, T.A. (2006). *Confirmatory Factor Analysis for Applied Research.* New York: Guilford Press.

Browne, M.B. (2001). An overview of analytic rotation in exploratory factor analysis. *Multivariate Behavioral Research*, *36*(1), 111–150.

Browne, M.W. & Cudeck, R. (1989). Single sample cross-validation indices for covariance structures. *British Journal of Mathematical and Statistical Psychology*, *37*, 62–83.

Box, G.E.P. (1976). Science and statistics. *Journal of the American Statistical Association*, *71*(356), 791–799.

Box, J.F. (1978). *R. A. Fisher: The Life of a Scientist.* New York: John Wiley & Sons.

Cattell, R.B. (1966). Psychological theory and scientific method (pp. 1–18). In R.B. Cattell (Ed.) *Handbook of Multivariate Experimental Psychology.* Chicago: Rand McNally & Co.

Cliff, N. & Krus, D. J. (1976). Interpretation of canonical variate analysis: Rotated vs. unrotated solutions. *Psychometrika*, *41*(1), 35–42.

Crouse, J. & Trusheim, D. (1988). *The Case Against the SAT.* Chicago: University of Chicago Press.

Cooley, J. & Lohnes, P. (1962, 1971, 1985). *Mutivariate Data Analysis.* Boston: Duxbury Press.

Dierker, L.C., Avenevoli, S., Goldberg, A., & Glantz, M. (2004). Defining subgroups of adolescents at risk for experimental and regular smoking. *Prevention Science*, *5*(3), 169–183.

Drezner, Z., Marcoulides, G.A., & Stohs, M.H. (2001). Financial application of a Tabu search variable selection model. *Journal of Applied Mathematics and Decision Sciences*, *5*(4), 215–234.

Dwinnell, W. (2000). Exploring MARS: An alternative to neural networks. *PC AI*, *14*(1), 21–24.

Efron, B. (1979). Bootstrap methods: Another look at the jackknife. *Annals of Mathematical Statistics*, *7*, 1–26.

Everitt, B.S. (1979). Unresolved problems in cluster analysis. *Biometrics*, *35*, 169–181.

Fielding, A. & O'Muircheaetaigh, C.A. (1977). Binary segmentation in survey analysis with particular reference to AID. *The Statistician*, *26*(1), 17–28.

Fisher, R.A. (1936). The use of multiple measurements in taxonomic problems. *Annals of Eugenics*, *7*, 179–188.

Feldesman, M.R. (2002). Classification trees as an alternative to linear discriminant analysis. *American Journal of Psychical Anthropology, 119*, 257–275.

Ferrer, E., Hamagami, F., & McArdle, J.J. (2004). Modeling latent growth curves with incomplete data using different types of structural equation modeling and multilevel software. *Structural Equation Modeling, 11*(3), 452–483.

Fox, J. (1997). *Applied Regression Analysis, Linear Models, and Related Methods.* Thousand Oaks, CA: SAGE.

Friedman, J.H. (1991). Multivariate adaptive regression splines. *Annals of Statistics, 19*(1), 1–67.

Fruend, Y. & Schapire, R.E. (1999) A short introduction to boosting. *Journal of the Japanese Society for Artificial Intelligence, 14*, pp. 771–780.

Gabadinho, A., Ritschard, G., Müller, N.S., & Studer, M. (2011). Analyzing and visualizing state sequences in R with TraMineR. *Journal of Statistical Software, 40*(4), 1–37.

Goldberger, L.R. (1965). Diagnotician vs. diagnostic signs: The diagnosis of psychosis vs neurosis from the MMPI. *Psychological Monographs* (602), p. 79.

Gordon, L. & Olshen, R. (1985). Tree-structured survival analysis. *Cancer Treatment Reports, 69*, 1065–1069.

Granger, C.V., Lackner, J.M., Kulas, M., & Russell, C.F. (2003). Outpatients with low back pain: An analysis of the rate per day of pain improvement that may be expected and factors affecting improvement. *American Journal of Physical Medicine & Rehabilitation, 82*(4), 253–260.

Grimm, K. J., McArdle, J. J., & Hamagami, F. (2007). Nonlinear growth mixture models in research on cognitive aging. In K. van Montfort, H. Oud, & A. Satorra (Eds.), *Longitudinal models in the behavioral and related sciences* (pp. 267–294). Mahwah, NJ: Erlbaum.

Gruenewald, T.L., Mroczek, D.K., Ryff, C.D., & Singer, B.H. (2008). Diverse pathways to positive and negative affect in adulthood and later life: An integrative approach using recursive partitioning. *Development Psychology, 44*(2), 330–343.

Hand, D.J., Blunt, G., Kelly, M.G., & Adams, N.M. (2000). Data mining for fun and profit. *Statistical Science, 15*(2), 111–131.

Hastie, R. & Dawes, R. (2010). *Rational choice in an uncertain world: The psychology of judgment and decision making* (2nd Ed). Los Angeles: SAGE.

Hastie, T., Tibshirani, R., & Freidman, J. (2001). *The Elements of Statistical Learning: Data Mining, Inference, and Prediction.* New York: Springer.

Hothorn, T., Hornik, K. & Zeileis, A. (2006). Unbiased recursive partitioning: A Conditional inference framework. *Journal of Computational and Graphical Statistics, 15*(3), 651–674.

Horn, J.L. (1965). A rationale and test for the number of factors in factor analysis. *Psychometrika, 30*, 179–185.

Horn, J.L. & Little, K.B. (1966). Isolating change and invariance in patterns of behavior. *Multivariate Behavioral Research, 1*(2), 219–228.

Horton, N.J. & Lipsitz, S.R. (2001). Multiple imputation in practice: Comparing software packages for regression models with missing values. *The American Statistician, 55*, 244–254.

Hosmer, D.W. & Lemeshow, S. (2000) *Applied Logistic Regression.* New York: John Wiley & Sons.

Ishwaran, H. (2007). Variable importance in binary regression trees and forests. *Electronic Journal of Statistics, 1*, 519–537.

Jöreskog, K.G. & Sörbom, D. (1979). *Advances in Factor Analysis and Structural Equation Models.* Cambridge, MA: Abt.

Jöreskog, K.G. and Wold, H. (1982). The ML and PLS techniques for modeling with latent variables: Historical and comparative aspects, in H. Wold and K. Jöreskog (Eds.), *Systems Under Indirect Observation: Causality, Structure, Prediction* (Vol. I), Amsterdam: North-Holland, 263–270.

Jung, T. & Wickrama, K.A.S. (2008). An introduction to latent class growth analysis and growth mixture modeling, *Social and Personality Psychology Compass*, 2/1, 302–317.

Keith, T.Z. (2006). *Multiple Regression Analysis and Beyond.* Boston: Allyn and Bacon.

Kerby, D.S. (2003). CART analysis with unit-weighted regression to predict suicide ideation from Big Five traits. *Personality and Individual Differences*, 35, 249–261.

King, D.W., King, L.A., Bachrach, P.S., & McArdle, J.J. (2001). Contemporary approaches to missing data: The glass is really half full. *PTSD Quarterly*, 12(2), 1–7.

Kline, R.B. (2010). *Principles and Practice of Structural Equation Modeling* (2nd Ed). New York: Guilford Press.

Lawley, D.N. & Maxwell, A.E. (1971). *Factor Analysis as a Statistical Method.* New York: Macmillan Publishing Co., Inc.

Lazersfeld, P.F. (1955). The interpretation of statistical relations as a research operation, in P.F. Lazersfeld & M. Rosenberg (Eds.), *The Language of Social Research.* Glencoe, IL: The Free Press.

Leamer, E.E. (1978). *Specification Searches: Ad Hoc Inferences with Nonexperimental Data.* New York: Wiley.

Lemon, S.C., Roy, J., Clark, M.A., Friedmann, P.D., & Rakowski, W. (2003). Classification and regression tree analysis in public health: Methodological review and comparison with logistic regression. *Annals of Behavioral Medicine*, 26(3), 172–181.

Liang, K.H., Krus, D.J., & Webb, J.M. (1995). K-fold cross-validation in canonical analysis. *Multivariate Behavioral Research*, 30, 539–545.

Loehlin, J.C. (2005). *Latent Variable Models: An Introduction to Factor, Path, and Structural Analysis* (5th edn). Mahwah, NJ: Lawrence Erlbaum Associates.

Loh, W-Y & Shih, Y-S. (1997). Split selection methods for classification trees. *Statistica Sinica*, 7, 815–840.

MacCallum R.C. (1986). Specification searches in covariance structure modeling. *Psychological Bulletin*, 100, 107–120.

MacCallum, R.C., Roznowski, M., & Necowitz, L.B. (1992). Model modifications in covariance structure analysis: The problem of capitalization on chance. *Psychological Bulletin*, 111, 490–504.

Marcoulides, G.A., Chin, W.W., & Saunders, C. (2009). A critical look at partial least squares modeling. *MIS Quarterly*, 33(1), 171–175.

Marcoulides, G.A. & Drezner, Z. (2004). Tabu search variable selection with resource constraints. *Communications in Statistics*, 33(2), 355–362.

McArdle, J.J. (1988). Dynamic but structural equation modeling of repeated measures data, in J.R. Nesselroade & R.B. Cattell (Eds.), *The Handbook of Multivariate Experimental Psychology*, (Vol. 2). New York, Plenum Press, 561–614.

McArdle, J.J. (1991). Principals versus principles in factor analysis. *Multivariate Behavioral Research*, 25(1), 81–87.

McArdle, J.J. (1994). Structural factor analysis experiments with incomplete data. *Multivariate Behavioral Research, 29*(4), 409–454.

McArdle, J.J. (1997). Modeling longitudinal data by latent growth curve methods. In G. Marcoulides (Ed.), *Modern Methods for Business Research*. Mahwah, NJ: Lawrence Erlbaum Associates, pp. 359–406.

McArdle, J.J. (1998). Test bias. In J.J. McArdle & R.W. Woodcock (Eds.), (1998). *Human Cognitive Abilities in Theory and Practice*. Mahwah, NJ: Lawrence Erlbaum Associates.

McArdle, J.J. (2009). Latent variable modeling of longitudinal data. *Annual Review of Psychology, 60*, 577–605.

McArdle, J.J. (2010). Ethical issues in factor analysis. In A. Panter (Ed.), *Statistics Through An Ethical Lens*. Washington, DC: APA.

McArdle, J.J. (2012). Exploratory data mining using CART in the behavioral sciences. In H. Cooper & A. Panter (Eds.), *Handbook of Methodology in the Behavioral Sciences* (Chapter 20). Washington, DC: APA.

McArdle, J.J. & Aber, M.S. (1990). Patterns of change within latent variable structural equation modeling. In A. von Eye (Ed.), *New Statistical Methods in Developmental Research*. New York: Academic Press, 151–224.

McArdle, J.J. & Cattell, R.B. (1994). Structural equation models of factorial invariance in parallel proportional profiles and oblique confactor problems. *Multivariate Behavioral Research, 29*(1), 63–113.

McArdle, J.J. & Epstein, D.B. (1987). Latent growth curves within developmental structural equation models. *Child Development, 58*(1), 110–133.

McArdle, J.J. & Hamagami, F. (1994). Logit and multilevel logit modeling studies of college graduation for 1984–85 freshman student-athletes. *The Journal of the American Statistical Association, 89*(427), 1107–1123.

McArdle, J.J. & Nesselroade, J.R. (1994). Using multivariate data to structure developmental change. In S.H. Cohen & H.W. Reese (Eds.), *Life-Span Developmental Psychology: Methodological Innovations*. Hillsdale, NJ: Erlbaum, pp. 223–267.

McArdle, J.J. & Wang, L. (2008). Modeling Age-Based Turning Points in Longitudinal Life-Span Growth Curves of Cognition (pp. 105–127). In P. Cohen (Ed.), *Applied data analytic techniques for turning points research, Multivariate applications series*. Mahwah, Erlbaum.

McCall, R.B., Appelbaum, M.I., Hogarty, P.A. (1973). Developmental Changes in Mental Performance. *Monographs of the Society for Research in Child Development*.

McDonald, R.P. (1985). *Factor Analysis and Related Methods*. Hillsdale, NJ: Erlbaum.

McDonald, R.P. (1999). *Test Theory: A Unified Treatment*. Mawah, NJ: Erlbaum.

McLachlan, G. & Peel, D. (2000). *Finite Mixture Models*. New York: Wiley.

McVicar, D. & Anyadike-Danes, M. (2002). Predicting successful and unsuccessful transitions from school to work by using sequence methods. *Journal of the Royal Statistical Society, Series A, 165*, Part 2, 317–334.

Meredith, W. & Horn, J.L. (2001). The role of factorial invariance in modeling growth and change. In L. Collins & A. Sayers (Eds.), *New Methods for the Analysis of Change*. Washington, DC: APA, pp. 203–240.

Miller, A.J. (1990). *Subset Selection in Regression*. London: Chapman & Hall.

Milligan, G.W. (1981). A review of Monte Carlo tests of cluster analysis. *Multivariate Behavioral Research, 16*, 379–407.

Mills, J.D., Olejnik, S.F., & Marcoulides, G.A. (2005). The Tabu search procedure: An alternative to the variable selection methods. *Multivariate Behavioral Research, 40*(3), 351–371.

Miloslava, K., Jiří, K., & Pavel, J. (2008). Application of decision trees in problem of air quality modeling in the Czech Republic locality. *WSEAS Transactions on Systems, 7*(10), 1166–1175.

Monahan, J., Steadman, H.J., Silver, E., Appelbaum, P.S., Robbins, P.C., Mulvey, E.P., Roth, L., Grisso, T., & Banks, S. (2001). *Rethinking Risk Assessment: The MacArthur Study of Mental Disorder and Violence.* Oxford: Oxford University Press.

Morgan, J.N. & Sonquist, J.A. (1963). Problems in the analysis of survey data: And a proposal. *Journal of the American Statistical Association, 58*, 415–434.

Mosteller, F. & Tukey, J.W. (1977). *Data Analysis and Regression: A Second Course in Statistics.* Reading, MA: Addison-Wesley.

Muthén, B. (2004). Latent variable analysis: Growth mixture modeling and related techniques for longitudinal data. In D. Kaplan (Ed.), *Handbook of Quantitative Methodology for the Social Sciences* (pp. 345–368). Newbury Park, CA: Sage Publications.

Nagin, D.S. (2005). *Group-Based Models of Development.* Cambridge: MA: Harvard University Press.

Nylund, K.L., Asparouhov, T., & Muthén, B.O. (2007). Deciding on the number of classes in latent class analysis and growth mixture modeling: A Monte Carlo simulation study. *Structural Equation Modeling, 14*(4), 535–569.

Osborne, R.T., & Suddick, D.E. (1972). A longitudinal investigation of the intellectual differentiation hypothesis. *Journal of Genetic Psychology, 121*, 83–89.

Penny, K., & Chesney, T. (2006). A comparison of data mining methods and logistic regression to determine associated with death following injury. In S. Zani, A. Cerioli, M. Riani, & M. Vichi (Eds.), *Data Analysis, Classification and the Forward Search.* Heidelberg: Springer-Verlag, pp. 417–423.

Put, R., Perrin, C., Questier, F., Coomans, D., Massart, D.L., & Vander Heyden, Y. (2002). Classification and regression tree analysis for molecular descriptor selection and retention prediction in chromatographic quantitative structure–retention relationship studies. *Journal of Chromatography A, 988*, 261–276.

Questier, F., Put, R., Coomans, D., Walczak, B., & Vander Heyden, Y. (2005). The use of CART and multivariate regression trees for supervised and unsupervised feature selection. *Chemometrics and Intelligent Laboratory Systems, 76*, 45–54.

Quinlan, J.R. & Cameron-Jones, R.M. (1995). Oversearching and layered search in empirical learning. In *Proceedings of the 14th International Joint Conference on Artificial Intelligence, Montreal,* Vol. 2 (Edited by M. Kaufman), pp. 1019–1024.

Riley, M.W. (1964). Sources and types of sociological data. In R.E. Faris (Ed.), *Handbook of Modern Sociology.* Chicago: Rand-McNally.

Rovlias, A. & Kotsou, S. (2004). Classification and regression tree for prediction of outcome after severe head injury using simple clinical and laboratory variables. *Journal of Neurotrauma, 21*(7), 886–893.

Sawitzki, G. (1994). Testing numerical reliability of data analysis systems. *Computational Statistics & Data Analysis, 18*(2), 269–286.

Scheffe, H. (1959). *The Analysis of Variance.* New York: Wiley.

Segal, M.R. (1992). Tree-structured methods for longitudinal data. *Journal of the American Statistical Association, 87*(418), 407–417.

Sephton, P. (2001). Forecasting recessions: Can we do better on MARS$^{(TM)}$? *Federal Reserve Bank of St. Louis Review, 83*(2), 39–49.

Seroczynski, A.D., Cole, D.A., Maxwell, S.E. (1997). Cumulative and compensatory effects of competence and incompetence on depressive symptoms in children. *Journal of Abnormal Psychology, 106*(4), 586–597.

Shmueli, G., Patel, N.R., & Bruce, P.C. (2007). *Data Mining for Business Intelligence*. New York: Wiley.

Sonquist, J. (1970). *Multivariate Model Building*. Ann Arbor, MI: Institute for Social Research.

Sonquist, J. & Morgan, J.N. (1964). *The Detection of Interaction Effects*. Ann Arbor, MI: Institute for Social Research.

Steadman, H., Mulvey, E., Monahan, J., Robbins, P., Appelbaum, P., Grisso, T., Roth, L., & Silver, E. (1998). Violence by people discharged from acute psychiatric inpatient facilities and by others in the same neighborhoods. *Archives of General Psychiatry, 55*, 393–401.

Steadman, H.J., Silver, E., Monahan, J., Applebaum, P.S., Robbins, P.C., Mulvey, E.P., Grisso, T., Roth, L.H., & Banks, S. (2000). A classification tree approach to the development of actuarial violence risk assessment tools. *Law and Human Behavior, 24*(1), 83–100.

Steinberg, D. & Colla, P. (1999). *MARS: An Introduction*. San Diego: Salford Systems.

Strobl, C. & Augustin, T. (2009). Adaptive selection of extra cutpoints—an approach towards reconciling robustness and interpretability in classification trees. *Journal of Statistical Theory and Practice, 3*(1), 119–135.

Strobl, C., Hothorn, T., & Zeileis, A. (2009). Party on! A new, conditional variable importance measure for random forests available in the party package. *The R Journal, 1*(2), 14–17.

Strobl, C., Malley, J., & Tutz, G. (2009). An introduction to recursive partitioning: Rationale, application and characteristics of classification and regression trees, bagging and random forests. *Psychological Methods, 14*(4), 323–348.

Strobl, C., Wickelmaier, F., & Zeileis, A. (2013). Accounting for individual differences in Bradley–Terry models by means of recursive partitioning. *Journal of Educational and Behavioral Statistics, 36*(2), 135–153.

Studer, M., Ritschard, G., Gabadinho, A., & Müller, N.S. (2011). Discrepancy analysis of state sequences. *Sociological Methods and Research, 40*(3), 471–510.

Su, X., & Tsai, C. (2005). Tree-augmented Cox proportional hazards models. *Biostatistics, 6*(3), 486–499.

Sullivan, P.F., Kovalenko, P., York, T.P., Prescott, C.A., & Kendler, K.S. (2003). Fatigue in a community sample of twins. *Psychological Medicine, 33*, 263–281.

Swets, J.A., Dawes, R.M., & Monahan, J. (2000). Psychological science can improve diagnostic decisions. *Psychological Science in the Public Interest, 1*(1), 1–26.

Tatsuoka, M.M. & Tiedeman, D.V. (1954). Discriminant analysis. *Review of Educational Research*, Washington, DC: AERA Press.

Temkin, N.R., Holubkov, R., Machamer, J.E., Winn, H.R., & Dikmen, S.S. (1995). Classification and regression trees (CART) for prediction of function at 1 year following head trauma. *Journal of Neurosurgery, 82*, 764–771.

Tibshirani, R.J. & Tibshirani, R. (2009). A bias correction for the minimum error rate in cross-validation. *The Annals of Applied Statistics, 3*(2), 822–829.

Tucker, L.R & Lewis, C. (1973). The reliability coefficient for maximum likelihood factor analysis. *Psychometrika, 38,* 1–10.

Tukey, J.W. (1962). The future of data analysis. *Annals of Mathematical Statistics, 33,* 1–67.

Tukey, J.W. (1977). *Exploratory Data Analysis.* Reading, MA: Addison-Wesley.

Ture, M., Tokatli, F., & Kurt, I. (2009). Using Kaplan–Meier analysis together with decision tree methods (C&RT, CHAID, QUEST, C4.5 and ID3) in determining recurrence-free survival of breast cancer patients. *Expert Systems with Applications, 36,* 2017–2026.

Vayssières, M.P., Plant, R.E., & Allen-Diaz, B.H. (2000). Classification trees: An alternative non-parametric approach for predicting species distributions. *Journal of Vegetation Science, 11,* 679–694.

Vermunt, J.K. & McCutcheon, A.L. (2004). *An Introduction to Modern Categorical Data Analysis.* CA: SAGE.

Wallace, K.A., Bergeman, C.S., & Maxwell, S.E. (2002). Predicting well-being outcomes in later life: An application of classification and regression tree (CART) analysis. In S. P. Shohov (Ed.), *Advances in Psychology Research* (Vol. 17). Hauppauge, NY: Nova Science Publishers, Inc., pp. 71–92.

Weber, G., Vinterbo, S., & Ohno-Machado, L. (2004). Multivariate selection of genetic markers in diagnostic classification. *Artificial Intelligence in Medicine, 31,* 155–167.

Wilcox, R. (2003). *Applying Contemporary Statistical Techniques.* Amsterdam: Academic Press.

Willingham, W.W., Lewis, C., Morgan, R., & Ramist, L. (1990*). Predicting College Grades: An Analysis of Institutional Trends over Two Decades.* New York: The College Board.

Witten, I.H. & Frank, E. (2005). *Data Mining: Practical Machine Learning Tools and Techniques* (2nd Edn). Amsterdam: Morgan Kaufman.

Yates, A. (1987). *Multivariate Exploratory Data Analysis.* Albany, NY: State University Press.

Zhang, H.P. (1997) Multivariate adaptive splines for longitudinal data. *Journal of Computational and Graphic Statistics, 6,* 74–91.

Zhang, H.P. (1999) Analysis of infant growth curves using MASAL. *Biometrics, 55,* 452–459.

Zhang, H.P. (2004) Multivariate adaptive splines in the analysis of longitudinal and growth curve data. *Statistical Methods in Medical Research, 13,* 63–82.

Zhang, H. & Singer, B. (1999). *Recursive Partitioning in the Health Sciences.* New York: Springer.

Zwick, W.R. & Velicer, W.F. (1986). Comparison of five rules for determining the number of components to retain. *Psychological Bulletin, 99*(3), 432–442.

2 CHAID and Earlier Supervised Tree Methods

Gilbert Ritschard

Introduction

The aim of this chapter is twofold. First we discuss the origin of tree methods. Essentially we survey earlier methods that led to CHAID (Kass, 1980; Biggs, De Ville, and Suen, 1991). The second goal is then to explain in detail the functioning of CHAID, especially the differences between the original method as described in Kass (1980) and the nowadays currently implemented extension that was proposed by Biggs et al. (1991).

Classification and regression trees, also known as recursive partitioning, segmentation trees or decision trees, are nowadays widely used either as prediction tools or simply as exploratory tools. Their interest lies mainly in their capacity to detect and account for non-linear effects on the response variable, and especially of even high order interactions between predictors. We demonstrate this interest while explaining the CHAID tree growing algorithm as it is implemented for instance in SPSS (2001) and by retracing the history of tree methods that led to it. We start with this latter point.

AID, THAID, ELISEE, and Other Earlier Tree Growing Algorithms

The first methods for inducing trees from data appeared in the survey analysis domain and were mainly developed by statisticians. We give here a short presentation of these earlier methods. See also the nice survey by Fielding and O'Muircheartaigh (1977).

Tree growing, also known as "hierarchical splitting," "partitioning," "group dividing," or "segmentation," finds its origin in the analysis of survey data. Perhaps the first published proposal is the one by Belson (1959). He addresses a matching issue that is in fact just a predictive one: Predicting the outcome for a second group given the outcome observed for the first one. Predictors as well as the outcome variable are dichotomized and the growing criterion used is the difference (for one of the two outcome categories) between the observed count and the number expected under the no association assumption. Other earlier proposals include those by Morgan and Sonquist (1963) who proposed the AID (Automatic Interaction Detector) algorithm for growing a binary regression tree, i.e., one in which the outcome variable is quantitative, and by Cellard, Labbé, and Savitsky (1967) who proposed ELISEE (Exploration of Links and Interactions through Segmentation of an Experimental Ensemble), which is a binary method for categorical dependent variables. The former regression tree method was popularized thanks to the AID computer programme developed at Ann Arbor by Sonquist, Baker, and Morgan (1971), while the latter segmentation method was popularized by Bouroche and Tenenhaus (1970, 1972). AID was certainly the most famous of these earlier programmes and Sonquist (1969) showed its interest as a complementary tool with multiple correlation analysis. See Thompson (1972) for a thorough comparison of AID and Belson's method. Messenger and Mandell (1972) and Morgan and Messenger (1973) extended AID for categorical outcomes using a so called theta criterion, which resulted in THAID (THeta AID). Gillo (Gillo, 1972; Gillo and Shelly, 1974) extended AID for multivariate quantitative outcome variables (MAID). Press, Rogers, and Shure (1969) developed an interactive tree growing tool allowing multibranching, IDEA (Interactive Data Exploration and Analysis). Independently from those developments in the survey data analysis framework, Hunt (see Hunt, Marin, and Stone, 1966) has proposed a series of decision tree induction algorithms called *Concept Learning Systems* (CLS-1 to CLS-9). Those algorithms were explicitly developed with an Artificial Intelligence perspective and are primarily intended for doing classification, i.e., for categorical response variables. CLS-1 to CLS-8 build binary trees, while the latter, CLS-9, allows multibranching.

Motivation behind the Earlier Methods

The motivation behind these first approaches is essentially to discover how the outcome variable is linked to the potential explanatory factors and more specifically to special configurations of factor values. If we except Hunt, authors are mainly interested in finding alternatives to the restrictions of the linear model, in which the effects of the explanatory variables are basically additive, i.e the effect of any variable is independent of the value taken by others. The primary concern is thus to detect important interactions, not to improve prediction, but just to gain better knowledge about how the outcome variable is linked to the explanatory factors.

> Particularly in the social sciences, there are two powerful reasons for believing that it is a mistake to assume that the various influences are additive. In the first place, there are already many instances known of powerful interaction effects—advanced education helps a man more

than it does a woman when it comes to making money, ... Second, the measured classifications are only proxies for more than one construct. ... We may have interaction effects not because the world is full of interactions, but because our variables have to interact to produce the theoretical constructs that really matter.

(Morgan and Sonquist, 1963, p. 416.)

Whenever these interrelationships become very complex—containing non-linearity and interaction—the usefulness of classical approaches is limited.

(Press, Rogers, and Shure, 1969, p. 364.)

It is interesting to look at the application domains considered in these earlier works. Indeed the need for such tree approaches followed the spread of survey analyses and the increasing place taken by empirical analyses within social sciences in the late 1950s. Belson (1959), for instance, developed his method for analysing survey data collected by the BBC for studying "the effects of exposure to television upon the degree to which individuals participate with others in the home in their various household activities." He thus grows a tree for "High" and "Low" degrees of joint activity. Later, in Belson (1978) he used a tree approach to investigate causal hypotheses concerning delinquent behavior. Morgan and Sonquist (1963) illustrate their AID method with an analysis of the differences in living expenses of families, on which "age and education can obviously not operate additively with race, retired status, and whether the individual is a farmer or not." Ross and Bang (1966) resort to AID for highlighting differences in family profiles that may explain differences in their chances of adoption. In the same vein, Orr (1972) makes an in depth study of transitions proportions between successive academic stages using AID on data from the National Survey of Health and Development. Interestingly in these latter applications, the response is a binary variable that is treated as quantitative. AID was also applied for instance in psychology (Tanofsky, Shepps, and O'Neill, 1969) and marketing (Armstrong and Andress, 1970; Assael, 1970). Press et al. (1969) illustrate their IDEA interactive technique with an example of questionnaire data about participation in an uprising. Gillo and Shelly (1974) discuss their multivariate tree method with an example where the outcome vector is the pattern of overall, job and leisure satisfactions. Cellard et al. (1967) is an exception in that their illustrations are not in the social science field. They present two applications, one in engineering where they are interested in the effects of different technical and environmental factors on the gripping of locomotive engines and one in marketing where they attempt to characterize standards expected by foreign hosts of French hotels. Nonetheless, their concern is still describing links and interactions and not making prediction or classification.

Table 2.1 Main earlier tree growing algorithms

Algorithm	Local Split	Dependent variable Quantitative	Dependent variable Categorical	Splitting criterion Association	Splitting criterion Purity	Splitting criterion p-value
Belson	binary		x	x		
AID	binary	x		x		
MAID	binary	x		x		
THAID	binary		x	x	x	
Hunt et al.	*n*-ary		x	x		
ELISEE	binary		x	x		
IDEA	*n*-ary	x	x	x		x
CHAID	*n*-ary	x	x	x		x

Splitting Criteria

The focus being on effects and interaction effects, the aim of these earlier methods (Table 2.1) was primarily to segment the data into groups with as different as possible distributions of the outcome variable. Therefore, the splitting criteria considered are naturally measures of the association strength between the outcome and split variables. This contrasts with more recent methods such as CART and C4.5 for instance, which are primarily oriented towards classification and prediction and attempt therefore to maximize the homogeneity of each group by means of purity measures. Note that the classification concern is, nevertheless, somehow implicitly present behind some of these earlier methods. The θ measure used in THAID is for instance just the classification error rate, and two (Shannon's entropy and Light and Margolin's (1971) variance of a categorical variable) among the three alternatives mentioned by Messenger and Mandell (1972) also are some kinds of purity measures.

The splitting criterion used depends indeed on the nature of the variables considered. Belson (1959) dichotomizes all variables. He looks at the 2×2 contingency table that cross-tabulates each explanatory factor with the outcome variable and compares this table with the one expected under the independence hypothesis. Except for its sign, the deviation is the same for all four cells. Belson's association criterion thus consists just in one of these deviances.

In AID, the dependent variable is quantitative and the splitting criterion proposed in Morgan and Sonquist (1963) is the largest reduction in unexplained sum of squares. The latter is commonly known as the residual or within sum of squares, WSS, in analysis of variance. It reads:

$$\text{WSS} = \sum_{j=1}^{g} \sum_{i=1}^{n_j} (y_{ij} - \bar{y}_j)^2$$

where $\bar{y}_j$ is the mean value of the y_{ij}s in node j. Here it is the WSS for the $g = 2$ groups that would be produced by the split. Maximizing this reduction is equivalent to maximizing the η^2 coefficient, i.e., the ratio BSS/TSS, where TSS is the total sum of squares (before the split)

$$\text{TSS} = \sum_{j=1}^{g} \sum_{i=1}^{n_j} (y_{ij} - \bar{y})^2,$$

which is independent of the split variable, and BSS $=$ TSS $-$ WSS the resulting between sum of squares. Hence, it is some sort of R^2 association measure. MAID (Gillo and Shelly, 1974) uses a generalized version of this criterion applicable in the multivariate case. The proposed generalization is indeed a variant of Wilks' Λ, namely $1 - k/\text{tr}(\mathbf{TW}^{-1}) = \text{tr}(\mathbf{BW}^{-1})/\text{tr}(\mathbf{TW}^{-1})$, where $\mathbf{T}$, $\mathbf{W}$ and $\mathbf{B} = \mathbf{T} - \mathbf{W}$ are respectively the total, within and between cross product matrices among the k dependent variables.

Kass (1975) introduces a statistical significance criteria for AID, namely the p-value of the BSS/TSS ratio that he evaluates through a distribution-free permutation test. A Chi-square approximation of this test is proposed in Scott and Knott (1976).

In the interactive IDEA system, Press et al. (1969) consider also a scaled indicator of the proportion of explained variation for the case of quantitative outcome. For categorical outcome variables they resort to the significance of the independence Pearson Chi-square for the table that cross-tabulates the predictor with the outcome discrete variable. ELISEE is also based on the Chi-square. Its authors indeed consider the squared ϕ, which is some normalized Chi-square obtained by dividing it by the number of cases. Since ELISEE considers only binary splits, the table that cross-tabulates at each node the split with the outcome always has only two columns. The ϕ criterion is in such cases equivalent to Cramer's v.

CHAID (Chi-square Automatic Interaction Detector), introduced by Kass (1980) as an evolution of AID and THAID, is certainly nowadays the most popular among these earlier statistical supervized tree growing techniques. We describe its functioning in detail hereafter.

CHAID

As indicated by its name, CHAID uses a Chi-square splitting criterion. More specifically, it uses the p-value of the Chi-square. In his 1980 paper in *Applied Statistics*, Kass discusses only the case of a categorical dependent variable. The method is, nevertheless, most often implemented with an option for also handling quantitative dependent variables. The criterion is in that case the p-value of the F statistic for the difference in mean values between the g nodes generated by the split:

$$F = \frac{\text{BSS}/(g-1)}{\text{WSS}/(n-g)} \sim F_{(g-1),(n-g)}.$$

An alternative could be using the Kass's (1975) permutation test or its χ^2 approximation (Scott and Knott, 1976).

The main characteristics of CHAID that contributed to its popularity are:

1 At each node, CHAID determines for each potential predictor the optimal n-ary split it would produce, and selects the predictor on the basis of these optimal splits.
2 CHAID uses p-values with a Bonferroni correction as splitting criteria.

Resorting to p-values as growing criteria provides stopping rules that automatically account for statistical significance. Thresholds are naturally set to the usual critical values considered for statistical significance, namely 1%, 5%, or 10%. Such p-value criteria are sensitive to the number of cases involved in the split and tend to avoid splitting into too small groups. Note that though CHAID popularized the idea of using p-values for selecting predictors, it was not the first attempt to do so. As mentioned above, p-values were also used by Press et al. (1969) in their IDEA system. The originality of the method proposed by Kass is, however, that for evaluating the significance of each split it applies first a Bonferroni correction on the p-value. We will explain later what this consists of.

The most original contribution of CHAID is no doubt the first point, i.e., the idea of looking at the optimal n-ary split for each predictor. First, as shown in Table 2.1, most methods considered only binary splits. Those that allow for n-ary splits, such as the method by Hunt et al. (1966), set the number of splits to the number of categories of the potential predictor, which could be particularly unsuited for predictors with many categories.

We explain how CHAID works by means of a real world example data set. Let us first describe these data.

Illustrative Data

We consider administrative data about the 762 first year students who were enrolled in fall 1998 at the Faculty of Economic and Social Sciences (ESS) of the University of Geneva (Petroff, Bettex, and Korffy, 2001). The data will be used to find out how the situation (1, eliminated; 2, repeating first year; 3, passed) of each student after the first year is linked to her/his available personal characteristics. The response variable is thus the student situation in October 1999. The predictors retained are birth year, year when first registered at the University of Geneva, chosen orientation

Table 2.2 Situation after first year by type of secondary diploma

	Classic Latin 1	Modern 2	Scientific 3	Economics 4	Technical 5	Non-Swiss 6	None 7	Missing 8	Total
Eliminated	23	44	18	37	5	76	6	0	209
Repeating	16	22	13	36	0	43	0	0	130
Passed	87	68	90	96	4	71	6	1	423
Total	126	134	121	169	9	190	12	1	762

(Social Sciences or Business and Economics), type of secondary diploma achieved (classic/Latin, scientific, economics, modern, technical, non-Swiss, none, missing), place where secondary diploma was obtained (Geneva, Switzerland outside Geneva, abroad), age when secondary diploma was obtained, nationality (Geneva, Swiss except Geneva, Europe, non-Europe) and mother's living place (Geneva, Switzerland outside Geneva, abroad).

Optimal n-ary Split

Consider the contingency (Table 2.2) between the dependent variable (situation after the first year) and the type of secondary diploma. The latter predictor has $c = 8$ categories, namely: classic or Latin, scientific, economics, modern, technical, non-Swiss, no secondary diploma, and missing. Ideally, we would like to look at all possibilities for segmenting the population by means of these eight categories. The number of such possibilities is given by the number of ways of splitting into two groups, plus the number of ways of splitting into three groups, and so on until the one way of splitting into c groups.

When, as in our case, the predictor is purely nominal, i.e., there is no restriction as to how to partition the c categories, the total number of ways of partitioning them is known as Bell's (1938) number $B(c)$ and may be obtained through the recursive formula:

$$B(c) = \sum_{g=0}^{c-1} \binom{c-1}{g} B(g)$$

with $B(0)$ set equal to 1. Alternatively, $B(c)$ may be expressed as

$$B(c) = \sum_{g=1}^{c} S(c,g)$$

where $S(c,g)$ is the Stirling number of the second kind giving the number of ways of splitting c values into g groups:

$$S(c,g) = \sum_{i=0}^{g-1} \frac{(-1)^i (g-i)^c}{i!(g-i)!}. \tag{1}$$

For our eight categories, this gives $B(8) = 4,140$ possibilities. We get the number of segmentation possibilities by subtracting 1 from this number since grouping all eight categories into a single class would not be a split. This gives 4,139 possibilities.

In the case of an ordinal predictor where the categories are naturally ordered, the groups should be constituted by contiguous categories only. The number of such partitions is:

$$G(c) = \sum_{g=0}^{c-1} \binom{c-1}{g} = 2^{(c-1)}.$$

For $c = 8$, we would have $G(8) = 128$, and hence $128 - 1 = 127$ possibilities of segmentation. This is much less than in the nominal case, but still increases exponentially with the number of categories c.

To avoid scanning all possibilities, Kass (1980, p. 121) suggests the following heuristic:

Step 1   Find the pair of categories of the predictor (only considering allowable pairs as determined by the type of the predictor) whose $2 \times r$ sub-table (r being the number of categories of the dependent variable) is least significantly different. If this significance does not reach a critical value, merge the two categories, consider this merger as a single compound category, and repeat this step.

Step 2   For each compound category consisting of three or more of the original categories, find the most significant binary split (constrained by the type of the predictor) into which the merger may be resolved. If the significance is beyond a critical value, implement the split and return to Step 1.

To illustrate our example, we provide in Table 2.3 the Pearson Chi-square computed for each pair of the original $c = 8$ categories.

The smaller Chi-square of .06 is obtained for the pair $\{5,7\}$, i.e., for technical diploma and no secondary diploma. The corresponding 3×2 table is shown as Table 2.4. The degrees of freedom are $df = (3-1)(2-1) = 2$, and the p-value of the obtained Chi-square is $p(\chi_2^2 \geq .06) = 96.7\%$. In the general case where the dependent variable has r categories, the p-values are obtained from a Chi-square distribution with $r - 1$ degrees of freedom.

Table 2.3 Chi-squares and their *p*-values by pair of categories

	1	2	3	4	5	6	7	8
1	0	9.62	0.87	5.25	7.53	30.64	7.37	.45
2	.008	0	15.66	4.79	2.81	5.88	2.92	.96
3	.647	.000	0	9.88	9.84	40.80	9.65	.34
4	.073	.091	.007	0	6.25	16.65	6.37	.76
5	.023	.245	.007	.044	0	2.66	.06	1.11
6	.000	.053	.000	.000	.264	0	3.47	1.66
7	.025	.232	.008	.041	.969	.177	0	.93
8	.800	.618	.842	.685	.574	.436	.629	0

Chi-squares are in the upper triangle and *p*-values in the lower triangle.

Table 2.4 3 × 2 contingency table with smallest Chi-square

	Technical Diploma 5	No secondary diploma 7	Merged {5,7}
Eliminated	5	6	11
Repeating	0	0	0
Passed	4	6	10
Total	9	12	21

Table 2.5 Chi-squares and their *p*-values by pair of categories

	1	2	3	4	{5,7}	6	8
1	0	9.62	0.87	5.25	12.98	30.64	.45
2	.008	0	15.66	4.79	5.44	5.88	.96
3	.647	.000	0	9.88	16.40	40.80	.34
4	.073	.091	.007	0	11.63	16.65	0.76
{5,7}	.002	.066	.000	.003	0	5.97	1.05
6	.000	.053	.000	.000	.0505	0	1.66
8	.800	.618	.842	.685	.592	.436	0

Chi-squares are in the upper triangle and *p*-values in the lower triangle.

We then repeat the same process on the 3 × 7 table obtained by replacing columns 5 and 7 of Table 2.2 with the single merged column {5,7}. The pairwise Chi-squares and their *p*-values are given in Table 2.5. The last significant Chi-square is .34 (*p*-value = 84.6%) and corresponds to the pair {3,8}, i.e., scientific and missing. We merge these two categories and iterate until we get significant Chi-squares for all remaining pairs.

Table 2.6 Successive merges

Iteration	Merge	Chi-square	p-value
1	{5,7}	.06	96.7%
2	{3,8}	.34	84.6%
3	{1,{3,8}}	.95	62.3%
4	{2,4}	4.79	9.1%
5	{6,{5,7}}	5.97	5.05%

Table 2.7 Final Chi-squares/p-values

	Classic/Latin scientific, missing {1,3,8}	Modern economics {2,4}	Technical, none non-Swiss {5,6,7}
{1,3,8}	0	18.04	52.94
{2,4}	.000	0	14.56
{5,6,7}	.000	.001	0

Chi-squares are in the upper triangle and p-values in the lower triangle.

The successive merges are recapitulated in Table 2.6. The process ends at iteration 6, with the Chi-squares and p-values depicted in Table 2.7.

The next step (Step 2) is to check that none of the compound categories formed by more than two categories ({1,3,8} and {5,6,7}) can be significantly dichotomized. This is a top-down way of testing the homogeneity of the groups obtained in Step 1 in a bottom-up manner. For group {1,3,8}, we have to check that splitting it into {1,3} and {8} is not significant. The other possibility has already been considered in the third row of Table 2.6. Likewise, for group {5,6,7}, we have to check that splitting into {5,6} versus {7} is not significant, the other possibility corresponding to the last row in Table 2.6. The results are

Split	Chi-square	p-value	
{1,3} versus {8}	.39	82.1%	non–significant
{5,6} versus {7}	2.28	19.4%	non–significant

Hence, there is no significant way of splitting the three groups obtained. The best way of segmenting the whole data by means of the type of secondary diploma is thus according to the heuristic: {classic/Latin, scientific, missing}, {modern, economics} and {technical, non-Swiss, no secondary diploma}.

Ordinal and Floating Predictors

We do not detail here the grouping process for all other predictors. It is worth mentioning, however, that "Birth year," "Year when first registered," and "Age at secondary diploma" have ordered categories. Such predictors, are called *ordinal* or *monotonic* and receive a special treatment allowing merges only of adjacent categories. In fact, beneath purely *nominal* (also called *free*) and *ordinal* (*monotonic*) predictors, Kass (1980) also distinguishes a third type of variable which he calls a *floating predictor*. This refers to variables that have ordered categories except for one of them (usually the missing value), which cannot be positioned on the ordinal scale. Such predictors are dealt with like ordinal ones except for the floating category, which is allowed to merge with any other category or compound category. "Year when first registered," which has one missing value, is such a floating predictor. It takes 11 different ordered year values and one missing category. "Age at secondary diploma" would also have been such a floating predictor because of those few students that were exceptionally accepted at the ESS Faculty without having obtained their secondary diploma. However, it was already recoded into four categories in the data set we used, students "without secondary diploma" being grouped with the older age class. Hence, this four category predictor is simply considered as an ordinal one.

To illustrate how floating predictors are handled, let us detail the merging process for the "Year when first registered" variable. Its cross tabulation with the target variable is shown in Table 2.8. For the first four older enrollment dates the distribution is the same: all concerned students were eliminated at the end of the first year. The four categories being adjacent, we merge them together. After this merge, the mere Chi-squares to be considered are shown in Table 2.9. Empty cells correspond to non-adjacent categories, and hence to unfeasible merges. Note that there are no such empty cells in the row and column corresponding to the "missing" category. The table clearly shows the reduction in the number of pairs to examine that is achieved by taking the order of categories into consideration.

The first suggested merge is to put the missing value with the group of the older registration dates (≤ 91). The Chi-square is 0, since the

Table 2.8 Situation after first year by "Year when first registered"

	77	85	89	91	92	93	94	95	96	97	98	*Missing*	*Total*
Eliminated	1	1	2	2	2	2	3	9	12	31	143	1	209
Repeating	0	0	0	0	0	1	0	2	7	10	110	0	130
Passed	0	0	0	0	2	3	4	4	24	39	347	0	423
Total	1	1	2	2	4	6	7	15	43	80	600	1	762

Table 2.9 Chi-squares and their *p*-values by pair of categories of the floating predictor "Year when first registered"

	≤ 91	92	93	94	95	96	97	98	*Missing*
≤ 91	0	3.75							0
92	0.153	0	.83						0.83
93		.659	0	1.27					1.56
94			.529	0	2.41				1.14
95				.300	0	5.18			0.64
96					.075	0	1.50		2.44
97						.472	0	8.53	1.55
98							.014	0	3.18
Missing	1	.659	.459	.565	.726	.295	.461	.204	0

Chi-squares are in the upper triangle and *p*-values in the lower triangle.

distribution is the same for "missing" and "≤ 91." Without showing here the Chi-squares recomputed to account for the previous merged categories, it seems quite obvious that the next merges will be "92" with "93," then the resulting group with "94," and so on. We leave it to the reader to check that the merging process ends up with two groups, namely "≤ 97 or missing" and "98," the latter being the group of those students who enrolled for the first time when they started their first year at the ESS Faculty. The final Chi-square is 18.72 for two degrees of freedom and its *p*-value 0.00863%, making it clear that the two final groups have dissimilar distributions.

Growing the Tree

Having explained how the categories of each predictor are optimally merged at each node, we can now describe the tree growing process. We start with a simple growing scheme based on the classical non-adjusted *p*-value of the independence Chi-square for the table that cross tabulates, at the concerned node, the target variable with the predictor. We will afterwards explain the Bonferroni adjustment proposed by Kass (1980) and show that using it may generate a different tree.

Table 2.10 summarizes the results obtained at the root node by applying the merging heuristic to the eight considered predictors. The Chi-square reported is the one for the cross-tabulation between the dependent target variable (*Status after first year*) and the concerned predictor with optimally merged categories. The number of the latter ones is indicated under #*splits*. Since we have three statuses (eliminated, repeating, passed) for the dependent variable, the number of degrees of freedom *df* is in each

Table 2.10 Summary of possible first level splits

Predictor	#categories	#splits	Chi-square	df	p-value
Type of secondary diploma	8	3	54.83	4	.000000000035
Birth year	25	3	53.01	4	.000000000085
Where secondary diploma	3	3	42.66	4	.0000000122
Mother living place	4	2	27.21	2	.00000123
Nationality	3	2	24.26	2	.00000540
Year when first registered	11+1	2	18.72	2	.0000863
Age at secondary diploma	4	4	21.86	6	.00128
Chosen orientation	2	2	1.39	2	.499

case $(3 - 1)(\#\text{splits} - 1)$. This set of information is the one considered for choosing the best first split in the tree growing process.

Table 2.10 reveals that the *Type of secondary diploma* grouped into the three groups found above has the smallest *p*-value. It is thus the most significantly linked with the *Status after first year* and hence the most discriminating factor. We select it for generating the first split.

We then iterate the process at each of the resulting nodes. Figure 2.1 shows the tree we obtain. At level 1 we notice that a different splitting variable is used for each node. The best splitting variable for the group formed by the 248 students who have a secondary diploma in the classic/Latin or scientific domain is the ordinal birth year variable with its values merged into three birth year classes, namely 1976 and before, 1977–78, and after 1978. These birth year classes correspond to age classes "22 and more," "20–21" and "less than 20" for students in 1998 when they started their first year at the ESS Faculty. The 303 students with a modern or economic oriented secondary diploma are split according to their chosen orientation in the ESS Faculty, which takes only two values. Finally, the remaining 211 students are distinguished according to their enrollment date. Indeed, the distinction is between those for whom the first year at the ESS Faculty was the first year at the University of Geneva, from those who spent already at least one year at the University.

We may wonder why the tree stopped growing after level 3. This is because each leaf (terminal node) met at least one stopping criterion. With CHAID, four stopping rules are classically considered:

- an α_{split} threshold for the splitting *p*-value above which CHAID does not split the node (was set to 5%);
- a maximal number of levels $\max_{level}$ (was set to 5);
- a minimal parent node size $\min_{parent}$ (was set to 100), meaning that CHAID does not try to split a node with less than $\min_{parent}$ cases;
- a minimal node size $\min_{node}$ (was set to 50), meaning that CHAID considers only splits into groups with each at least $\min_{node}$ cases.

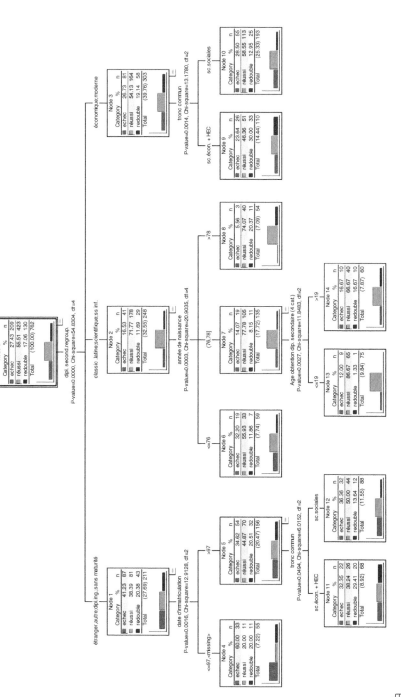

Figure 2.1 CHAID, without Bonferroni adjustment.

In the tree of Figure 2.1 there are seven out of nine leaves with less than 100 cases, which thus meet our $\min_{node} = 100$ constraint. For the two remaining ones, CHAID found no further significant split, all p-values exceeding the $\alpha_{split} = 5\%$ threshold.

Bonferroni Adjustment

As already mentioned in the introduction of this section, one of the main characteristics of CHAID is the use of Bonferroni adjusted p-values. The aim of the Bonferroni adjustment is to account for multiple testing. For instance, at level 1 we have seen that the p-value for the optimally merged 'Type of secondary diploma' predictor is 0.000000000035. Since this corresponds to the optimal n-ary grouping of the categories, the p-value for any other possible partition should be greater than this value. This supposes that we have also tested these other solutions, hence the multiple tests. For the best split to be non-statistically significant, all other possibilities should also be non-significant. Assuming independence between the m tests and the same type I error probability α for each of them, the total type I error probability is $m\alpha$. To ensure a total probability of α, the Bonferroni correction consists thus in lowering the critical value α for the sole test considered, by dividing it by the number m of underlying tests, or alternatively, what CHAID does, by multiplying the p-value of the optimal solution by m. Since it ignores dependences between tests, the Bonferroni correction is conservative and known to be often much too restrictive (Abdi, 2007). Furthermore, for our splitting issue, it is not so evident to determine the number of tests to take into consideration.

Kass (1980) proposes, as an approximation, to set the Bonferroni multiplier m to the number of ways a c category predictor can be reduced to g groups, g being the final number of optimally merged categories. The way of calculating m depends indeed on the nature of the predictor. We give hereafter the formulae used by the classical CHAID algorithm for each of the three types of predictors.

Purely nominal (free) predictors.   The number m of ways of partitioning c categories into g groups is given by the Stirling number of the second kind (Equation 1):

$$m = \sum_{i=0}^{g-1} \frac{(-1)^i (g-i)^c}{i!(g-i)!}.$$

For instance for partitioning $c = 4$ categories a, b, c, d into $g = 2$ groups, there are $m = (2^4/2) + (-1) = 7$ possibilities, namely the seven ones depicted in Table 2.11.

Ordinal (monotonic) predictors.   Here the groups are non-overlapping subsequences of categories. The first such subsequence starts necessarily

Table 2.11 The seven ways of grouping a, b, c, d into two groups

i	Group 1	Group 2	i	Group 1	Group 2
1	a	b, c, d	5	a, b	c, d
2	b	a, c, d	6	a, c	b, d
3	c	a, b, d	7	a, d	b, c
4	d	a, b, c			

with the first category. Hence m is the number of ways we can choose the starting points of the other $g - 1$ subsequences among the $c - 1$ remaining categories, that is

$$m = \binom{c-1}{g-1}.$$

For $c = 4$ ordered categories a, b, c, d, there are thus $m = 3!/2 = 3$ possibilities to partition them into two groups, namely $\{a, bcd\}$, $\{ab, cd\}$, and $\{abc, d\}$.

Floating predictors

$$m = \binom{c-2}{g-2} + g\binom{c-2}{g-1} = \frac{g-1+g(c-g)}{c-1}\binom{c-1}{g-1}.$$

Thus three ordered categories a, b, c and one floating value f can be grouped in $m = (1 + 2 \cdot 2)/3)3 = 5$ ways into two groups, namely the five possibilities shown in Table 2.12.

Table 2.13 gives the Bonferroni adjusted p-value for each of our eight predictors at the root node. The table should be compared with the values in Table 2.10. The *rank* column in Table 2.13 reveals changes in the ranking of the predictors as compared with what resulted from the unadjusted p-values. With a multiplier of 966 the 'Type of secondary diploma' is now only the third best predictor, while the retained splitting variable will here be 'Where secondary diploma', which has three categories only.

This example shows clearly that predictors with many categories tend to be more penalized by this Bonferroni correction than predictors with

Table 2.12 The five ways of grouping three ordered values a, b, c and one float f into two groups

i	Group 1	Group 2	i	Group 1	Group 2
1	a, f	bc	4	ab, f	c
2	a	bc, f	5	ab	c, f
3	abc	f			

Table 2.13 Bonferroni adjusted *p*-values, Kass's method

Predictor		p-value	Multiplier	Adjusted p-value	Rank
Type of secondary diploma	Nominal	.000000000063	966	.0000000341	3
Birth year	Ordinal	.000000000085	276	.0000000234	2
Where secondary diploma	Nominal	.0000000122	1	.0000000122	1
Mother living place	Nominal	.00000123	7	.00000864	4
Nationality	Nominal	.00000540	3	.0000162	5
Year when first registered	Floating	.0000863	19	.00164	7
Age at secondary diploma	Ordinal	.00128	1	.00128	6
Chosen orientation	Nominal	.499	1	.499	8

few categories. Also, nominal predictors are more penalized than ordinal or floating predictors. This is somehow justified since the more there are split possibilities, the greater the chance to find the best split among them. Nevertheless, the correction often seems to be excessive, and hence reverses the bias in favor of a predictor offering few splitting alternatives.

Using the Kass proposition of the Bonferroni correction, we get the grown tree shown in Figure 2.2. The tree is quite different from that (Figure 2.1) obtained with uncorrected *p*-values. First, it looks less complex. This is indeed a direct consequence of the implicit reinforcement of the α_{split} stopping rule that follows from the Bonferroni adjustment. Second, as depicted by the detailed table for the first split (Table 2.13), applying the correction may change at each node the ranking of the predictors, and hence the splitting predictor used.

Exhaustive CHAID

Biggs et al. (1991) propose two important improvements to the CHAID method.

1   A more thorough heuristic for finding, at each node, the optimal way of grouping the categories of each predictor.
2   A better suited approximation for the Bonferroni correction factor that avoids discriminating nominal variables with a large number of categories.

The improved CHAID method is commonly known as Exhaustive CHAID. The name is confusing however, since though it is more thorough, the search is still not exhaustive. Furthermore, it refers only to the first of

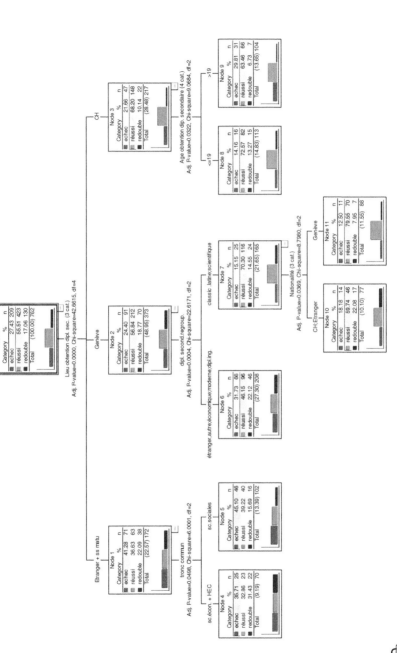

Figure 2.2 CHAID, with Bonferroni adjustment.

Table 2.14 p-values for each of the best split into $g = 2, \ldots, c$ groups

g	Best χ^2	df	p-value	Smallest group size
8	67.85	14	.000000004725	1
7	67.76	12	.000000000839	1
6	67.50	10	.000000000135	21
5	66.72	8	.000000000022	21
4	61.92	6	.000000000018	21
3	54.83	4	.000000000035	211
2	39.64	2	.000000002473	248

the two improvements, while differences between CHAID and Exhaustive CHAID trees more often result from the alternative Bonferroni adjustment.

Extended Search for the Optimal g-way Split

Kass's (1980) heuristic merges iteratively pairs of similar distributed columns, considering indeed only pairs allowed by the nominal, ordinal or floating nature of the predictor, until no pair can be found with a non-significant associated Chi-square. Biggs et al. (1991) propose pursuing the merging process until we obtain only two groups. The best g-way split is then determined by seeking among the thus determined successive best $c - 1$ groupings into $c, c - 1, \ldots, 2$ classes, the one for which the cross-tabulation with the response variable has the most significant Chi-square.

The floating category of a floating predictor is also handled differently. Kass's heuristic considers merges involving the floating category at each iteration, while Biggs et al.'s proposition is to first seek the best merge of the ordered categories and only afterwards choose between merging the floating category with the most alike group or keeping it as a category per se.

To illustrate, consider again our example of the "Type of secondary diploma" at the root node. With Kass's heuristic we stopped with Table 2.7, in which all Chi-squares are statistically significant. Biggs et al.'s heuristic goes one step further and merges {2, 4} with {5, 6, 7}, that is, it also considers the best grouping into two classes, namely {classic/Latin, scientific, missing} and {modern, economics, technical, non-Swiss, none}. It looks then at the p-value for the cross tabulation of each grouping with the response variable. There are $c - 1$ such tables with respectively $c, c - 1, \ldots, 2$ columns. Table 2.14 reports these values. The lowest p-value is attained for $g = 4$ groups, which differs from the partitioning into three groups suggested by Kass's heuristic. The difference concerns the group {technical, non-Swiss, none}, which in this four group solution is broken down into {non-Swiss} and {technical, none}.

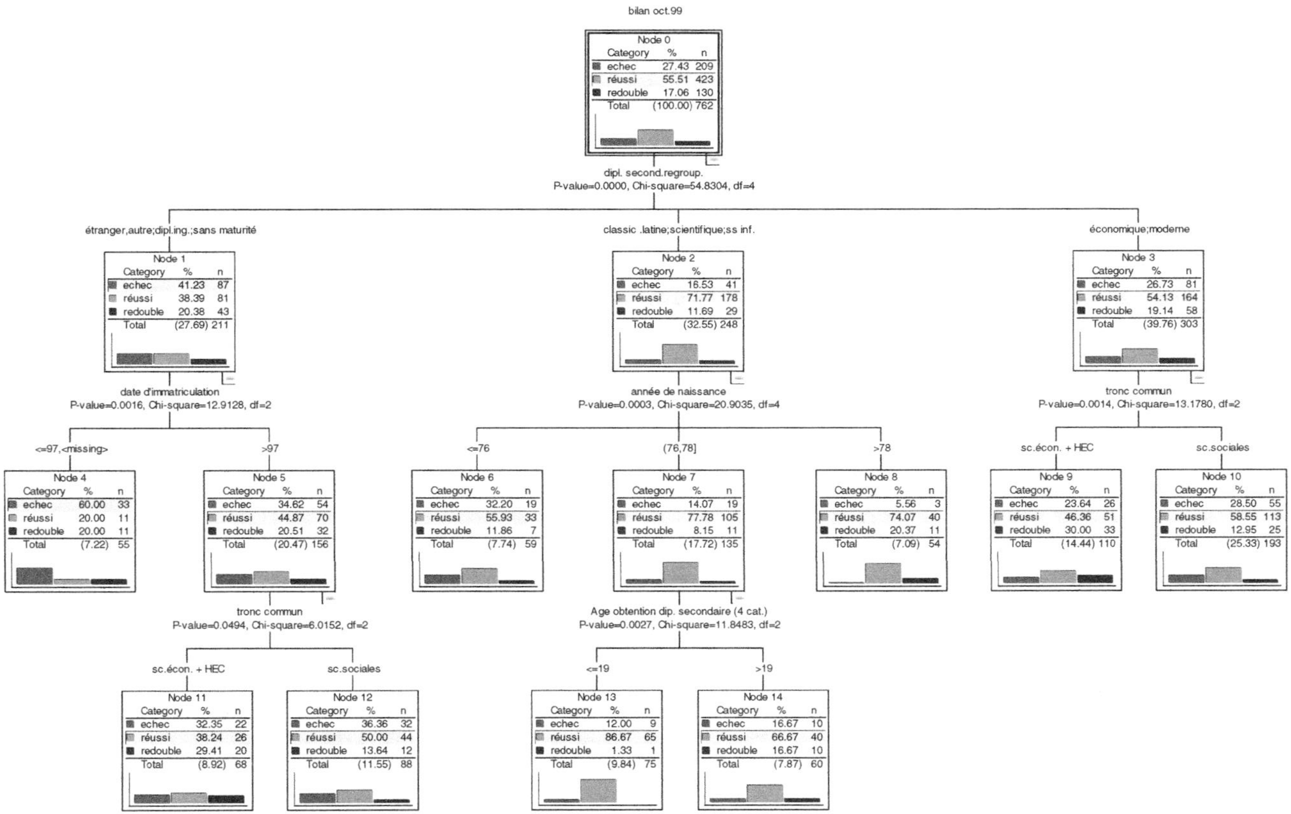

Figure 2.3 XCHAID, without Bonferroni adjustment.

Note that the best g-way grouping must indeed comply with the minimal node size constraint if we want to use it effectively for splitting the node. From Table 2.2 it is readily shown, for example, that the {technical, none} group of the four group solution contains only 21 cases and does not fit our $\min_{node}$ constraint that was set at 50. Hence this solution cannot be retained. Furthermore, since the solution into g groups is derived from that into $g - 1$ groups, this same stopping rule would indeed also preclude using any of the 'best' solutions into $g \geq 4$ groups. Hence, in our example, the only choice is to split into two or three groups. The solution into two groups has a much higher p-value, which makes it less interesting. Eventually, we end up again with the same partition into three groups as before.

Growing the tree on our data with this refined best g-way grouping method—but without using Bonferroni adjustments—we get the tree shown in Figure 2.3. Note that it is the same as that (Figure 2.1) obtained with Kass's best g-way grouping heuristic and no Bonferroni adjustment. This illustrates the small impact of this first refinement by Biggs et al.'s.

Revised Bonferroni Correction

We have seen that the Bonferroni multiplier proposed by Kass excessively penalizes predictors with many categories. The penalization is at its highest when the number of formed groups is close to $c/2$ and favors for instance splits into two groups rather than three, since the multiplier is much smaller for $g = 2$ than for $g = 3$. Moreover, we may also wonder why Kass considers only the alternative ways of grouping into g groups, and disregards those into $g + 1, g + 2, \ldots, c$ groups. These are the main weaknesses addressed by Biggs et al. (1991).

Those authors propose a Bonferroni correction taking explicitly into account the fact that when looking for the best split into $k - 1$ groups, one only explores groupings that result from the optimal solution into k groups. For non-floating predictors with c categories, their Bonferroni multiplier thus reads:

$$m_B(c) = 1 + \sum_{k=2}^{c} m(k, k-1) \tag{2}$$

where $m(k, k - 1)$ denotes the number of ways of grouping k categories into $k - 1$ groups and the '1' stands for the trivial partitioning into c values. Unlike Kass's solution, this formula is related to the iterative working of the merging heuristic and is therefore better sounded. In formula (2) the value of $m(k, k - 1)$ indeed depends on the either ordinal or nominal nature of the predictor.

Table 2.15 Biggs et al.'s Bonferroni multipliers

c	Nominal $m_B^{nom}(c)$	Ordinal $m_B^{ord}(c)$
3	4	3
4	10	6
5	20	10
6	35	15
7	56	21
8	84	28
9	120	36
10	165	45

- For nominal (free) predictors it is given by the corresponding Stirling number of the second kind:

$$m(k, k-1) = S(k, k-1) = \sum_{i=0}^{k-2} (-1)^i \frac{(k-1-i)^k}{i!(k-1-i)!}.$$

- For ordinal (monotonic) predictors it is just:

$$m(k, k-1) = k - 1.$$

The Bonferroni multiplier $m_B^{float}(c)$ for a *floating predictor* with c values is the one $m_B^{ord}(c-1)$ for an ordinal variable with $c-1$ categories plus the number of ways of placing the floating category. There are at most $c-1$ groups into which the floating category can be placed, or it can be kept as a separate group. Hence, the multiplier for floating predictors is:

$$m_B^{float}(c) = m_B^{ord}(c-1) + (c-1) + 1 = m_B^{ord}(c),$$

which is the same as for ordinal (monotonic) predictors.

An important point is that whatever the nature of the predictor, Biggs et al.'s proposition is, unlike Kass's solution, independent of the final number of groups retained. It depends upon c only. Table 2.15 reports the values of the multiplier until $c = 10$ categories. Though increasing exponentially with c, the correction factors grow much more slowly than Kass's solution. For the "Type of secondary diploma" that is selected as the best splitting attribute at the root node, the multiplier is for instance 84 with Biggs et al.'s proposition, while it was 966 with Kass's solution for the retained partition into three groups. This is more than a factor of 10 difference.

To illustrate the use of Biggs et al.'s Bonferroni adjustment, we consider again the possibility of splitting the root node by using the best k-way groupings of each predictor. The latter groupings are the same as those

Table 2.16 Bonferroni adjusted *p*-values, Biggs et al.'s method

Predictor		*c*	p-*value*	$m_B(c)$	*Adj* p-*value*	*Rank*
Type of secondary diploma	Nominal	8	.000000000035	84	.0000000030	1
Birth year	Ordinal	25	.000000000085	300	.0000000254	2
Where secondary diploma	Nominal	3	.0000000122	4	.0000000487	3
Mother living place	Nominal	4	.00000123	10	.0000123	4
Nationality	Nominal	3	.00000540	4	.0000216	5
Year when first registered	Floating	11	.0000863	55	.00475	6
Age at secondary diploma	Ordinal	4	.00128	6	.00770	7
Chosen orientation	Nominal	2	.499	1	.499	8

obtained with Kass's methods, which allows us to start again from the information gathered in Table 2.10. The *p*-values without adjustment and the adjusted values are reported in Table 2.16. We may notice that the ranking of the predictor remains unchanged, meaning that here and unlike what happened with Kass's Bonferroni adjustment, the same predictor "Type of secondary diploma" is retained with and without Bonferroni adjustment.

Iterating the same process at each new node obtained, we end up with the tree shown in Figure 2.4. Comparing this tree with the one in Figure 2.3 reveals that the use of Biggs et al.'s Bonferroni adjustment did not change the retained splitting predictor at any node. We end up however with a slightly less complex tree. The first level node corresponding to those students who had none or a technical or non-Swiss secondary diploma is not split, while it was in the first tree. This is because there is now at this node no possible split for which the critical adjusted *p*-value is below the critical value (set at 5%).

Conclusion

Despite their many advantages and obvious efficiency for detecting interactions that matter, and more generally as exploratory tools, tree methods were also more or less severely criticized. Einhorn (1972, 1973) for instance warned about easy tree misuse. He draw attention to the nowadays well known over-fitting problem (i.e., the danger of getting a too-complex tree when focusing only on the training sample), and suggested resorting to cross-validation. More generally, he was skeptical of the reliability of tree outcomes. The discussion with Morgan and Andrews (1973) relates mainly to the meaning of exploratory approaches, Einhorn claiming that only model building based on clearly enounced assumptions

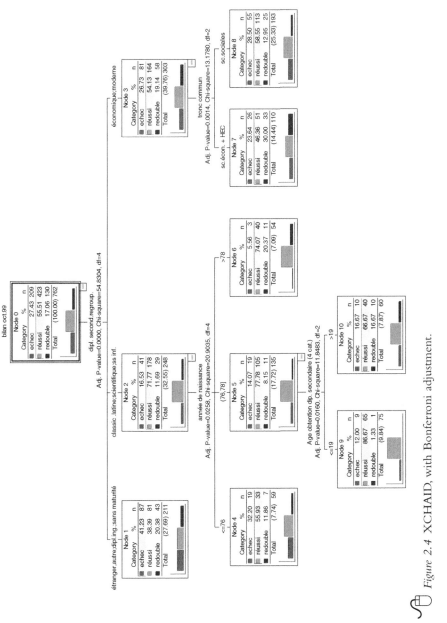

Figure 2.4 XCHAID, with Bonferroni adjustment.

may usefully exploit quantitative data. Doyle (1973) also advanced a series of criticisms that were more specifically addressed to the application of AID that Heald (1972) ran on a sample of only 70 data. He recalls that trees are mainly intended for large samples (a minimum of about 1,000 cases). He stresses that no general goodness of fit measure is provided with trees, and that trees cannot determine the global importance of the factors that intervene in the tree. Indeed, the tree does not provide effects controlled by all other covariates. Instead, it shows the additional information brought by the covariate conditional to the splits already made. The tree structure obtained from a sample data set is known to be unstable in the sense that a very small change in the data may considerably affect it.

Most of these criticisms have since then received effective answers. Cross-validation and test of generalization on test data are nowadays currently used with trees. Ensemble methods, and especially forest trees introduced by Breiman (2001) offer solutions to the robustness issue when the concern is prediction. The measures of the variable importance in forest trees proposed for instance by Strobl, Malley, and Tutz (2009) also answer one of Doyle's criticisms. As for the goodness of fit of trees, we proposed ourself (Ritschard and Zighed, 2003; Ritschard, 2006) deviance-based measures for investigating statistical information brought by the obtained segmentation. Nonetheless, trees remain exploratory tools. Although they have demonstrated good prediction capacities, they are not intended for testing causality hypotheses.

It also is worth mentioning that, although we focused in this chapter on early tree methods, their development did not stop with CHAID. More recent important milestones are among others the CART method of Breiman, Friedman, Olshen, and Stone (1984), which boosted the use of trees in the 1980s, and the ID3 (Quinlan, 1986) and C4.5 (Quinlan, 1993) algorithms. These methods are clearly oriented towards prediction and have in common the use of split criteria based on entropy reduction (Gini, i.e., quadratic entropy, for CART and Shannon's entropy for ID3 and C4.5) aiming at finding pure nodes. In the statistical area, recent developments have been oriented towards solving the bias favoring predictors with many different values. Here, milestones are QUEST (Loh and Shih, 1997), GUIDE (Loh, 2007) and 'party' (Hothorn et al., 2006a,b).

References

Abdi, H. (2007). Bonferroni and Sidak corrections for multiple comparisons. In N. Salkind (Ed.), *Encyclopedia of Measurement and Statistics*. Thousand Oaks, CA: Sage.

Armstrong, J. S. and J. G. Andress (1970). Exploratory analysis of marketing data: Tree vs regression. *Journal of Marketing Research* 7, 487–492.

Assael, H. (1970). Segementing markets by group purchasing behavior. *Journal of Marketing Research* 7, 153–158.

Bell, E. T. (1938). The iterated exponential numbers. *Ann. Math. 39*, 539–557.

Belson, W. A. (1959). Matching and prediction on the principle of biological classification. *Applied Statistics 8*(2), 65–75.

Belson, W. A. (1978). Investigating causal hypotheses concerning delinquent behaviour, with special reference to new strategies in data collection and analysis. *The Statistician 27*(1), 1–25.

Biggs, D., B. De Ville, and E. Suen (1991). A method of choosing multiway partitions for classification and decision trees. *Journal of Applied Statistics 18*(1), 49–62.

Bouroche, J.-M. and M. Tenenhaus (1970). Quelques méthodes de segementation. *Revue française d'informatique et de recherche opérationnelle 4*(2), 29–42.

Bouroche, J.-M. and M. Tenenhaus (1972). Some segmentation methods. *Metra 7*, 407–418.

Breiman, L. (2001). Random forest. *Machine Learning 45*, 5–32.

Breiman, L., J. H. Friedman, R. A. Olshen, and C. J. Stone (1984). *Classification and Regression Trees*. New York: Chapman and Hall.

Cellard, J. C., B. Labbé, and G. Savitsky (1967). Le programme ELISEE, présentation et application. *Metra 3*(6), 511–519.

Doyle, P. (1973). The use of automatic interaction detector and similar search procedures. *Operational Research Quarterly 24*(3), 465–467.

Einhorn, H. J. (1972). Alchemy in the behavioral sciences. *The Public Opinion Quarterly 36*(3), 367–378.

Einhorn, H. J. (1973). Reply to Morgan and Andrews. *The Public Opinion Quarterly 37*(1), 129–131.

Fielding, A. and C. A. O'Muircheartaigh (1977). Binary segmentation in survey analysis with particular reference to AID. *The Statistician 26*(1), 17–28.

Gillo, M. W. (1972). MAID, a Honeywell 600 program for an automatized survey analysis. *Behaviorial Science 17*(2), 251–252.

Gillo, M. W. and M. W. Shelly (1974). Predictive modeling of multivariable and multivariate data. *Journal of the American Statistical Association 69*(347), 646–653.

Heald, G. I. (1972). The application of the automatic interaction detector programme and multiple regression techniques to the assessment of store performance and site selection. *Operational Research Quarterly 23*(4), 445–457.

Hothorn, T., K. Hornik, and A. Zeileis (2006a). Party: A laboratory for recursive part(y)itioning. User's manual.

Hothorn, T., K. Hornik, and A. Zeileis (2006b). Unbiased recursive partitioning: A conditional inference framework. *Journal of Computational and Graphical Statistics 15*(3), 651–674.

Hunt, E. B., J. Marin, and P. J. Stone (1966). *Experiments in induction*. New York and London: Academic Press.

Kass, G. V. (1975). Significance testing in automatic interaction detection (A.I.D.). *Applied Statistics 24*(2), 178–189.

Kass, G. V. (1980). An exploratory technique for investigating large quantities of categorical data. *Applied Statistics 29*(2), 119–127.

Light, R. J. and B. H. Margolin (1971). An analysis of variance for categorical data. *Journal of the American Statistical Association 66*(335), 534–544.

Loh, W.-Y. (2007). GUIDE (version 5). User manual. Technical report, Department of Statistics, University of Wisconsin, Madison.

Loh, W.-Y. and Y.-S. Shih (1997). Split selection methods for classification trees. *Statistica Sinica* 7, 815–840.

Messenger, R. and L. Mandell (1972). A modal search technique for predictive nominal scale multivariate analysis. *Journal of the American Statistical Association* 67(340), 768–772.

Morgan, J. N. and F. M. Andrews (1973). A comment on Einhorn's "Alchemy in the behavioral sciences." *The Public Opinion Quarterly* 37(1), 127–129.

Morgan, J. N. and R. C. Messenger (1973). THAID: a sequential analysis program for analysis of nominal scale dependent variables. Survey Research Center, Institute for Social Research, University of Michigan, Ann Arbor.

Morgan, J. N. and J. A. Sonquist (1963). Problems in the analysis of survey data, and a proposal. *Journal of the American Statistical Association* 58, 415–434.

Orr, L. (1972). The dependence of transition proportions in the education system on observed social factors and school characteristics. *Journal of the Royal Statistical Society. Series A (General)* 135(1), 74–95.

Petroff, C., A.-M. Bettex, and A. Korffy (2001). Itinéraires d'étudiants à la Faculté des sciences économiques et sociales: le premier cycle. Technical report, Université de Genève, Faculté SES.

Press, L. I., M. S. Rogers, and G. H. Shure (1969). An interactive technique for the analysis of multivariate data. *Behavioral Science* 14(5), 364–370.

Quinlan, J. R. (1986). Induction of decision trees. *Machine Learning* 1, 81–106.

Quinlan, J. R. (1993). *C4.5: Programs for Machine Learning*. San Mateo: Morgan Kaufmann.

Ritschard, G. (2006). Computing and using the deviance with classification trees. In A. Rizzi and M. Vichi (Eds.), *COMPSTAT 2006—Proceedings in Computational Statistics*, pp. 55–66. Berlin: Springer.

Ritschard, G. and D. A. Zighed (2003). Goodness-of-fit measures for induction trees. In N. Zhong, Z. Ras, S. Tsumo, and E. Suzuki (Eds.), *Foundations of Intelligent Systems, ISMIS03*, Volume LNAI 2871, pp. 57–64. Berlin: Springer.

Ross, J. A. and S. Bang (1966). The AID computer program, used to predict adoption of family planning in Koyang. *Population Studies* 20(1), 61–75.

Scott, A. J. and M. Knott (1976). An approximate test for use with AID. *Applied Statistics* 25(2), 103–106.

Sonquist, J. A., E. L. Baker, and J. N. Morgan (1971). Searching for structure (Alias–AID–III). Survey Research Center, Institute for Social Research, University of Michigan, Ann Arbor.

Sonquist, J. A. (1969). Finding variables that work. *The Public Opinion Quarterly* 33(1), 83–95.

SPSS (2001). *Answer Tree 3.0 User's Guide*. Chicago: SPSS Inc.

Strobl, C., J. Malley, and G. Tutz (2009). An introduction to recursive partitioning:. *Psychological Methods* 14(4), 323–348.

Tanofsky, R., R. R. Shepps, and P. J. O'Neill (1969). Pattern analysis of biographical predictors of success as an insurance salesman. *Journal of Applied Psychology* 53(2, Part 1), 136–139.

Thompson, V. R. (1972). Sequential dichotomisation: Two techniques. *The Statistician* 21(3), 181–194.

3 The Potential of Model-based Recursive Partitioning in the Social Sciences

Revisiting Ockham's Razor

Julia Kopf, Thomas Augustin, and Carolin Strobl

Introduction

A variety of new statistical methods from the field of machine learning have the potential to offer new stimuli for research in the social, educational, and behavioral sciences. In this chapter we focus on one of these methods: model-based recursive partitioning. This algorithmic approach is reviewed and illustrated by means of instructive examples and an application to the Mincer equation, which is commonly used to describe the association between education, job experience, and income in econometric and sociological research. For readers unfamiliar with algorithmic methods, the explanation starts with the introduction of the predecessor method classification and regression trees. As opposed to classification and regression trees that search for groups of observations that differ in the values of a response variable, model-based recursive partitioning searches for groups differing in their estimated parameters of a postulated statistical model. With respect to the application and interpretation of model-based recursive partitioning, we highlight the principle of parsimony and Ockham's Razor. To facilitate applicability in the social sciences, we close with a section on model-based recursive partitioning software available in the free R system for statistical computing. In addition, a supplement with worked examples for classification and regression trees as well as model-based recursive partitioning is provided on the book's website as a hands-on tutorial.

The aim of this chapter is to demonstrate the potential of model-based recursive partitioning (Zeileis, Hothorn, & Hornik 2008; related approaches have previously been suggested by Loh 2002; Li, Lue, & Chen 2000; Chaudhuri, Lo, Loh, & Yang 1995; Wang & Witten 1997), a statistical method adopted from the field of machine learning, for applications in the social sciences. In particular, we will point out that this algorithmic method provides a powerful tool to evaluate whether relevant covariates have been omitted in a statistical model and, therefore, whether a theoretically postulated model is in conflict with Ockham's Razor.

As a prototypical example, the method is employed to evaluate the appropriateness of the so called Mincer equation (Mincer 1974), which explains different income levels through rates of return from schooling and work experience by means of a linear model. The analysis relies on data from the German Socio-Economic Panel Study (SOEP) from 2008, provided by DIW Berlin (the German Institute for Economic Research).

Model-based recursive partitioning can be regarded as a powerful synthesis between nonparametric partitioning methods and parametric regression models. In contrast to standard multiple regression approaches, model-based recursive partitioning is based on the successive segmentation of the sample used: the data are split further as long as different groups of observations still display substantially different values of the estimated parameters of the statistical model of interest. Hence, the objective of model-based recursive partitioning is related to the objective of latent class or mixture models, where different regression parameter estimates are permitted between subgroups of the data set (see e.g., Vermunt 2010; Leisch 2004, for general introductions to mixture or latent class models, and Ünlü 2011, for a specific application to knowledge structures). In latent class regression models, these groups are unobserved, whereas in model-based recursive partitioning the groups are determined from combinations of observed covariates.

For example, in our investigation of the Mincer equation we will see that the intercept and the estimated coefficient for further education vary across groups of men and women working full-time in east or west Germany. Thus additional sociological and economic theories, such as discrimination in labor markets (e.g., Aigner & Cain 1977; Phelps 1972), need to be considered to explain these differences.

The model-based recursive partitioning method constitutes an advance in classification and regression trees, which are widely used in life sciences (cf., e.g., Hannöver, Richard, Hansen, Martinovich, & Kordy 2002; Kitsantas, Moore, & Sly 2007; Romualdi, Campanaro, Campagna, Celegato, Cannata, Toppo, Valle, & Lanfranchi 2003; Zhang, Yu, Singer, & Xiong 2001) and have recently been applied in social and behavioral sciences (e.g., Berk 2006). Classification and regression trees will be summarized briefly in the following section, beginning with an informal description of the resulting tree-structure. After reviewing some technical details of classification and regression trees, the advanced method of model-based recursive partitioning is addressed, first by pointing out the main differences with and similarities to classification and regression trees. The review of model-based recursive partitioning will be continued by interpreting an instructive example and recapitulating the statistical background. To facilitate the use of this powerful algorithmic method in the social sciences, the chapter highlights interpretation with regard to the principle of parsimony in the context of model construction. Moreover, application to the Mincer equation demonstrates the potential of model-based recursive partitioning in empirical research. Finally, for further research, software available in the R system for statistical computing is described.

In summary, in this chapter we show how model-based recursive partitioning allows us to decide whether a postulated model fails to describe the whole sample in a suitable way, because the method may detect varying parameter estimates in different subgroups of the sample. Model-based recursive partitioning therefore offers a synthesis of the theory-based and data-driven approaches. In particular, it can be used to detect violations of Ockham's Razor. If subgroups with different parameter estimates are found, the postulated model is too simple and not appropriate for the entire sample.

Classification and Regression Trees

Classification and regression trees (cf., e.g., Breiman, Friedman, Olshen, & Stone 1984) are based on a purely data-driven paradigm. Without referring to a concrete statistical model, they search recursively for groups of observations with similar values of the response variable by building a tree structure. If the response is categorical, one refers to classification trees; if the response is continuous, one refers to regression trees. The basic principles of this approach will be explained by means of an example application in the following section.

Basic Principles of Classification and Regression Trees

As a first example, we consider the respondents in the SOEP study 2008 (see Wagner, Frick, & Schupp 2007 for details about SOEP). In this data set, groups of subjects vary with respect to whether they participate in full-time labor or not (the latter including all categories like part-time or marginal employment, civil or military service, vocational training, and unemployment labeled here as "other").

These groups can be described by means of covariates, such as age and gender. The covariates (here: age and gender), together with the response variable (full time or other), are handed over to the algorithm. The resulting tree-structure is displayed in Figure 3.1.

From the entire sample of about 19,553 respondents living in private households, the covariate with the highest association (for technical details see the next section) to the response is chosen for the first split. It is the participant's gender, and, thus, 9,318 male respondents (represented in the left branch) are separated from the rest of the sample (10,235 female respondents, represented in the right branch). In the next step the male group is further diversified: it is split into two new subgroups, over the age of 62 or not (node 3 and node 4). Figure 3.1 also shows that the majority of women in the lower age group respond differently (node 6) compared to those in node 7. Here we stop the algorithm for simplicity.

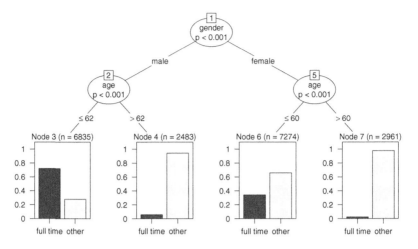

Figure 3.1 Classification tree: assessing different frequencies of full-time jobs in Germany (SOEP 2008). The resulting tree-structure shows varying participation rates in full-time labor in three splits according to the covariates gender and age.

The respective cutpoint for these splits depends on the type of the covariate: while gender has only two categories—male and female—and thus offers only one cutpoint, referring to age the algorithm must also find the "best" cutpoint within this variable. This optimal cutpoint turns out to be located at the threshold of 62 years for the male subsample and 60 years for the female subsample (technical details are given below).

The resulting tree-structure is easily interpreted and shows groupwise frequencies for full-time and non-full-time workers in the end nodes: the left node indicates that the majority of men up to 62 years in Germany work full-time, while the majority of women up to 60 do not. Women over 60 years are hardly ever employed full-time. The tree-structure in this example represents an interaction effect between gender and age (see e.g., Strobl et al. 2009, for details on the interpretation of the main effects and interactions in classification trees).

In contrast to this classification problem, regression trees focus on continuous response variables. Instead of looking at the frequencies of the categories, groups with different average response values are separated and visualized, e.g. by means of box plots (as in Figure 3.2). These groups are again detected automatically.

In this example the regression tree searches for different patterns of the (outlier adjusted) requested income at which 950 unemployed respondents would take a job (outliers are defined as participants with a requested income higher than the third quartile plus 1.5 times interquartile range).

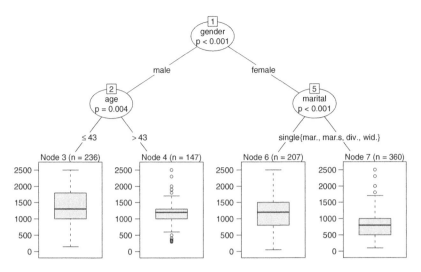

Figure 3.2 Regression tree: assessing different requested incomes of unemployed respondents (SOEP 2008). Three different levels are obtained in groups related to gender, age and marital status.

Additional covariates that are handed over to the algorithm are gender, age, nationality (nation) and marital status (marital). Figure 3.2 again shows the first split in the variable gender (node 1). The second split in the male subsample is again related to age (node 3 and node 4), while the third split in the female subsample is associated with marital status (node 6 and node 7). The cutpoint in a categorical variable is chosen automatically in an optimal way from all possible combinations of categories. Here the requested mean income is associated with marital status, in particular the request of female singles differs from the other categories (married, married but separated, divorced, and widowed). The latter categories have smaller values of the requested income (median $x_{med} = 800$, mean $\bar{x} = 870$) than female singles ($x_{med} = 1,200$, $\bar{x} = 1,166$), who seem to be of the same magnitude as men over 43 years ($x_{med} = 1,200$, $\bar{x} = 1,159$). The highest average of requested income occurs within the male subsample up to the age of 43 years (node 3, $x_{med} = 1,300$, $\bar{x} = 1,349$). After these three splits, all of the determined groups are homogeneous enough to let the algorithm come to a stop, without further splitting, e.g. according to the nationality of the respondent. This exemplifies another attractive feature of partitioning methods: they implicitly perform a flexible variable selection. A more detailed description of the technical procedure underlying classification and regression trees is given in the next section.

Some Technical Details

Classification trees search for different patterns in the response variable according to the available covariates. Since the sample is divided in rectangular partitions defined by values of the covariates and since the same covariate can be selected for multiple splits, classification trees can assess even complex interactions, non-linear and non-monotone patterns. The structure of the underlying data-generating process is not specified in advance, but is determined in an entirely data-driven way. These are the key distinctions between classification and regression trees, and classical regression models. The approaches differ, first, with respect to the functional form of the relationship that is limited to, for example linear influence of the covariates in most parametric regression models, and, second, with respect to the pre-specification of the model equation in parametric models.

Historically, the foundations for classification and regression trees were first developed in the 1960s as Automatic Interaction Detection (Morgan & Sonquist 1963). Later the most popular algorithms for classification and regression trees were developed by Quinlan (1993) and Breiman et al. (1984). Here we concentrate on a more recent framework by Hothorn, Hornik, & Zeileis (2006b), which is based on the theory of conditional inference developed by Strasser & Weber (1999). The major advantage of this approach is that it avoids two fundamental problems of earlier algorithms for classification and regression trees: variable selection bias and overfitting (cf., e.g., Strobl, Malley, & Tutz 2009).

The algorithm of Hothorn, Hornik, & Zeileis (2006b) for binary recursive partitioning can be described in three steps: first, beginning with the whole sample, the global null hypothesis that there is no relationship between any of the covariates and the response variable is evaluated. If no violation of the null hypothesis is detected, the procedure stops. If, however, a significant association is discovered, the variable with the largest association is chosen for the split. Second, the best cutpoint in this variable is determined and used to split the sample into two groups according to values of the selected covariate. Then the algorithm recursively repeats the first two steps in the subsamples until there is no further violation of the null hypothesis, or a minimum number of observations per node is reached.

In the following, we briefly summarize which covariates can be analyzed using classification and regression trees, how variables are selected for splitting and how the cutpoint is chosen.

The Response Variable in the End Nodes

As outlined in the previous section, classification trees search for groups of similar response values with respect to a categorical dependent variable, whereas regression trees focus on continuous variables. Hothorn et al.

(2006b) stress that their conditional inference framework can be applied beyond that to situations of ordinal, censored survival times and multivariate response variables.

Within the resulting tree-structure, all respondents with the same covariate values—represented graphically in one end node—obtain the same prediction for the response, i.e. the same class membership for categorical responses or the same value for continuous response variables.

Selection of Splitting Variables

The next question is how the variables for the potential splits are chosen and how the related cutpoints can be obtained. As outlined above, Hothorn et al. (2006b) provide a statistical framework for tests applicable to various data situations. In the binary recursive partitioning algorithm, each iteration is related to a current data set (beginning with the whole sample), where the variable with the highest association is selected by means of permutation tests as described in the following. Use of permutation tests allows evaluation of the global null hypothesis H_0 that none of the covariates has an influence on the dependent variable. If H_0 holds (in other words, if the independence between any of the covariates Z_j $(j = 1, \ldots, l)$ and the dependent variable Y cannot be rejected), the algorithm stops. Therefore, the statistical test acts both for variable selection and as a stopping criterion.

Otherwise the strength of the association between the covariates and the response variable is measured in terms of the p-value that corresponds to the test of the null hypothesis that the specific covariate is not associated with the response. Thus, the variable with the smallest p-value is selected for the next split. The advantage of this approach is that the p-value criterion guarantees an unbiased variable selection regardless of the scales of measurement of the covariates (cf., e.g., Hothorn et al. 2006b; Strobl, Boulesteix, & Augustin 2007; Strobl et al. 2009).

Permutation tests are constructed by evaluating the test statistic for the given data under H_0. Monte-Carlo or asymptotic approximations of the exact null-distribution are employed for the computation of the p-values (see Hothorn, Hornik, van de Wiel, & Zeileis 2006a; Hothorn et al. 2006b; Strasser & Weber 1999, for more details).

Selection of the Cutpoints

After the variable for the split has been selected, we need a cutpoint within the range of the variable to find the subgroups that show the strongest difference in the response variable. In the procedure described here, the selection of the cutpoint is also based on the permutation test statistic: the idea is to compute the two-sample test statistic for all potential splits within the covariate. In the case of continuous variables all potential cutpoints between any two successive observations are investigated (except for a certain percentage of the smallest and largest observations to avoid too small nodes). In the case of ordinal variables the ordering of the categories is accounted for. The resulting split is located where the binary separation of two data sets leads to the highest test statistic. This reflects the largest discrepancy in the response variable with respect to the two groups.

In the case of missing data, the algorithm proceeds as follows: observations that have missing values in the currently evaluated covariate are ignored in the split decision, whereas the same observations are included in all other steps of the algorithm. The class membership of these observations can be approximated by means of so-called surrogate variables (Hothorn et al. 2006b; Hastie, Tibshirani, & Friedman 2008).

From Classification and Regression Trees to Model-based Recursive Partitioning

Model-based recursive partitioning was developed as an advance from classification and regression trees. Both methods originate from the field of machine learning, which is influenced by both statistics and computer sciences.

The algorithmic rationale behind classification and regression trees is described by Berk (2006, p. 263) in the following way:

> With algorithmic methods, there is no statistical model in the usual sense; no effort has been made to represent how the data were generated. And no apologies are offered for the absence of a model. There is a practical data analysis problem to solve that is attacked directly with procedures designed specifically for that purpose.

In that sense, classification and regression trees are purely data-driven and exploratory—and thus mark the entire opposite of the theory-based approach of model specification that is prevalent in the empirical social sciences.

The advanced model-based recursive partitioning method, however, brings together the advantages of both approaches: at first, a parametric model is formulated to represent a theory-driven research hypothesis. Then this parametric model is handed over to the model-based recursive partitioning algorithm that checks whether other relevant covariates have been omitted which would alter the parameters of the model of interest. Note that, as opposed to latent class regression, the groups yielding differing parameter estimates are explained by covariates and not by a latent class approach.

Technically, the tree-structure obtained from classification and regression trees remains the same for model-based recursive partitioning. However, instead of splitting for different patterns of the response variable, now we

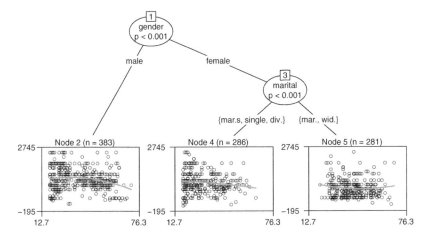

Figure 3.3 MOB: assessing different relationships between age and requested income of unemployed respondents in Germany (SOEP 2008). The line shows the estimated relationship in the current subsample and indicates the varying parameters according to groups related to age and marital status.

search for different patterns of the association between the response variable and other covariates, that have been pre-specified in the parametric model. Therefore the end nodes in the model-based tree represent statistical models, such as linear models, and no longer mere values of the response variable. The execution of a split in the model-based tree then indicates a parameter instability in the original model, i.e. the postulated model is too simple to explain the data.

Basic Principles of Model-based Recursive Partitioning

As an instructive example for a partitioned model, Figure 3.3 shows the tree-structure for a sample of unemployed respondents in the SOEP study. The model of interest here is the relationship between the requested income, at which respondents would take on a new job, and age. The functional form of this relationship is fixed to a quadratic polynomial as often found intuitively for models relating age and income:

$$\text{requested income} = \beta_0 + \beta_1 \cdot \text{age} + \beta_2 \cdot \text{age}^2 + \varepsilon.$$

Additional covariates passed over to the algorithm are marital status, gender, and nationality.

Beginning with the whole outlier adjusted sample of 950 unemployed respondents the model with the linear and quadratic term is fitted, where

Table 3.1 Estimated coefficients of the models in the end nodes in Figure 3.3 and standardized coefficients of regression models estimated from the data sets that correspond to the end nodes in parentheses

Node	$\widehat{\beta_0}$	$\widehat{\beta_1}$	$\widehat{\beta_2}$
2	1014.5837	22.5446	−0.3618
		(0.6524)	(−0.8302)
4	1212.6621	−0.0236	−0.0871
		(−0.0006)	(−0.1479)
5	1390.9708	−25.3983	0.2737
		(−0.6111)	(0.5642)

the estimated coefficients $\widehat{\beta_0}$, $\widehat{\beta_1}$, and $\widehat{\beta_2}$ indicate parameter instability. The highest instability is related to gender and thus a split in this variable is performed. While in the sample of the male respondents (node 2) no more instabilities are detected, the female subset is again divided into two subgroups with differing parameter estimates. The end node in the middle shows the result for married but separated, single, or divorced women (node 4). The rightmost end node contains the linear model for married and widowed women (node 5). Interestingly, even the direction of the relationship changes from a parabola on the left, where men of higher age tend to request less income, to a slight u-shape on the right, where married or widowed women request more income with higher age. The attractive feature of implicit variable selection is also maintained in model-based recursive partitioning: the nationality does not occur in any split decision in this example.

In Table 3.1 the parameter estimates for the different groups— represented in the end nodes of the tree-structure—are displayed. The varying signs of the coefficients confirm what is illustrated in Figure 3.3: the inverse u-shape holds only for part of the sample and is reversed for other parts. Thus, the example illustrates that model-based recursive partitioning is indeed able to detect different functional forms that might be masked when a single model is fitted to the data.

The example also shows that, as opposed to classification and regression trees, the end nodes in model-based recursive partitioning do not contain values of a response variable, but represent a statistical model for each specific subpopulation. Between these groups the estimated parameters of the common underlying model vary significantly, but the postulated basic functional form (here polynomial) stated by the researcher is fixed. Within the subgroups no significant parameter instability is present.

Hence, the interpretation of a tree without any split is quite simple: there are no significant parameter instabilities found in any of the covariates handed over to the algorithm. If, however, a tree-structure is displayed, it reveals that the postulated model is not appropriate for describing the entire sample. The variation of the parameters highlights structural differences in the obtained subgroups, which can be easily interpreted by examining the estimates or the graphical output.

In the next section, important steps of the model-based partitioning algorithm are outlined. Then we take a closer look at the interpretation in social or behavioral sciences in the following section.

Some Technical Details

The model-based recursive partitioning algorithm maintains the fundamental steps of the partitioning method reviewed above, but coherently extends them in the light of the model-based paradigm. According to this paradigm, the recursive process now estimates the basic statistical model beginning with all available observations. The result of this step is the estimated parameter vector from the optimization of the objective function, typically the (log-)likelihood. In almost the same manner as classification trees, the recursive process starts: instead of testing the association, now the parameter instability is assessed using so called generalized M-fluctuation tests (Zeileis 2005; Zeileis & Hornik 2007). If the data indicate parameter instability, the split of the parent node in two daughter nodes is executed. Relying on the data points in the new subgroups only, the algorithm again searches for parameter instability until no further significant instability is found, or another stopping criterion is fulfilled. The resulting tree structure can be visualized as illustrated in the examples presented below, so that the different groups can be compared. Note, however that statistical tests conducted after model selection—such as significance tests for group differences after recursive partitioning—may be affected by the effects described by Leeb & Pötscher (2005) and Berk (2010), and should thus be based on new data.

This brief overview of the similarities and differences in the algorithms leaves some questions that have yet to be answered: Which models can be partitioned recursively? How can we assess parameter instability and where are the optimal cutpoints in the covariates in model-based recursive partitioning? These questions are addressed in the next subsections, which are structured in the same way as above.

The Statistical Model in the End Nodes

The foundation of a general statistical framework for model-based recursive partitioning by Zeileis et al. (2008) allows the use of variety of underlying statistical models, such as linear and logistic regression models. The wide range of applications emerges from the inclusion of several widely used test statistics in a unified approach (Zeileis 2005) called generalized M-fluctuation tests.

Technically, the generalized M-fluctuation test used for the split decisions relies on the objective function $\Psi(.)$ of the parameter estimation, like least-squares and maximum–likelihood-estimation:

$$\widehat{\theta} = \arg\max_{\theta} \sum_{i=1}^{n} \Psi(y_i, \theta), \tag{1}$$

where y_i $(i = 1, \ldots, n)$ symbolizes the vector of all values of the dependent and independent variables in the postulated model for subject i, and θ represents the (potentially vector-valued) parameter. To keep the notation simple, here we use the full sample notation and do not distinguish whether the underlying observations are the entire sample or a specific subgroup arising from the recursive application of the procedure.

The estimation process is based on the individual contributions of each subject i to the score function

$$\psi(y_i, \theta) = \frac{\partial \Psi(y_i, \theta)}{\partial \theta},$$

as outlined below.

In addition to the model specification, the algorithm requires categorical or numeric covariates—denoted as Z_j $(j = 1, \ldots, l)$—for potential splits in the model-based tree.

Selection of Splitting Variables

After the first step of the algorithm—fitting the underlying model for the whole sample and obtaining a preliminary estimate $\widehat{\theta}$—a test of parameter instability is performed. It is based on the statistical framework developed by Zeileis & Hornik (2007) to detect structural changes by fluctuation tests. In econometrics, these tests for structural changes are widely used to detect, for example, a drop in the expected value of a time series for a stock exchange due to an economic crisis.

To detect a systematic change in the parameter over the range of a covariate Z_j, the observations are ordered according to their values of Z_j. Under the null hypothesis of parameter stability, no systematic structural change is present. The null hypothesis is rejected if one or more parameters of the postulated model change significantly over the ordering induced by the covariate Z_j.

The construction of general test statistics relies on the partial derivatives of the objective function, e.g. of the log-likelihood. The contributions of each individual observation i to the derivative of the objective function (i.e., to the score function) evaluated at the current parameter estimate, $\psi(y_i, \widehat{\theta})$, are ordered with respect to the potential splitting variable Z_j. The individual contributions $\psi(y_i, \widehat{\theta})$ are depicted as vertical dashed lines for an instructive example in Figure 3.4 (left panel).

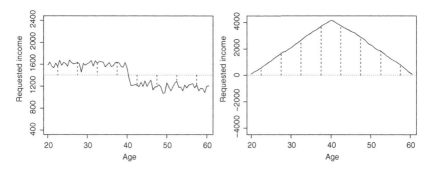

Figure 3.4 Structural change in the mean over age (artificial data). The left plot displays the mean income over all age groups (dotted line) and the individual deviations (dashed lines), the right shows the cumulated deviations over the variable, age.

Under the null hypothesis, the individual contributions $\psi\left(y_i, \widehat{\theta}\right)$ should fluctuate randomly around the mean zero, whereas in Figure 3.4 (left panel) a clear structural change can be detected. To grasp this structural change statistically, we turn from the individual contributions to their cumulative sums in Figure 3.4 (right panel). Zeileis & Hornik (2007) proved the convergence of the cumulative sum process (also termed decorrelated empirical fluctuation process)

$$W_j\left(t\right) = \widehat{J}^{-\frac{1}{2}} n^{-\frac{1}{2}} \sum_{i=1}^{\lfloor nt \rfloor} \psi\left(y_i, \widehat{\theta}\right)$$

against a k-dimensional Brownian bridge. The first part of the formula, $\widehat{J}^{-\frac{1}{2}}$, denotes an estimator of the covariance $Cov\left(\psi(Y, \widehat{\theta})\right)$. The summation over all $\lfloor nt \rfloor$ refers to the first $n \cdot t$ (with $t \in [0; 1]$) observations according to the order with respect to covariate Z_j (for example the first 50%, where the $\lfloor . \rfloor$ indicates that the integer part of $n \cdot t$, i.e. the lower whole number, is used).

The instructive example in Figure 3.4 can be interpreted as the variation of income before and after the simulated threshold of 40 years. The path of the cumulative sum process increases until the age of 40 and decreases after that threshold, with a sharp peak at the change-point of 40. The strength of this peak is used as a statistical measure for the strength of the parameter instability.

The asymptotic properties of the cumulative sum process allow for the construction of test statistics that are used for detecting the structural change. The test statistic for numeric variables is directly built from the empirical fluctuation process $W_j\left(t\right)$, while the test statistic for categorical variables takes into account that the categories and the observations within the category are not ordered. The result of Zeileis et al. (2008) also permits the computation of *p*-values and thus the statistical decision whether the parameters differ significantly from parameter stability. If parameter instability is detected, the algorithm selects the variable with the smallest *p*-value. Splitting continues until there is no further instability in any current node.

Selection of the Cutpoint

In case of a splitting decision the cutpoint can be sought by a criterion that also includes the maximization of the objective function in the two potential subsamples. In the case of ordered or numeric covariates, these subsamples can easily be defined as $L(\zeta) = \{i \mid z_{ij} \leq \zeta\}$ and $R(\zeta) = \{i \mid z_{ij} > \zeta\}$ for a candidate cutpoint ζ and the component z_{ij} of z_j.

The optimal cutpoint $\zeta^\star$ is determined by maximizing

$$\sum_{i \in L(\zeta)} \Psi\left(y_i, \hat{\theta}^{(L)}\right) + \sum_{i \in R(\zeta)} \Psi\left(y_i, \hat{\theta}^{(R)}\right) \tag{2}$$

over all candidate cutpoints ζ. $\hat{\theta}^{(L)}$ and $\hat{\theta}^{(R)}$ are the estimated parameters in the subsets. In the case of unordered categorical covariates all potential binary partitions need to be evaluated and the partition with the highest criterion is chosen for the split (Zeileis et al. 2008).

Both parts of the binary split generate new parent nodes. Unless there is no further parameter instability found or another stopping criterion is satisfied (such as a minimum sample size in the current node) the algorithm continues searching for instability and splitting the current (sub-)data set into daughter nodes.

Potential in the Social Sciences

The application of model-based recursive partitioning offers new impetus for research in the social, educational, and behavioral sciences. For the interpretation of model-based recursive partitioning, we would like to point out the connection to the principle of parsimony: following the fundamental research paradigm that theories developed in the social sciences should yield falsifiable hypotheses, the latter are translated into statistical models. The aim of model construction is thus to simplify the complex reality.

The decision on the complexity of the formulated model can be guided by "*a working rule known as Occam's Razor whereby the simplest possible descriptions are to be used until they are proved to be inadequate*" (Richardson 1958, p. 1247). This rule implies the objective of parsimonious model formulation: a model should be no more complex than necessary, but it also needs to be complex enough to describe the empirical data. In the regression context the usage of sparse and simple models with few variables explaining the response is usually propagated (e.g., Gujarati 2003) – as long as no relevant explanatory variables are omitted. The strength of model-based recursive partitioning in this context lies in the power to let the data decide this question. Indeed, it offers the possibility of detecting whether the suggested model is inadequate because relevant covariates are missing and it explicitly selects these relevant covariates. If the algorithm executes at least one split, we obtain the statistical decision that the parameters are unstable and the data are too heterogeneous to be explained by the postulated model. In this case, the presumed functional form does not describe the entire sample in an appropriate way and thus subgroups have to be constructed. Moreover, the tree-structured results provide information about which subgroups differ in their association patterns. This information can either be integrated into a revision of the substantial theory and the formulation of a new parametric model, or it should be pointed out in the interpretation that the postulated model applies only to a limited range of subjects.

Consequently, model-based recursive partitioning can identify different shapes of a parametric model stated by the researcher in different subgroups of the sample. Model-based recursive partitioning offers a synthesis of the theory-based and the data-driven approach that can be used for evaluating violations of the "working rule" Ockham's Razor. If the method detects no instability of the model parameters, the model is not rejected based on the additional covariates provided to the algorithm. If, however, the method does detect instability, the postulated model is too simple.

Empirical Example

To illustrate the potential of model-based recursive partitioning further, we turn to another example, based on an extension of the so called Mincer equation. In the seminal econometric work of Mincer (1974) the logarithmic income is described as a function of the variables: years of schooling (time_edu) and full-time experience (included in linear and squared terms, full_ex, full_ex^2).

The Mincer equation owes its popularity to the straightforward interpretation of the coefficients as approximated rates of return from education (cf. Björklund & Kjellström 2002, for a critical discussion). We focus on the following extension of the Mincer equation that also includes a dummy variable for further education on the job (further_edu):

$$\ln(\text{income}) = \beta_0 + \beta_1 \, \text{time\_edu} + \beta_2 \, \text{full\_ex} + \beta_3 \, \text{full\_ex}^2 + \beta_4 \, \text{further\_edu} + \varepsilon,$$

with ε i.i.d. $N(0, \sigma^2)$. Here we restrict the observations from the SOEP study to over 6,000 respondents in full-time employment who are not in vocational training and earn more than 500 Euros monthly.

The examination of the Mincer equation, which is driven by the principle of parsimony, via model-based recursive partitioning is illustrated in Figure 3.5. The model formulation involves the effects (which are displayed as symbols in the end nodes) of years of schooling, further education, and work experience in full-time jobs (linear and squared term) on the logarithmic gross income of fully employed respondents in Germany. Again, further potentially influencing variables are passed over to the algorithm, namely the location of the employer in east or west Germany, gender, and the size of the company. The results show significantly different parameter estimates related to each of the additional covariates. These estimated coefficients of the Mincer equation are approximated rates of return, e.g. from schooling. A closer look at the estimated parameters for the detected subgroups (Table 3.2) shows quite similar effects on the logarithmic income for some covariates from the original Mincer equation, such as the percentage change for time of education on earnings ($\widehat{\beta}_1$). However, the estimated coefficients for further education ($\widehat{\beta}_4$) and the intercept ($\widehat{\beta}_0$) differ more strongly between the groups. In particular, the effect of further education ($\widehat{\beta}_4$) is higher for employers in east as opposed to west Germany.

Our results are in accordance with current empirical social and economic research on heterogeneous effects of further education for men in Germany by Kuckulenz & Zwick (2005). One reason for the violation of a joint model for all respondents may lie in the strong assumption of the Mincer equation that there is no relevant change in the economy under research. In the SOEP study, this assumption is clearly violated by the reunification of the eastern and western parts of Germany. As a consequence, we find a split according to the location of the employer in east or west Germany in Figure 3.5.

Our findings imply that more elaborate theories explaining the different income levels in these subgroups, such as discrimination theories (e.g., Aigner & Cain 1977; Phelps 1972), and a more specific investigation of the differential effects of further education may be necessary to explain the observed group differences.

Software

The data analysis presented here uses the R system for statistical computing (R Development Core Team 2009), which is freely available under terms of the GNU General Public Licence (GPL) from the Comprehensive R Archive Network at http://CRAN.R-project.org/. Methods for classification, regression, and model-based trees are provided in the package "party." The conditional inference framework is implemented in the function `ctree()` (Hothorn et al. 2006b), while the model-based recursive partitioning algorithm is available via the function `mob()` (Zeileis et al. 2008). At present the algorithm can be applied to various types of

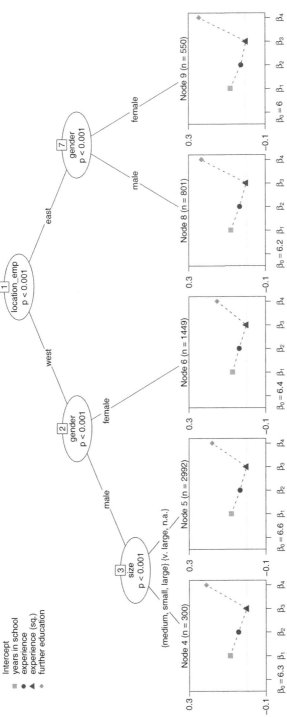

Figure 3.5 Model-based recursive partitioning of the extended Mincer equation (SOEP 2008). The symbols in the end nodes illustrate the estimated coefficients in the subgroups related to the location of the employer, gender, and size of the company.

Table 3.2 Estimated coefficients of the models in the end nodes in Figure 3.5 and standardized coefficients of regression models estimated from the data sets that correspond to the end nodes in parentheses

Node	$\hat{\beta}_0$	$\hat{\beta}_1$	$\hat{\beta}_2$	$\hat{\beta}_3$	$\hat{\beta}_4$
4	6.2743	0.0860	0.0430	−0.0009	0.2110
		(0.4757)	(0.8681)	(−0.6382)	(0.2065)
5	6.5620	0.0796	0.0335	−0.0005	0.1785
		(0.4683)	(0.7241)	(−0.4930)	(0.1732)
6	6.4486	0.0718	0.0369	−0.0007	0.1520
		(0.4477)	(0.8634)	(−0.5802)	(0.1605)
8	6.1543	0.0801	0.0340	−0.0006	0.2332
		(0.4104)	(0.7175)	(−0.5557)	(0.2207)
9	6.0173	0.0817	0.0258	−0.0004	0.2454
		(0.4321)	(0.5795)	(−0.3718)	(0.2394)

generalized linear models, survival models, or linear models. Moreover, the authors allow the users to build their own model classes and pass them on to the existing mob() function. A vignette explaining the use of the software for linear regression and logistic regression trees including the R-code is also available (Zeileis, Hothorn, & Hornik 2010). Additionally, we provide worked examples for classification and regression trees and model-based recursive partitioning as a supplement. The examples are similar in spirit to the examples presented in this chapter, but they rely on simulated data since the usage of the SOEP data requires a data distribution contract (further information on http://www.diw.de/). Ongoing research expands the unbiased recursive partitioning approach presented here to psychometric models such as the Bradley–Terry model for detecting different preference structures (Strobl, Wickelmaier, & Zeileis 2011) as well as the Rasch model (Strobl, Kopf, & Zeileis 2010; Strobl, Kopf, & Zeileis 2013) and factor analytic and structural equation models (Merkle & Zeileis 2013) for the assessment of measurement invariance.

Conclusion

Algorithmic procedures, such as classification and regression trees, have become popular and widely used tools in many scientific fields. Our aim here was to highlight that the recent development of model-based recursive partitioning allows us to combine the power of these algorithmic methods with that of theory-based parametric models by means of enhancing the purely data-driven approach towards a segmentation procedure for postulated models. We have highlighted the relation between this approach and the principle of parsimonious model construction. The tree-structured results allow straightforward interpretations of potential parameter instabilities that have been detected via empirical fluctuation tests. The detection of parameter instability leads to the interpretation that the statistical model under investigation cannot describe the whole sample appropriately, because relevant covariates have been omitted. Thus, model-based recursive partitioning can be used as a diagnostic check for inadequately simple descriptions of the relationship between response and explanatory variables.

The application in social science research is eased by the freely accessible and well documented packages provided in the statistical software R.

Note

Julia Kopf is supported by the German Federal Ministry of Education and Research (BMBF) within the project "Heterogeneity in IRT-Models" (grant ID 01JG1060). The authors would like to thank Achim Zeileis and Torsten Hothorn for their expert advice.

References

Aigner, D. J., & Cain, G. G. (1977). Statistical theories of discrimination in labor markets. *Industrial and Labor Relations Review*, 30, 175–187.

Berk, R. A. (2006). An introduction to ensemble methods for data analysis. *Sociological Methods & Research*, 34, 263–295.

Berk, R. A. (2010). An introduction to statistical learning from a regression perspective. In A. R. Piquero & D. Weisburd (Eds.), *Handbook of Quantitative Criminology*, New York: Springer (pp. 725–740).

Björklund, A., & Kjellström, C. (2002). Estimating the return to investments in education: How useful is the standard Mincer equation? *Economics of Education Review*, 21, 195–210.

Breiman, L., Friedman, J. H., Olshen, R. A., & Stone, C. J. (1984). *Classification and Regression Trees*. New York: Chapman and Hall.

Chaudhuri, P., Lo, W.-D., Loh, W.-Y., & Yang, C.-C. (1995). Generalized regression trees. *Statistica Sinica*, 5, 641–666.

Gujarati, D. N. (2003). *Basic Econometrics*. Boston: McGraw-Hill, 4th. edn.

Hannöver, W., Richard, M., Hansen, N. B., Martinovich, Z., & Kordy, H. (2002). A classification tree model for decision-making in clinical practice: An application based on the data of the German multicenter study on eating disorders, project tr-eat. *Psychotherapy Research*, 12, 445–461.

Hastie, T., Tibshirani, R., & Friedman, J. H. (2008). *The Elements of Statistical Learning. Data Mining, Inference and Prediction*. New York: Springer, 2nd. edn.

Hothorn, T., Hornik, K., van de Wiel, M., & Zeileis, A. (2006a). A Lego system for conditional inference. *The American Statistician*, 60, 257–263.

Hothorn, T., Hornik, K., & Zeileis, A. (2006b). Unbiased recursive partitioning: A conditional inference framework. *Journal of Computational and Graphical Statistics*, 15, 651–674.

Kitsantas, P., Moore, T., & Sly, D. (2007). Using classification trees to profile adolescent smoking behaviors. *Addictive Behaviors*, 32, 9–23.

Kuckulenz, A., & Zwick, T. (2005). Heterogene Einkommenseffekte betrieblicher Weiterbildung. *Die Betriebswirtschaft*, 65, 258–275.

Leeb, H., & Pötscher, B. M. (2005). Model selection and inference: Facts and fiction. *Econometric Theory*, 21, 21–59.

Leisch, F. (2004). Flexmix: A general framework for finite mixture models and latent class regression in R. *Journal of Statistical Software*, 11, 1–18.

Li, K.-C., Lue, H.-H., & Chen, C.-H. (2000). Interactive tree-structured regression via principal Hessian directions. *Journal of the American Statistical Association*, 95, 547–560.

Loh, W.-Y. (2002). Regression trees with unbiased variable selection and interaction detection. *Statistica Sinica*, 12, 361–386.

Merkle, E. C., & Zeileis, A. (2013). Tests of measurement invariance without subgroups: A generalization of classical methods. *Psychometrika*, 78(1), 59–82.

Mincer, J. A. (1974). *Schooling, Experience, and Earnings*. New York: National Bureau of Economic Research.

Morgan, J. N., & Sonquist, J. A. (1963). Problems in the analysis of survey data, and a proposal. *Journal of the American Statistical Association*, 58, 415–434.

Phelps, E. S. (1972). The statistical theory of racism and sexism. *The American Economic Review*, 62, 659–661.

Quinlan, J. R. (1993). *C4.5: Programms for Machine Learning*. San Francisco: Morgan Kaufmann Publishers Inc.

R Development Core Team (2009). *R: A Language and Environment for Statistical Computing*. R Foundation for Statistical Computing, Vienna, Austria.

Richardson, L. F. (1958). Mathematics of war and foreign politics. In J. R. Newman (Ed.), *The World of Mathematics*, New York: Simon and Schuster.

Romualdi, C., Campanaro, S., Campagna, D., Celegato, B., Cannata, N., Toppo, S., Valle, G., & Lanfranchi, G. (2003). Pattern recognition in gene expression profiling using DNA array: A comparison study of different statistical methods applied to cancer classification. *Human Molecular Genetics*, 12, 823–836.

Strasser, H., & Weber, C. (1999). On the asymptotic theory of permutation statistics. *Mathematical Methods of Statistics*, 8, 220–250.

Strobl, C., Boulesteix, A.-L., & Augustin, T. (2007). Unbiased split selection for classification trees based on the Gini index. *Computational Statistics & Data Analysis*, 52, 483–501.

Strobl, C., Kopf, J., & Zeileis, A. (2010). Wissen Frauen weniger oder nur das Falsche? Ein statistisches Modell für unterschiedliche Aufgaben-Schwierigkeiten in Teilstichproben. In S. Trepte & M. Verbeet (Eds.), *Wissenswelten des 21. Jahrhunderts–Erkenntnisse aus dem Studentenpisa-Test des SPIEGEL*, Wiesbaden: VS Verlag.

Strobl, C., Kopf, J., & Zeileis, A. (2013). Rasch trees: A new method for detecting differential item functioning in the Rasch model. *Psychometrika*. (Accepted.)

Strobl, C., Malley, J., & Tutz, G. (2009). An introduction to recursive partitioning: Rationale, application and characteristics of classification and regression trees, bagging and random forests. *Psychological Methods*, 14, 323–348.

Strobl, C., Wickelmaier, F., & Zeileis, A. (2011). Accounting for individual differences in Bradley–Terry models by means of recursive partitioning. *Journal of Educational and Behavioral Statistics*, 36, 135–153.

Ünlü, A. (2011). A note on the connection between knowledge structures and latent class models. *Methodology: European Journal of Research Methods for the Behavioral and Social Sciences*, 7, 63–67.

Vermunt, J. (2010). Latent class models. In *International Encyclopedia of Education*, vol. 7, Oxford: Elsevier, 3rd. edn. (pp. 238–244).

Wagner, G. G., Frick, J. R., & Schupp, J. (2007). The German Socio-Economic Panel Study (SOEP)–scope, evolution and enhancements. *Schmollers Jahrbuch*, 127, 139–169.

Wang, Y., & Witten, I. H. (1997). Induction of model trees for predicting continuous classes. In *Proceedings of the European Conference on Machine Learning*, Prague: University of Economics, Faculty of Informatics and Statistics.

Zeileis, A. (2005). A unified approach to structural change tests based on ML scores, F statistics, and OLS residuals. *Econometric Reviews*, 24, 445–466.

Zeileis, A., & Hornik, K. (2007). Generalized M-Fluctuation tests for parameter instability. *Statistica Neerlandica*, 61, 488–508.

Zeileis, A., Hothorn, T., & Hornik, K. (2008). Model-based recursive partitioning. *Journal of Computational and Graphical Statistics*, 17, 492–514.

Zeileis, A., Hothorn, T., & Hornik, K. (2010). *Party with the Mob: Model-based Recursive Partitioning in R*. R package version 0.9-9999.

Zhang, H., Yu, C.-Y., Singer, B., & Xiong, M. (2001). Recursive partitioning for tumor classification with gene expression microarray data. *Proceedings of the National Academy of Sciences of the United States of America*, 98, 6730–6735.

4 Exploratory Data Mining with Structural Equation Model Trees

Andreas M. Brandmaier, Timo von Oertzen,
John J. McArdle, and Ulman Lindenberger

Introduction

Structural Equation Model Trees (SEM Trees) combine Structural Equation Models (SEM) and decision trees. SEM Trees are tree structures that partition a dataset recursively into subsets with significantly different sets of parameter estimates. The method allows the detection of heterogeneity observed in covariates and thereby offers the possibility to automatically discover non-linear influences of covariates on model parameters in a hierarchical fashion. The methodology allows an exploratory approach to SEM by providing a data-driven but hypothesis-constrained exploration of the model space. We summarize the methodology, show applications on empirical data, and discuss Hybrid SEM Trees, an extension of SEM Trees that allows the finding of subgroups that differ with respect to model parameters and model specification.

In this chapter, we present an overview and selected applications of a multivariate statistical framework, Structural Equation Model Trees (SEM Trees; Brandmaier, von Oertzen, McArdle, & Lindenberger, 2013), that combines benefits from confirmatory and exploratory approaches to data analysis. SEM Trees allow a data-driven refinement of models reflecting prior hypotheses about the data. Confirmatory aspects are provided by using Structural Equation Modeling as the framework, and exploratory aspects arise from the incorporation of decision trees, also known as *classification and regression trees* or *recursive partitioning*. This combined approach yields trees representing a recursive partitioning of a dataset into subgroups maximally differing with respect to their model-predicted distributions. SEM Trees allow an exploratory approach to finding variables that influence the model parameters for any model that can be described as a linear combination of observed and latent variables. This class of models includes, for instance, regression models (McArdle & Epstein, 1987), factor analytic models (Jöreskog, 1969), autoregressive models (Jöreskog, 1979; McArdle & Aber, 1990), latent growth curve models (McArdle & Epstein, 1987), latent difference score models (McArdle &

Hamagami, 2001), or latent differential equation models (Boker, Neale, & Rausch, 2004).

A typical workflow for empirical research in the behavioral sciences can be described by the following steps. First, hypotheses about the population are derived as tentative explanations of observed phenomena. Then, a study is designed and conducted to collect a dataset with variables representing concrete observations of the phenomena. Finally, each hypothesis is formalized in a model and inference-statistical methods are used to gauge the evidence of the data for or against the hypotheses. This is often called the *confirmatory* approach of data analysis. Unfortunately, it is often found that models describe the data inadequately. Consequently, researchers move to an exploratory phase, in which they adapt hypotheses and models, in order to find a better representation of the observed phenomena, for instance, by adding variables or removing variables from their models. As an alternative approach to improving the model as a description of the complete dataset, a second approach can be followed: The dataset is partitioned into groups that differ with respect to the parameter estimates of the model. This multi-group approach assumes that the model is a valid description of the phenomena, however, it does not require the sample to be homogeneous with respect to the parameters of the model. SEM Trees realize this approach by recursively partitioning a dataset into subgroups that maximally differ in the model-predicted distributions.

In this chapter, we summarize the algorithm for inducing SEM Trees and highlight details such as the estimation of parameters in the models, the evaluation of split candidates, and the incorporation of measurement invariance in the SEM Tree framework. We draw attention to the dual motivation of the candidate selection procedure from a statistical and an information-theoretic point of view. As a methodological innovation, we discuss Hybrid SEM Trees that allow the specification of a set of different SEMs representing competing hypotheses about the data. Hybrid SEM Trees allow the retrieval of partitions of the dataset that are best described by different models and, ultimately, by different hypotheses. In order to illustrate the utility of SEM Trees, we conclude with an application of regular SEM Trees and Hybrid SEM Trees on empirical datasets.

All reported analyses are based on the freely available *semtree* package (Brandmaier, 2012b) for the statistical computing language R (Ihaka & Gentleman, 1996). The package is based on OpenMx (Boker et al., 2011) for defining and estimating SEMs.

Decision Trees

Decision trees are classifiers that discriminate between states of a response variable based on a hierarchy of decisions on a set of covariates. Put in a statistical context, a decision tree represents partitions of the covariate space that are associated with significant differences in the response variable.

The earliest representative, called the *Automatic Interaction Detector* (AID), was devised by Sonquist and Morgan (1964). The paradigm gained popularity through the seminal works of Breiman, Friedman, Olshen, and Stone (1984) and Quinlan (1986). Many aspects of decision trees have been developed since then. We restrict ourselves to mentioning only a small selection of the various approaches available today: ID3 (Quinlan, 1986), C4.5 (Quinlan, 1993), CART (Breiman et al., 1984), and QUEST (Loh & Shih, 1997). A more recent development are model-based trees. These trees maximize differences of outcome variables with respect to a hypothesized model, e.g., logistic regression trees (Chan & Loh, 2004), multivariate adaptive regression splines (MARS; Friedman, 1991), or the comprehensive model-based partitioning framework by Zeileis, Hothorn, and Hornik (2008). Decision trees are increasingly used in the context of ensemble learning. In this paradigm, multiple decision trees, typically based on resampled subsets of the original dataset, are aggregated into a decision forest, e.g., random forests (Breiman, 2001) or conditional random forests (Strobl, Boulesteix, Zeileis, & Hothorn, 2007), thereby trading increased computation time and decreased interpretability for increased stability and predictive accuracy of the results.

In the machine learning community, the popularity of tree methods gradually diminished in favor of more recent learning machine methods that allowed learning about more complex decision boundaries. Nevertheless, decision trees can be helpful in the process of knowledge discovery and scientific theory building due to their clear advantage in depicting the predictive structure visually. Figure 4.1 illustrates decision boundaries of a two-class problem obtained from a logistic regression as a representative of linear discriminating models and the corresponding boundaries produced by a decision tree. From a decision-tree perspective, a linear model can be thought of as an *oblique tree* restricted to a height of one. Oblique trees (Murthy, Kasif, & Salzberg, 1994), sometimes referred to as *multivariate trees* (Brodley & Utgoff, 1995), allow decisions to be represented as linear combinations of covariates in inner nodes of the tree. While oblique trees allow a larger number of possible splits, traditional decision tree approaches with axis-parallel splits have the advantage of allowing a straightforward interpretation of the decision tree as simple decision rules in natural language, specifically in rule sets describing conditions on covariates, e.g., "IF a participant is younger than 25 AND works out regularly, THEN she/he has a low risk of heart disease."

The result of a recursive partitioning algorithm is typically visualized in a dendrogram (cf. Figure 4.1). A dendrogram is a pictorial description of the hierarchical differences in the model-implied predictions and the predictors that determine these differences. Ovals represent decision nodes. The label of a decision node contains the name of the covariate that is subject to partitioning. If the partitioning is based on a statistical test, a corresponding p-value or a test statistic is shown in the label. Each oval has two or more

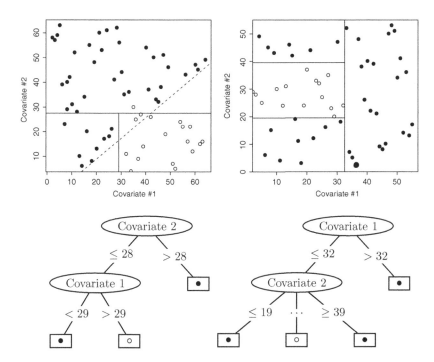

Figure 4.1 The top row shows two-dimensional decision boundaries on two hypothetical datasets with two classes depicted as empty and solid dots. On the left, the solid lines depict the axis-parallel decision boundaries of the tree below that discriminates the two classes. The dotted line depicts the decision boundary of a logistic regression discriminating between both classes. On the right, the solid lines represent axis-parallel decision boundaries of the decision tree below. In the example on the right, adequate discrimination is not possible using a linear model.

outgoing edges that represent the partitioning of the dataset into subsets corresponding to whether the covariate value of the oval node matches the condition that is depicted on the respective edge. Leaf nodes of the trees are depicted as rectangles that contain information about the predicted outcome; in the case of SEM Trees, they contain parameter estimates for the chosen SEM.

Structural Equation Model Trees

Hyafil and Rivest (1976) showed that finding an optimal tree is NP-hard (which can only be solved in non-deterministic polynomial time), whereas optimality is defined as a minimization of the expected number of decisions, and the underlying problem is computationally demanding to solve. This motivates the widespread application of heuristics for the induction of decision trees. When applying decision trees, the general idea is to choose the covariate from a set of candidates that divides the dataset into groups that maximally increase the predictability of the outcome variable. This process is recursively applied to each resulting partition of the dataset as long as meaningful covariates are found. The "goodness-of-split" can be formalized in various ways, for example, based on information-theoretic or statistical tests.

SEM Trees are based on the idea that datasets can not only be partitioned into subgroups that are homogeneous with respect to a single outcome variable, but into subgroups that are homogeneous with respect to the parameters of an SEM. In the behavioral and social sciences, Structural Equation Models (SEMs) have become widely accepted as a statistical tool for modeling the relations between latent and observed variables. SEMs are based on an isomorphism between (a) a set of linear equations for observed and latent variables and distributional assumptions about these variables and (b) a graphical representation of these equations and assumptions.

SEM Trees recover decision boundaries in the covariate space dividing the dataset into multiple groups that are each represented by a different parameter set for an SEM. If Figure 4.1 represented covariate space boundaries of an SEM Tree, one could imagine the solid and empty dots representing participants with two different associated parameter sets.

The algorithm for the induction of an SEM Tree is geared to the traditional decision tree algorithms. Thus, it is a greedy, top-down, recursive partitioning procedure that chooses the locally best split of the covariate space. In each recursion step, it chooses the covariate that is maximally informative about the model-predicted distribution and, according to this choice, permanently splits the dataset. Inputs of the algorithm are: (1) an SEM that formalizes the researchers' hypotheses about the data, which is also referred to as *template model*; (2) an empirical dataset that is modeled by the SEM; and (3) a set of covariates whose influence on the SEM is to be explored. In a first step, parameters of the template model are estimated from the complete dataset. This parameter set is associated with the root of a decision tree. For each covariate, the dataset is temporarily partitioned into subgroups according to the values of the covariate. Then, parameters are estimated for each partitioned dataset. We refer to the model that estimates parameters on the unsplit dataset as the *pre-split model*. The set of models resulting from splitting the dataset into subgroups can be seen as a single multiple-group model, which we refer to as the *post-split model*. If the parameter estimates are obtained with maximum likelihood estimation, the pre-split model and the post-split model are algebraically nested models and their log-likelihood ratio is asymptotically χ^2-distributed under the null hypothesis that the covariate is uninformative about the model-predicted distribution (Brandmaier, 2012a). At each level of the tree, the covariate with the maximum log-likelihood ratio is chosen. This selection procedure is recursively continued in each resulting partition. The known

Algorithm 4.1 The elementary algorithm for the induction of SEM Trees with discrete covariates and multi-way splits. Typically, a preprocessing step reduces all covariates to sets of binary covariates yielding a binary tree.

InduceSEMTree (*Dataset*, *Covariates*, *Model*)

1. Create a new *node*
2. Estimate free parameters $\hat{\theta}$ from *Dataset*
3. If *Covariates* is empty, return *node*
4. For each covariate $C_i \in$ *Covariates*,
    a. For each value v_{ij} of the covariate C_i,
        i. Estimate $\hat{\theta}_{v_i}$ from the subset $D_{v_{ij}} \subset D$ for which covariate C_i has value v_{ij}
    b. Calculate the log-likelihood ratio statistic Λ_{C_i} for covariate C_i as:
$$\Lambda_{C_i} = -2\mathcal{LL}\left(M\left(\hat{\theta}\right)|D\right) + \sum_j 2\mathcal{LL}\left(M\left(\hat{\theta}_{v_{ij}}\right)|D_{v_{ij}}\right)$$
5. Find best covariate candidate $b = \arg\max_i \Lambda_{C_i}$
6. If Λ_{C_b} is above the critical value,
    a. For each v_{bj} in C_b create a child node by recursively calling InduceSEMTree $(D_{v_{bj}},\text{*Covariates*}\backslash C_b,\text{*Model*})$ and create an edge between *node* and the new child node with label v_{bj}
7. Otherwise return *node*

---

distributional properties of the estimator under the null hypothesis allow the usage of a hypothesis-testing framework to determine when to stop splitting the dataset. If there is no covariate having a significantly large test statistic such that the null hypothesis can be rejected, the induction algorithm terminates. The elementary algorithm for the induction of an SEM Tree with discrete covariates is shown in Algorithm 4.1. This algorithm allows multi-way splits according to the levels of discrete covariates. In the following, a generalization to continuous covariates is described.

Multi-valued Attributes

Drawing upon ideas of Breiman et al. (1984) and Loh and Shih (1997), SEM Trees perform dichotomous splits in the covariates leading to binary tree representations. Generally, trees with multiway splits can be represented as binary trees. However, Kim and Loh (2001) illustrate an example that yields different results depending on whether a multiway and a binary tree are used.

To allow continuous, and multi-valued ordinal and discrete covariates, all multi-valued covariates are transformed to sets of dichotomous covariates. The conversion depends on the type of variable. Ordinal variables having values from an ordered set are transformed into a set of dichotomous variables, in which each variable represents a "smaller or equal" relation on one of the possible split points. Continuous variables with values from an ordered set imply cutpoints in the center between pairs of sorted observed values. Let N be the number of observations in a dataset. With the proposed procedure, ordinal and continuous variables can imply a maximum of $N - 1$ dichotomous covariates. Categorical covariates imply dichotomous covariates representing splits corresponding to partitions in all possible pairs of non-empty subsets yielding a maximum of $2^{N-1} - 1$ dichotomous covariates.

Model Estimation

Parameters in models of an SEM Tree are estimated using maximum likelihood estimation, which is a common technique in SEM (e.g., von Oertzen, Ghisletta, & Lindenberger, 2009). Let X be a dataset. Let M be a template SEM that encodes researchers' prior hypotheses about the data. Under the assumption of independence of the observations in the sample, the likelihood of the model given the dataset with a total of N observations is the product of the likelihoods of observing the model given individual observations. Let Σ be the model-implied covariance matrix and let μ be the model-implied mean vector. Under the assumption that the variables are normally distributed, the likelihood of the model given a single datum x is defined by the multivariate Gaussian distribution:

$$\mathcal{L}(\mu, \Sigma | x) = ((2\pi)^p |\Sigma|)^{-\frac{1}{2}} \exp\left[-\frac{1}{2}(x-\mu)^T \Sigma^{-1}(x-\mu)\right] \qquad (1)$$

Given a set of observations, let m be the sample mean vector and S be the sample covariance matrix. Furthermore, let θ be a vector parametrizing μ and Σ. The simplified log-likelihood function derived from Equation 1 is:

$$\begin{aligned} &-2\mathcal{LL}(\mu, \Sigma, \theta | m, S) \\ &= N\left[\text{const} + \log|\Sigma| + tr\left(\Sigma^{-1}S\right) + (m-\mu)^T \Sigma^{-1}(m-\mu)\right] \end{aligned} \qquad (2)$$

The maximum likelihood estimate of the parameters $\hat{\theta}$ given the data m, S, and the model μ, Σ is found by minimizing $-2\mathcal{LL}$:

$$\hat{\theta} = \arg\min_{\theta} -2\mathcal{LL}(\mu, \Sigma, \theta | m, S) \qquad (3)$$

Because it is difficult to find a closed-form solution for general models, numeric procedures are employed to find the maximum of the likelihood function or the minimum of the negative function. There are a variety of methods for the numerical solution of the problem. Most solutions revolve around the *Newton method* in order to find the minimum of the function numerically. The gradient, i.e., the partial derivatives of the likelihood function with respect to the parameters, describes the rate of increase or decrease of the likelihood function depending on an infinitesimally small change of the free parameters. *Gradient descent* methods propose iteratively calculating the gradient, and climbing or descending the likelihood function until they arrive at a maximum or minimum. An improvement of this approach also considers the matrix of the partial second derivatives, the *Hessian matrix*. This comprises the class of Newton methods. If the Hessian matrix is iteratively approximated rather than fully approximated at each iteration, this is called a *quasi-Newton method*. An important and widely used representative of the latter method is the *Broyden, Fletcher, Goldfarb, and Shanno method* (BFGS), which was independently suggested by each of the four authors. A comprehensive overview of optimization algorithms is given by Fletcher (1994). OpenMx (Boker et al., 2011), on which the *semtree* package is based, uses a general-purpose optimization scheme that involves numerical estimation of the gradient and the Hessian matrix. Von Oertzen et al. (2009) report that a dampened Newton method, which fully calculates the gradient and the Hessian at each step and adapts the step width by a line search, works well in practice.

Split Candidate Evaluation

A recursive tree-inducing algorithm proceeds by selecting the best candidate at each step of the recursion. There are natural stopping criteria for a recursive tree-inducing algorithm, including the following: (1) there are no remaining covariates to split the dataset; (2) the number of observations in a leaf node is below a certain threshold; (3) a pre-determined height of the tree has been reached; and (4) the best split candidate is not good enough. The fourth criterion is introduced in order to avoid overfitting, i.e., to avoid choosing an apparently adequate split candidate although its aptitude is only due to random fluctuation in the sample. In SEM Trees, split candidate selection can be based on the log-likelihood ratio that allows a statistical test to determine when splitting the dataset and growing the tree should be stopped.

Given a θ-parametrized model $M(\theta)$ and a dataset D, let $\hat{\theta}$ be a parameter vector that minimizes the *negative two log-likelihood* of seeing the model given the data, i.e., the maximum-likelihood estimate of M given D. Let $D_{v_{ij}}$ be the partitions of a dataset with respect to the $j = 1, \ldots, k$ values of the ith covariate in the dataset and let $\hat{\theta}_{v_{ij}}$ be the parameter vector that minimizes $-2\mathcal{LL}(M(\hat{\theta}_{v_{ij}})|D_{v_{ij}})$. For covariate i, the likelihood ratio of the pre-split model (left summand) and the post-split model of the $j = 1, \ldots, k$ resulting partitions (right summand) is:

$$\Lambda_i = -2\mathcal{LL}\left(M(\hat{\theta})|D\right) + \sum_{j=1}^{k} 2\mathcal{LL}\left(M(\hat{\theta}_{v_{ij}})|D_{v_{ij}}\right) \tag{4}$$

Following Wilks's (1938) theorem, Λ_i is χ^2-distributed with $(k-1)m$ degrees of freedom with m being the number of free parameters in the template SEM. Given the distributional properties of Λ_i and a chosen α level, a critical value c can be calculated such that $Pr(\Lambda > c) = \alpha$. Only split candidates for which $\Lambda_i > c$ holds are considered potential split candidates. The best split is chosen by selecting the covariate with the maximum log-likelihood ratio, that is, the covariate with the largest evidence against the pre-split model.

It has been pointed out that any tree-structured algorithm carrying out an exhaustive search in the split attribute space suffers from a multiple comparison problem (Jensen & Cohen, 2000) that can lead to overfitting (i.e., an over-representation of apparent structure in the sample that is merely due to sampling fluctuations). Solutions for tackling this problem include correcting the critical value under the assumption of independence of covariates, known as Bonferroni correction, or using cross-validation to obtain estimates of the expected log-likelihood ratio. The first approach corrects the sampling distribution under independence assumptions at low computational costs but is known to be overly conservative. The latter approach yields a score that can be treated as an individual score instead of a maximum score but requires additional computations.

Several authors (Dobra & Gehrke, 2001; Jensen & Cohen, 2000; Loh & Shih, 1997; Shih, 2004; Zeileis et al., 2008) have cautioned about the problem of variable selection bias in the context of decision trees. By definition, split candidate selection procedures that preferably select certain types of variables over others under the null hypothesis that all variables are uninformative, suffer from variable selection bias. A typical observation is that categorical variables with a larger number of categories are more likely to be chosen under the null hypothesis than those with less. An unbiased variable selection algorithm is expected to have no preference for any variable under the null hypothesis. Brandmaier et al. (2013) showed that using the likelihood-ratio-based split selection procedure as described above suffers from selection bias. This bias can be reduced to a neglible amount if variable cutpoints and the selection between variables are estimated in a two-step procedure (Kim & Loh, 2001). As an alternative approach, Zeileis et al. (2008) suggested a unified framework for unbiased model-based recursive partitioning that is based on tests for parameter instability.

Missing Values

Under the assumption that variable values are *missing at random* (Rubin, 1976), SEM Trees can handle missing values in the observed variables of the model as well as in the covariates. If values are missing in the covariates, the likelihood calculation is performed with Full Information Maximum Likelihood (FIML; Finkbeiner, 1979), which is equivalent to Equation 2 under the assumption of no missingness:

$$-2\mathcal{LL}(\mu, \Sigma | x_1, \ldots, x_N)$$

$$= N \cdot p \cdot \ln(2\pi) + \sum_{i=1}^{N} \left[\ln |\Sigma_i| + (x_i - \mu_i)^T \Sigma_i^{-1} (x_i - \mu_i) \right] \quad (5)$$

where Σ_i is the model-implied covariance matrix with rows and columns deleted according to the pattern of missingness in the ith observation x_i, and μ_i the model-implied mean vector with elements deleted according to the respective pattern of missingness.

Missing values in covariates can be handled by removing the respective missing rows in the dataset during the evaluation of a split candidate, effectively modifying only Equation 4 based on the patterns of missingness in the dataset. Others, e.g., Hastie, Tibshirani, and Friedman (2001), employ a surrogate approach, which is based on finding a surrogate covariate that most closely describes the same partition of the dataset as the variable with the missing value.

Measurement Invariance

A fundamental issue in psychometrics is measurement invariance. A measurement is invariant if "under different conditions of observing and studying phenomena, measurement operations yield measures of the same attribute" (Horn & McArdle, 1992, p. 117). Measurement invariance is traditionally examined through a sequence of hypothesis tests. Typically, a set of statistical tests are carried out to determine what level of measurement invariance is tenable. These tests are administered in the following order: (1) *configural invariance* (also *configuration invariance* or *pattern invariance*) requires the invariance of the pattern of zero and non-zero factor loadings across groups; (2) *metric invariance, weak invariance,* or *factor pattern invariance* requires the invariance of the values of factor loadings across groups; (3) *strong factorial invariance* or *scalar invariance* requires intercepts of all indicators and all factor loadings to be equal across groups; (4) *strict invariance* establishes the additional restriction that the residual error variances are equal across groups, in order to allow the interpretation of standardized coefficients across groups.

SEM Trees incorporate the concept of measurement invariance if desired by the researcher. Models with measurement invariance are integrated in the process of split candidate evaluation in the following way. Valid split candidates must fulfill the user-specified level of measurement invariance *and* their log-likelihood ratio must exceed the chosen critical value. By construction, a measurement-invariant post-split model of configural, metric, strong, or strict invariance is algebraically nested in a non-invariant post-split model. Furthermore, the invariant model is nested in the pre-split model (Brandmaier, 2012a). Consolidating these observations, one can determine valid split candidates by first performing a set of likelihood-ratio tests in order to assure measurement invariance for the split candidates. With the set of candidates for which measurement invariance could not be rejected, the normal procedure of split candidate evaluation (see above) is then performed.

Information-Theoretic Interpretation

From a machine learning perspective, split candidate selection procedures for traditional decision trees with a categorical outcome variable maximize the predictability of the outcome variable conditional on the knowledge of the state of the selected predictors. This is often formalized as follows. Let $H(X)$ be the Shannon entropy of a random variable X, let $x_1, \ldots, x_N$ be observed outcomes of X, and let N_y be the number of observations for which $Y = y$. The information gain about X when knowing the state of Y is:

$$Gain\,(x_1, \ldots, x_N, Y) = H\,(X) - \sum_{y \in \mathit{Values}(Y)} \frac{N_y}{N} H\,(x_1, \ldots, x_n | Y = y) \quad (6)$$

Let M be a model. The duality between the Gaussian log-likelihood and the entropy of the model-predicted distribution can be formulated as:

$$\lim_{N \to \infty} \frac{1}{N} \sum_{i=1}^{N} \mathcal{LL}\,(x_i | M) = -H\,(X | M) \quad (7)$$

Estimating the parameters of an SEM by maximizing transformations of the multivariate Gaussian fit function seen in Equation 2 minimizes the entropy of the model-implied distribution. The likelihood-ratio test statistic that is used to determine the significance of a split candidate is the difference between the log-likelihood of the parent model and the sum of the log-likelihoods of the potential child models. It can be shown that the likelihood-ratio test statistic from Equation 4 is proportional to the information gain shown in Equation 6 (see Brandmaier, 2012a, for a detailed proof). This important relation motivates the variable selection approach in

SEM Trees from both a statistical and an information-theoretic perspective. At each level, an SEM Tree chooses the covariate that maximizes the likelihood ratio statistic of the pre-split and the post-split model. This is equivalent to choosing the covariate that maximizes the information gain about the model-predicted distribution. The expected information gain is the mutual information. By using cross-validation as a variable selection procedure, the expected likelihood-ratio is estimated and, following the same reasoning as above, the expected information gain is maximized at each level of the decision tree.

Hybrid Trees

Thus far, SEM Trees were defined such that their goal is a recursive partitioning of a dataset into subsets that maximize the difference of the model-implied distributions. As an extension to that, Brandmaier et al. (2013) suggested that it can make sense to allow not only a single template model but a set of competing template models. This implies that leaf nodes in the tree are associated with subgroups represented by both different SEM and individual parameter estimates. We call these types of trees *Hybrid SEM Trees*. Hybrid SEM Trees can answer research questions that involve a choice between multiple candidate models for the representation of a dataset. Instead of providing a single choice of a "best" model for the complete dataset, Hybrid SEM Trees retrieve different models for different subgroups in the sample.

Suppose a group of researchers is interested in cognitive development over the life span. They suspect that some participants have a linear increase in cognitive abilities that saturates at some point, whereas a competing hypothesis assumes a simple linear change process without saturation, and a third hypothesis proposes a drastic exponential decline in cognitive change that was observed for very old participants. Rather than fitting a single model to the complete dataset, which might imply an unfavorable trade-off between the different observed phenomena of change, a Hybrid SEM Tree can recover subgroups that are best described by individual models.

We will outline the underlying model selection problem in Hybrid SEM Trees in more detail. Comparing a set of different models for subgroups renders the likelihood ratio test for model selection inappropriate because, in general, the set of competing models is not algebraically nested. However, the evaluation of split candidates by an estimate of their predictive performance is still feasible. As suggested before in the context of SEM Trees with a single template model, we employ k-fold cross-validation to obtain an estimate of the expected likelihood ratio of a pair of candidate models. The cross-validated test statistic $\hat{CV}$ is obtained by averaging the k test statistics of a k-fold cross-validation. Let M_1 and M_2 be two competing models and let x be an observation, and n be the sample size.

We obtain:

$$\lim_{n\to\infty} \hat{CV} = E\left[\log P(x|M_1) - \log P(x|M_2)\right] = E\left[\log \frac{P(x|M_1)}{P(x|M_2)}\right]$$

where $E()$ is the expectation. For model selection, we require a procedure to determine the evidence for the model. Using Bayes' formula, we can rewrite the test statistic as:

$$E\left[\log \frac{P(x|M_1)}{P(x|M_2)}\right] = E\left[\log \frac{P(M_1|x) P(M_2) P(x)}{P(M_2|x) P(M_1) P(x)}\right]$$

Under the assumption that, a priori, all candidate models are equally likely, $P(M_1) = P(M_2)$, we obtain a model selection procedure selecting covariates that maximize the expected posterior probability of the post-split model:

$$\lim_{n\to\infty} \hat{CV}_{P(M_1)=P(M_2)} = E\left[\log \frac{P(M_1|x)}{P(M_2|x)}\right]$$

Under the assumptions of non-identical priors for the models, a correction term can be added to the cross-validation statistic that represents the models' prior ratio. Effectively this is the sum of the estimated log-likelihood ratio and the difference of the log-priors.

Hybrid SEM Trees allow model selection between a set of competing SEMs representing competing hypotheses about the data, whereas recursive partitioning elicits a hierarchy of covariates associated with differences in the dataset. In the process of induction of a Hybrid SEM Tree, covariates are selected that not only maximize differences with respect to a single model-implied distribution but also select between parametrized distributions that are the best representation for the observed phenomena.

With hybrid trees, we distinguish between heterogeneity with respect to models and heterogeneity with respect to parameters. Common SEM Trees find heterogeneity with respect to parameters assuming that the model holds for all subgroups of the dataset. Additionally, Hybrid SEM Trees are able to choose between competing models. When applying Hybrid SEM Trees with a single template model, this is equivalent to applying a common SEM Tree with cross-validation for variable selection.

Case Studies

In this section, we present applications of SEM Trees to selected datasets, including a univariate SEM Tree, a regression SEM Tree, and a longitudinal SEM Tree.

Univariate SEM Tree and Regression SEM Tree

In order to illustrate the usage of univariate SEM Trees, we show an analysis based on a freely available dataset from the *psych* package (Revelle, 2011) for R. The dataset includes scores of 700 participants on the Scholastic Aptitude Test (SAT) and the American College Test (ACT) that were collected as part of the Synthetic Aperture Personality Assessment (Revelle, Wilt, & Rosenthal, 2009). Three additional covariates are included for each participant: sex (male/female), education (high school, high school graduate, college, college graduate, or graduate degree), and age ($\mu = 25.6 \pm 9.5$).

Suppose a researcher is interested in performance differences on the SAT verbal (SATV) scale. Therefore, the researcher sets up a fully saturated SEM that measures the mean of the SATV score as μ_{SATV} and the variance of the score as σ^2_{SATV}.

For the analysis, the SAT scores were standardized by subtracting the mean and dividing by the standard deviation. The SEM Tree that was generated from these data with Bonferroni-corrected p values and an alpha level of .01, is shown in Figure 4.2. The tree chooses education as the first partition in a group depending on whether participants graduated from college or not. For those who graduated from college, a second split according to their age is the optimal partition. In order to estimate the difference between the two age groups for graduates, we first calculated the pooled standard deviation of both age groups $\sigma^2 = 0.70$ and obtained an effect size of $d = 0.59$ for a split at an age of 28. Naturally, the question arises whether this particular age reflects a fixed change point of the investigated phenomenon. In general, researchers should be careful

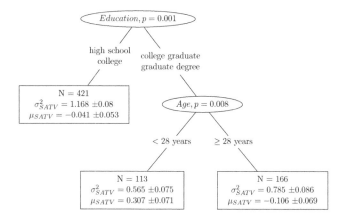

Figure 4.2 SEM Tree for the SAT dataset. Variables "education" and "age" are chosen to split the dataset into subgroups that describe differences in the SAT scores. Parameter estimates are given as point estimates with standard error.

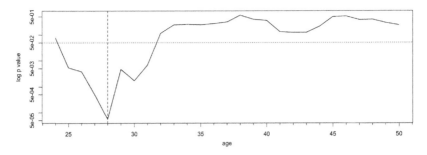

Figure 4.3 The plot shows *p* values for the splits according to the variable "age" conditioned on people with an education equivalent to college degree or higher. This corresponds to the right subtree of the SEM Tree in Figure 4.2. The *p* values are shown on a logarithmic scale. The horizontal, dotted line indicates a significance level of 5%. Values below the threshold are potential split candidates. The minimum value is attained when a split at an age of 28 years is chosen (marked by the vertical, dashed line). The set of significant split points of the age variable ranges from ages 25 to 31.

when reifying continuous splits. We advise further inspection of the tree and the dataset in order to determine an appropriate range of a continuous covariate that supports significant splits. In this example, we inspected the *p* values of all possible age-related splits. The *p* values were already corrected for multiple testing. Figure 4.3 shows the log *p* values versus possible age-related splits. A clear minimum is visible at an age of 28. However, the range from 25 to 31 is below a significant threshold of .05. We conclude that this age range should be interpreted as a fuzzy partitioning or fuzzy decision rule for describing the age-related subgroups instead of cresting at 28. This type of ex-post analysis can be carried out for any covariate with more than two ordered levels.

In the following, we use the same dataset to investigate a regression model with SEM Trees. Beyond the verbal score of the SAT, the dataset contains self-reports of the quantitative score on the SAT. Again, both scores were normalized to obtain zero mean and unit variance. A regression model between the two scores can reveal insights about how strongly these two scores are correlated or how well one score can predict the other score assuming a linear relationship between the two variables. The model contains five parameters. A plot of the model's path diagram is shown in Figure 4.4.

The correlation between the verbal and the quantitative score was $r = 0.64$, $p < 2.2 \times 10^{-16}$. Based on this measure, the proportion of variance shared between the two variables in the sample is about 41% under the assumption that the population is homogeneous with respect to the correlation of these abilities. Applying an SEM Tree to this model allows the

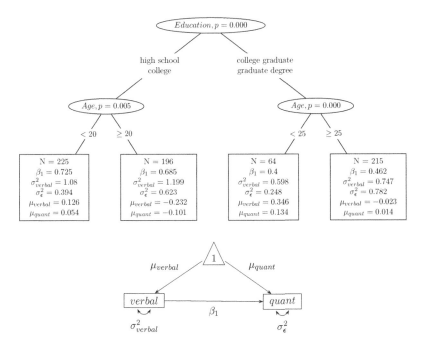

Figure 4.4 SEM Tree (upper panel) for a regression model (lower panel) of the quantitative and the verbal SAT score.

detection of subgroups that differ in the strength of the linear relationship between the two scores. Differences in the parameter β_1 between groups indicate a different strength of the linear relation between the two variables. Differences between the parameter μ_{verbal} or μ_{quant} indicate differences in the expected values of the scores between subgroups. Differences in the residual error term, σ_ϵ^2, between groups indicate differences in the model fitness, particularly, differences in the amount of variance unexplained by the linear model. The SEM Tree was induced using a significance threshold of .01. The resulting tree is shown in Figure 4.4. The subgroups that are implied by the first-level split of the SEM Tree are shown as scatterplots in Figure 4.5, in which solid and dotted regression lines indicate the linear relation between the two variables as retrieved by the SEM Tree. The variable "education" that was chosen as a first split explains differences in the linear relation between the quantitative and the verbal SAT scores. The better-educated subgroup shows a smaller linear relation between the scores. Splits with respect to age are found in both education-related subgroups. Generally, we can observe a decrease in the average performance on both scales in each of the subgroups with higher age. The difference in the cutpoint of the continuous variable "age" in the left and right subtree

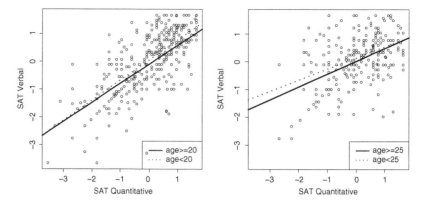

Figure 4.5 Regression plots of the quantitative and verbal SAT scores for the subgroups implied by the first split of the SEM Tree. The split variable was education and the two partitions were high school, high school graduation or college (left plot) and college graduation or graduate degree (right plot). The regression lines depict the linear relation between both scores for the second-level splits of the SEM Tree, both with respect to the covariate age.

might indicate a lag in the underlying age-related change process that is explained by "education."

Without SEM Trees, a researcher might have stopped when finding a highly significant correlation between test scores. SEM Trees proved useful here because they recovered subsets that differ with respect to the model parameters. In the tree, we found hints that the covariate "age" predicts lower average scores on both tests, whereas "education" predicts a difference in the shared variance of the scores.

Longitudinal SEM Tree

Brandmaier et al. (2013) showed illustrations of SEM Trees based on a factor model and a latent growth curve model. In the following, we further demonstrate how SEM Trees can be employed to explore structure in the data. We present an application of SEM Trees to data from the Berlin Aging Study (BASE; cf. Baltes & Mayer, 1999; Lindenberger, Smith, Mayer, & Baltes, 2010; Lövdén, Ghisletta, & Lindenberger, 2004). BASE is a multidisciplinary study of aging with extensive measurements from psychology, sociology and social policy, internal medicine and geriatrics, and psychiatry. The first wave of measurements started in 1990. The sample consisted of 516 participants who were recruited from the city registry of Berlin, Germany. The initial sample was stratified by sex and by age with a mean age of 84.9 years and an age range of 69.7–103.1 years. It was followed up longitudinally in seven further waves until 2009. For our analysis, we relied on data from the first six waves, of which the last was completed in 2005.

For illustration purposes, we focus the analysis on the digit letter test, a measure of perceptual speed as a marker of cognitive functioning. Participants performed the task on 11 occasions in six waves spread over fifteen years. Ghisletta and Lindenberger (2004) have modeled the digit letter task with latent growth curve models before. They report that the digit letter task displayed both reliable fixed and random linear time effects but no statistically significant quadratic effects of time for the mean and variance of a quadratic slope factor. Therefore, a linear latent growth curve model was used to model changes in the performance of individuals over time. A latent intercept variable I with mean μ_I and variance σ_I^2 models the baseline performance on the task, and a latent linear slope variable S with mean μ_S and variance σ_S^2 models the increase or decrease of performance over time in the study. A correlation between the latent intercept and the latent slope was estimated as σ_{IS}^2. The exact individual time points of measurement were available for each participant and for each occasion of measurement. This allowed accurate modeling of the cognitive change process with individual but fixed slope loadings for each participant by employing these individual time points as definition variables on the slope's loadings. Put differently, the SEM can be thought of as a multi-group model with each participant being its own group, and each group being characterized by individual but fixed slope loadings, whereas each freely estimated parameter is restricted to be equal across groups. Furthermore, the slope loadings were individually centered at the mean age of the observed time span for each participant. Also, participants' age was controlled for at the latent level by adding age as a covariate with loadings onto intercept and slope. The corresponding latent growth curve model is depicted in Figure 4.6.

Due to the age of the participants, mortality led to high attrition and 61.67% of the measurements are missing. Therefore, FIML estimation was used to estimate the parameters in the model.

Covariates included education and newspaper reading and book reading habits, of each participant's father and mother. The reading habit variables were encoded on an ordinal scale with the values "often," "sometimes," "seldom," and "never." The education index was encoded on an ordinal scale with three values: "elementary school without apprenticeship," "elementary school with apprenticeship," and "secondary school certificate."

The resulting SEM Tree is depicted in Figure 4.7. The first chosen covariate is the education variable. The decline in cognitive score is lower for the higher-educated group (first-level split, left subtree). For the lower-educated group, there is a difference based on their fathers' reading behavior. For participants whose fathers read newspapers sometimes or often, the decline in perceptual speed is comparable to the decline in the better-educated group. However, if fathers did not or seldom read newspapers,

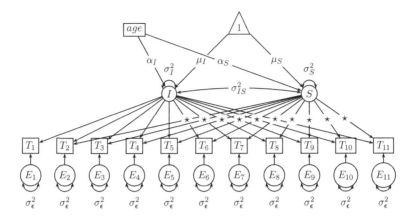

Figure 4.6 Linear latent growth curve model with individual time points for the measurement occasions. Participants were measured on 11 occasions spread over six waves. The residual error for each measurement has a variance of σ_ϵ^2. Repeated measurements of the digit letter score are represented as T_i with i being the measurement occasion. The measurement error for each observation is accordingly named E_i. The latent trajectory is modeled with an intercept $I \sim N\left(\mu_I, \sigma_I^2\right)$ and a slope $S \sim N\left(\mu_S, \sigma_S^2\right)$. The correlation between intercept and slope is modeled as σ_{IS}^2. The factor loadings of the slope are marked with stars to indicate that they are individual but fixed for each participant. The variable *age* is controlled for with loadings α_I and α_S on the latent level.

the decline is about 27% stronger than in the other group. Possibly, the fathers' reading behavior acts as a proxy for parents' education and the tree might depict an interaction of children's and fathers' educational level in predicting cognitive decline in old age.

Factor-analytic SEM Tree

For this example, we analyze a personality dataset available as "bfi" in the *psych* package (Revelle, 2011) for R. The dataset includes 25 personality self-report items taken from the International Personality Item Pool for 2,700 participants. For an illustration, we set up a single factor model as an SEM that models the personality trait "extraversion" as follows: five observed variables $X_1, X_2, \ldots, X_5$ model five items related to the factor extraversion with items including "make friends easily" or "find it difficult to approach others." Each score has an individual measurement error $E_1, E_2, \ldots, E_5$ with individual residual variances $\sigma_{\epsilon_1}^2, \sigma_{\epsilon_2}^2, \ldots, \sigma_{\epsilon_5}^2$. The means of the items were modeled as $\mu_1, \mu_2, \ldots, \mu_5$. The latent factor *"ext"*

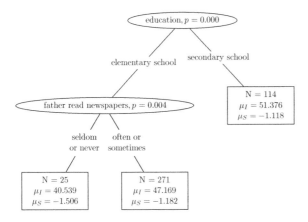

Figure 4.7 An SEM Tree based on a linear latent growth curve model. Candidate covariates included four variables of parents' reading behavior and a variable about the participants' education.

was modeled to have a mean of μ_{ext} and a variance of σ_{ext}^2. Figure 4.8 depicts the factor model.

The resulting SEM Tree was constructed with the requirement of weak invariance, that is, factor loadings were required to be equal in subgroups, and by using the Bonferonni-corrected variable selection procedure. Covariates include the variables gender, education, and age. The tree is shown in Figure 4.9. It has two levels. The first split is with respect to the age covariate, the second split is conditional on whether participants were younger than thirty years. On the second level, only the younger group is partitioned according to gender, whereas the older group is not. Comparing the estimates of the nodes, we find a difference in the means of the latent variable. Participants older than 30 years and females younger than 30 have comparable scores on the extraversion factor, both above average, whereas the young males have a below-average value of "extraversion." The interpretation of the differences can be guided by inspecting the set of estimates in the leaf nodes. For example, a comparison of the residual errors of items $(\sigma_{\epsilon 1}^2, \ldots, \sigma_{\epsilon 5}^2)$ might show that variance in certain items is less well explained by the common factor in one group than in the other. In this example, this does not seem to be the case. Differences in the mean values for the individual items $(\mu_{x2}, \ldots, \mu_{x5})$ could hint at systematic differences. Note that some researchers might recommend a stricter level of measurement invariance to draw conclusions from the differences. This could be a level of measurement error that does not allow differences in the expected values of the items. Nevertheless, building a tree with weaker constraints can give insights into which covariates induce subgroups that maximally break this requirement.

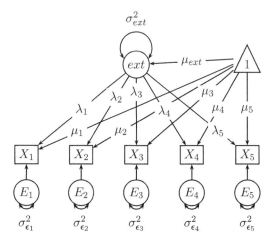

Figure 4.8 A single factor SEM for the personality factor "extraversion." The factor *ext* is measured by five items, X_1 to X_5.

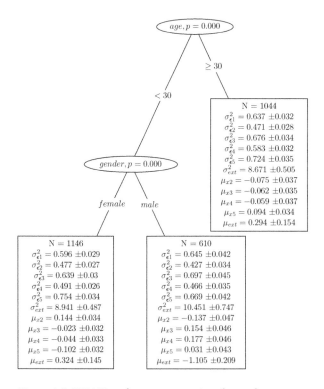

Figure 4.9 SEM Tree for an extraversion factor from a personality questionnaire.

A Hybrid SEM Tree simulation

For an illustration of a Hybrid SEM Tree, we first simulated a dataset measuring hypothetical cognitive decline in younger and older age. In this hypothetical sample, participants are described by two dichotomous covariates that carry the following meanings: (a) participants were sampled from young and older adults, and (b) participants were either part of a training program or a control group. Assume a group of researchers is interested in finding a model that describes cognitive development in their sample. Two template latent growth curve models were constructed for the Hybrid SEM Tree, one describing a model of linear change and one describing a model of quadratic change. The linear model assumes that a datum $x_{i,t}$ describing the score of individual i at time point t is an observation of the following generative model:

$$x_{i,t} = I_i + (t - t_0) S_i + E_i$$

with I being an intercept term that is distributed with mean μ_I and σ_I^2, S a slope term being distributed with μ_S and σ_S^2, and E being a residual error term that is distributed with zero mean and a variance σ_ϵ^2. The quadratic growth curve model is analogously created with a quadratic instead of a linear growth term:

$$x_{i,t} = I_i + (t - t_0)^2 S_i + E_i$$

In our simulated dataset, younger adults have an approximately linear decline in their cognitive abilities that is mitigated by the cognitive training program, whereas older adults have an accelerating, quadratic decline over time. Participants were either male or female, which had no effect on cognitive development for any subgroup. The dataset was simulated with the following values. For all participants the residual error had a variance of $\sigma_\epsilon^2 = .01$ and the covariance between intercept and slope was set at $\sigma_{IS} = 0$. The younger participants without treatment were simulated with an intercept of $\mu_I = 0$ and $\sigma_I^2 = 1$, and a slope of $\mu_I = -0.8$ and $\sigma_S^2 = .25$. The younger participants with treatment were simulated with an intercept of $\mu_I = 0$ and $\sigma_I^2 = 1$, and a slope of $\mu_I = -1.6$ and $\sigma_S^2 = .25$. Independently of the received treatment, older participants were simulated with an intercept of $\mu_I = 0$ and $\sigma_I^2 = 1$, and a quadratic slope component of $\mu_I = -0.8$ and $\sigma_S^2 = .25$.

An example SEM Tree that was obtained from randomly creating a dataset with the given values is depicted in Figure 4.10. A graphical representation of the expected growth curves of the subgroups, as they were detected by the tree, is also shown in Figure 4.10. The tree finds two subgroups with respect to age in the first split. In the young group,

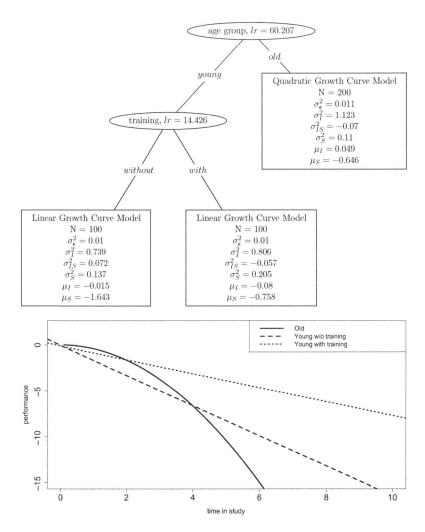

Figure 4.10 Analysis of a simulated dataset using a hybrid SEM Tree with two
template models: a linear latent growth curve model and a quadratic
latent growth curve model. Upper panel: the SEM Tree shows a first
partition with respect to the age group. Older participants experience
a quadratic decline, while younger participants are better described
by a linear decline. Treatment has an influence on the slope of the
change process for younger participants while no significant parameter
differences for the older group were found on the second level of the
tree. Lower panel: expected growth curves over time in the study for
the three subgroups of the simulated dataset that were recovered by
the tree.

the tree partitions the sample according to treatment. Most noteworthy, the hybrid tree correctly chose different models for the young and old subtrees. Young participants are represented with a linear model of change and old participants' accelerating decline is represented by a quadratic change. Of course, this example can also be phrased in a confirmatory setting, in which group differences with respect to the polynomial growth are expected and tested in a multi-group model. We expect Hybrid SEM Trees to work well when (1) there are a number of competing hypotheses, (2) models are complex, and (3) the number of covariates is large and their influence on the individual models and the nature of their interactions is not known (e.g., if researchers have additional sets of behavioral, cognitive, or genetic covariates).

Hybrid SEM Tree with a Developmental Latent Growth Curve Model: Wechsler Intelligence Score for Children

The data for the following illustration of an application of Hybrid SEM Trees were originally obtained by Osborne and Suddick (1972) between 1961 and 1965. These data have been analyzed in depth before (e.g., McArdle & Epstein, 1987; McArdle, 1988). An analysis with SEM Trees was performed by Brandmaier et al. (2013) using a linear latent growth curve model. We extend this analysis to a Hybrid SEM Tree.

The dataset was created from measurements for 204 children on eleven different items from the *Wechsler Intelligence Scale for Children* (WISC; Wechsler, 1949). The children were repeatedly measured on four occasions, at the ages of six, seven, nine, and eleven years. The raw scores of four "verbal" subscales and four "performance" subscales were aggregated into a composite score, which was rescaled to a range between 0 and 100. In this analysis, the covariates included the dichotomous variables "sex," the continuous variable "age," and the continuous variable "years of education" of each mother and father. The covariate "father's education" had missing values. We set up an equivalent latent growth curve model for an analysis with SEM Trees based on the description by Brandmaier et al. (2013). Then, we created modified versions of this model representing competing hypotheses about the underlying growth curve of cognitive development. All candidate models were derived from the following baseline model (see Figure 4.11), which we refer to as *BASELINE*. This model has four observed variables representing the test scores at the four occasions of measurement. Each occasion has independent errors of measurement. Five freely estimated parameters describe the distribution at the latent level: The intercept is assumed to be distributed as $I \sim \mathcal{N}\left(\mu_I, \sigma_I^2\right)$, the slope term is assumed to be distributed according to $S \sim \mathcal{N}\left(\mu_S, \sigma_S^2\right)$, and the covariance between both is modeled as σ_{IS}^2. The first slope loading was fixed at zero and the remaining slope loadings are parametrized as $\lambda_1 \lambda_2$, and λ_3. The residual variance σ_ϵ^2 is assumed to be equal for each occasion

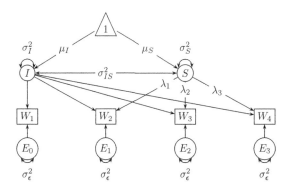

Figure 4.11 Longitudinal latent growth curve model for the WISC dataset. A composite score of WISC is measured longitudinally on four occasions represented by variables W_1, W_2, W_3, and W_4 with corresponding errors of measurement $E_{1,\ldots,4}$. Model parameters represent the latent intercept $I \sim \mathcal{N}(\mu_I, \sigma_I^2)$, the latent slope $S \sim \mathcal{N}(\mu_S, \sigma_S^2)$, the covariance between both σ_{IS}^2, the residual error terms at each occasion with variance σ_ϵ^2 and the slope loadings beyond the first occasion of measurement λ_1, λ_2, and λ_3.

of measurement. We derived four candidate models that are nested in the template model.

1   A linear latent growth curve model represents the hypothesis that cognitive development in this time period is approximately linear. In order to obtain this model from BASELINE, we set $\lambda_1 = 1$, $\lambda_2 = 3$, and $\lambda_3 = 5$. This model is referred to as *LINEAR*.
2   A hypothesis of no cognitive change is formalized in the *FLAT* model, by removing the slope component from BASELINE and thereby eliminating the parameters $\mu_S, \sigma_S^2, \sigma_{IS}^2, \lambda_1, \lambda_2$, and λ_3.
3   We assume that cognitive improvement stops after an individual maximum level of improvement that is attained at the third occasion. This STOPPING model is achieved by setting $\lambda_1 = 1$ and adding the constraint $\lambda_2 = \lambda_3$ to BASELINE.
4   The fourth model SATURATION is a variant of the previous model that assumes a linear trajectory of cognitive development on the first three occasions and allows a different slope for the fourth occasion of measurement that can represent either a saturation of change or a boost in change. This is achieved by setting $\lambda_1 = 1$, $\lambda_2 = 3$ and freely estimating λ_3.

A Hybrid SEM Tree was induced from the dataset using the set of all four described template models, which reflect competing hypotheses about the expected trajectories of cognitive development. The resulting SEM Tree with Bonferroni-corrected *p* values is shown in Figure 4.12 and the expected growth trajectories that are implied by the models associated with the four leaf nodes of the tree are plotted in Figure 4.13. The covariate "father's education" constitutes the primary partition. In each subset, "mother's education" is the second split covariate. The selection of the split point is the same as reported by Brandmaier et al. (2013).

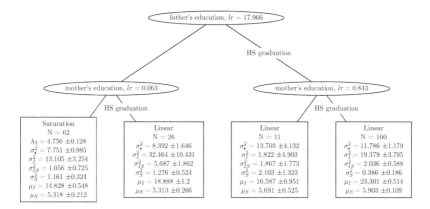

Figure 4.12 Hybrid SEM Tree on the WISC dataset. The LINEAR model representing linear change throughout all occasions of measurement is chosen for all subgroups with the exception of the subgroup in which both parents did not graduate from high school (HS; leftmost leaf). Parameter estimates are given with their standard deviations. *lr* indicates the expected log-likelihood ratio.

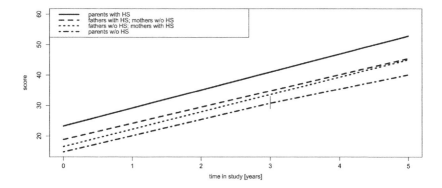

Figure 4.13 Expected growth trajectories for the subgroups retrieved by an SEM Tree with a height of two. The trajectories differ in their intercepts depending on whether parents graduated from high school (HS) or not. The linear growth of the dot-dashed trajectory exhibits a slight saturation at time point 3, as marked by a vertical line.

Among the ordinal values representing different qualities of education, the maximally informative split for both variables is related to whether parents graduated from high school or not. The difference in effect size of the slope difference between both extreme groups, the group with the higher-educated parents and the group with the less-educated parents, is quite high with an effect size of $d = .76$. However, they seem to be dominated by differences in the mean trajectories (see Figure 4.13). While previous analyses have focussed on the assumption of an approximately linear developmental change, we discovered an interesting effect with the hybrid analysis. The cognitive trajectory of children with parents that did not graduate from high school (leftmost leaf in the tree, see Figure 4.12) is better represented by the SATURATION model that allows a change of slope between the last two occasions of measurement. This hints that the cognitive development of the children from this subgroup saturates earlier than for those from the remaining groups, for which the linear model assumes a fixed $\lambda_3 = 5$, while it is estimated as $\lambda_3 = 4.76$ for the low-performing group, representing a less pronounced increase between the last two occasions. This can be observed as a slight flattening of the expected developmental trajectory of this subgroup, shown in Figure 4.13.

Conclusion

In this chapter we have reviewed the SEM Tree methodology and high-lighted important methodological and algorithmic aspects. Furthermore, we have contributed an extension to the paradigm: outlining a Hybrid SEM Tree methodology that not only allows model parameters to differ across subgroups but also across model specification. We have concluded with empirical examples to demonstrate how SEM Trees can be used to find influences of covariates on models and model parameters.

We have provided several examples of how SEM Trees can be applied to empirical data. Based on SAT scores, we have shown trees with a univariate and a bivariate regression model as template models. In the former tree, the continuous covariate "age" was chosen as the split variable. We presented an analysis of the different possible age-related split points that discourages reification of the binary age-split but rather suggests reporting a range of changes. Furthermore, we have presented a longitudinal SEM Tree based on cognitive data from the BASE study and showed the influence of parents' reading behavior, presumably as a proxy of educational background, on children's cognitive development in old age. In addition, we have shown a factor-analytic SEM Tree that identifies an interaction of age and gender in the extraversion score on a personality test. We have concluded with a demonstration of Hybrid SEM Trees that allow a set of potentially non-nested template models instead of a single model. First, we illustrated the method with simulated data and proceeded to demonstrate it on an empirical dataset of the Wechsler Intelligence Score for Children. Between a set of longitudinal SEMs that represent competing hypothesis of cognitive change, the SEM Tree selects a linear model that has been similarly described before but also hints that there is a subgroup for which a change point model could be a better representation.

In the context of longitudinal data, SEM Trees are related to longitudinal recursive partitioning, as introduced by Segal (1992), including the extension presented by Zhang and Singer (1999). Su, Wang, and Fan (2004) suggested building trees based on models estimated with maximum-likelihood procedures. An alternative framework for SEM Trees is provided in the R package *pathmox* (Sanchez & Aluja, 2012) by Sanchez (2009) that is based on partial least squares estimation of linear models. Zeileis et al. (2008) proposed a general framework for recursive partitioning based on permutation tests available in the package *party* (Hothorn, Hornik, Strobl, Zeileis, & Hothorn, 2011). Merkle and Zeileis (2011) suggested use of the R package *strucchange* (Zeileis, Leisch, Hornik, & Kleiber, 2002) to create trees based on factor models that recover subgroups that maximally break specified invariance assumptions. Further research comparing the different estimation methods and covariate selection procedures remains to be done.

Not all datasets are equally suited for the application of recursive partitioning. Due to the successive partitioning of the dataset, the resulting subsets quickly reduce in size. Generally, the larger a dataset, the better it is suited to recursive partitioning. As a rule of thumb, Hawkins (1999) has suggested that the sample size should have at least three digits in order to apply a recursive partitioning algorithm. However, the required sample size depends on many factors including number and type of free parameters, effect size, missingness, normality of the data, and choice of estimator. Also, if decision boundaries are linear but not orthogonal to the axes, or if decision boundaries are complex, decision trees tend to yield large and overly complex tree descriptions. Furthermore, the framework assumes that heterogeneity in the dataset is observed, that is, that covariates were obtained that elicit meaningful group difference with respect to the model. If unobserved heterogeneity is assumed, latent mixture models (McLachlan & Peel, 2000; Lee & Song, 2003) might provide a good starting point for further analyses. However, the clear advantage of trees is their straightforward interpretability as a "white box" model and the possibility of searching a large space of covariates and covariate interactions for influences on the model-predicted distribution.

Critics tend to allege that researchers reporting exploratory results cannibalize chance and are simply dredging data instead of carefully excavating reliable patterns. They claim that the exploitation of a large hypothesis space generates results that are merely random fluctuations and that exploration compromises their confirmatory results. Addressing the former problem, SEM Trees are equipped with procedures that support the generalizability of findings, for example, by controlling statistical error or employing cross-validation for model and variable selection. The latter criticism is addressed in the propositions of ethical data analysis by McArdle (2010), in which he advocates performing exploratory analysis *after* the confirmatory analysis.

To conclude, SEM Trees provide a versatile exploratory data analysis tool for SEM given that a set of covariates is available whose influence on the model is as yet unclear. The method combines exploratory detection of influences of these covariates on parameter estimates for observed variables, latent variables, and their relations, and formal confirmatory mechanisms to ensure generalizability. An implementation of SEM Trees providing a range of features described in this article is available as the *semtree* package (Brandmaier, 2012b).

Note

We would like to thank Julia Delius for her helpful assistance in language and style editing.

References

Baltes, P. B., & Mayer, K. U. (Eds.), (1999). *The Berlin Aging Study: Aging from 70 to 100.* New York: Cambridge University Press.

Boker, S., Neale, M., Maes, H., Wilde, M., Spiegel, M., Brick, T., et al. (2011). OpenMx: An open source extended structural equation modeling framework. *Psychometrika, 76*(2), 306–317.

Boker, S., Neale, M., & Rausch, J. (2004). Latent differential equation modeling with multivariate multi-occasion indicators. In K. van Montfort, J. Oud, & A. Satorra (Eds.), *Recent developments on structural equation models: Theory and applications* (pp. 151–174). Dordrecht, The Netherlands: Kluwer Academic Publishers.

Brandmaier, A. M. (2012a). *Permutation distribution clustering and structural equation model trees.* Dissertation, Saarland University, Saarbrücken.

Brandmaier, A. M. (2012b). semtree: Recursive partitioning of Structural Equation Models in R [Computer software manual]. Available from http://www.brandmaier.de/semtree

Brandmaier, A. M., von Oertzen, T., McArdle, J. J., & Lindenberger, U. (2013). Structural equation model trees. *Psychological Methods, 18*(1), 71–86.

Breiman, L. (2001). Random forests. *Machine Learning, 45*(1), 5–32.

Breiman, L., Friedman, J. H., Olshen, R. A., & Stone, C. J. (1984). *Classification and regression trees.* Belmont, CA: Wadsworth International.

Brodley, C., & Utgoff, P. (1995). Multivariate decision trees. *Machine Learning, 19*, 45–77.

Chan, K., & Loh, W. (2004). Lotus: An algorithm for building accurate and comprehensible logistic regression trees. *Journal of Computational and Graphical Statistics, 13*(4), 826–852.

Dobra, A., & Gehrke, J. (2001). Bias correction in classification tree construction. In *Proceedings of the Eighteenth International Conference on Machine Learning* (pp. 90–97). San Francisco, CA: Morgan Kaufmann.

Finkbeiner, C. (1979). Estimation for the multiple factor model when data are missing. *Psychometrika*, *44*(4), 409–420.

Fletcher, R. (1994). An overview of unconstrained optimization. In E. Spedicato (Ed.), *Algorithms for continuous optimization: The state of the art* (pp. 109–143). Dordrecht, The Netherlands: Kluwer Academic Publishers.

Friedman, J. (1991). Multivariate adaptive regression splines. *Annals of Statistics*, *19*(1), 1–67.

Ghisletta, P., & Lindenberger, U. (2004). Static and dynamic longitudinal structural analyses of cognitive changes in old age. *Gerontology*, *50*, 12–16.

Hastie, T., Tibshirani, R., & Friedman, J. (2001). *The elements of statistical learning.* Berlin: Springer.

Hawkins, D. (1999). *Firm: Formal inference-based recursive modeling* [Tech. Rep.]. Department of Applied Statistics, University of Minnesota.

Horn, J., & McArdle, J. (1992). A practical and theoretical guide to measurement invariance in aging research. *Experimental Aging Research*, *18*(3), 117–144.

Hothorn, T., Hornik, K., Strobl, C., Zeileis, A., & Hothorn, M. (2011). *party: A laboratory for recursive partitioning.* Available from http://cran.r-project.org/web/packages/party/index.html

Hyafil, L., & Rivest, R. (1976). Constructing optimal binary decision trees is NP-complete. *Information Processing Letters*, *5*(1), 15–17.

Ihaka, R., & Gentleman, R. (1996). R: A language for data analysis and graphics. *Journal of Computational and Graphical Statistics*, *5*(3), 299–314.

Jensen, D., & Cohen, P. (2000). Multiple comparisons in induction algorithms. *Machine Learning*, *38*(3), 309–338.

Jöreskog, K. G. (1969). A general approach to confirmatory maximum likelihood factor analysis. *Psychometrika*, *34*(2), 183–202.

Jöreskog, K. G. (1979). Statistical models and methods for analysis of longitudinal data. In K. Jöreskog, D. Sörbom, & M. J. (Eds.), *Advances in factor analysis and structural equation models* (pp. 129–169). Cambridge, MA: Abt Books.

Kim, H., & Loh, W. (2001). Classification trees with unbiased multiway splits. *Journal of the American Statistical Association*, *96*(454), 589–604.

Lee, S., & Song, X. (2003). Maximum likelihood estimation and model comparison for mixtures of structural equation models with ignorable missing data. *Journal of Classification*, *20*(2), 221–255.

Lindenberger, U., Smith, J., Mayer, K., & Baltes, P. (Eds.), (2010). *Die Berliner Altersstudie* [The Berlin Aging Study]. Berlin: Akademie Verlag.

Loh, W., & Shih, Y. (1997). Split selection methods for classification trees. *Statistica Sinica*, *7*, 815–840.

Lövdén, M., Ghisletta, P., & Lindenberger, U. (2004). Cognition in the Berlin Aging Study: The first ten years. *Aging, Neuropsychology, and Cognition*, *11*, 104–133.

McArdle, J. (1988). Dynamic but structural equation modeling of repeated measures data. In J. Nesselroade & R. Cattell (Eds.), *Handbook of multivariate experimental psychology* (Vol. 2, pp. 561–614). New York: Plenum Press.

McArdle, J. (2010). Some ethical issues in factor analysis. In A. Panter & S. Sterber (Eds.), *Quantitative methodology viewed through an ethical lens* (pp. 313–339). Washington, DC: American Psychological Association Press.

McArdle, J., & Aber, M. (1990). Patterns of change within latent variable structural equation modeling. In A. von Eye (Ed.), *New statistical methods in developmental research* (pp. 151–224). New York: Academic Press.

McArdle, J., & Epstein, D. (1987). Latent growth curves within developmental structural equation models. *Child Development, 58*(1), 110–133.

McArdle, J., & Hamagami, F. (2001). Latent difference score structural models for linear dynamic analyses with incomplete longitudinal data: New methods for the analysis of change. In L. M. Collins & A. G. Sayer (Eds.), *New methods for the analysis of change.* Washington, DC: American Psychological Association.

McLachlan, G., & Peel, D. (2000). *Finite mixture models.* New York: John Wiley and Sons.

Merkle, E. C., & Zeileis, A. (2011). *Generalized measurement invariance tests with application to factor analysis* (Working Paper No. 2011-09). Universität Innsbruck: Working Papers in Economics and Statistics, Research Platform Empirical and Experimental Economics. Available from http://EconPapers.RePEc.org/RePEc:inn:wpaper:2011-09

Murthy, S., Kasif, S., & Salzberg, S. (1994). A system for induction of oblique decision trees. *Journal of Artificial Intelligence Research, 2,* 1–32.

Osborne, R., & Suddick, D. (1972). A longitudinal investigation of the intellectual differentiation hypothesis. *Journal of Genetic Psychology, 121*(pt 1), 83–89.

Quinlan, J. (1986). Induction of decision trees. *Machine Learning, 1*(1), 81–106.

Quinlan, J. (1993). *C4.5: Programs for machine learning.* San Francisco, CA: Morgan Kaufmann.

Revelle, W. (2011). *psych: Procedures for psychological, psychometric, and personality research* [Computer software manual]. Evanston, Illinois. Available from http://personality-project.org/r/psych.manual.pdf (R package version 1.01.9)

Revelle, W., Wilt, J., & Rosenthal, A. (2009). Personality and cognition: The personality-cognition link. In A. Gruszka, G. Matthews, & B. Szymura (Eds.), *Handbook of individual differences in cognition: Attention, memory and executive control.* New York: Springer.

Rubin, D. (1976). Inference and missing data. *Biometrika, 63*(3), 581–592.

Sanchez, G. (2009). *PATHMOX approach: Segmentation trees in partial least squares path modeling.* Dissertation, Departament Estadistica i Investigacio Operativa. Universitat Politecnica de Catalunya.

Sanchez, G., & Aluja, T. (2012). *pathmox: Segmentation trees in partial least squares path modeling* [Computer software manual]. Available from http://CRAN.R-project.org/package=pathmox (R package version 0.1-1)

Segal, M. (1992). Tree-structured methods for longitudinal data. *Journal of the American Statistical Association, 87*(418), 407–418.

Shih, Y. (2004). A note on split selection bias in classification trees. *Computational Statistics and Data Analysis, 45,* 457–466.

Sonquist, J., & Morgan, J. (1964). *The detection of interaction effects. A report on a computer program for the selection of optimal combinations of explanatory variables* (No. 35). Ann Arbor, MI: Survey Research Centre, The Institute for Social Research, University of Michigan.

Strobl, C., Boulesteix, A., Zeileis, A., & Hothorn, T. (2007). Bias in random forest variable importance measures: Illustrations, sources and a solution. *BMC Bioinformatics*, *8*(1), 25.

Su, X., Wang, M., & Fan, J. (2004). Maximum likelihood regression trees. *Journal of Computational and Graphical Statistics*, *13*(3), 586–598.

von Oertzen, T., Ghisletta, P., & Lindenberger, U. (2009). Simulating statistical power in latent growth curve modeling: A strategy for evaluating age-based changes in cognitive resources. In M. Crocker & J. Siekmann (Eds.), *Resource-adaptive cognitive processes* (pp. 95–117). Heidelberg: Springer.

Wechsler, D. (1949). *Wechsler Intelligence Scale for Children: Manual*. New York: Psychological Corporation.

Zeileis, A., Hothorn, T., & Hornik, K. (2008). Model-based recursive partitioning. *Journal of Computational and Graphical Statistics*, *17*(2), 492–514.

Zeileis, A., Leisch, F., Hornik, K., & Kleiber, C. (2002). strucchange: An R package for testing for structural change in linear regression models. *Journal of Statistical Software*, *7*(2), 1–38.

Zhang, H., & Singer, B. (1999). *Recursive partitioning in the health sciences*. New York: Springer Verlag.

5 Validating Tree Descriptions of Women's Labor Participation with Deviance-based Criteria

Gilbert Ritschard, Fabio B. Losa, and Pau Origoni

Introduction

This chapter presents a full scale application of induction trees for non-classificatory purposes. The grown trees are used to highlight regional differences in women's labor participation, by using data from the Swiss Population Census. Hence, the focus is on their descriptive rather than predictive power. A first tree provides evidence for three separate analyses for non-mothers, married or widowed mothers, and divorced or single mothers. For each group, trees grown by language regions exhibit fundamental cultural differences supporting the hypothesis of cultural models in female participation. From the methodological standpoint, the main difficulties with such a non-classificatory use of trees have to do with their validation, since the classical classification error rate does not make sense in this setting. We comment on this aspect and propose deviance-based solutions that are both consistent with our non-classificatory usage and easy to compute.

Induced decision trees have become, since Breiman et al. (1984), popular multivariate tools for predicting continuous dependent variables and for classifying categorical variables from a set of predictors. They are called *regression trees* when the outcome is quantitative and *classification trees* when it is categorical. Though their primary purpose is to predict and to classify, trees can be used for many other relevant purposes: such as exploratory methods for partitioning and identifying local structures in datasets, as well as alternatives to statistical descriptive methods like linear or logistic regression, discriminant analysis, and other mathematical modeling approaches (Fabbris, 1997).

This contribution demonstrates such a *non-classificatory* use of classification trees by presenting a full scale application (Losa et al., 2006) to female labor market data from the Swiss 2000 Population Census (SPC). The use of trees for our analysis was dictated by our primary interest in discovering the interactions effects of predictors of the women's labor participation. Since the goal is no longer to extract classification rules, but to understand—from a cross-cultural perspective—the forces that drive women's participation behavior, misclassification rates do not make sense when they are used to validate the trees. We therefore rely on best suited alternative fit criteria that we initially introduced in Ritschard and Zighed (2003) and Ritschard (2006). Our experiment brings insight into the limits and practicability of these criteria for large scale applications.

Apart from these methodological aspects, the practical experiment discussed in this paper is original in at least two respects: (1) the use of trees for microeconomic analysis, which does not appear to be a common domain of application; (2) the use of induction trees for a complete population census dataset.

The next section briefly recalls the principle of classification trees. Then we present the socio-economic research objectives and discuss the main findings. The following section is devoted to the validation issue and then we apply the introduced deviance-based measures to the trees grown for studying womens' participation. Finally, we conclude with an overall evaluation of the experience and of the application of classification trees for non-classificatory purposes.

Principle of Classification Trees

Classification trees are grown by seeking, through successive splits of the learning data set, some optimal partition of the predictor space to predict the outcome class. Each split is carried out according to the values of one predictor. The process is greedy. At the first step, it tries all predictors to find the "best" split. Then, the process is repeated at each new node until some stopping rule is reached. This requires a local criterion to determine the "best" split at each node. The choice of the criterion is the main difference between the various tree growing methods that have been proposed in the literature, of which CHAID (Kass, 1980), CART (Breiman et al., 1984), C4.5 (Quinlan, 1993) and "party" (Hothorn et al., 2006) are perhaps the most popular. For our application, we used CART which only builds binary trees by choosing at each step the split that maximizes the gain in purity measured by the Gini index. CART uses relatively loose stopping rules, but proceeds to a pruning round after the preliminary growing phase.

One of the striking features of induction trees is their ability to provide results in a visual form that allows straightforward interpretations. This visual feature, when compared with the outcome of regression models for instance, has exceptional advantages in terms of user-friendliness and in supporting the knowledge discovery process. Furthermore, by their very nature, trees provide a unique description of the predictor interaction effects on the response variable. These advantages remain true as long as the tree does not become too complex. That is why we chose CART for our analysis, despite the gain in purity seeming less appropriate for a non-classificatory purpose than, for example, the strength of association criterion used by CHAID. Indeed, the great readability of the binary CART trees was decisive when compared with the *n*-ary CHAID trees that had, even at the first level, much too high a number of nodes to allow any useful interpretation.

As with other statistical modeling approaches, it is essential to assess the quality of the obtained tree before drawing any conclusion from it. Our point is to make it clear that the validation criteria are largely dependent on the pursued goal. In particular, it is worth mentioning that the misclassification rate, which is most often the only validation criterion provided by software programs, is of little help in non-classificatory settings.

The Applied Study

We begin by setting out the applied research framework, then we sketch our global analysis procedure, and, finally, we present selected findings.

The Topic: Female Labor Market Participation in Switzerland

Female labor market participation shows significant differences across countries. In Europe, scholars often identify at least two general models: a Mediterranean one (Italy, Greece, Portugal, etc.) versus a model typical of Central and Northern Europe (Reyneri, 1996). The first is represented by an inverse L-shaped curve of the *activity* or *participation rate* by age, where, after a short period of high participation rate (at entry to the labor market), the proportion of women working or seeking work begins to steadily decline up to retirement. The same graph is depicted by an M-shaped curve in Central and Northern European countries, characterized by high participation at entry, followed by a temporary decline during the period of motherhood and childbearing, and a subsequent return to work, up to a certain age where the process of definite exit begins.

In this respect, Switzerland is an interesting case. First, Switzerland is situated across the Alps, which are considered to be one of the cleavages dividing Southern Europe from Central and Northern Europe. Second, there are three main languages spoken by people living in three geographically distinct regions: French in the western part on the border with France, German in the northern and eastern parts on the border with Germany and Austria, and Italian south of the Alps in a region leading to Italy.

The existence of three regions, with highly distinctive historical, social, and cultural backgrounds and characters, and the fact that the Italian-speaking region is divided from the other two regions by the Alps highlight the very specific particularity of this country for a cross-cultural analysis of female participation in the labor market. Moreover, the fact that the comparative analysis is performed amongst regions of the same country guarantees, despite differences stemming from the Swiss federal system, a higher degree of comparability on a large series of institutional, political, and other factors than one would obtain with cross-country studies.

The idea of the research project was to verify the existence of differing cultural models of female labor market participation, by analysing activity rates and hours worked per week—in terms of proportions of full-timers and part-timers—across the three linguistic regions in Switzerland, by using the SPC 2000 data.

To briefly describe the data, we can say that the Federal Statistical Office made available a clean census dataset covering the approximately 7 million inhabitants of Switzerland. For our study, only the about 3.5 million women were of interest. In the preprocessing step we disregarded young ($< 20, 23\%$) and elderly ($> 61, 18\%$) women, as well as non-Swiss women not born in Switzerland (1.6%), i.e. about 43% of the women. This left us with about 2 million cases. Finally, we dropped about 350,000 cases with missing values, and hence included 1,667,494 cases in the analysis.

The Empirical Research Design

The research procedure used classification trees at two different stages, with differing but complementary purposes.

A tree was first grown in what we refer to as the *preliminary step*. Its main goal was, in the spirit of structured induction (Shapiro, 1987) and local pattern detection (Hand, 2002; Rüping, 2005), to find a sound partition of the analysed population into a limited number of homogeneous groups—homogeneous female labor supply behavior in terms of activity and choice between full-time and part-time employment—over which a tailored analysis could be performed. In other words, in order to avoid an "average" analysis at global level, classification trees have been used to structure the research and to identify those groups of the population which could be used to guide subsequent analysis.

This first step was run on the whole Swiss female population aged 20 to 61, using their *labor market status*[1] as outcome variable, and general socio-demographic characteristics (civil status, mother/non-mother ...) as predictive attributes. From this, a robust partition into three groups was chosen, as the best compromise between level of details for the subsequent analysis and population size of each group. The three groups are the *non-mothers*, the *married or widowed mothers*, and the *divorced or single mothers*. The first group is composed of 609,861 women (36.6%), the second one of 903,527 women (54.2%), and the third one of 154,106 women (9.2%).

The second application of classification trees took place in the analysis of cross-cultural female labor supply behavior for each selected group. Here again the outcome variable was the *labor market status* of the women. A much broader series of predictive variables was retained, however: age, profession, educational level, mother/non-mother, number of children, age of last-born child, type of household, etc. Before growing trees, we carried out a series of simple bivariate analyses between the labor market status and each selected predictive attribute. This helped to identify the most relevant attributes for the retained cross-cultural perspective. The analysis of their raw impact on the labor status provided useful indications of how important each one is when it comes to explaining female labor supply behavior.

Classification trees have been produced separately for each region and then compared, as described in the next section, in order to analyse cultural patterns in the participation behavior of the main language regions in Switzerland. At this stage, classification trees and traditional analyses have been used in a complementary way allowing for interplay between them. This proved to be highly productive in stimulating the knowledge discovery process as well as in analysis and understanding of relevant phenomena.

It is worth mentioning here that the final trees retained are simplified versions of those that resulted from the stopping and pruning criteria. They were selected on the basis of comprehensibility and stability. We checked for instance that the splits retained stayed the same when randomly removing 5% of the cases from the learning data set.

Results

Definition of the Groups of Analysis

The three groups identified in the preliminary local pattern detection step appear to exhibit a high degree of inter-group diversity combined with a significant intra-group homogeneity. Inter-group diversity is highlighted by the very specific participation rates by age depicted in Figure 5.1.[2]

Comparison of part-time versus full-time employment reinforces the picture by highlighting the very different choices made by working women of the three groups: a majority of the non-mothers choose full-time employment throughout their professional life, divorced and single mothers switch from part-time jobs during motherhood and early childbearing to full-time (or long-hours part-time) jobs, and the married and widowed mothers prefer short-hours part-time employment in the majority of cases.

The Determinants of Labor Supply Behavior of Divorced and Single Mothers

In order to identify cultural models of female labor supply, three trees (one per region) were generated for each group. These—in combination with the results of the traditional bivariate analyses—were compared and thoroughly analyzed in terms of structure and results. We give here a very brief overview of the main results for the third group, i.e. divorced or single mothers.[3] For details, interested readers may consult the research report by

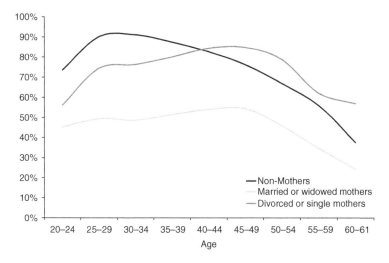

Figure 5.1 Activity rates by age of the three groups selected.

Losa and Origoni (2005). In Figures 5.2 and 5.3, a white background is used for nodes with a majority of non-active women, light gray for a majority of part-timers, and dark gray for a majority of full-timers. We see that opting for inactivity seems to be much more frequent in the Italian speaking region.

Profession and *age of the mother* point out specific groups with particular distinct behaviors. The former sets apart a group of professions—in the fields of health, education, science, etc.—which are known to be characterized by high proportions of part-time jobs. Age plays a central role in the Swiss Italian (Figure 5.2) and in the (not shown) Swiss German tree by clearly splitting the period of active life (up to age 54–55), from that of definite withdrawal from the labor market.

The *age of the last-born child* appears as the most discriminative factor in all the regions, demonstrating the very central role within the family–work conflict of being a mother for the women of this group, who live mainly in single-parent households. The most significant differences across the Swiss language regions appear in this variable, namely in its position in the tree, its split values, and the distribution in the classes of the resulting partition. There is a high proportion of inactivity among Swiss Italian women living in single-parent households when last-born child is 2 years old or younger. This proportion decreases for the first time when the child is 3 (access to public kindergarten) and for the second time when the child reaches 6 (access to primary school). Swiss German women also quit the labor market, but re-enter sooner, while Swiss French are almost indifferent to this factor, showing constant activity rates per age of last-born child.

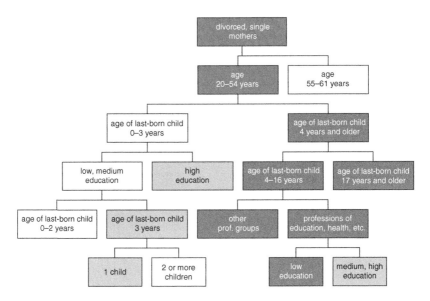

Figure 5.2 Tree for participation of divorced or single mothers, Italian speaking region.

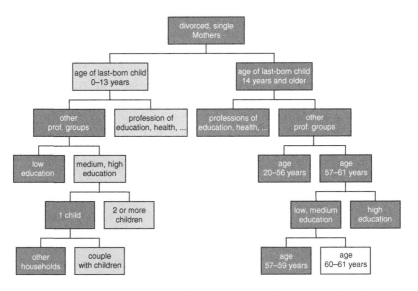

Figure 5.3 Tree for participation of divorced or single mothers, French speaking region.

In all three regions, *educational level* has a strong influence on female labor supply. The higher the educational level, the higher the proportion of active mothers and the lower the proportion of full-timers. This double effect is particularly evident in the Italian speaking and German speaking regions, when the last-born child is very young (less than 4 and 6 years, respectively). Mothers with elementary or intermediate level education decide in the majority of cases to quit their jobs and to stay at home during this period, while mothers of higher education level work on a part-time basis.

The *presence of the partner* and the *number of children*, which strongly influence the behavior of married women have only a limited effect on divorced women.

Validating the Tree Descriptive Ability

For the reliability of the description, individual predictions do not matter. Rather, we focus on the posterior distribution of the response variable (i.e., on the distribution conditioned by the values of the predictors). These posterior distributions are the columns of the target table (see the Terminology subsection below). Our concern is thus to measure how well a tree may predict this target table. This is a goodness-of-fit issue very similar to that encountered in the statistical modeling of multiway cross tables. According to our knowledge, however, it has not been addressed so far for induced trees. Textbooks, like Han and Kamber (2006) and Hand et al. (2001) for example, do not mention it, and, as far as this model assessment issue is concerned, statistical learning focuses almost exclusively on the statistical properties of the classification error rate (see, e.g. Berk 2009 or Hastie et al. 2001, chap. 7).

In statistical modeling (e.g. linear regression, logistic regression or more generally generalized linear models (GLM)), the goodness-of-fit is usually assessed by using two kinds of measures. On the one hand, indicators such as the coefficient of determination R^2 or pseudo R^2s tell us how much better the model does than some naive baseline model. On the other hand, we measure, usually with divergence or deviance statistics, how well the model reproduces some target or, in other words, how far we are from the target.

Our contribution is a trick that permits to use this statistical machinery with induced trees. The trick allows us to propose, among others, an adapted form of the Likelihood Ratio deviance statistic with which we can test statistically the significance of any expansion of a tree. Other criteria discussed are R^2-like measures and the powerful model selection AIC and BIC criteria.

Before describing the deviance, we start by introducing an illustrative example data set that will be used throughout the section. We then specify the notations, terminology, and concepts that we use.

Table 5.1 Example: the data set

Civil status	Gender	Activity sector	Number of cases
Married	Male	Primary	50
Married	Male	Secondary	40
Married	Male	Tertiary	6
Married	Female	Primary	0
Married	Female	Secondary	14
Married	Female	Tertiary	10
Single	Male	Primary	5
Single	Male	Secondary	5
Single	Male	Tertiary	12
Single	Female	Primary	50
Single	Female	Secondary	30
Single	Female	Tertiary	18
Divorced/widowed	Male	Primary	5
Divorced/widowed	Male	Secondary	8
Divorced/widowed	Male	Tertiary	10
Divorced/widowed	Female	Primary	6
Divorced/widowed	Female	Secondary	2
Divorced/widowed	Female	Tertiary	2

Illustrative Example

We consider a fictional example where we are interested in predicting the civil status (married, single, divorced/widowed) of individuals from their gender (male, female) and sector of activity (primary, secondary, tertiary). The civil status is the outcome or response variable, while gender and activity sector are the predictors. The data set is composed of the 273 cases described in Table 5.1.

Terminology and Notations

Classification trees are grown by seeking, through recursive splits of the learning data set, some optimal partition of the predictor space to predict the outcome class. Figure 5.4 shows the tree grown for our illustrative data.

A *leaf* is a terminal node. There are four leaves in Figure 5.4.

In the machine learning community, predictors are also called attributes and the outcome variable is called the predicted attribute. The values of the outcome variable are called the classes. We prefer using "outcome values" to avoid confusion with the classes of the population partition defined by the leaves.

We call *profile* a vector of predictor values. For instance, (female, tertiary) is a profile in Table 5.1.

We call *target table* and denote by T the contingency table that cross classifies the outcome values with the set of possible profiles. As shown in Table 5.2, there are six possible profiles for our data.

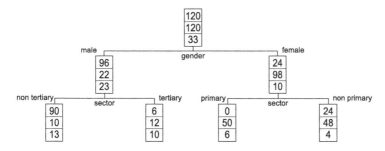

Figure 5.4 Example: induced tree for civil status (married, single, divorced/widowed).

Table 5.2 Target table

	Male			Female			
	Primary	Secondary	Tertiary	Primary	Secondary	Tertiary	Total
Married	50	40	6	0	14	10	120
Single	5	5	12	50	30	18	120
Divorced/ widowed	5	8	10	6	2	2	33
Total	60	53	28	56	46	30	273

Note that the root node contains just the marginal distribution of the outcome variable. It is also useful to point out that the columns of the target table are just the leaves of a maximally developed tree (see the right side of Figure 5.5). We call this maximally developed tree *saturated tree*.

The count in cell (i, j) of the target table T is denoted n_{ij}. We designate by $n_{.j}$ and $n_{i.}$ the total of respectively the jth column and ith row.

The Deviance

Having defined the target table, we propose using the deviance to measure how far the induced tree is from this target (Figure 5.5). By comparing this with the deviance between the root node and the target, we should also be able to evaluate the overall contribution of the predictors, i.e. what is gained over not using any predictor.

The general idea of the deviance of a statistical model m is to measure how far the model is from the target, or more specifically how far the values predicted by the model are from the target. In general (see for instance McCullagh and Nelder, 1989), this is measured by minus twice the log-likelihood of the model ($-2\text{LogLik}(m)$) and is just the log-likelihood ratio

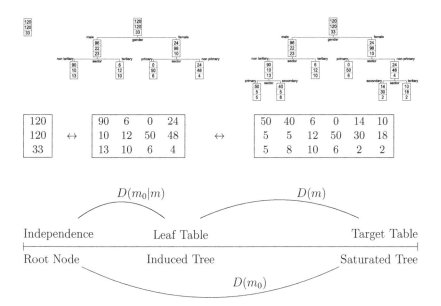

Figure 5.5 Deviance.

Chi-square in the modeling of multiway contingency tables (Agresti, 1990). For a two way $r \times c$ table, it reads for instance

$$D(m) = 2 \sum_{i=1}^{r} \sum_{j=1}^{c} n_{ij} \ln \left(\frac{n_{ij}}{\hat{n}_{ij}} \right), \tag{1}$$

where $\hat{n}_{ij}$ is the estimation of the expected count provided by the model for cell (i, j). The likelihood is obtained by assuming simply a multinomial distribution which is in no way restrictive. Under some regularity conditions (see for instance Bishop et al., 1975, chap. 4), the Log-Likelihood Ratio statistic has an approximate Chi-square distribution when the model is correct. The degrees of freedom d are given by the difference between the number of cells and the number of free parameters of the model.

The advantage of the deviance over, for instance, the Pearson Chi-square is an additivity property that permits us to test the difference between a model m_1 and a restricted version m_2 with the difference $D(m_2|m_1) = D(m_2) - D(m_1)$. This difference has indeed also an approximate Chi-square distribution when the restricted model is correct. Its number of degrees of freedom equals the difference $d_2 - d_1$ in degrees of freedom for each model.

Deviance for a Tree

We have already defined the target table for a classification tree with discrete attributes. Hence, we should be able to compute a deviance for the tree. We face two problems, however:

1 How do we compute the predicted counts $\hat{n}_{ij}$ from the induced tree?
2 What are the degrees of freedom?

To answer these questions we postulate a (non-restrictive) multinomial distribution of the outcome variable for each profile. More specifically, we assume a discrete distribution

$$\mathbf{p}_j = (p_{1|j}, \ldots, p_{r|j}),$$

where $p_{i|j}$ is the probability of being in state i of the outcome variable for a case with profile $\mathbf{x}_j$.

A tree with $q \leq c$ leaves can be seen as a model of the target table. It states that the probability $p_{i|j}$ of being in the ith value of the outcome variable is equal for all profiles j belonging to the same leaf k, i.e.

$$p_{i|j} = p^*_{i|k}, \quad \text{for all } \mathbf{x}_j \in \mathcal{X}_k, k = 1, \ldots, q,$$

where $\mathcal{X}_k$ stands for the set of profiles of leaf k. The tree parameterizes the rc probabilities $p_{i|j}$ in terms of rq parameters $p^*_{i|k}$, which leaves

$$d = (r-1)(c-q) \text{ degrees of freedom.} \tag{2}$$

The probabilities $p^*_{i|k}$s are estimated by the observed proportions, i.e $\hat{p}^*_{i|k} = n_{ij}/n_{.j}$. Estimates of the probabilities $p_{i|j}$ are derived from those of the $p^*_{i|k}$s, i.e. $\hat{p}_{i|j} = \hat{p}^*_{i|k}$ when $\mathbf{x}_j \in \mathcal{X}_k$.

For given n_js and given distributions $\mathbf{p}_j$, the expected count for a profile $\mathbf{x}_j$ is $n_{.j}p_{i|j}$, for $i = 1, \ldots, r$. Now, replacing the $p_{i|j}$s by their estimates, we get estimates $\hat{n}_{ij}$ of the expected counts:

$$\hat{n}_{ij} = n_{.j}\hat{p}^*_{i|k} \quad \text{for all } \mathbf{x}_j \in \mathcal{X}_k, k = 1, \ldots, q. \tag{3}$$

Table 5.3 shows the counts predicted this way from the tree in Figure 5.4.

Considering the counts of the target table and the estimates (3), the deviance $D(m)$ of a tree m can be computed using formula (1). For our example we find $D(m) = 1.69$. The number of degrees of freedom is $d(m) = (3-1)(6-4) = 4$. The obtained deviance being much less than $d(m)$, it is clearly not statistically significant, indicating that the induced tree fits well the target T.

Table 5.3 Predicted counts

| | Male | | | Female | | | |
	Primary	Secondary	Tertiary	Primary	Secondary	Tertiary	Total
Married	47.8	42.2	6	0	14.5	9.5	120
Single	5.3	4.7	12	50	29.1	18.9	120
Divorced/ widowed	6.9	6.1	10	6	2.4	1.6	33
Total	60	53	28	56	46	30	273

Using the Deviance

The approximated Chi-square distribution of the deviance holds when the expected counts per cell are all, say, greater than five. This is rarely the case when the number of predictors is large. Hence, the deviance will not be so useful for testing the goodness-of-fit. Note that we have exactly the same problem with, for instance, logistic regression.

Nevertheless, the difference in the deviance for two nested trees will have a Chi-square distribution, even when the deviances themselves do not.

$$D(m_2|m_1) = D(m_2) - D(m_1) \sim \chi^2 \text{ with } d_2 - d_1 \text{ degrees of freedom.}$$

Thus, the main interest of the deviance is to test differences between nested trees. A special case is testing the difference with the root node with $D(m_0|m)$, which is the equivalent of the usual Likelihood Ratio Chi-square statistic used in logistic regression.

For our example, we have $D(m_0|m) = 167.77$ for six degrees of freedom. This is clearly significant and demonstrates that the tree describes the outcome significantly better than independence (root node). The predictors bring significant information.

As a further illustration, let us test if pruning the branches below "female" in the tree of Figure 5.4 implies a significant change. The reduced tree m_1 has a deviance $D(m_1) = 32.4$ for six degrees of freedom. This is statistically significant, indicating that the reduced tree does not fit the target correctly. The difference with the induced tree m is $D(m_1|m) = 32.4 - 1.7 = 30.7$ for two degrees of freedom. This is also significant and demonstrates that pruning the branch significantly deteriorates the deviance.

Deviance-based Quality Measures

It is very convenient to measure the gain in information in relative terms. Pseudo R^2s, for instance, represent the proportion of reduction in the root node deviance that can be achieved with the tree. Such pseudo R^2s come in different flavors. McFadden (1974) proposed simply

$(D(m_0) - D(m))/D(m_0)$. A better choice is the improvement of Cox and Snell's (1989) proposition suggested by Nagelkerke (1991):

$$R^2_{\text{Nagelkerke}} = \frac{1 - \exp\left\{\frac{2}{n}\left(D(m_0) - D(m)\right)\right\}}{1 - \exp\left\{\frac{2}{n}D(m_0)\right\}}.$$

The McFadden pseudo R^2 is 0.99, and with the Nagelkerke formula we get 0.98.

We may also consider the percent reduction in uncertainty of the outcome distribution for the tree as compared with the root node. The uncertainty coefficient u of Theil (1970), which reads $u = D(m_0|m)/(-2\sum_i n_i \cdot \ln(n_i/n))$ in terms of the deviance, and the association measure τ of Goodman and Kruskal (1954) are two such measures. The first is the proportion of reduction in Shannon's entropy and the second the proportion of reduction in quadratic entropy. These two indexes generally produce very close values. They evolve almost in a quadratic way from no association to perfect association (Olszak and Ritschard, 1995). Their square root is therefore more representative of the position between these two extreme situations. For our induced tree, we have $\sqrt{u} = 0.56$, and $\sqrt{\tau} = 0.60$, indicating that we are a bit more than half way to full association. For the reduced tree m_1 (pruning branch below female), these values are smaller $\sqrt{u} = 0.51$, and $\sqrt{\tau} = 0.57$, indicating that the pruned branch bears some useful information about the distribution.

From the deviance, we can derive AIC and BIC information criteria. For instance, the BIC value for a tree m is

$$\text{BIC}(m) = D(m) - d\ln(n) + \text{constant},$$

where n is the number of cases and d the degrees of freedom in the tree m. The constant is arbitrary, which means that only differences in BIC values matter. Recall that following Raftery's rules of thumb (Raftery, 1995), a difference in BIC values greater than 10 provides strong evidence for the superiority of the model with the smaller BIC in terms of trade-off between fit and complexity.

Computational Aspects

Though the deviance could easily be obtained in our simple example, its practical use for real life data raises two major issues.

1   Existing software packages for growing trees do not provide the deviance. Furthermore, most of them do not provide the data needed

to compute the target table and the estimates $\hat{p}_{i|j}$ in an easily usable form.

2   The number of possible distinct profiles which defines the number c of columns of the target table rapidly becomes excessively large when the number of predictors increases. Theoretically, denoting by c_v the number of values of the variable $x_v, v = 1, \ldots, V$, the number of distinct profiles may be as large as $\prod_v c_v$, which may become untractable.

Regarding the *first point*, we need to compute the "profile" variable, i.e., assign to each case a profile value. The profile variable can be seen as a composite variable x_{prof} with a unique value for each cell of the cross classification of all predictors x_v. Assuming that each variable has less than 10 values, we can compute it, for example, by using successive powers of 10

$$x_{prof} = \prod_{v=1}^{V} 10^{v-1} x_v.$$

We need also a "leaf" variable x_{leaf} that indicates to which leaf each case belongs. Here we have to rely on tree growing software packages that either directly produce this variable ("rpart," Therneau and Atkinson, 1997, or "party," Hothorn et al., 2006), or like AnswerTree (SPSS, 2001) for instance, generate rules for assigning the leaf number to each case.

The next step is to compute the counts of the target table and those of the leaf table resulting from the cross tabulation of the outcome variable with the leaf variable. This can be done by resorting to software that directly produces cross tables. However, since the number of columns, especially that of the columns of the target table, may be quite large and the tables very scarce, a more careful coding that would take advantage of the scarcity is a real concern. A solution is to aggregate cases by profiles and outcome values, which is for instance easily done with software such as SPSS. Creating a similar file by aggregating by leaves and outcome values, the resulting files can then be merged together so as to assign the leaf data to each profile. From here, it is straightforward to get the estimated counts with formula (3) and then compute the deviance $D(m)$ with formula (1). Figure 5.6 shows the SPSS syntax we used for getting the deviance of our example induced tree.

An alternative solution that can be used by those who do not want to write code, is to use the Likelihood Ratio Chi-square statistic that most statistical packages provide for testing the row–column independence in a contingency table. For the target table this statistic is indeed the deviance $D(m_0)$ between the root node m_0 and the target, while for the leaf table it is the deviance $D(m_0|m)$ between the root node and the leaf table associated

```
GET FILE='civst_gend_sector.sav'.
compute profiles
  = ngender*10^1 + nsect.
**Rules generated by AnswerTree**.
IF (ngender NE 2) AND (nsect NE 3)
  leaf = 3.
IF (ngender NE 2) AND (nsect EQ 3)
  leaf = 4.
IF (ngender EQ 2) AND (nsect EQ 1)
  leaf = 5.
IF (ngender EQ 2) AND (nsect NE 1)
  leaf = 6.
END IF.
**Computing the deviance**.
SORT CASES BY profiles .
AGGREGATE
  /OUTFILE='profiles.sav'
  /PRESORTED
  /BREAK=profiles
  /prof_mar = PIN(ncivstat 1 1)
  /prof_sgl = PIN(ncivstat 2 2)
  /prof_div = PIN(ncivstat 3 3)
  /leaf = first(leaf)
  /nj=N.
SORT CASES BY leaf.
AGGREGATE
  /OUTFILE='leaves.sav'
  /PRESORTED
  /BREAK=leaf
  /leaf_mar = PIN(ncivstat 1 1)
  /leaf_sgl = PIN(ncivstat 2 2)
  /leaf_div = PIN(ncivstat 3 3)
  /nj=N.
GET FILE='profiles.sav'.
SORT CASES BY leaf.
MATCH FILES /FILE=*
  /TABLE='leaves.sav'
  /RENAME (nj = d0)
  /DROP d0
  /BY leaf.
COMPUTE pre_mar=leaf_mar*nj/100.
COMPUTE pre_sgl=leaf_sgl*nj/100.
COMPUTE pre_div=leaf_div*nj/100.
COMPUTE n_mar=prof_mar*nj/100.
COMPUTE n_sgl=prof_sgl*nj/100.
COMPUTE n_div=prof_div*nj/100.

**Restructuring data table**.
VARSTOCASES
  /MAKE count
    FROM n_mar n_sgl n_div
  /MAKE pre
    FROM pre_mar pre_sgl pre_div
  /INDEX= Index1(3)
  /KEEP = profiles leaf
  /NULL = DROP
  /COUNT= nclass .

SELECT IF count > 0.
COMPUTE
  deviance=2*count*ln(count/pre).
SORT CASES BY leaf profiles.
COMPUTE newleaf = 1.
IF (leaf=lag(leaf,1))
  newleaf = 0.
COMPUTE newprof = 1.
IF (profiles=lag(profiles,1))
  newprof = 0.
COMPUTE one = 1.
FORMAT one (F2.0)
  /newleaf newprof (F8.0).
**Results in one row table**.
AGGREGATE
  /OUTFILE='deviance.sav'
  /PRESORTED
  /BREAK=one
  /deviance = sum(deviance)
  /nprof    = sum(newprof)
  /nleaves  = sum(newleaf)
  /nclass   = first(nclass)
  /ncells   = N.
GET FILE='deviance.sav'.
**DF and Significance**.
COMPUTE
  df=(nclass-1)*(nprof-nleaves).
COMPUTE
  sig=CDF.CHISQ(deviance,df).
EXECUTE.
```

Figure 5.6 SPSS syntax for computing the deviance of the tree.

to the induced tree. The deviance for the model is then just the difference between the two deviances (see Figure 5.5)

$$D(m) = D(m_0) - D(m_0 \mid m).$$

For our example, we obtain with SPSS $D(m_0) = 169.46$ and $D(m_0|m) = 167.77$, from which we deduce $D(m) = 169.46 - 167.77 = 1.69$. This is indeed the value we obtained by directly applying formula (1). Note that this approach is limited by the maximal number of columns (or rows) accepted for cross tables. This is for instance 1,000 in SPSS 13, which makes this approach inapplicable when the number of possible profiles exceeds this number.

Let us now turn to the *second issue*, i.e. the possibly excessive number of a priori profiles. The solution we propose is to consider partial deviances. The idea is to define the target table from the mere predictors retained during the growing process. This will reduce the number of variables. We could go even further and group the values of each predictor according to the splits used in the tree. For instance, if the induced tree leads to the three leaves "male," "female and primary sector," "female and non-primary sector," we would not distinguish between secondary and tertiary sectors. There would thus be four profiles—instead of six—for the target table, namely "male and primary sector," "male and non-primary sector," "female and primary sector," "female and non-primary sector."

The resulting target table T^* is clearly somewhat arbitrary. The consequence is that the partial deviance, i.e. the deviance $D(m|m_{T*})$ between the tree m and T^*, has no real meaning by itself. However, we have $D(m) = D(m|m_{T*}) + D(m_{T*})$ thanks to the additivity property of the deviance. It follows that $D(m_2) - D(m_1) = D(m_2|m_{T*}) - D(m_1|m_{T*})$. The difference in the partial deviance of two nested trees m_1 and m_2 remains unchanged, whatever target m_{T*} is used. Thus, all tests based on the comparison of deviances, between the fitted tree and the root node for example, remain applicable.

The partial deviance can also be used for defining AIC and BIC criteria, since only differences in the values of the latter matter. Pseudo R^2s, however, are not very informative when computed from partial deviances, due to the arbitrariness of the target table. It is preferable to consider the percentage reduction in uncertainty, which does not depend on the target table, and to look at the square root of Theil's u or Goodman and Kruskal's τ.

Validating the Women's Participation Trees

We must first define the target table in order to compute the deviance for our three regional trees. As explained above, this is quite easy as long as only a limited number of attributes each with a limited number of values are used. For our real full-scale application, it happened, nevertheless, to be a virtually unmanageable task. Indeed, cross tabulating the observed values of the attributes considered gives rise to more than a million different profiles, i.e., columns for the target table.

We therefore considered only a partial deviance $D(m|m_{T*})$ that measures the departure from the partition m_{T*} defined by the mere split values used in the tree. In other words, we compare the partition defined by the tree with the finest partition that can be achieved by combining the groups of values defined by the splits.

For our application, we obtained the partial deviances with SPSS. Two deviances were computed, namely $D(m_0|m_{T*})$ and $D(m_0|m)$, where m_0 is the root node and m the fitted tree. We first recoded the attributes so as to group the values that remain together all over the tree. It was then easy to build a profile variable taking a different value for each observed combination of the recoded values. The target table m_{T*} results from the cross tabulation of this profile variable with the outcome variable, i.e., the type of participation in the labor market.

The deviance $D(m_0|m_{T*})$ is finally simply the independence Log Likelihood Ratio Chi-square statistic (LR) for this target table. Likewise, the deviance $D(m_0|m)$ between the root node and the fitted tree is the LR statistic for the table that cross tabulates the leaf number with the response variable. Since the trees were grown with Answer Tree (SPSS, 2001), we readily obtained the leaf number of each case with the SPSS code generated by this software. The deviance $D(m|m_{T*})$ that measures how far the tree is from the target, is obtained as the difference between those two computed deviances:

$$D(m|m_{T*}) = D(m_0|m_{T*}) - D(m_0|m).$$

Similar relations hold for the degrees of freedom. Recall, however, that the partial deviance has no real meaning by itself. Its interest lies in that it permits us to test statistically differences between nested trees.

We also derive BIC values from the partial deviance. This is not restrictive since only differences in the values of the latter matter. We thus compute the BIC value for a tree m as

$$\mathrm{BIC}(m) = D(m|m_{T*}) - \ln(n)(c^* - q)(\ell - 1),$$

where n is the number of cases, c^* is the number of different profiles in the target table m_{T*}, q the number of leaves of the tree, and ℓ the number of outcome classes, i.e., in our application, the four types of participation in the labor market. The product $(c^* - q)(\ell - 1)$ gives the degrees of freedom associated with the partial deviance.

It is also very convenient to measure the gain in information in relative terms. Pseudo R^2s are not very informative when computed from partial deviances, due to the arbitrariness of the target table. It is preferable to consider the percent reduction in uncertainty about the outcome distribution achieved with the tree when compared to the root node such as that measured, for instance, by the uncertainty coefficient u of Theil (1970).

Table 5.4 reports some of the quality figures we have computed for each of the three regional trees: CHI for the Italian speaking region, CHF for the

Table 5.4 Tree quality measures

	q	c^*	p	n	$D(m_0|m)$	d	*sig.*	ΔBIC	u	$\sqrt{u}$
CHI	12	263	299	5770	822.2	33	.00	536.4	.056	.237
CHF	10	644	674	35239	4293.3	27	.00	4010.7	.052	.227
CHG	11	684	717	99641	16258.6	30	.00	15913.3	.064	.253

French speaking region, and CHG for the German speaking region. The deviances $D(m_0|m)$ are all very large for their degrees of freedom. This tells us that the grown trees clearly improve the description as compared to the root node. The deviances $D(m|m_{T*})$, not shown here, are also very large, indicating that there remains room for improving the fit. The difference ΔBIC in the BIC values between the root node and the grown trees lead to a similar conclusion. They are largely superior to 10, providing evidence of the superiority of the grown trees over the root node. For CHI and CHF, the BIC values of the grown trees are also much smaller than those of the associated saturated trees. This is not the case, however, for CHG. There is thus definite room for improvement in this last case. Remember, however, that we are interested in pointing out the main forces that drive female participation in the labor market. Hence, we have a comprehension purpose, for which increasing complexity would undoubtedly be counter productive. This is typical in socio-economic modeling, where we cannot let the modeling process be entirely driven by purely statistical criteria. Indeed, the trees need to make sense.

The Theil uncertainty coefficient u seems to exhibit a low proportion of gain in uncertainty. However, looking at its square root, we see that we have covered about 25% of the distance to perfect association. Furthermore, the values obtained should be compared with the maximal values that can be achieved with the attributes considered. For the target table, which retains a partition into c^* classes, the u is, respectively, .28, .24 and .23. The square root of these values is about .5, i.e. only about twice the values obtained for the trees. Thus, with the grown trees that define a partition into q classes only, we are about half-way from the target table.

To illustrate how these measures can be used for tree comparison, consider the simplified tree in Figure 5.2 obtained from CHI by pruning a branch grown from the node "age 55–61 years." The original tree CHI has $q = 12$ leaves, while the simplified tree has only nine terminal nodes. For the latter, we get $D(m_0|m) = 799.4$ with $d = 24$, which leads to a significant difference in deviances of 22.8 for nine degrees of freedom. The ΔBIC between the two trees is however 55.1 in favor of the simplified tree, the loss in fit being more than compensated by the complexity reduction. This grounds the retained simplification statistically.

Conclusion

The experiment reported here demonstrates the great potential of classification trees as an analytical tool for investigating socio-economic issues. Especially interesting is the visual tree outcome. For our study, this synthetic view of the relatively complex mechanisms that steer the way women decide about their participation in the labor market provided valuable insight into the issue studied. It allowed us to highlight regional cultural differences in the interaction effects of attributes like age of last-born child, number of children, profession, and education level that would have been hard to uncover through regression analysis, for example.

It is worth mentioning that generating reasonably sized trees is essential when the purpose is to describe and understand underlying phenomenon. Indeed, complex trees with many levels and hundred of leaves, even with excellent classification performance in generalization, would be too confusing to be helpful. Furthermore, in a socio-economic framework, like that considered here, the tree should make sense from the social and economic standpoint. The tree outcomes should therefore be confronted with other bivariate analyses and modeling approaches. Our experience benefited a great deal from this interplay.

Now, as end users, we had to face the lack of suitable validation measures provided by the tree growing software programs for our non-classificatory purpose. The main novelty proposed here is the partial deviance and the efficient way to compute it. The relevance of the partial deviance is based on the additivity property of the deviance. Alternative chi-square divergence measures (Pearson for example) could be considered. However, since they do not share the additivity property, we could not as easily derive partial forms of them.

Although we were able to obtain relevant indicators and statistics afterwards by means of classical cross tabulation outcomes, we would urge software developers to include such validation measures in their software output. Even more, we are convinced that better descriptive trees can be generated when maximal change in overall deviance or BIC values is used as a criterion for growing trees.

Notes

1 Labor market status is a categorical variable with four values: full-time active (at least 90% of standard hours worked per week), long part-time active (50% to 90%), short part-time active (less than 50%) and non-active, where active means working or seeking for a job.

2 Figure 5.1 demonstrates that the M- or L-shaped curves encountered in cross-country studies may result from the superposition of group specific curves.

3 For space reasons, only the (slightly simplified) trees of the Italian and French speaking regions are presented.

References

Agresti, A. (1990). *Categorical Data Analysis.* New York: Wiley.

Berk, R. A. (2009). *Statistical Learning from a Regression Perspective.* New York: Springer.

Bishop, Y. M. M., S. E. Fienberg, and P. W. Holland (1975). *Discrete Multivariate Analysis.* Cambridge MA: MIT Press.

Breiman, L., J. H. Friedman, R. A. Olshen, and C. J. Stone (1984). *Classification and Regression Trees.* New York: Chapman and Hall.

Cox, D. R. and E. J. Snell (1989). *The Analysis of Binary Data* (2nd ed.). London: Chapman and Hall.

Fabbris, L. (1997). *Statistica multivariata: analisi esplorativa dei dati.* Milan: McGraw Hill.

Goodman, L. A. and W. H. Kruskal (1954). Measures of association for cross classifications. *Journal of the American Statistical Association 49*, 732–764.

Han, J. and M. Kamber (2006). *Data Mining: Concept and Techniques* (2nd ed.). San Francisco: Morgan Kaufmann.

Hand, D. J. (2002). The framing of decisions as distinct from the making of decisions. In A. M. Herzberg and R. W. Oldford (Eds.), *Statistics, Science, and Public Policy VI: Science and Responsibility*, pp. 157–161. Kingston, Ontario, Queen's University.

Hand, D. J., H. Mannila, and P. Smyth (2001). *Principles of Data Mining. Adaptive Computation and Machine Learning.* Cambridge MA: MIT Press.

Hastie, T., R. Tibshirani, and J. Friedman (2001). *The Elements of Statistical Learning.* New York: Springer.

Hothorn, T., K. Hornik, and A. Zeileis (2006). Unbiased recursive partitioning: A conditional inference framework. *Journal of Computational and Graphical Statistics 15*(3), 651–674.

Kass, G. V. (1980). An exploratory technique for investigating large quantities of categorical data. *Applied Statistics 29*(2), 119–127.

Losa, F. B. and P. Origoni (2005). The socio-cultural dimension of women's labour force participation choices in Switzerland. *International Labour Review 44*(4), 473–494.

Losa, F. B., P. Origoni, and G. Ritschard (2006). Experiences from a socio-economic application of induction trees. In L. Todorovski, N. Lavrač, and K. P. Jantke (Eds.), *Discovery Science, 9th International Conference, DS 2006, Barcelona, October 7–10, 2006, Proceedings*, Volume LNAI 4265, pp. 311–315. Berlin, Heidelberg: Springer.

McCullagh, P. and J. A. Nelder (1989). *Generalized Linear Models.* London: Chapman and Hall.

McFadden, D. (1974). The measurement of urban travel demand. *Journal of Public Economics 3*, 303–328.

Nagelkerke, N. J. D. (1991). A note on the general definition of the coefficient of determination. *Biometrika 78*(3), 691–692.

Olszak, M. and G. Ritschard (1995). The behaviour of nominal and ordinal partial association measures. *The Statistician 44*(2), 195–212.

Quinlan, J. R. (1993). *C4.5: Programs for Machine Learning.* San Mateo: Morgan Kaufmann.

Raftery, A. E. (1995). Bayesian model selection in social research. In P. Marsden (Ed.), *Sociological Methodology*, pp. 111–163. Washington, DC: The American Sociological Association.

Reyneri, E. (1996). *Sociologia del mercato del lavoro*. Bologna: Il Mulino.

Ritschard, G. (2006). Computing and using the deviance with classification trees. In A. Rizzi and M. Vichi (Eds.), *COMPSTAT 2006—Proceedings in Computational Statistics*, pp. 55–66. Berlin: Springer.

Ritschard, G. and D. A. Zighed (2003). Goodness-of-fit measures for induction trees. In N. Zhong, Z. Ras, S. Tsumo, and E. Suzuki (Eds.), *Foundations of Intelligent Systems, ISMIS03*, Volume LNAI 2871, pp. 57–64. Berlin: Springer.

Rüping, S. (2005). Learning with local models. In K. Morik, J.-F. Boulicaut, and A. Siebes (Eds.), *Local Pattern Detection*, Volume 3539 of *LNCS*, pp. 153–170. Berlin: Springer.

Shapiro, A. D. (1987). *Structured Induction in Expert System*. Wokingham: Adison-Wesley.

SPSS (2001). *Answer Tree 3.0 User's Guide*. Chicago: SPSS Inc.

Theil, H. (1970). On the estimation of relationships involving qualitative variables. *American Journal of Sociology 76*, 103–154.

Therneau, T. M. and E. J. Atkinson (1997). An introduction to recursive partitioning using the RPART routines. Technical Report Series 61, Mayo Clinic, Section of Statistics, Rochester, Minnesota.

6 Exploratory Data Mining Algorithms for Conducting Searches in Structural Equation Modeling

A Comparison of Some Fit Criteria

George A. Marcoulides and Walter Leite

This chapter reports the results of a study examining the performance of two heuristic algorithms commonly used for conducting specification searches in structural equation modeling (SEM): the ant colony optimization (ACO) and the Tabu search algorithms. A secondary goal of the study was to determine which goodness of fit criteria perform best with the ACO and Tabu algorithms in specification searches. A number of fit indices were examined: the chi-square statistic, the Bayesian Information Criterion (BIC), the Comparative Fit Index (CFI), the Tucker–Lewis Index (TLI), and the Root Mean Square Error of Approximation (RMSEA). These indices were selected because of their widespread popularity in evaluating model goodness of fit within SEM. Using data with known structure, the algorithms and fit criteria were examined under different model misspecification conditions. In all conditions, the Tabu search procedure outperformed the ACO algorithm. Recommendations are provided for researchers to consider when undertaking specification searches.

Model fitting techniques are commonly used by researchers in the behavioral sciences. Most of these techniques focus on the minimization or maximization of some model fit function with respect to a set of free and constrained parameters for a given collection of data. If the specified model does not fit the data, a research may accept this fact and leave it at that, or may explore modifying the model in order to improve its overall fit to the data.

Attempts to automate these exploratory model fitting activities have received much attention in the methodological literature over the past few decades, particularly as they relate to structural equation modeling applications. These model fitting activities have been referred to by a variety of names, such as "discovering structure in data," "learning from data," and "data mining techniques" (Marcoulides, 2010; Marcoulides & Ing, 2012). Another commonly accepted taxonomy includes "directed data mining" versus "non-directed data mining," and "supervised," "semi-supervised," or "unsupervised" methods (Larose, 2005; Marcoulides, 2005). In supervised learning (also referred to as directed data mining) the learning process is directed by previous knowledge about the variables under consideration. Unsupervised learning (which is often referred to as non-directed data mining) is much closer to the exploratory spirit of data mining, as there is no previously known result to guide the algorithm in building the model. From a multivariate statistical perspective the differences between supervised learning and unsupervised learning are the same as those distinguishing, for example, the commonly used techniques of discriminant analysis from cluster analysis. Supervised learning requires that the target variable is adequately well defined and that a sufficient number of its values are given. In the case of a discriminant analysis, this would imply specifying *a priori* the number of groups to be considered and focusing on which variables can best be used for the maximum discrimination or differentiation of the groups. For unsupervised learning, typically the target variable is unknown, which in the case of cluster analysis corresponds to the number of groupings or clusters of observations based upon the selectively considered variables. Semi-supervised is a combination of both—the algorithms are provided with limited information concerning the must and/or cannot-link constraints on the model.

Within SEM, the differences between supervised learning and unsupervised learning are the same as those distinguishing, for example, between confirmatory factor analysis (CFA; Jöreskog, 1969) and exploratory factor analysis (EFA; Spearman, 1904). Semi-supervised learning is represented by the newly proposed so-called exploratory structural equation modeling (ESEM; Asparouhov & Muthén, 2009; Marsh, Muthén, Asparouhov, Lüdtke, Robitzsch, Morin & Trautwein, 2009) approach, which attempts to initially incorporate some partial prior knowledge into the determination of the appropriate best fitting model. For example, in the common factor analytic model $\Sigma_{xx} = \Lambda \Phi \Lambda' + \Theta$, where Σ_{xx} is the covariance matrix of the observed $\mathbf{x}$ variables, Λ the factor loading matrix, Φ the covariance matrix of the common factors, and Θ is the covariance matrix of the unique factors (commonly assumed to be diagonal), the ESEM approach (and somewhat similarly to EFA) would be set up with fewer restrictions imposed on Λ and Φ than in CFA. Thereby, in an ESEM with correlated factors, the necessary m^2 restrictions would be imposed by a priori fixing the diagonal of Φ to a value of 1 and the elements above the diagonal in Λ to a value of 0 (for further details on model restrictions in factor analysis, see Hayashi & Marcoulides, 2006). In contrast, as a supervised learning approach, CFA would impose more than m^2 restrictions on the model by additionally fixing other aspects of these particular matrices.

Automated model fitting activities within structural equation modeling that integrate all three learning approaches into a single framework have also received some attention in the literature (e.g., Marcoulides & Drezner, 2001, 2003; Marcoulides & Ing, 2012; Marcoulides, Drezner & Schumacker, 1998; Scheines, Spirtes, Glymour, Meek, & Richardson, 1998). Another term used to describe such model fitting activities is *specification search* (Long, 1983).

A specification search is deemed necessary whenever a researcher is interested in detecting and correcting any potential specification errors between an initially proposed model and the true model characterizing the population and variables under study. To date, most currently available SEM programs (e.g., AMOS, EQS, LISREL, *Mplus*) provide researchers with some rudimentary options to conduct specification searches that can be used to try to improve model to data fit. Although research has demonstrated that specification errors can have serious consequences and that one should try to correct those errors, to date no optimal procedure or single ideal strategy has been proposed for conducting such searches (MacCallum, 1986; Marcoulides & Drezner, 2001; Marcoulides & Ing, 2012).

The most common approach for conducting specification searches is to successively change individual parameter restrictions (e.g., to either free up or constrain parameters) in the initially proposed model using either Lagrange Multiplier and/or Wald tests and evaluate whether a particular restriction is statistically inconsistent with the data (Sörbom, 1989). Bentler (1995), however, indicated that perhaps a preferable alternative to successively changing individual parameters restrictions in an initially proposed model is to consider an all possible subset selection of parameters. Unfortunately, a serious concern with any all possible subset selection strategy is that a very large number of prospective solutions may need to be examined before deciding on a final model.

For many considered models there may just be a small number of possibilities to consider. For example, with only two observed variables, there are only four possible models to examine. For three variables, there are 64 total possible models to examine. However, what happens when the number of possible model combinations becomes prohibitively large? For example, even with just six observed variables there are 1,073,741,824 possible model combinations to examine. One way to consider the total number of models among p investigated variables is the number of possible ways each pair can be connected, to the power of the number of pairs of variables and is determined by $4^{[p(p-1)/2]}$ (Glymour et al., 1987).

In cases where enumeration procedures for examining all possible models are simply impractical, various heuristic optimization data–mining-type search algorithms are available to examine the parameter space (Marcoulides et al., 1998). Heuristic algorithms are specifically designed to determine the best possible solution, but do not guarantee that the optimal solution is found—though their performance using empirical testing or worst cases analysis indicates that in many situations they seem to be the only way forward to produce any viable results (Salhi, 1998). So as the models become more complicated, automated data mining procedures can at the very least make "chaotic situation(s) somewhat more manageable by narrow(ing) attention to models on a recommendation list" (Marcoulides & Drezner, 2001, p. 266).

To date, heuristic procedures have been successfully applied to tackle specification searches in SEM. Examples of such numerical heuristic procedures in SEM include: ant colony optimization (Leite, Huang, & Marcoulides, 2008; Leite & Marcoulides, 2009; Marcoulides & Drezner, 2003), genetic algorithms (Marcoulides & Drezner, 2001), ruin-and-recreate (Marcoulides, 2009), simulated annealing (Marcoulides & Drezner, 1999), and Tabu search (Marcoulides et al., 1998; Marcoulides & Drezner, 2004)— and over the years a great variety of modifications have been proposed to these heuristic search procedures (e.g., Drezner & Marcoulides, 2003; Marcoulides, 2010). Indeed, the literature on these algorithms is quite extensive and technically involved. For a recent overview of all the various available algorithms, we refer the reader to Marcoulides and Ing (2012). In summary, all of these methods focus on the evaluation of an objective function, which is usually based upon some aspect of model to data fit and can even provide users with a candidate list of, say, the top 10 or 20 feasible well-fitting models.

Among the several proposed optimization algorithms in the SEM literature, the ant colony optimization (ACO) and Tabu search algorithms have the advantage of being relatively easy to implement. In this chapter, we provide an overview of these two automated SEM strategies and elaborate on some conceptual and methodological details related to their performance in conducting specification searches. Although there are various criteria that can be used to evaluate comparatively the performance of these search procedures (e.g., the quality of the solutions provided and the computational effort in terms of central processing time needed to obtain a solution), to date most studies within the SEM field have focused mainly on implementations and empirical tests of the different algorithms to perform various types of model searches. The studies by Leite and Marcoulides (2009), Leite, Huang and Marcoulides (2008), Marcoulides et al. (1998), and Whittaker and Marcoulides (2007) appear to be some of the few that have examined goodness of fit performance issues, but concentrated solely on just one algorithm at a time (ACO or Tabu). The results reported by Leite and Marcoulides (2009) indicated that when faced with specification searches for highly complex SEM models, non-convergence of ACO solutions occurs quite frequently. However, in a prior study of complex models using ant colony algorithms, Leite et al. (2008) actually found that non-convergence rarely occurred. Similar good convergence research results were also found for the Tabu search procedure (see Marcoulides et al., 1998). As such, at least with regard to SEM specification search and goodness of fit issues, it would appear that much further research is needed before definitive conclusions can be drawn about preferring one algorithm and model fit criterion over another.

The effectiveness of an algorithm used for specification searches may indeed depend on the selection of one or more goodness of fit criteria to be optimized (e.g., chi-square statistic, or other fit indices). However, whether one criterion or combination of criteria performs best for automated specification searches has also not been adequately examined in previous studies. As a consequence, we examined which goodness of fit criteria perform best with the ACO and Tabu algorithms for specification searches. Given the plethora of fit indices available in the SEM literature, we focused on examining the performance of these algorithms using only a few of the fit information provided by the popular programs *Mplus* (Muthén & Muthén, 2008) and LISREL (Jöreskog & Sorböm, 2005), namely: the chi-square test statistic, the Bayesian Information Criterion (BIC), the Comparative Fit Index (CFI), the Tucker–Lewis Index (TLI), and the Root Mean Square Error of Approximation (RMSEA)—although we note that any of the other commercially available programs would be expected to provide similar results. These particular indices were selected because of their widespread popularity in evaluating model goodness of fit within the SEM field (Raykov & Marcoulides, 2008). For obvious space limitations, we selectively introduce just a few illustrative model example results. We also emphasize that because each application of the described heuristic search requires problem and model specific construction, programming design and setup, only overviews of the procedural steps are described throughout the chapter (for more extensive computer and programming details, see Resenda & Sousa, 2003). Throughout the chapter we use a notational system and equations generally considered to be consistent with the so-called Jöreskog–Keesling–Wiley (Jöreskog & Sörbom, 2005) framework, although this choice is quite arbitrary, as specialized variants of the equations (e.g., the Bentler–Weeks model; Bentler & Weeks, 1980) can also be readily used).

Ant Colony Optimization

Ant colony optimization (ACO) is a class of optimization algorithms based on the foraging behavior of ants (Dorigo & Stützle, 2004). Using a colony of Argentine ants (*Linepithema humile*, formerly *Iridomyrmex humili*) Deneubourg and his associates (1989; 1983) determined how ants are able to establish a shortest route path from their colony nest to food sources and back. Ants are able to find the shortest path between the nest and the food source using a process that starts with the ants randomly attempting alternative paths and leaving feedback on the ground in the form of a chemical substance called a pheromone. The amount of pheromone accumulates faster on the shortest path, which repeatedly stimulates the other ants to follow the same path. Consequently, a positive feedback mechanism occurs as more and more ants follow the particularly attractive path based upon the deposited pheromone level. After some time, enough pheromone accumulates along the shortest path that ants strongly prefer it and rarely choose alternative longer paths.

Marcoulides and Drezner (2003), Leite et al. (2008), and Leite and Marcoulides (2009) demonstrated implementations of an ACO algorithm to perform various types of model searches in structural equation modeling (SEM) under a variety of conditions. In order to appropriately implement the ACO based search procedure for conducting specification searches in SEM, a key feature concerns the determination of the pheromone level, which is essentially based upon the criterion for the fit and selection of a model. Marcoulides and Drezner (2003) implemented the ACO algorithm with the pheromone level based upon the minimization of the non-centrality parameter (NCP = χ^2- df) while Leite et al. (2008) used a pheromone level that was based upon obtaining different fit indices (i.e., CFI, TLI, RMSEA) that indicated adequate model fit.

In both the Marcoulides and Drezner (2003) and the Leite et al. (2008) implementations of the ACO algorithm, parameters are sampled from all possible parameters to be freely estimated in a model, while the other parameters are fixed at zero. The sampling of parameters is performed given sampling weights based on the pheromone levels that may be initially set to be equal for all parameters, which correspond to having no initial hypothesis about model specification and therefore an initial model is randomly selected. Alternatively, initial sampling weights can be specified to reflect the quality of model fit of a user specified theoretical model, and therefore the algorithm will be more likely to sample models that are similar to the initially proposed model during the first iterations. Specifying initial sampling weights according to an *a priori* proposed theoretical model rather than equal weights may result in faster convergence of the algorithm. The sampling weights are updated after each successful iteration based on the fit criteria being optimized. If better fit criteria are obtained with a given model configuration, the pheromone levels of the parameters freely estimated increase, while the pheromone levels of the parameters fixed at zero decrease. Consequently, the probability of the same configuration of parameters being sampled again increases. This process is repeated until parameters that provide the best model fit when freely estimated have very large pheromone levels, while the other parameters have small pheromone levels. When the pheromone levels stabilize, the same model specification is selected at each iteration. As it turns out, a variety of different convergence criteria can be used: Marcoulides and Drezner (2003) chose to finish the search when an NCP of zero was found or after a fixed number of iterations, whereas Leite et al. (2008) stopped the search after the solution did not improve whenever a pre-specified number of iterations was reached.

As indicated above, in the Marcoulides and Drezner (2003) ACO, the actual NCP value is not used directly as the pheromone level. Instead, the function *int*[100/(NCP + 1)] (where *int* refers to an integer) is used to provide the pheromone level to be added after each iteration. The use of this function is necessary because the pheromone level should increase as NCP decreases in order to provide the feedback mechanism described previously. For example, consider a simple CFA model with the following known two factor loading matrix Λ with five observed indicators:

$$\Lambda = \begin{bmatrix} \lambda_{11} & 0 \\ \lambda_{21} & 0 \\ \lambda_{31} & 0 \\ 0 & \lambda_{42} \\ 0 & \lambda_{52} \end{bmatrix}$$

In order to simplify matters for the ACO algorithm, this matrix is binary coded as the vector 1110000011 (for ease of presentation we do not discuss here the Φ correlation between the factors, or the Θ error variances normally considered in a factor model). We note that when the value of the factor loading is "set to 0" it implies that the particular element is fixed to zero in the model examined, whereas "set to 1" implies that the particular element is freely estimated. Let us assume that we did not know the specific structure of this factor loading matrix and wish to use the ACO approach to data to examine an initially proposed mis-specified model, which is represented by the vector 1100000111—in other words, we assume that the first factor is only measured by indicators #1 and #2, whereas the second factor is measured by indicators #3, #4, and #5. To implement the ACO algorithm for the specification search, a pheromone table can be used (Table 6.1; for complete details see Marcoulides & Drezner, 2003). We note that the initial pheromone level in the entire table is initially set at a value of 1 for each variable in the Λ matrix of the model.

Now assuming that the initial mis-specified model provides an NCP = 46.47, based on the above provided pheromone level function formula $int[100/(\text{NCP} + 1)]$, the obtained pheromone level function yields a value of 2. This obtained pheromone level value is added to the second column of the table for the parameters set to 1 and to the third column for the parameters set to 0, resulting in an updated table (Table 6.2).

Table 6.1

Loading	Pheromone level for loading = 1	Pheromone level for loading = 0	Sampling weight
λ_{11}	1	1	1/1
λ_{21}	1	1	1/1
λ_{31}	1	1	1/1
λ_{41}	1	1	1/1
λ_{51}	1	1	1/1
λ_{12}	1	1	1/1
λ_{22}	1	1	1/1
λ_{32}	1	1	1/1
λ_{42}	1	1	1/1
λ_{52}	1	1	1/1

Note: λ_{ij} is the loading of item i on factor j.

Table 6.2

Loading	Pheromone level for loading = 1	Pheromone level for loading = 0	Sampling weight
λ_{11}	3	1	3/1
λ_{21}	3	1	3/1
λ_{31}	1	3	1/3
λ_{41}	1	3	1/3
λ_{51}	1	3	1/3
λ_{12}	1	3	1/3
λ_{22}	1	3	1/3
λ_{32}	3	1	3/1
λ_{42}	3	1	3/1
λ_{52}	3	1	3/1

Note: λ_{ij} is the loading of item i on factor j.

Table 6.3

Loading	Pheromone level for loading = 1	Pheromone level for loading = 0	Sampling weight
λ_{11}	140	1	140/1
λ_{21}	140	1	140/1
λ_{31}	134	7	134/7
λ_{41}	34	107	34/107
λ_{51}	37	104	37/104
λ_{12}	1	140	1/140
λ_{22}	3	138	3/138
λ_{32}	39	102	39/102
λ_{42}	140	1	140/1
λ_{52}	138	3	138/3

Note: λ_{ij} is the loading of item i on factor j.

The process of updating the pheromone level based upon the obtained NCP for each randomly generated model continues for a number of iterations until either NCP = 0 or a fixed number of iterations are completed. With respect to the above considered example model this occurs after only seven iterations, whereupon the pheromone table displayed in Table 6.3 would emerge.

Based upon Table 6.3, one can quickly discern that the model 1110000011 is preferred (i.e., it is dominated in terms of its pheromone levels), which is of course the correct known Λ in the model given earlier as:

$$
\Lambda = \begin{bmatrix}
\lambda_{11} & 0 \\
\lambda_{21} & 0 \\
\lambda_{31} & 0 \\
0 & \lambda_{42} \\
0 & \lambda_{52}
\end{bmatrix}
$$

The Marcoulides and Drezner (2003) ACO algorithm is aimed at maximizing model fit in terms of converging on the correct model. Leite et al. (2008) used a slightly different approach to the ACO algorithm by suggesting the use of a multiple-ant search strategy with a "best-so-far" pheromone update. This notion of a "best-so-far" pheromone update is somewhat similar to that employed in the Tabu search (for a complete description see the section below), where the best solution is maintained in a list and updated whenever better solutions are encountered. A multiple-ant search strategy consists of evaluating n potential model fit solutions before actually updating the pheromone level and then moving to the next iteration. This process is equivalent to sending collectively a whole group of ants, instead of just a single ant at a time, to go looking for food and evaluating the pheromone trails after all the ants from that group return. The multiple ants essentially form the ant colony that searches for a solution simultaneously (Dorigo & Stützle, 2004). After the n solutions are evaluated, the best solution from the group of ants is subsequently selected. Then, pheromone is added to the search weights if the pheromone level of the current best solution exceeds the largest pheromone level obtained with the previous best solution (i.e., the "best-so-far" pheromone update). We also note that the number of potential solutions examined at a time can be set to any pre-specified number (e.g., $n = 25$ suggests the use of a colony of 25 ants at a time). Thereby the performance of the ACO algorithm can be uniquely tuned for each problem by varying the size of the selected ant colony. Although increasing n ultimately makes the algorithm take longer to converge, it also ensures that fewer sub-optimal solutions are obtained. Conversely, although reducing n may make the algorithm converge faster, it can also increase the number of non-optimal solutions obtained. Additionally, a process of pheromone evaporation can also be implemented in which, before the pheromone is updated, the pheromone associated with each component of the pheromone table can be proportionately reduced (e.g., by 5%, 10%, 15%, or even higher percentage values). Pheromone evaporation has been shown in some research to reduce the influence of the solutions obtained at earlier stages of the search, when poor-quality solutions are more likely to be selected (Dorigo & Stützle, 2004). A maximum neighborhood size for the search can be specified, which constrains the maximum number of parameters that are allowed to change from the best-so-far solution at each iteration. Also, any number of constraints to the acceptability of a solution selected by each ant can be specified (for example, the variance estimates may not contain negative variances). It is important to note that the ACO algorithm can be programmed to take into account any number of pre-specified modeling criteria of fit qualities simultaneously. All that is necessary is a modification of the definition of the pheromone level in the ACO algorithm. The implementation of the ACO algorithm to conduct any search activity is thereby to a great degree problem and model specific.

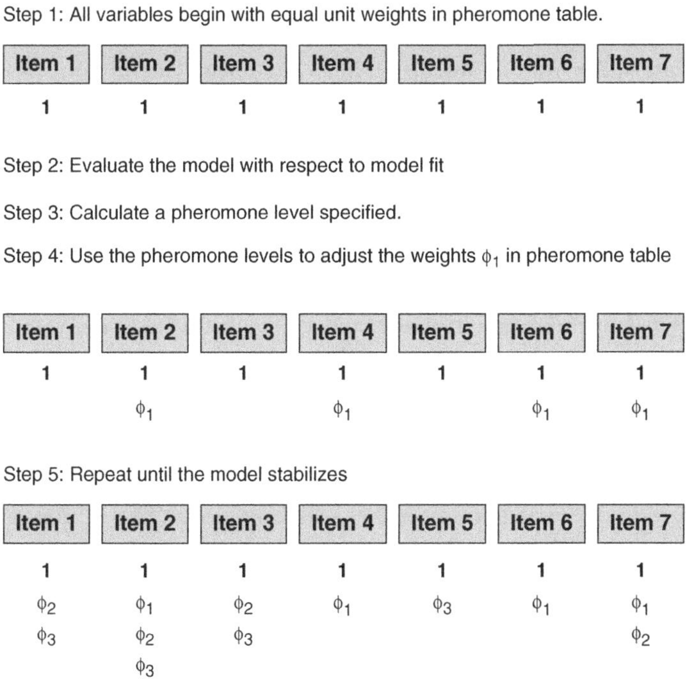

Step 1: All variables begin with equal unit weights in pheromone table.

Item 1	Item 2	Item 3	Item 4	Item 5	Item 6	Item 7
1	1	1	1	1	1	1

Step 2: Evaluate the model with respect to model fit

Step 3: Calculate a pheromone level specified.

Step 4: Use the pheromone levels to adjust the weights ϕ_1 in pheromone table

Item 1	Item 2	Item 3	Item 4	Item 5	Item 6	Item 7
1	1	1	1	1	1	1
	ϕ_1		ϕ_1		ϕ_1	ϕ_1

Step 5: Repeat until the model stabilizes

Item 1	Item 2	Item 3	Item 4	Item 5	Item 6	Item 7
1	1	1	1	1	1	1
ϕ_2	ϕ_1	ϕ_2	ϕ_1	ϕ_3	ϕ_1	ϕ_1
ϕ_3	ϕ_2	ϕ_3				ϕ_2
	ϕ_3					

Figure 6.1 Schematic presentation of the ant colony optimization algorithm.

To obtain the results reported in this chapter, the ACO algorithm was programmed using the R statistical package and implemented alongside the *Mplus* program[1] (although other SEM programs could also be used, because our implementation of the ACO algorithm does not compute the parameter estimates and fit statistics but uses the values generated by the utilized SEM program). The R program implementation basically passes information back and forth between the ACO algorithm and *Mplus*. At the first iteration, all items are assumed to have equal pheromone levels (initially set to a value of 1, as illustrated previously in Table 6.1 above). Item pheromone levels work as sampling weights for model parameter selection. After a model is selected, it is fit to the data using *Mplus*. Once the results from the model fit are obtained, the algorithm calculates a pheromone level based on the quality of the solution (for a schematic of this process, see Figure 6.1).

Tabu Search

Tabu search is a memory-based search strategy to guide the optimization of the objective function away from parts of the solution space that have already been explored. This is usually achieved by prohibiting the re-examination of solutions already visited and stored in the Tabu list. Tabu search procedures are closely tied to the field of artificial intelligence in which intelligent uses of memory help to exploit useful historical information concerning interrelationships within data (for a complete discussion see Salhi, 1998). Tabu search procedures are basically local search strategies that proceed by examining a neighborhood of the current solution. Unlike steepest descent type procedures where the search terminates when there is no further improvement with respect to the objective function examined, Tabu allows the search to exploit inferior solutions. This flexibility enables the search to actually get out of local optimality when making uphill moves, and to avoid cycling Tabu search imposes a sort of off-limits status (i.e., a "tabu status") to those attributes recently involved in the choice of the new solution.

In order to implement a Tabu search procedure in SEM, several definitions and parameters must be carefully considered (Drezner, Marcoulides, & Salhi, 1999).[2] These include: (a) the criterion for the selection of a model; (b) the definition of the neighborhood; (c) a starting model; (d) a definition of the Tabu list, the Tabu size (the length of the Tabu list), and admissible models; (e) the search parameters; and finally (f) a stopping criterion. Each of the above is described next, followed by a list of the Tabu search procedure steps for implementation with any type of structural equation model.

Criterion for the Selection of a Model

Numerous criteria have been proposed in the SEM literature to evaluate the goodness of fit of a specified model (Marsh, Balla, & Hau, 1996; Raykov & Marcoulides, 2008). Most criteria define goodness of fit in terms of the discrepancy between the observed and the model implied covariance matrices. Given the plethora of fit indices available in the SEM literature and because there is no known "best" index, one can essentially choose to rely on any one (or more) fit index as the criterion for the selection of a model when using a Tabu search. As indicated above, for the purposes of this study, we will use some of the fit information provided by the popular programs *Mplus* (Muthén & Muthén, 2008) and LISREL (Jöreskog & Sörbom, 2005), namely: the chi-square statistic, the Bayesian Information Criterion (BIC), the CFI, TLI, and the RMSEA.

Definition of the Neighborhood

A variety of neighborhood definitions can be utilized. For example, two CFA models can be considered neighbors if the set of free terms in the Λ matrix differs by the definition of just one term. The two CFA models would also be considered neighbors if the set of free terms in the factor correlation matrix Φ differed by just one term. This implies that the neighborhood of a considered model K is a model for which either one fixed parameter is freed, or one free parameter is fixed to zero (or any other value).

A Starting Model

A variety of proposed models can be utilized as the starting model. For example, a user specified theoretical model, a randomly generated initial model, or even a model where all parameters are constrained to zero (or all parameters are set free—a null model). Although any number of choices can be selected as the starting model, a starting model of some sort must be given.

The Tabu List, Tabu Size, Admissible Models, and Search Parameters

The Tabu list contains a list of all Tabu moves. When a move is performed (e.g., a constrained term in the model is freed), reversing this action is added to the Tabu list. The Tabu list contains the names of the terms whose status has been changed. A pre-specified maximum length of the Tabu list is termed the "Tabu size." When the Tabu size is exceeded, the last member in the Tabu list is discarded in a "First In First Out" manner. A model in the neighborhood is termed *admissible* if the term changed in the current model (i.e., one that is freed or fixed) is not in the Tabu list. The Tabu size used in the search procedure can be set to any number (e.g., five times the number of observed variables). When a new best model is found, the Tabu list is emptied as if it is a new starting solution. Entries in the Tabu list commonly used can either be short term or long term Tabu memory. Drezner and Marcoulides (2009) considered a modified robust Tabu and showed that increasing the range (in the sense of the defined list) for the random generation of the Tabu tenure improves the performance of the algorithm.

Stopping Criterion

The stopping criterion for a Tabu search can be set to any number according to the complexity of the model examined (e.g., when 100 consecutive iterations do not produce a new best solution). Specifying higher numbers of consecutive iterations that do not produce a new solution inevitably increases the convergence time of the algorithm.

An Overview of the Tabu Search Procedure

1. An initial model K is specified.
2. The best current model K_{best} is set to K.
3. The iteration counter is set to iter $= 0$ (current iteration).

4. The neighborhood $N(K)$ of the model K is created.
5. The objective function $F(K')$ for all K' in $N(K)$ are evaluated.
6. If $F(K') < F(K_{best})$ for any K' in $N(K)$, set $K_{best} = K'$ (if there are several K' that fulfill this condition, select the best one). Go to step 8.
7. If for all K' in $N(K)$: $F(K') \geq F(K_{best})$, choose the best admissible model K' in $N(K)$.
8. Set $K = K'$ and iter $=$ iter $+ 1$.
9. The Tabu list is updated. Go to Step 4 unless the stopping criterion is met.

Data Analysis Method

The approach used in this study to examine the ACO and Tabu algorithms for conducting specifications searches in SEM was the following: (a) utilize data for which there is a known correct population model, (b) initially fit a mis-specified model to the data, (c) determine whether a specification search leads to the correct model, (d) determine which fit criteria performed best in identifying the correct model. The population model was generated based on common practice in Monte Carlo studies (Hoogland & Boomsma, 1998) and a review of models commonly used in empirical studies within social and behavioral science research.

A confirmatory factor analytic model based on 12 observed variables with three common factors was used for the current study. Empirically based factor loadings, factor covariances and error variances similar to those obtained in a factor model of the Teachers' Sense of Efficacy Scale—Short Form (Tschannen-Moran & Woolfolk Hoy, 2001) were used in this analysis. The factor loading matrix Λ has the following structure:

$$\Lambda = \begin{bmatrix} 0.802 & 0.000 & 0.000 \\ 0.839 & 0.000 & 0.000 \\ 0.861 & 0.000 & 0.000 \\ 0.574 & 0.000 & 0.000 \\ 0.000 & 0.632 & 0.000 \\ 0.000 & 0.716 & 0.000 \\ 0.000 & 0.648 & 0.000 \\ 0.000 & 0.821 & 0.000 \\ 0.000 & 0.000 & 0.786 \\ 0.000 & 0.000 & 0.849 \\ 0.000 & 0.000 & 0.790 \\ 0.000 & 0.000 & 0.856 \end{bmatrix}$$

For this model, variances of the factors were set to 1.0 and covariances among the three factors in the matrix Φ were set at $\sigma_{12} = 0.605$, $\sigma_{13} = 0.626$, and $\sigma_{23} = 0.621$. To evaluate the effect of model complexity on the performance of the algorithms, two true models were generated: The first true model has the structure shown in the factor loading matrix above and the following diagonal Θ covariance matrix of measurement errors:

$$\Theta = \begin{bmatrix}
0.382 \\
0 & 0.357 \\
0 & 0 & 0.297 \\
0 & 0 & 0 & 0.259 \\
0 & 0 & 0 & 0 & 0.601 \\
0 & 0 & 0 & 0 & 0 & 0.279 \\
0 & 0 & 0 & 0 & 0 & 0 & 0.376 \\
0 & 0 & 0 & 0 & 0 & 0 & 0 & 0.268 \\
0 & 0 & 0 & 0 & 0 & 0 & 0 & 0 & 0.488 \\
0 & 0 & 0 & 0 & 0 & 0 & 0 & 0 & 0 & 0.580 \\
0 & 0 & 0 & 0 & 0 & 0 & 0 & 0 & 0 & 0 & 0.670 \\
0 & 0 & 0 & 0 & 0 & 0 & 0 & 0 & 0 & 0 & 0 & 0.326
\end{bmatrix}$$

The second true model has the same factor loading matrix Λ, but the covariance matrix of errors Θ has a structure commonly found in multi-trait/multi-method (MTMM) instruments where the errors of items that share a common method correlate. The second true model is identical to the correlated trait/correlated uniqueness model described by Marsh (1989). For the sake of simplicity, correlations between items sharing the same method were set to about 0.2 in the population. For the second model, the Θ covariance matrix of errors has the following structure:

$$\Theta = \begin{bmatrix}
0.54 \\
0 & 0.95 \\
0 & 0 & 0.49 \\
0 & 0 & 0 & 0.60 \\
0.12 & 0 & 0 & 0 & 0.70 \\
0 & 0.12 & 0 & 0 & 0 & 0.35 \\
0 & 0 & 0.11 & 0 & 0 & 0 & 0.58 \\
0 & 0 & 0 & 0.09 & 0 & 0 & 0 & 0.32 \\
0.13 & 0 & 0 & 0 & 0.15 & 0 & 0 & 0 & 0.82 \\
0 & 0.15 & 0 & 0 & 0 & 0.09 & 0 & 0 & 0 & 0.62 \\
0 & 0 & 0.19 & 0 & 0 & 0 & 0.21 & 0 & 0 & 0 & 1.88 \\
0 & 0 & 0 & 0.11 & 0 & 0 & 0 & 0.08 & 0 & 0 & 0 & 0.50
\end{bmatrix}$$

Although any number of possible specification errors can be made for the population models presented above, for the current study we will use just two mis-specified models: (1) A model with incorrect loadings was fit to data generated with the first population model; (2) A model with incorrect cross-loadings and no correlated errors was fit to data generated with the second population model. The first mis-specified model contains the following three incorrect loadings in the Λ matrix:

$$\Lambda = \begin{bmatrix} X & 0 & 0 \\ X & 0 & 0 \\ X & 0 & 0 \\ 0 & X & 0 \\ 0 & X & 0 \\ 0 & X & 0 \\ 0 & X & 0 \\ 0 & 0 & X \\ 0 & 0 & X \\ 0 & 0 & X \\ 0 & 0 & X \\ X & 0 & 0 \end{bmatrix};$$

The second mis-specified model contains four incorrect cross-loadings and twelve omitted error covariances. The incorrect cross-loadings in the Λ matrix are as follows:

$$\Lambda = \begin{bmatrix} X & 0 & 0 \\ X & 0 & X \\ X & 0 & 0 \\ X & 0 & 0 \\ X & X & 0 \\ 0 & X & X \\ 0 & X & 0 \\ 0 & X & 0 \\ X & 0 & X \\ 0 & 0 & X \\ 0 & 0 & X \\ 0 & 0 & X \end{bmatrix}.$$

In order to examine the performance of the ACO and Tabu algorithms, 50 datasets of sample size $n = 1{,}000$ were simulated from the known population models. Data generation was performed in *Mplus* (Muthén & Muthén, 2010). Data were drawn from a multivariate normal distribution $(0, \Sigma)$. In all, there were 2 (models) $\times$ 50 (replications) $= 100$ examples examined. Sample size was not varied because it is not expected to affect the performance of the ACO or Tabu algorithms, but the complexity of the true model may be expected to have an impact.

For each replication, the ACO and Tabu algorithms were run five times for each of three fit criteria: the chi-square test statistic, Bayesian Information Criterion (BIC) and fit indices reported by *Mplus* and LISREL. The pheromone level maximized for these criteria were $100/(\chi^2 - df + 1)$, $100/\mathrm{BIC}$ and $\max(\mathrm{CFI}, \mathrm{TLI}, 1\text{-RMSEA})$, respectively (all with positive values). Final solutions with each criterion were obtained by taking the parameters that were selected in at least three of the five runs performed for each dataset. The solutions obtained with single runs of the algorithms were also recorded for comparison with solutions obtained with five runs.

Results

For each of the 100 datasets simulated with a CFA model, fifteen runs of the ACO and Tabu search algorithm were performed (i.e., 5 runs of each of 3 criteria), producing a total of 1500 solutions. Descriptive statistics about the success of both the ACO and Tabu algorithms for specification searches using the different criteria (i.e., chi–square statistics, fit indices, and BIC) are shown in Table 6.4. An examination of this table immediately reveals that the Tabu search completely outperformed the ACO algorithm in all conditions examined. In contrast, the ACO algorithm struggled in a variety of ways to identify the correct model. Single runs of the ACO algorithm minimizing the chi–square statistic produced solutions that included all of the correct parameters in only 36.8% of the replications. Furthermore, it consistently produced over parameterized solutions. Although the correct model only had 12 factor loadings, the mean number of factor loadings in the solutions selected was 20.912 (SD = 2.366). None of the solutions contained only the correct parameters. Running the ACO algorithm for each dataset five times and choosing parameters present in at least 3 solutions increased the number of solutions containing all of the correct parameters to 56%, but did not reduce the

Table 6.4 Summary of the performance of the ACO and Tabu algorithms for specification searches of a CFA model with diagonal error covariance matrix

Optimization criteria	Chi-square		Fit indices		BIC	
	Single run	Five runs	Single run	Five runs	Single run	Five runs
Proportion of solutions containing all of the correct loadings	0.368[a] *1.000[b]*	0.560 *1.000*	0.912 *1.000*	0.900 *1.000*	0.936 *1.000*	0.940 *1.000*
Proportion of solutions containing only correct loadings	0.000 *1.000*	0.000 *1.000*	0.000 *1.000*	0.000 *1.000*	0.232 *1.000*	0.460 *1.000*
Mean number of loadings selected	20.912 *12*	21.940 *12*	20.016 *12*	19.940 *12*	14.604 *12*	13.260 *12*
Standard deviation of number of loadings selected	2.366 *0*	2.676 *0*	2.653 *0*	2.721 *0*	2.614 *0*	1.794 *0*
Mean number of irrelevant loadings in each solution	9.620 *0*	10.440 *0*	8.104 *0*	8.120 *0*	2.668 *0*	1.400 *0*

Note: [a]ACO results are provided in plain text; [b]Tabu results are italicized.

over-parameterization problem. The mean number of parameters selected was still high at 21.940 (SD = 2.676).

Single runs of the ACO algorithm for specification searches of CFA models optimizing fit indices were found to obtain solutions with mean CFI, TLI, and RMSEA of 0.998 (SD = 0.007), 0.999 (SD = 0.012), and 0.005 (SD = 0.016) respectively. With single ACO runs maximizing fit indices, 91.2% of the solutions contained all of the correct parameters. However, these runs were also found to consistently over-parameterize the model. The mean number of parameters selected was 20.016 (SD = 2.653), and none of the solutions contained only correct parameters. A close examination of the over-parameterized solutions indicated that they included parameters with very small estimated values. It was observed that 78% of the individual ACO runs resulted in CFI = 1.000 and RMSEA = 0.000, which demonstrates that the presence of extra parameters in the solution did not substantially disturb the fit indices. Furthermore, from the solutions where all 12 correct loadings were included, 75.6% had a CFI = 1.000 and RMSEA = 0.000. When a final solution was selected containing only parameters present in at least three of five solutions, 90% of the solutions included all of the correct parameters. The over-parameterization problem persisted, with an average number of parameters selected of 19.940 (SD = 2.721).

The solutions selected with single runs of the ACO algorithm minimizing the BIC included all of the correct parameters in 93.6% of the iterations. The resulting models were frequently over-parameterized, but to a smaller extent than when the chi-square statistic or fit indices were optimized. The mean number of parameters selected was 14.640 (SD = 2.614). Furthermore, 23.2% of the solutions only contained correct parameters. The selection of parameters present in the at least three of each five solutions resulted in similar proportion of solutions including all the correct parameters (i.e. 94%), but there was a strong improvement in the number of solutions that included only correct parameters (i.e. 46%).

With the specification search for the population model with a non-diagonal error covariance matrix, the ACO algorithm optimizing the chi-square statistic, fit indices, or BIC resulted in even worse model over-parameterization than with the population model with diagonal error covariance matrix (see Table 6.5). In contrast, the Tabu algorithm again performed superbly in all modeling situations considered. The population model contained a total of 24 non-zero factor loadings and error covariances.

Single runs of the ACO algorithm selected a mean number of parameters equal to 56.960, (SD = 4.369), 58.124 (SD = 4.205), and 43.652 (SD = 8.435), with optimizations of the chi-square statistic, fit indices, and BIC, respectively. Optimization of the BIC resulted in selection of the correct loadings in 58.8% of runs, while optimization of the chi-square statistic and fit indices resulted in 40% and 33.2%, respectively, of solutions containing

Table 6.5 Summary of the performance of the ACO and Tabu algorithms for specification searches of a CFA model with non-diagonal error covariance matrix

Optimization criteria	Chi-square		Fit indices		BIC	
	Single run	Five runs	Single run	Five runs	Single run	Five runs
Proportion of solutions containing all of the correct loadings	0.400[a]	0.620	0.332	0.640	0.588	0.620
	1.000[b]	*1.000*	*1.000*	*1.000*	*1.000*	*1.000*
Proportion of solutions containing all of the corrected error covariances	0.000	0.000	0.000	0.040	0.012	0.040
	1.000	*1.000*	*1.000*	*1.000*	*1.000*	*1.000*
Proportion of solutions containing only correct parameters	0.000	0.000	0.000	0.000	0.000	0.000
	1.000	*1.000*	*1.000*	*1.000*	*1.000*	*1.000*
Mean number of parameters selected	56.960	59.500	58.124	60.840	43.652	38.880
	24	*24*	*24*	*24*	*24*	*24*
Standard deviation of number of parameter selected	4.369	8.105	4.205	6.932	8.435	12.706
	0	*0*	*0*	*0*	*0*	*0*
Mean number of irrelevant parameters in each solution	38.944	39.960	40.364	41.600	24.832	19.240
	0	*0*	*0*	*0*	*0*	*0*

Note: [a]ACO results are provided in plain text; [b]Tabu results are italicized.

all of the correct parameters. However, the ACO algorithm runs optimizing the BIC, chi-square statistic, and fit indices selected all of the correct error covariances in 1.2%, 0%, and 0% of the replications.

Over-parameterization also occurred when five runs of the ACO algorithm were used to create solutions based on parameters selected at least three times. The mean number of parameters selected was 59.500 (SD = 8.105), 60.840 (SD = 6.932), and 38.880 (SD = 12.706) with optimizations of the chi-square statistic, fit indices, and BIC, respectively. All of the correct loadings were selected in 62%, 64%, and 62% of the runs optimizing the chi-square statistic, fit indices, and BIC, respectively, but the proportion of solutions containing all of the correct error covariances was only 4% with the BIC and fit indices and 0% with the chi-square statistic. None of the solutions obtained for the model with non-diagonal error covariance matrix with either single runs or five runs of the ACO algorithm contained only correct parameters for any of the criteria optimized.

Conclusion

The performance of various fit criteria for conducting specification searches in SEM was examined based on the ACO and Tabu search algorithms. The results indicate that the Tabu search had absolutely no difficulties in identifying the correct model in any of the conditions examined (regardless of fit criteria used), whereas the ACO algorithm consistently selected models with too many parameters. This could be due to an insufficient convergence criterion (i.e., 500 runs without a change), excessively aggressive search parameters (i.e., pheromone evaporation = 0.95, pheromone comparisons after each solution as opposed to after $n > 1$ solutions), or because all of the functions being maximized (i.e., $100/(\chi^2 - df + 1)$, $100/BIC$ and max(CFI, TLI, 1-RMSEA) did not produce large enough differences in pheromone levels between models with correct parameters, plus irrelevant parameters and models with just correct parameters. One pervasive problem of the ACO algorithm that hampers their performance in specification searches for complex SEM models is the fact that non-convergence of solutions occurs too frequently. This problem slows down the algorithm considerably and makes it difficult to enforce stricter convergence criteria. In other applications of the ACO algorithm, such as item selection for scale short-forms (Leite et al., 2008), non-convergence rarely occurred. Perhaps one issue to consider for the future is the specifics of the conditions that might lead to non-convergence issues.

For optimizations of fit indices and BIC, the performance of the ACO algorithm with the CFA model with diagonal error covariances was substantially better than with the CFA model containing error covariances. With the simpler CFA model, all of the correct parameters were selected in more than 90% of the runs. Although some incorrect parameters were always present in the solutions obtained, they were very small and had large standard errors. Therefore, a researcher could easily identify the irrelevant parameters by inspecting the estimates and their standard errors. Interestingly, the ACO algorithm with the chi-square statistic performed considerably worse than with the fit indices and BIC for the simpler CFA model, but the performances of the three criteria were very similar with the more complex model.

Despite the uneven performance of the ACO algorithm across models, the results indicate that when using such an approach the BIC is a better optimization criterion than the chi-square statistic and fit indices. This is not surprising because it is well known that the BIC rewards model parsimony. The performance of the optimizations of fit indices may have been reduced by the fact that the CFA has a maximum of 1 and the RMSEA has a minimum of 0. Therefore, it is recommended that future studies by specification searchers for SEM models that are not based on a Tabu search consider the use of the BIC as the optimization criterion. It is quite evident that the Tabu search (Marcoulides et al., 1998; Marcoulides & Drezner, 2004) (and perhaps even some of the other available algorithms mentioned above) may overcome the problems encountered with the ACO algorithm. For example, Dorigo and Stützle (2004) indicated that ACO algorithms do not perform as well as genetic algorithms in general optimization problems.

It is important to keep in mind that specification searches for structural equation models are completely "data-driven exploratory model fitting" and, as such, can capitalize on chance (MacCallum, 1986). We believe that such automated specification searches are best used in cases where all available information has been included in a specified model and this information is not enough to obtain an acceptable model fit. All final generated models must be cross-validated before their validity can be claimed. More research on optimization algorithms is needed to establish which one works best with structural equation models. For now, we believe that the Tabu search is one of the best available automated specification search procedures to use in structural equation modeling applications. Based on its performance for conducting specification searches, we feel that is can provide very valuable assistance in modeling complex educational and psychological phenomena.

Notes

1 The R program that implements the ACO algorithm alongside with *Mplus* is provided at the second author's website: http://education.ufl.edu/leite/publications/
2 A FORTRAN program that implements the Tabu search algorithm is available upon request from the first author.

References

Asparouhov, T., & Muthén, B. O. (2009). Exploratory structural equation modeling. *Structural Equation Modeling, 16*, 397–438.

Bandalos, D. L. (2006). The use of Monte Carlo studies in structural equation modeling research. In R. C. Serlin (Series Ed.) & G. R. Hancock & R. O. Mueller (Vol. Eds.), *Structural equation modeling: A second course* (pp. 385–462). Greenwich, CT: Information Age.

Bentler, P. M. (1995). *EQS structural equation program manual.* Encino, CA: Multivariate Software Inc.

Bentler, P. M., & Weeks, D. G. (1980). Linear structural equations with latent variables, *Psychometrika, 45*, 289–308.

Deneubourg, J. L., & Goss, S. (1989). Collective patterns and decision making. *Ethology, Ecology, and Evolution, 1*, 295–311.

Deneubourg, J. L., Pasteels, J. M., & Verhaeghe, J. C. (1983). Probabilistic behaviour in ants: A strategy of errors? *Journal of Theoretical Biology, 105*, 259–271.

Dorigo, M., & Stützle, T. (2004). *Ant colony optimization.* Cambridge, MA: The Massachusets Institute of Technology Press.

Drezner, Z., & Marcoulides, G. A. (2003). A distance-based selection of parents in genetic algorithms. In M. Resenda & J. P. Sousa (Eds.), *Metaheuristics: Computer decision-making* (pp. 257–278). Boston, MA: Kluwer Academic Publishers.

Drezner, Z., & Marcoulides, G. A. (2009). On the range of Tabu tenure in solving quadratic assignment problems. In P. Petratos & G.A. Marcoulides (Eds.),

Recent advances in computing and management information systems (pp. 157–167). ATINER SA Publishers.

Drezner, Z., Marcoulides, G. A., & Salhi, S. (1999). Tabu search model selection in multiple regression analysis. *Communications in Statistics – Computation and Simulation, 28*, 349–367.

Glymour, C., Schienes, R., Spirtes, P., & Kelly, K. (1987). *Discovering causal structure: Artificial intelligence, philosophy of science, and statistical modeling.* San Diego, CA: Academic Press.

Hayashi, K., & Marcoulides, G. A. (2006). Identification issues in factor analysis. *Structural Equation Modeling, 13*, 631–645.

Hoogland, J. J., & Boomsma, A. (1998). Robustness studies in covariance structure modeling: An overview and a meta-analysis. *Sociological Methods & Research, 26*, 329–367.

Jöreskog, K. G. (1969). A general approach to confirmatory maximum likelihood factor analysis. *Psychometrika, 34*, 183–202.

Jöreskog, K. G., & Sörbom, D. (2005). *LISREL 8 user's reference guide.* Chicago, IL: Scientific Software International, Inc.

Larose, D. T. (2005). *Discovering knowledge in data: An introduction to data mining.* Hoboken, NJ: Wiley.

Leite, W. L., Huang, I.-C., & Marcoulides, G. A. (2008). Item selection for the development of short forms of scales using an Ant Colony Optimization algorithm. *Multivariate Behavioral Research, 43(3)*, 411–431.

Leite, W. L., & Marcoulides, G. A. (2009, April). *Using the ant colony optimization algorithm for specification searches: A comparison of criteria.* Paper presented at the Annual Meeting of the American Education Research Association, San Diego: CA.

Long, J. S. (1983). *Covariance structure models: An introduction to LISREL.* Beverly Hills, CA: Sage Publications.

MacCallum, R. (1986). Specification searches in covariance structure modeling. *Psychological Bulletin, 100(1)*, 107–120.

Marcoulides, G. A. (2005). Review of *Discovering knowledge in data: An introduction to data mining. Journal of the American Statistical Association, 100(472)*, 1465–1465.

Marcoulides, G. A. (2009, June). *Conducting specification searches in SEM using a ruin and recreate principle.* Paper presented at the Annual Meeting of the American Psychological Society, San Francisco, CA.

Marcoulides, G. A. (2010, July). *Using heuristic algorithms for specification searches and optimization.* Paper presented at the Albert and Elaine Borchard Foundation International Colloquium, Missillac, France.

Marcoulides, G. A., & Drezner, Z. (1999). Using simulated annealing for model selection in multiple regression analysis. *Multiple Linear Regression Viewpoints, 25*, 1–4.

Marcoulides, G. A., & Drezner, Z. (2001). Specification searches in structural equation modeling with a genetic algorithm. In G. A. Marcoulides & R. E. Schumacker (Eds.), *Advanced structural equation modeling: New developments and technique* (pp. 247–268). Mahwah, NJ: Lawrence Erlbaum Associates, Inc.

Marcoulides, G. A., & Drezner, Z. (2003). Model specification searches using Ant Colony Optimization algorithms. *Structural Equation Modeling, 10(1)*, 154–164.

Marcoulides, G. A., & Drezner, Z. (2004). Tabu Search Variable Selection with Resource Constraints. *Communications in Statistics: Simulation & Computation, 33,* 355–362.

Marcoulides, G. A., Drezner, Z., & Schumacker, R. E. (1998). Model specification searches in structural equation modeling using Tabu search. *Structural Equation Modeling, 5(4),* 365–376.

Marcoulides, G.A., & Ing, M. (2012). Automated structural equation modeling strategies. In R. Hoyle (Ed.), *Handbook of structural equation modeling.* New York, NY: Guilford Press.

Marsh, H. W. (1989). Confirmatory factor analyses of multitraitmultimethod data: Many problems and a few solutions. *Applied Psychological Measurement, 13(4),* 335–361.

Marsh, H. W., Balla, J. R., & Hau, K. (1996). An evaluation of incremental fit indices: A clarification of mathematical and empirical properties. In G.A. Marcoulides & R.E. Schumacker (Eds.), *Advanced structural equation modeling: Issues and techniques* (pp. 315–353). Mahwah, NJ: Lawrence Erlbaum Associates, Inc.

Marsh, H. W., Muthén, B. O., Asparouhov, T., Lüdtke, O., Robitzsch, A., Morin, A. J., & Trautwein, U. (2009). Exploratory structural equation modeling, integrating CFA and EFA: Application to students' evaluations of university teaching. *Structural Equation Modeling, 16,* 439–476.

Muthén, B., & Muthén, L. (2010). *Mplus user's guide.* Los Angeles, CA: Muthén & Muthén.

Raykov, T., & Marcoulides, G. A. (2008). *An introduction to applied multivariate analysis.* New York: Taylor & Francis.

Resenda, M., & Sousa, J. P. (2003). (Eds.), *Metaheuristics: Computer decision-making.* Boston, MA: Kluwer Academic Publishers.

Salhi, S. (1998). Heuristic search methods. In G. A. Marcoulides (Ed.), *Modern Methods for Business Research* (pp. 147–175). Mahwah, NJ: Lawrence Erlbaum Associates, Inc.

Sörbom, D. (1989). Model modification. *Psychometrika, 54(3),* 371–384.

Scheines, R., Spirtes, P., Glymour, C., Meek, C., & Richardson, T. (1998). The TETRAD Project: Constraint based aids to causal model specification. *Multivariate Behavioral Research, 33,* 65–117.

Spearman, C. (1904). "General intelligence" objectively determined and measured. *American Journal of Psychology, 5,* 201–293.

Tschannen-Moran, M., & Woolfolk Hoy, A. (2001). Teacher efficacy: Capturing an elusive construct. *Teaching and Teacher Education, 17,* 783–805.

Whittaker, T. A., & Marcoulides, G. A. (2007). Model specification searches using cross-validation indices. Paper presented at the Annual Meeting of the American Educational Research Association, Chicago, IL.

7 A Simulation Study of the Ability of Growth Mixture Models to Uncover Growth Heterogeneity

Kevin J. Grimm, Nilam Ram, Mariya P. Shiyko, and Lawrence L. Lo

Introduction

Growth mixture models (GMMs) are often used to explore whether initially undetermined subgroups of participants exhibit distinct patterns of change. A major issue in the practical application of GMMs is model selection and identification of the proper number of latent classes. The choice is heavily based on the comparative fit indices that compare models with a different number of classes because there are no formal statistical tests for model comparison. We carried out a simulation study to examine the effects of the separation of group means, relative group sizes (mixing proportions), the overall sample size, and the density of observations on model selection. Results indicated that the magnitude of group differences was most strongly associated with the model selection. Surprisingly, the growth mixture model was only favored when class differences were extremely large (~3 standard deviations).

Structural equation models (SEMs) are often fit to data under the assumption that the study sample is drawn from a single population—as might be indicated by the data being distributed in a multivariate normal manner. Recently, the finite mixture model has been incorporated into the SEM framework to model unobserved heterogeneity in the outcome (Muthén & Shedden, 1999; Nagin, 1999). Often, results from finite mixture models are interpreted as evidence that the study sample is drawn from multiple (latent) populations. Frequently referred to as a *person-centered* method (Magnusson & Bergman, 1988), mixture models shift the focus from the study of associations among variables to the identification of homogeneous groups/classes of persons (Muthén & Muthén, 2000). One explicitly longitudinal *person-centered* model that has gained substantial traction in the study of development is the GMM, a model that integrates the latent growth curve (McArdle, 1986; McArdle & Epstein, 1987; Meredith & Tisak, 1990; McArdle & Nesselroade, 2003) within a finite mixture modeling framework (McLachlan & Basford, 1988; McLachlan & Peel, 2000). Applications of the GMM now span many areas of behavioral inquiry, including education (Muthén, 2001; Muthén, Khoo, Francis, & Boscardin, 2003), substance use (Jacob, Koenig, Howell, Wood, & Randolphaber, 2009; Martino, Ellickson, & McCaffrey, 2009; Muthén & Muthén, 2000), psychological well-being (Chen & Simons-Morton, 2009; Crocetti, Klimstra, Keijsers, Hale, & Meeus, 2009; Lincoln & Takeuchi, 2010) and intervention efficacy (Gueorguieva, Mallinckrodt, & Krystal, 2011; Jo & Muthén, 2003; Leoutsakos, Muthén, Breitner, & Lyketsos, 2011), in order to identify and categorize similar individuals into groups/classes based on their observed change trajectories.

A fundamental issue in the application of GMMs to longitudinal data has been model selection. There are no formal statistical tests to compare models with a different number of classes, which has led researchers to utilize a variety of fit indices (e.g., Bayesian Information Criteria) in their quest to determine how many groups are hiding in their sample. Various fit indices have been recommended (see Nylund, Asparouhov, & Muthén, 2007; Tofighi & Enders, 2008), indicating that recommendations are dependent on the simulation conditions and that no single fit index is universally preferred. In the present simulation study, we examine how the separation of group means and the relative sizes of the groups (mixing proportions) are related to accurate identification of true mixtures using a variety of different fit criteria (e.g., Bayesian Information Criteria, Lo–Mendell–Rubin likelihood ratio test) and study procedures (density of repeated measures and sample size). In the following sections we provide an overview of the latent growth, multiple group growth, and growth mixture modeling frameworks, highlight the importance of separation of group means and mixing proportion for identifying mixtures, and introduce the simulation study. In discussing the findings we highlight some implications for design and interpretation of growth mixture modeling results.

Latent Growth Model

The latent growth model (LGM; Meredith & Tisak, 1990) emerged from foundational work of Tucker (1958) and Rao (1958) as a formal model of the means and covariances of repeated measures data based on the common factor model. Assume we have repeatedly measured $n = 1$ to N individuals on a single variable, Y, on $t = 1$ to T occasions. Following McArdle (1986), the latent growth curve model can be written as

$$Y_{nt} = g_{0n} + A_t \cdot g_{1n} + e_{nt} \tag{1}$$

where g_{0n} is a latent intercept, g_{1n} is a latent slope, A is a vector of basis coefficients (factor loadings) indicating the functional relationship of the changes in Y with *time*, and e_{nt} is a time dependent residual that is assumed to be normally distributed with a mean of zero, uncorrelated with g_{0n} and g_{1n}, and restricted to have a single variance ($e_{nt} \sim N\left(0, \sigma_e^2\right)$). The intercept and slope (random) coefficients are assumed to be multivariate normal distributed with estimated means and covariance structure

$$g_{0n}, g_{1n} \sim N\left(\begin{matrix} \mu_0 \\ \mu_1 \end{matrix}, \begin{matrix} \sigma_0^2 \\ \sigma_{10} \ \sigma_1^2 \end{matrix}\right).$$

The elements of the vector of basis coefficients, A, can be fixed to specific values to test whether specific functions of time (e.g., linear) provide a good representation of the data, or they can be estimated from the data in an exploratory way using only minimal constraints (as in typical applications of confirmatory factor analysis).

Figure 7.1 is a path diagram of the latent growth curve model (Equation 1) with five repeated measures of a single variable (Y1–Y5) and two growth factors (g_0 and g_1) that have means (μ_0 and μ_1), variances (σ_0^2 and σ_1^2) and a covariance (σ_{10}). The time-specific residuals (e1–e5) are assumed to have means of zero and homogeneous variance (σ_e^2). The factor loadings from the intercept factor to the manifest measures are all set equal to 1. The pattern of factor loadings (A_t) for the slope determines the shape of the growth curve. For example, linear changes would be captured by a model where A_t would be fixed to linear series (e.g., 0.0, 0.25, 0.50, 0.75, 1.0). More complex patterns of change are

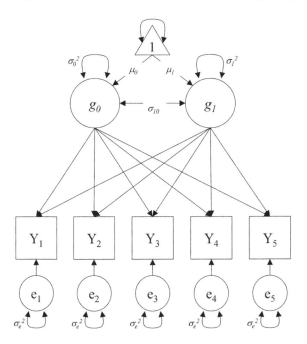

Figure 7.1 Path diagram of a latent growth curve.

articulated by fixing the elements of A_t to values that map onto the specific nonlinear functions (see Grimm & Ram, 2009; Grimm, Ram & Hamagami, 2011; Ram & Grimm, 2007). The latent growth curve model provides a flexible framework for articulating basic developmental theory describing interindividual differences in intraindividual change (Baltes & Nesselroade, 1979).

Multiple Group Growth Model

The multiple group growth model (MGGM; McArdle & Epstein, 1987; McArdle & Anderson, 1990; McArdle & Hamagami, 1996) extends the latent growth curve model as a multiple group structural equation model (Jöreskog, 1971; Sörbom, 1974; Muthén & Christoffesson, 1981; Tisak & Meredith, 1989) and is often used to evaluate differences in the developmental trajectories of two or more pre-defined groups. Assume we have repeatedly measured $n = 1$ to N subjects from $g = 1$ to G groups on a single variable, Y, on $t = 1$ to T occasions, the multiple group growth model is

$$Y_{nt}^{(g)} = g_{0n}^{(g)} + A_t^{(g)} \cdot g_{1n}^{(g)} + e_{nt}^{(g)}, \tag{2}$$

where the aspects of the growth model are group specific. In multiple group models, every estimated parameter $(\mu_0, \mu_1, \sigma_0^2, \sigma_1^2, \sigma_{10}, \sigma_e^2, A_t)$ can differ across groups.

Finite Mixture Modeling

In complement to the multiple group structural equation model, the finite mixture model provides a rigorous statistical foundation for the identification of multiple (latent) subpopulations. In brief, the density of the observed distribution is represented as a weighted sum of two or more other distributions. Thus, the finite mixture model provides an exploratory representation of heterogeneity (between-person differences). Although more general, we concentrate our brief overview on mixtures of normal distributions. The density of the mixture distribution $f(Y)$ of Y, can be written as

$$f(Y) = \sum_{k=1}^{K} \pi^{(k)} f^{(k)}(Y), \tag{3}$$

where $\pi^{(1)}, ..., \pi^{(k)}$ are mixing proportions and $f^{(k)}(Y)$ are the component densities of the group-specific normal distributions that are being *mixed* together. The mixing proportions are non-negative quantities, $0 \le \pi^{(k)} \le 1$, that sum to one, $\sum_{k=1}^{K} \pi^{(k)} = 1$. The number of components, k, is considered to be fixed a priori in any given model run (McLachlan & Peel, 2000), but the optimal number of components is inferred from the data by fitting and comparing multiple models that differ in the number of groups/classes (2 classes, 3 classes, etc.). Formally, when $\pi^{(1)} = 1$ the model reduces to a single group model described by a single density function, $f(Y) = f^{(1)}(Y)$ and when $\pi_n^{(k)} = 0$ or 1, where $\pi^{(k)} = \sum_n^N \left(\frac{\pi_n^{(k)}}{N} \right)$, then the model is equivalent to a multiple groups model as though group membership was observed, which has the same density function with $\pi^{(k)}$ representing the proportion of individuals within each group.

Growth Mixture Model

The growth mixture model combines the statistical foundation of the finite mixture model with the multiple group growth curve model. Combining Equations 2 and 3 and assuming we have repeatedly measured $n = 1$ to N subjects from $k = 1$ to K *latent classes* on a single variable, Y, on $t = 1$ to T occasions, the growth mixture model is

$$Y_{nt} = \sum_{k=1}^{K} \pi_n^{(k)} \left(g_{0n}^{(k)} + A_t^{(k)} \cdot g_{1n}^{(k)} + e_{nt}^{(k)} \right)$$

where $\pi_n^{(k)}$ is individual n's probability of membership in latent class k given $0 \leq \pi_n^{(k)} \leq 1$ and $\sum_{k=1}^{K} \pi_n^{(k)} = 1$. The latent growth factors ($g_{0n}^{(k)}$ and $g_{1n}^{(k)}$), basis coefficients ($A_t^{(k)}$), and time dependent residual ($e_{nt}^{(k)}$) all have the same meaning as in the multiple group growth curve model. As before, model parameters ($\mu_0, \mu_1, \sigma_0^2, \sigma_1^2, \sigma_{10}, \sigma_e^2, A_t$) can, but may or may not, vary over classes. As such, the growth mixture model can be thought of as a form of multiple group growth model in which the grouping variable is *unobserved* as opposed to *observed* (Ram & Grimm, 2009; Shiyko, Ram & Grimm, 2012).

In the (most used) basic linear growth mixture model, the elements of A_t are fixed to follow a specific linear form (e.g., 0, .25, .5, .75, 1.0) and are assumed invariant across classes. Differences between classes are restricted to the mean (μ_0 and μ_1) and covariance ($\sigma_0^2, \sigma_1^2, \sigma_{10}$, and σ_e^2) parameters. Although not necessary, in many applications the covariance parameters are held invariant across classes. A path diagram representation of a growth mixture model is presented in Figure 7.2. In this representation, the latent class variable (C) has an effect on the growth model. The means of the growth factors are superscripted with the latent class to denote that these parameters are free to vary across the latent classes, and all of the other parameters do not have superscripts denoting equality constraints. When all

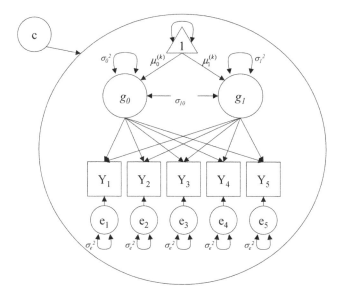

Figure 7.2 Visual representation of a growth mixture model with factor means allowed to vary over latent classes.

the group differences (are assumed to) manifest in the means of the growth factors, we are in the exact space outlined by Preston (1953), where the skewness and kurtosis of the observed distributions are used to infer the separation and mixing proportions of unobserved component distributions.

Simulation Motivation

Ever since applications of growth mixture models began appearing in the literature, questions have been raised about what fit indices to use when determining the number of latent classes and how prone the method is to class overextraction (Bauer & Curran, 2003a, 2004). In a landmark study, Bauer and Curran (2003a) simulated repeated measures data from a single population with a normal distribution, transformed the data to be slightly skewed and kurtotic (the types of distributions commonly seen in social science research) and applied growth mixture models. Across a variety of conditions, they consistently obtained results that would be interpreted as evidence for multiple classes, even though there was only a single (non-normal) population. In a simulation study examining the effectiveness of various fit statistics for identifying the correct number of classes, Tofighi and Enders (2008) also noted how often the wrong number of classes was obtained even "relatively mild levels of nonnormality" were imposed on the within-class distributions (skew = 1 and kurtosis = 1). As with all mixtures, skew and kurtosis (the higher order moments) are at the heart of growth mixture modeling inferences.

In commenting on Bauer and Curran's (2003a) work, Ringskopf (2003) further highlighted some of the difficulties faced when trying to determine whether an observed distribution is actually a mixture of multiple distributions. He plotted several mixtures of two groups that varied in the degree of the mean separation and showed that, when mixed together, two groups that have quite different means (two standard deviations apart) appear as a single, unimodal, more or less normal distribution. Few would suspect that the data were actually obtained from two distinguishable groups, even though a two standard deviations difference is a large separation of group means given traditional measures of effect size. In a simulation study of the effectiveness of various fit indices for identification of the correct number of classes (in three-class data), Tofighi and Enders (2008) also noted the rather dramatic impact of class separation in performance of growth mixture models. Identifying the correct number of classes was easier in a "high separation" condition (i.e., 2.96 SD difference between two of three classes) compared to a "low separation" condition (i.e., 2.30 SD difference). Worrisome for use in the real world is that the "low separation" condition is rather uncommon, especially when considered in terms of Cohen's *d* effect size.

General rules of thumb for study design suggest obtaining groups of equal size for better power. However, as Rindskopf (2003) illustrated, the growth mixture model has difficulty in identifying the observed distribution as a mixture when the classes are equal in size. Previously, Preston (1953) explicitly expressed the skew and kurtosis of a normal distribution as a function of the separation between two normal distributions (assumed to have the same variance, σ^2) and a mixing proportion. Specifically, skew is

$$\gamma_1 = \frac{a(a-1)\Delta^3}{\{a\Delta^2 + (a+1)^2\}^{3/2}}, \tag{4}$$

and kurtosis is

$$\gamma_2 = \frac{a(a^2 - 4a + 1)\Delta^4}{\{a\Delta^2 + (a+1)^2\}^2}, \tag{5}$$

where Δ is the Mahalanobis distance between the means of the two distributions (μ_1 and μ_2), $\Delta = \frac{|\mu_1 - \mu_2|}{\sigma}$ (i.e., separation), and a is the relative ratio of the mixing proportions of the two groups (proportion of the sample in the larger group/proportion of the sample in the smaller group). Thus, the separation of means and mixing proportions is directly related to the higher order moments that are interpreted as indicators of the presence of sub-populations. Equations 4 and 5 illustrate how the shape of the density distribution of a univariate mixture of two normal distributions depends on *both* the magnitude of separation between the two groups' means (e.g., 2 vs. 3 SDs) and the mixing proportion (e.g., .50/.50 vs. .80/.20). For example, consider the case when two groups/classes with very different means ($\mu_1 - \mu_2 = 3$ SDs) and equal size (e.g., $\pi^{(1)} = .5$) are mixed, the combined density distribution $f(Y)$ will visually appear bimodal. However, when the two groups/classes are of unequal size (e.g., $\pi^{(1)} = .8$), the combined density distribution may simply appear as a unimodal distribution with a fat tail (skew). Given that skew and kurtosis are at the heart of growth mixture modeling inferences, it will often be easier to identify a mixture when the groups are of unequal sizes. However, this contrast with typical notions of power-promoting equal sample sizes only goes so far. Tofigi and Enders (2008) noted the difficulty in identifying the true number of classes when one class (of three) only comprised 7% of the sample. Extending the cases given above, when the relative mixing proportion gets very large (e.g., $\pi^{(1)} = .95$) the tail of the distribution is no longer fat enough to be noticed as something different than a few potential outliers.

Following the previous work on performance of GMMs, the purpose of this simulation is to examine how different levels of separation and mixing proportion ratios are related to accurate identification of true mixtures when using a variety of different fit criteria (i.e., Bayesian Information Criteria [BIC], Akaike's Information Criteria [AIC], Sample Size Adjust BIC [ABIC], and the Lo–Mendell–Rubin likelihood ratio test) and study designs features (density of repeated measures, sample size). Specifically, we evaluate the accuracy of identification of groups with three different levels of separation in growth factor means (1, 2, or 3 SD differences in their average intercepts or average linear slopes) and three levels of relative sample size (ratio of mixing proportions = .5/.5, .8/.2, or .95/.05).

Method

Data Generation

Repeated measures data were simulated based on a two-group linear growth model. Each group followed a linear growth curve with non-zero mean intercept and slope parameters with corresponding random effect to reflect inter-individual differences, a zero correlation between the intercept and slope, and a single residual variance. Population parameter values for the first class are given in Table 7.1. Parameter values for the second class were identical except for the intercept or slope mean, which was 1, 2, or 3 standard deviations higher.

The simulation was designed with manipulations of five factors: (1) separation between group means ($\Delta = 1.0$, 2.0, or 3.0 SD); (2) separation located either in the intercept or slope mean; (3) mixing proportions of the two groups ($a = .50/.50$, $.80/.20$, or $.95/.05$); (4) sample size ($N = 200$, 500, 1,000); and (5) data density (i.e., number of repeated measures; $T = 5$,

Table 7.1 Fixed effect growth parameters in the simulation

Parameter	Class 1 population value
Intercept mean (μ_0)	20
Slope mean (μ_1)	20
Basis coefficients	
5 occasions	0, .25, .5, .75, 1
7 occasions	0, .167, .333, .5, .667, .833, 1
9 occasions	0, .125, .25, .375, .5, .625, .75, .875, 1
Intercept variance (σ_0^2)	25
Slope variance (σ_1^2)	16
Intercept–slope covariance (σ_{10})	0
Residual variance (σ_e^2)	9

7, or 9). Of note, across data density conditions, the first and last occasions remained the same so that the total amount of observed change remained identical across conditions. Only the number or density of observations between those occasions differed. Two hundred data sets were generated for each cell in a fully crossed design using the SAS system macro language.

Model Fitting and Evaluation

First, a baseline single-group linear growth model was fit to each data set. Next, a two-class growth mixture model was fit to each data set. Intercept and slope means were allowed to differ between classes. Covariance parameters and the basis coefficients were held invariant. Fit statistics and parameter estimates from each model for each data set were obtained using Mplus (Muthén & Muthén, 1998–2011) and the accompanying RUNALL utility interface (Nguyen, Muthén, & Muthén, 2001). Specifically we obtained: (1) whether or not a viable solution was obtained (i.e., if the model converged or not): and (2) the relative fits of the one class growth model and the two class growth mixture model as indicated by (a) Bayesian Information Criteria (BIC), (b) Akaike's Information Criteria (AIC), (c) sample size Adjusted BIC (ABIC), and (d) Lo–Mendell–Rubin (2001) Likelihood Ratio Test (LMR–LRT)

The BIC, AIC, and ABIC are calculated based on the -2 loglikelihood with differing penalties for the number of estimated parameters. The BIC equals $-2LL + \ln(N) \cdot p$, where $-2LL$ is twice the negative of the loglikelihood, N is the sample size, and p is the number of independently estimated parameters. The AIC equals $-2LL + 2p$ and the ABIC equals $-2LL + \ln\left(\frac{N+2}{24}\right) \cdot p$. For all information criteria, there is a penalty for the number of estimated parameters. Thus, the model with the smallest information criteria is considered the best representation of the data and is the chosen model.

One likelihood ratio test was also examined. Likelihood ratio tests compare the fitted model with k classes to a model with $k - 1$ classes. Lo et al. (2001) derived the appropriate reference distribution for the likelihood ratio test along with an ad hoc adjustment. A low p-value for the likelihood ratio test indicates the model with k classes fits significantly better than the model with $k - 1$ classes. In this simulation, a p-value of .05 was used as the criteria for model selection—if the two-class growth mixture model had a likelihood ratio test p-value less than .05, then this model was the selected model.

Results

We obtained viable solutions for the one class latent growth curve model for 100% of the data sets. Fit indices for the two-class growth mixture model were obtained for 99.6% (32,273) of the data sets. However, an additional 18% of the growth mixture runs led to inadmissible solutions (e.g., negative estimated variance) or corresponding to the choice of a single-class model (mean of categorical variable > |10| indicating more than 99.99% of the sample was in one of the two classes). For every growth mixture model run that produced fit indices, we determined whether the proper two-class solution was selected following usual model selection procedures (i.e., selection of the model with better comparative fit). Overall, the true model (two-class growth mixture) was the favored representation of the data relatively infrequently. The growth mixture model was selected in only 10% (3,209) of the data sets based on the BIC; using AIC, only in 23% (7,408); using ABIC, only in 17% (5,511); and using LMR–LRT, only in 18% (5,894). Simulated datasets for which fit indices were not obtained were excluded from subsequent analyses. However, simulated datasets for which the GMM resulted in inadmissible and one-class solutions were included, because this information would indicate preference of the one-class solution (and fit indices agreed with this conclusion).

Logistic multiple regression models were then fit to evaluate if and how each factor contributed to the probability of selecting the true two-class model (only main effects for design factors are reported because interactions accounted for less than an additional 3% of explained variance). Overall, five manipulated factors explained 40% of variance in model selection based on the BIC, and even less based on other indices (13% for the AIC, 19% for the ABIC, and 17% for the Lo–Mendell–Rubin LRT p-value).

As can be seen in Table 7.2, most simulation design factors were significantly related to model selection. Of note, data density (measurement occasions) and a mixing proportion of .50/.50 compared to .95/.05 tended to be unrelated to model choice. The remaining effects were generally consistent across the fit indices. Greater separation of group means, differences in intercept versus the slope, larger sample sizes, and when the mixing proportion was .80/.20 versus .95/.05 were each related to

Table 7.2 Odds ratios and confidence intervals for the effects of the manipulated factors on model selection based on the Bayesian Information Criteria (BIC), Akaike Information Criteria (AIC), sample size Adjusted BIC (ABIC), and Lo–Mendell–Rubin Likelihood Ratio Test (LMR–LRT)

Fit criteria	Density		Sample size		Location of differences	Mixing proportion		Mean distance	
	5 vs. 9	7 vs. 9	200 vs. 1000	500 vs. 1000	Intercept vs. slope	.50 vs. .95	.80 vs. .95	1 SD vs. 3 SD	2 SD vs. 3 SD
BIC	**1.18**	1.02	**.14**	**.50**	**2.57**	**1.14**	**2.47**	**<.01**	**.10**
	(1.01–1.24)	(.92–1.13)	(.13–.16)	(.46–.55)	(2.36–2.81)	(1.02–1.27)	(2.23–2.74)	(<.01–.01)	(.09–.11)
AIC	1.02	**.89**	**.37**	**.69**	**1.34**	1.01	**1.36**	**.22**	**.52**
	(.96–1.09)	(.83–.95)	(.34–.39)	(.64–.73)	(1.27–1.41)	(.95–1.08)	(1.27–1.45)	(.21–.24)	(.49–.55)
ABIC	1.06	.93	**.53**	**.70**	**1.64**	1.02	**1.63**	**.09**	**.36**
	(.98–1.14)	(.86–1.00)	(.49–.57)	(.65–.76)	(1.54–1.74)	(.94–1.10)	(1.51–1.75)	(.08–.10)	(.34–.39)
LMR–LRT	**1.10**	.99	**.29**	**.63**	**1.55**	1.01	**1.52**	**.17**	**.35**
	(1.02–1.18)	(.92–1.07)	(.27–.31)	(.59–.67)	(1.46–1.64)	(.93–1.09)	(1.42–1.64)	(.15–.18)	(.33–.38)

Note: The experimental condition listed second is the reference group for the comparisons; odds ratios that are in bold type are significantly different from 1, $p < .05$.

greater likelihood that the growth mixture model was selected. Differences in separation between group means accounted for the majority (63–82%) of the explained variance, especially for ABIC (which is inherently less affected by differences in sample size).

Looking at the pattern of results across fit statistics, the BIC was most sensitive to differences in the five design factors. For example, moving from a sample size of 500 to a sample of 1,000 increased the likelihood of correctly identifying the two-class solution by a factor of two; having intercept differences versus slope differences increased this likelihood by a factor of 2.57; having an .80/.20 mixing proportion versus a .95/.05 mixing proportion increased the likelihood by a factor of 2.47 (a similar sized effect was found when comparing .80/.20 with .50/.50); and going from a 2 to a 3 standard deviation mean difference increased the likelihood by a factor of 10.00.

Discussion

Summary of Findings

Growth mixture models are often used to identify and model heterogeneity in longitudinal data. The purpose of this simulation was to examine how different levels of group separation and mixing proportions were related to accurate identification of true mixtures across a set of fit criteria and study design factors (density of repeated measures, sample size). Generally, performance of the procedure was relatively low, with the true two-class model being identified, at best, in 23% of our simulated data sets. As expected, greater sample sizes and greater separation between groups were strongly related to model selection. Group differences in intercept facilitated identification of group differences to a greater extent than did group differences in slopes. In contrast to typical notions of power in balanced designs, the true two class model was identified correctly more often when the groups were unbalanced (80% vs. 20%), but not severely so (95% vs. 5%). Specifically, the GMM procedures were 1.34 to 2.17 times more likely to correctly identify the presence of two groups when they were mixed .80/.20 compared to when they were mixed .50/.50, depending on the chosen fit index.

In one of the prominent discussions about the viability and dangers of growth mixture modeling, Bauer and Curran (2003a) illustrated how the extent of skew in the observed data facilitated identification of a mixture (even in cases where no mixture was present). Variables may be skewed for a variety of reasons, including variables' inherent measurement properties, non-normal distribution of interindividual differences in the population, and sampling or mixing of multiple populations. Fifty years apart, and in quite different settings, Preston (1953) and Rindskopf (2003) noted explicitly how unequal sample sizes can contribute to higher levels of skew. For example, when two groups whose means are separated by 3 standard deviations are mixed in a balanced way (50% vs. 50%), the resulting distribution's skew is 0.0. However, when mixed in an unbalanced way (80% vs. 20%), skew is 0.68. A visualization of this association is presented in Figure 7.3(a), where increases in the mixing proportion ratio (*a*) and Mahalanobis Distance (delta) are related to increasing skew.

The mixing proportion ratio is also related to kurtosis. When two groups whose means are separated by 3 standard deviations and mixed in a balanced way (50% vs. 50%), the resulting distribution has kurtosis of

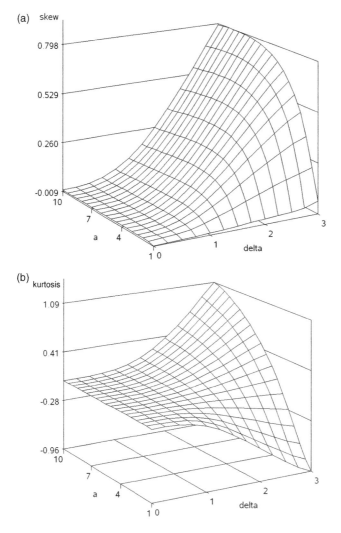

Figure 7.3 Associations between Mahlanabis' distance (delta), mixing proportion ratio (*a*), and (a) skew and (b) kurtosis.

−0.95 ("skinny tails"). However, when mixed in an unbalanced (80% vs. 20%) way, kurtosis is .09 (relatively normal tails). Figure 7.3(b) demonstrates how increases in the mixing proportion ratio (a) and Mahalanobis Distance (delta) are related to changes in kurtosis. Largely reflecting tail behavior, high kurtosis is sometimes used to identify the presence of outliers (DeCarlo, 1997). As the mixing proportion ratio increases, and skew and kurtosis get higher, the second group looks more like a small set of outliers and may be identified more easily. The implication is that, when looking for mixtures, it is better to have sampled fewer people from the minority class.

As well, it seems that the minority class may need to be outliers. In the best case scenario of large samples ($N = 1,000$) with .80/.20 mixing proportions of groups that were separated by 3 standard deviations (a typical cutoff used to identify outliers), GMM procedures correctly identified two groups only 60% of the time (regardless of the fit index). This is a bit disheartening, especially when considering the ease with which the human eye can discriminate differences of 3 standard deviations in raw data plots. However, these results corroborate differences noted between "low separation" (+2.3 SD) and "high separation" (+2.9 SD) conditions in previous simulations (Tofighi & Enders, 2008). When the groups were separated by only a small amount, 1 SD, the presence of two groups was only identified 10% of the time. Our general conclusion is that care should be taken in interpretation of GMM results from simple linear growth models, as it is not clear that the results are replicable, even under ideal conditions.

Outlook

Although our results are somewhat pessimistic, they also suggest that there are many avenues to explore regarding how and when growth mixture models are effective. Further work is needed on the effectiveness of growth mixture models for identifying more than two latent classes, for examining heterogeneity of change in non-continuous manifest variables (e.g., categorical and count data), along nonlinear change trajectories (i.e. exponential, latent basis, latent difference score models), and in the covariances among growth factors. For example, our simulations were simply generated with intercepts and slopes variances equal across groups and unrelated to each other within classes (i.e., zero covariance). This was done to distinguish between intercept and slope differences (if intercept and slope are correlated, then intercept differences would leak into slope differences and vice versa). Generalizations should consider how differences in the covariances (and variances) of the growth factors facilitate or hinder identification of classes. It would also be extremely informative to know how differences in groups' patterns of change (as opposed to magnitude of change) are related to model selection (small unpublished simulation studies indicated this may greatly facilitate identification).

While we did not highlight the differences and relative strengths of the various fit indices used in selection among models with different numbers of classes, future work might follow some of the early work in exploratory factor analysis related to choosing the correct number of factors. For example, Muthén (2001) advocated plotting of the BIC (or other fit index) against the number of classes to locate a break point after which additional classes might be considered "noise" classes, just as scree plots are used to determine the number of factors in exploratory factor analysis.

Development of analogues of parallel analysis, for selecting the number of factors in exploratory factor analysis, and other techniques seems warranted. Simulations and applications should consider the many reasons why a distribution might exhibit skew and kurtosis, including restrictions on measurement range, the presence of outliers, imbalanced sampling, and the like. Growth mixture modeling provides researchers with exciting opportunities to discover systematic patterns in repeated measures data. With careful replication of results across data sets and variables, we are sure that much can be learned. In the meantime, keeping an open mind about discoveries in the data and consideration of a variety of reasons for emerging data patterns is warranted.

References

Baltes, P. B., & Nesselroade, J. R. (1979).History and rational of longitudinal research. In J. R. Nesselroade & P. B. Baltes (Eds.), *Longitudinal research in the study of behavior and development*. New York: Academic Press.

Bauer, D. J. & Curran, P. J. (2003a). Distributional assumptions of growth mixture models: Implications for overextraction of latent trajectory classes. *Psychological Methods, 8*, 338–363.

Bauer, D. J. & Curran, P. J. (2003b). Overextraction of latent trajectory classes: Much ado about nothing? Reply to Rindskopf (2003), Muthén (2003), and Cudeck and Henly (2003). *Psychological Methods, 8*, 384–393.

Bauer, D. J. & Curran, P. J. (2004). The integration of continuous and discrete latent variable models: Potential problems and promising opportunities. *Psychological Methods, 9*, 3–29.

Chen, R., & Simons-Morton, B. (2009). Concurrent changes in conduct problems and depressive symptoms in early adolescents: A developmental person-centered approach. *Development and Psychopathology, 21*, 285–307.

Crocetti, E., Klimstra, T., Keijsers, L., Hale, W., III, & Meeus, W. (2009). Anxiety trajectories and identity development in adolescence: A five-wave longitudinal study. *Journal of Youth and Adolescence, 38*, 839–849.

DeCarlo, L. T. (1997). On the meaning and use of kurtosis. *Psychological Methods, 2*, 292–307.

Grimm, K. J., & Ram, N. (2009). Non-linear growth models in M*plus* and SAS. *Structural Equation Modeling, 16*, 676–701.

Grimm, K. J., Ram, N., & Hamagami, F. (2011). Nonlinear growth curves in developmental research. *Child Development, 82*, 1357–1371.

Gueorguieva, R., Mallinckrodt, C., & Krystal, J. (2011). Trajectories of depression severity in clinical trials of Duloxetine. *Archives of General Psychiatry, 68*, 1227–1237.

Jacob, T., Koenig, L. B., Howell, D. N., Wood, P. K., & Randolphaber, J. (2009). Drinking trajectories from adolescence to the fifties among alcohol-dependent men. *Journal of Studies on Alcohol and Drugs, 70*, 859–869.

Jo, B., & Muthén, B. (2003). Longitudinal studies with intervention and noncompliance: Estimation of causal effects in growth mixture modeling. In S. P. Reise & N. Duan (Eds.), *Multilevel modeling: Methodological advances, issues, and applications* (pp. 112–139). Mahwah, NJ: Erlbaum.

Jöreskog, K. G. (1971). Simultaneous factor analysis in several populations. *Psychometrika, 36*, 409–426.

Leoutsakos, J. S., Muthén, B. O., Breitner, J. C. S., & Lyketsos, C. G. (2011). Effects of non-steroid anti-inflammatory drug treatments on cognitive decline vary by phase of pre-clinical Alzheimer disease: Findings from the randomized controlled Azheimer's Disease Anti-Inflammatory Prevention Trial. *International Journal of Geriatric Psychiatry, 27*, 364–374.

Lincoln, K. D., & Takeuchi, D. (2010). Variation in the trajectories of depressive symptoms: Results from the Americans' changing lives study. *Biodemography & Social Biology, 56*, 24–41.

Lo, Y., Mendell, N. R., & Rubin, D. B. (2001). Testing the number of components in a normal mixture. *Biometrika, 88*, 767–778.

Magnusson, D., & Bergman, L. R. (1988). Individual and variable-based approaches to longitudinal research on early risk factors. In M. Rutter (Ed.), *Studies of psychosocial risk: The power of longitudinal data* (pp. 45–61). Cambridge: Cambridge University Press.

Martino, S. C., Ellickson, P. L., & McCaffrey, D. F. (2009). Multiple trajectories of peer and parental influence and their association with the development of adolescent heavy drinking. *Addictive Behaviors, 34*, 693–700.

McArdle, J. J. (1986). Latent variable growth within behavior genetic models. *Behavior Genetics, 16*, 163–200.

McArdle, J. J., & Anderson, E. (1990). Latent variable growth models for research on aging. In J. E. Birren & K. W. Schaie (Eds.), *Handbook of the psychology of aging*, 3rd Edn. (pp. 21–44). San Diego, CA: Academic Press Inc.

McArdle, J. J., & Epstein, D. (1987). Latent growth curves within developmental structural equation models. *Child Psychology, 58*, 110–133.

McArdle, J. J., & Hamagami, F. (1996). Multilevel models from a multiple group structural equation perspective. In G. Marcoulides & R. Schumacker (Eds.), *Advanced structural equation modeling techniques* (pp. 89–124). Hillsdale, NJ: Erlbaum.

McArdle, J. J., & Nesselroade, J. R. (2003). Growth curve analysis in contemporary psychological research. In J. A. Schinka & W. F. Velicer (Eds.), *Handbook of psychology: Research methods in psychology*, Vol. 2. (pp. 447–480). New York: John Wiley & Sons, Inc.

McLachlan, G. J., & Basford, K. E. (1988). *Mixture models: Inference and applications to clustering*. New York: M. Dekker.

McLachlan, G. J., & Peel, D. (2000). *Finite mixture models*. New York: John Wiley & Sons, Inc.

Meredith, W., & Tisak, J. (1990) Latent curve analysis. *Psychometrika, 55,* 107–122.

Muthén, B. (2001). Second-generation structural equation modeling with a combination of categorical and continuous latent variables: New opportunities for latent class–latent growth modeling. In L. M. Collins & A. G. Sayer (Eds.), *New methods for the analysis of change* (pp. 179–199). Washington, DC: American Psychological Association.

Muthén, B., & Christofferson, A. (1981). Simultaneous factor analysis of dichotomous variables in several groups. *Psychometrika, 46,* 407–419.

Muthén, B., Khoo, S. T., Francis, D. J., & Boscardin, C. K. (2003). Analysis of reading skills development from kindergarten through first grade: An application of growth mixture modeling to sequential processes. In S. P. Reise & N. Duan (Eds.), *Multilevel modeling: Methodological advances, issues, and applications* (pp. 71–89). Mahwah, NJ: Erlbaum.

Muthén, B., & Muthén, L. K. (2000). Integrating person-centered and variable-centered analysis: Growth mixture modeling with latent trajectory classes. *Alcoholism: Clinical and experimental research, 24,* 882–891.

Muthén, L. K., & Muthén, B. (1998–2011). *Mplus User's Guide,* 6th Edn. Los Angeles, CA: Muthén & Muthén.

Muthén, B., & Shedden, K. (1999). Finite mixture modeling with mixture outcomes using the EM algorithm. *Biometrics, 55,* 463–469.

Nagin, D. S. (1999). Analyzing developmental trajectories: A semiparametric, group-based approach. *Psychological Methods, 4,* 139–157.

Nguyen, T., Muthén, L. K., & Muthén, B. (2001). RUNALL (Versión 1.1) [Computer Software]. http://statmodel.com/runutil.html.

Nylund, K. L., Asparhouv, T., & Muthen, B. (2007). Deciding on the number of classes in latent class analysis and growth mixture modeling. A Monte Carlo simulation study. *Structural Equation Modeling, 14,* 535–569.

Preston, E. J. (1953). A graphical method for the analysis of statistical distributions into two normal distributions. *Biometrika, 40,* 460–464.

Ram, N., & Grimm, K. J. (2007). Using simple and complex growth models to articulate developmental change: Matching method to theory. *International Journal of Behavioral Development, 31,* 328–339.

Ram, N., & Grimm, K. J. (2009). Growth mixture modeling: A method for identifying differences in longitudinal change among unobserved groups. *International Journal of Behavioral Development, 33,* 565–576.

Rao, C. R. (1958). Some statistical methods for comparison of growth curves. *Biometrika, 14,* 1–17.

Rindskopf, D. (2003). Mixture or homogenous? Comment on Bauer and Curran (2003). *Psychological Methods, 8,* 364–368.

Shiyko, M. P., Ram, N., & Grimm, K. J. (2012). An overview of growth mixture modeling: A simple nonlinear application in OpenMx. In R. H. Hoyle (Ed.), *Handbook of structural equation modeling* (pp. 532–546). New York: The Guilford Press.

Sörbom, D. (1974). A general method for studying differences in factor means and factor structure between groups. *British Journal of Mathematical and Statistical Psychology, 27,* 229–239.

Tisak, J., & Meredith, W. (1989). Exploratory longitudinal factor analysis in multiple populations. *Psychometrika, 54,* 261–281.

Tofighi, D., & Enders, C. (2008). Identifying the correct number of classes in growth mixture models. In G. R. Hancock & K. M. Samuelsen (Eds.), *Advances in latent variable mixture models* (pp. 317–341). Charlotte, NC: Information Age Publishing, Inc.

Tucker, L. R. (1958). Determination of parameters of a functional relation by factor analysis. *Psychometrika, 23*, 19–23.

8   Mining for Association Between Life Course Domains

Raffaella Piccarreta and Cees H. Elzinga

Introduction

We explore three kinds of association between life courses. First, there is case-based association, which relies on the average contingency between states in different domains. This kind of association turns out not to be very practical: measures based on entropy of state contingencies are insensitive to order of states. However, a new measure based on contingency of subsequences turns out to be a numerical anomaly. Second, we investigate global association between dissimilarities or distances in either domain and apply Escouffier's R_V and Mantel's coefficient to life-course data. These coefficients are based upon the strong assumption of monotonicity of distances across domains. Finally, we focus on local association, exploiting neighborhoods. It turns out that this approach leads to a coefficient that has a nice interpretation in terms of predictability, is numerically tractable and seems well behaved at the full range of associations.

In most human life courses, entering a first job precedes entering parenthood. Apparently, these events are associated although they come from quite different life course domains: the labor market career and the family formation trajectory. It is generally held that these domains are associated and that people's paths or trajectories in these domains are related.

In this chapter, we deal with the problem of quantifying the strength of association between trajectories in different life course domains. In our approach, the trajectories are represented as sequences. Thus, a life course is represented as an ordered collection of encoded life course events (states) experienced by a given individual in a specified period. The set of the states used to describe a domain is called an alphabet: different life course domains will be characterized by different alphabets.

For example, if one considers the family formation trajectory and the work career, the alphabet for the former domain will consist of states like being single, marriage, or unmarried cohabitation, while the alphabet of the latter will include states like employment, vocational training, and unemployment.

In an abstract sense, quantifying the strength of association between two domains amounts to quantifying the association between the states visited by individuals in the two domains.

If two domains are associated, it seems natural to study the trajectories jointly. Multiple sequence analysis (see e.g., Pollock, 2007; Gauthier et al., 2010) has been recently introduced with this aim. Quantification of the strength of association between domains in this case can be useful as a screening procedure aiming at individuating which domains are worth being studied jointly and which instead should be analyzed individually, sharing little with the others.

Alternatively, one may adopt an asymmetric perspective, and study how the life course domain of one person affects the same life course of another person. For example, Liefbroer and Elzinga (2012) relate family formation trajectories of children to family formation trajectories of their parents in an attempt to investigate intergenerational transfer of such patterns. Another example is in Mooi-Reci (2008): she relates labor market careers of children and parents to see if scarring effects from parental unemployment affect labor market careers of their children. In these cases, it can be interesting to understand if and to what extent the knowledge of one career permits us to draw conclusions about the other (see e.g., Piccarreta and Lior, 2010). This can be particularly relevant in cases when only information on one career is available.

A first approach to exploring if and to what extent two domains are associated is to focus on each individual separately. As discussed below, for each individual the association between the sequences in the two domains is measured, and the results obtained for all the cases are averaged to summarize the association in the sample. To measure this *case-based* association we refer to indices introduced in the literature for categorical variables. These criteria cannot be optimal when focusing on sequences, since they do not take into proper account the ordering of states in the two sequences. To deal with this issue, we also consider a case-based association measure properly taking into account this relevant feature of trajectories.

Case-based measures are reasonable when one assumes that the association indices at the individual level come from the same population, so that the amount of association can be reasonably estimated through the (assumed to be common) mean of the individual indices.

Nonetheless, as discussed in below, in many applications one can be interested in evaluating whether a *relation* exists between domains, by adapting the ideas underlying regression models to sequences. In this context, no measurements are available, and the standard techniques cannot be applied. Nonetheless, criteria exist to properly measure the dissimilarities among all the possible pairs of cases. One possibility widely used in social sciences is Optimal Matching (OM) (see e.g., Abbott, 1995; Abbott and Forrest, 1986; Abbott and Hrychak, 1990; Anyadike-Danes and McVicar,

2010) or one of its variants (Halpin, 2010; Hollister, 2009; Lesnard, 2010), but alternatives were also successfully applied (Elzinga and Liefbroer, 2007; Bras et al., 2010). Therefore, attention can be focused on the relation between dissimilarities in the two domains, and to the measurement of the extent to which a variation of the dissimilarity in one domain *impacts* on the dissimilarity in the other one. This can be done by referring to the correlation between the dissimilarities or to measures based upon the same intuition but suitably modified to deal with dissimilarity data. This *global* approach is based upon the idea that two domains are associated if cases which are similar in one domain are also similar in the other one *and* cases which are dissimilar in one domain are also dissimilar in the other one, the dissimilarities in the two domains thus increasing together.

From the literature on regression, it is known that when the assumption of linearity or of monotone association is too far fetched (as is possible in the dissimilarity case) smoothed (or *local*) regression (see e.g., Härdle, 1992) can be used. Instead of fitting a global model ideally holding for each data point, these methods rely on a more flexible relation, fitted locally, allowed to vary case-wise. We will describe how to extend this method to dissimilarities below. For now, it is worth mentioning that this approach is based upon the idea that two domains are associated when sequences which are close in one domain *also* are similar in the other domain, without making assumptions about the reaction of dissimilarities in one domain to an increase or a decrease of the dissimilarity in the other domain.

To illustrate the association measures discussed, we will apply them to real data and therefore the next section briefly discusses these data. In addition, we concisely deal with the concept of Sequence Index Plot, which plays an important role in illustrating our reasoning about association between careers and the application of our ideas to real data.

Data

The data used in this paper come from the Panel Study on Social Integration in The Netherlands (PSIN; Liefbroer and Kalmijn (1997)). PSIN was designed to investigate social integration in two life domains: personal relations and family formation and educational and occupational career. Today, PSIN consists of seven waves of data collection from individuals born in 1961, 1965, and 1969. In this example, we use data on work and family trajectories covering the age span of 15 to 34 years (229 consecutive months) from persons born in 1961 and 1965, confined to those individuals from whom we had data covering the whole age span chosen. Table 8.1 describes the composition of the sample according to gender and birth-year cohort.

For each individual, the status in the family and work careers is coded for each of the $T = 229$ months during which they are observed. We encoded the family formation histories distinguishing between subjects

Table 8.1 Partition of PSIN data, covering
229 months starting at the age of 15, by
gender and year of birth

	1961	1965	Total
Females	187	139	326
Males	171	117	288
Total	358	256	614

who are single (the subject has no partner but may live with parents or not), or who cohabitate with a partner, being married or not. This information was combined with the number of children. A preliminary analysis revealed that few cohabiting individuals have children. For this reason, we only focused on the number of children and disregarded the partnership situation. Thus, the following mutually exclusive codes were used: S, U, M, (single, unmarried cohabitation, married without children) and C1, C2, C3 (S, U, or M with 1, 2, 3, or more children). As for the educational/occupational careers, we used three mutually exclusive levels of secondary education—E1, E2, E3 (Lower secondary education, Higher secondary education, and Tertiary education)—and three mutually exclusive states on the labor market—WP (Working Part time), WF (Working Full time), U (Unemployed; does not necessarily imply that the subject would be available for the labor market). In the next sections, we will use these data to illustrate some of the properties of the pertaining association measures.

From a substantive point of view, we expect association between the mutual life domains for a number of reasons. First, entry into a partner relationship and entry into parenthood is often expected to occur only after finishing one's education and establishing oneself on the labor market (Blossfeld and Huinink, 1991). To the extent that a male breadwinner model is operative, this dependence of events in the family formation career on events in the work career is expected to be stronger for men than for women. At the same time, given the difficulties in combining motherhood and a career in the Netherlands (Liefbroer and Corijn, 1999), women's work career is expected to be more strongly dependent on events in the family life career than men's.

Among cohorts born since the 1950s, a general process of de-standardization of the life course has been observed (Elzinga and Liefbroer, 2007). As individuals became more autonomous in their life choices, the standard order in which events in the work and family life domains were expected to happen, started to break down. If this process is observable among these birth cohorts in the Netherlands, one would expect a reduction of the association between events in both careers.

Sequence Index Plots

To visualize sequences, we use sequence index plots (SIP), first introduced in Scherer (2001). In these plots individuals are placed on the horizontal axis, and time on the vertical axis. Individuals' careers are represented by stacked, colored bars, each stack representing a career. Distinct colors (see the book's web site for a color version of the figure) of the bars denote distinct states from the career alphabet and the lengths of the bars denote the duration of the states in the careers. To ease analysis of the plots, the careers are ordered along the horizontal axis according to some property of the individuals or of the careers themselves, so that similar careers are placed close to each other.

The visual inspection of such plots may reveal one or more general career patterns in the data. As an example, in Figure 8.1, we present the SIP for the work careers of PSIN data. For a more detailed account of SIPs and their application, see Piccarreta and Lior (2010) or Gabadinho et al. (2011).

Case-Based Association Measures

A first approach to exploring association between two domains is to focus on each individual separately. Hence, for each individual the association between the sequences in the two domains is measured. One possibility is to refer to one of the association indices introduced in the literature for two categorical variables. Here we refer to the concept of "proportional

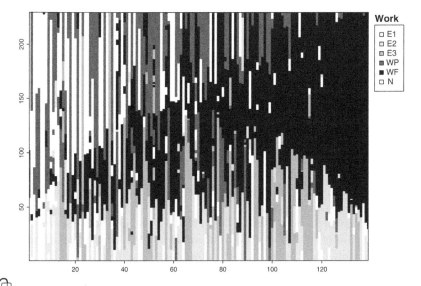

Figure 8.1 Sequence index plot of the careers of females born in 1965, taken from the PSIN data. In the plot, cases are placed on the horizontal axis and time on the vertical axis.

reduction of variation" introduced by Guttman (1941). Consider a variate Y and let $V(Y)$ denote the amount of its variation. If $V(Y)$ is reduced to some extent by knowing the value of another variate, X, then the two variables are said to be associated. Let $V(Y|X)$ be the average of the variation of Y conditioned to the values of X. If $0 \leq V(Y|X) \leq V(Y)$, then the quantity

$$0 \leq R(Y|X) = \frac{V(Y) - V(Y|X)}{V(Y)} = 1 - \frac{V(Y|X)}{V(Y)} \leq 1 \qquad (1)$$

equals the proportional reduction in variation of Y as a result of conditioning on X. If X and Y are independent, knowledge of X will not result in a reduction of variation in Y, and we will have $R(Y|X) = 0$. Instead, if X fully determines Y, $V(Y|X)$ will equal zero and hence $R(Y|X)$ will be maximal. Note that the association measure R is not necessarily symmetric.

Indeed, here we discuss two association measures having the structure of Equation (1), based on two alternative measures of dispersion: Shannon's (1948) entropy H and Gini's (1912, 1939) concentration G. The two measures are very well known in the context of categorical data analysis and are also used in the analysis of log-linear models (Haberman, 1982).

Both the measures start from the general idea that a categorical variable has no variation if all the cases present the same category, and has maximal variation if all possible categories are equally frequent. Even though the measures we are going to present are very well known in the literature on categorical variables, in the next section we briefly review some of their more relevant characteristics, and show how they can be adapted to the case when the association between two domains is of interest.

Association as Reduction of Entropy

Shannon's entropy $H(Y)$ quantifies the uncertainty about the realizations of a variable Y with J possible outcomes, $\{y_1, \ldots, y_J\}$, and is defined as:

$$0 \leq H(Y) = -\sum_j p_j \log_2 p_j \leq \log_2 J, \qquad (2)$$

where $p_j = Fr(Y = y_j)$ is the proportion of cases characterized by the jth level of Y. $H(Y)$ is minimal when one of the p_js equals 1, i.e. when only one outcome is observed. Indeed, setting $0 \cdot \log 0 = 0$, one easily obtains that in this case $H(Y) = 0$. On the other hand, when $p_j = (1/J)$ for each j, uncertainty is maximal and $H(Y) = \log_2 J$.

Consider now a variable X with K possible outcomes, $\{x_1, \ldots, x_K\}$, and let $H(Y|X = x_k)$ denote the entropy of Y conditioned to the kth level of X. The conditional entropy $H(Y|X)$ is defined as the average of the entropies $H(Y|X = x_k)$:

$$H(Y|X) = \sum_{k=1}^{K} p_k H(Y|X = x_k) = -\sum_{k=1}^{K}\sum_{j=1}^{J} p_{kj} \log_2 p_{j|k}, \qquad (3)$$

with $p_{kj} = Fr(X = x_k, Y = y_j)$ and $p_{j|k} = Fr(Y = y_j|X = x_k) = p_{kj}/p_k$. It can be shown that $H(Y|X) \leq H(Y)$ (e.g., Cover and Thomas, 1991). Using H as a measure of variation in (1), one obtains the so called *uncertainty coefficient* (Theil, 1972; Cover and Thomas, 1991):

$$R_H(Y|X) = 1 - \frac{H(Y|X)}{H(Y)}, \qquad (4)$$

measuring what fraction, on the average, of the uncertainty about Y is reduced when we have information about X. If this reduction is complete, so that $H(Y|X) = 0$, the association is perfect and $R_H(Y|X)$ reaches its maximum value, 1. Analogously, $R_H(X|Y)$ can be defined. It is important to emphasize that $R_H(Y|X)$ and $R_H(X|Y)$ reach their maximal values independently. The measure of association we are considering is consequently not symmetric.

As for the minimum value of R_H, it is attained when Y and X are statistically independent, so that $p_{kj} = p_k p_j$, and $p_{j|k} = p_j$. In this case, it is $H(Y|X) = H(Y)$ and $H(X|Y) = H(X)$, so that $R_H(Y|X) = R_H(X|Y) = 0$. Indeed, the R_Hs also measure the degree to which the conditional distributions differ from the ones we would expect under full independence of the two variables.

We now refer to (4) to measure the association between two domains, X and Y, characterized by the alphabets (set of states) $\mathcal{A}_x = \{x_1, x_2, \ldots, x_K\}$ and $\mathcal{A}_y = \{y_1, y_2, \ldots, y_J\}$ respectively. Consider the ith individual and the two paired T-long sequences, x and y, describing her careers in the two considered domains (the subscript i is omitted for the sake of simplicity). Consider the joint distribution of the visited states: now p_{kj} denotes the number of periods when the ith individual experienced the kth state in $\mathcal{A}_x$ and the jth state in $\mathcal{A}_y$ simultaneously, whilst p_k and p_j denote the marginal proportions characterizing the two states respectively. For the sake of clarity, we consider two toy-sequences constructed from the alphabets $\mathcal{A}_x = \{a, b, c\}$ and $\mathcal{A}_y = \{p, q, r\}$ respectively. In particular, $x = (a/20; b/10; a/20; c/50)$ and $y = (p/30; r/10; q/20; r/40)$. In Table 8.2 the joint distribution of the states is reported, together with the marginal frequencies. From the latter, one can easily obtain $H(x) = 1.361$ and that $H(y) = 1.49$. Calculating the entropy within each row (column) and taking the weighted average, the conditional entropies are obtained, $H(x|y) = 0.84$ and $H(y|x) = 0.96$. The corresponding associations indices between the two toy sequences are $R_H(x|y) = 0.39$ and $R_H(y|x) = 0.35$ respectively.

Table 8.2 Marginal and joint distributions of $x = (a/20; b/10; a/20; c/50)$ and $y = (p/30; r/10; q/20; r/40)$

$x; y$	p	q	r	Total
a	20	10	10	40
b	10			10
c		10	40	50
Total	30	20	50	100

The same procedure can be applied to all the n observed sequences. The individual R_Hs can be averaged across cases, to obtain a summary of the entropy-based association.

This procedure was applied to the PSIN data over all the individual pairs of careers in family formation (for short, the "f-career") and in education/employment (for short, the "e-career"). In Table 8.3, the averages of the R_Hs conditioned to the combinations of gender and cohort are reported, together with their standard errors. From the literature on three-way contingency tables and on log-linear models (see e.g., Agresti, 2002), the quantities in the table can be interpreted (for each combination of gender and cohort) as measures of the association between the f and the e domain when the interactions between each domain and individuals are removed.

From the table we observe that the association between the e-career on the f-career (in both directions) is systematically lower for men than for women, even when the cohorts are separated. In addition, the differences between cohorts are relatively high when considering $R_H(e|f)$ for females and $R_H(f|e)$ for males. So, we conclude that R_H detects differences in state contingencies between men and women but, apparently, it does not detect

Table 8.3 PSIN data: average of the individual $R_H(e|f)$ and $R_H(f|e)$ conditioned to gender and cohort (standard errors in parentheses)

$R_H(e|f)$	*1961*	*1965*	*Whole sample*
Females	0.49 (0.018)	0.42 (0.020)	0.46 (0.013)
Males	0.32 (0.016)	0.29 (0.018)	0.31 (0.012)
Whole sample	0.41 (0.013)	0.36 (0.014)	0.39 (0.010)

$R_H(f|e)$	*1961*	*1965*	*Whole sample*
Females	0.40 (0.015)	0.41 (0.019)	0.41 (0.012)
Males	0.22 (0.012)	0.29 (0.019)	0.25 (0.011)
Whole sample	0.32 (0.011)	0.36 (0.014)	0.33 (0.08)

Table 8.4 PSIN data: observed combinations of *e*-states and *f*-states, for women born in 1961

	S	U	M	C1	C2	C3	Total
E1	1546	7	50	23	26	108	1760
E2	2776	37	14	34	73	121	3055
E3	4167	384	127	78	59	19	4834
N	896	329	633	2318	4004	940	9120
WP	1357	569	866	1194	1570	343	5899
WF	8303	3359	5078	1042	349	24	18155
Total	19045	4685	6768	4689	6081	1555	42823

differences between cohorts that are the result of increasing complexity of the life course.

To gain some insights about the results found, the most frequent combinations of *f* and *e* states (for each gender–cohort combination) can be inspected using the contingency tables between the *f* and the *e* states collapsed across cases.

For example, in Table 8.4 the joint frequencies of the *e*- and *f*-states are reported. The last column and the last row ("Total") show the marginal counts of the states. Note that the grand total is the total number of observed paired *e* and *f* states, that is $(n \cdot T) = (187 \cdot 229) = 42{,}823$.

As expected, we see that the children and non-children states are often combined with specific states in the *e*-career. More precisely, having a partner but not children (states U and M) is often experienced together with full-time employment (WF). Instead, once motherhood is entered, the C-states are mostly associated to non-availability (N) in the labor market.

Weaker combinations are observed in Table 8.5, which pertains to men born in 1961. Here the distributions of the *f*-states, given the state they are combined to in the *e*-career are displayed. The last row ("Marg.") shows the marginal distributions of the *e*-states for these men.

Here, strong association is observed between being enrolled in any type of education and being single: these states are almost always experienced together. However, enrollment in education only makes up less than 25%

Table 8.5 PSIN data: distributions (on the rows) of the *f*-states. Given the *e*-careers' state, for men born in 1961

	S	U	M	C1	C2	C3	Marg.
E1	0.98	0.01	0.00	0	0.01	0	0.04
E2	0.96	0.02	0.02	0.01	0.00	0	0.08
E3	0.93	0.04	0.01	0.02	0.01	0	0.12
N	0.56	0.10	0.07	0.12	0.14	0.02	0.07
WP	0.60	0.11	0.04	0.12	0.09	0.05	0.02
WF	0.41	0.16	0.21	0.10	0.10	0.03	0.67

of all observed *e*-states. The most relevant *e*-state is full-time employment (WF), which comprises 67% of the observed *e*-states. Precisely, conditioned on being in that state, the distribution of the non-empty *f*-states is quite flat. As mentioned above, Table 8.5 is collapsed across cases, so that it is not possible to make reasonable guesses on the distributions of the *f* and the *e* states for the individuals. Nonetheless, from the table one may guess that after education most of the men work full time (since this state is the most visited one) and they build their family (and thus experience the *e*-states) while remaining in this state, so that knowing the *e*-state a man is in, does not help to make any reasonable guess about the *f*-state. Also, working full-time is much less dependent on the status in the *f*-career for men than it is for women. This helps one to understand why the association between the two domains turns out to be lower for men.

Association as Reduction of Gini's Heterogeneity

Another index of individual-based association can be obtained by measuring the heterogeneity of a variable Y using Gini's heterogeneity index (Gini, 1912, 1939):

$$G(Y) = 1 - \sum_{j=1}^{J} p_j^2 = \sum_{j=1}^{J} p_j(1 - p_j) \tag{5}$$

instead of entropy. It is not difficult to see that $G(Y) = 0$ when all observations take the same value and that, when all of the labels are equally frequent, $G(Y) = \frac{J-1}{J}$. So, just like entropy H, dispersion G increases with the number of categories when the p_js are all equal.

As for the entropy, the Gini heterogeneity for Y can be calculated conditioned upon the categories of another variable, X, and $G(Y|X)$ can be derived by averaging. Applying (1), one obtains the so-called Goodman and Kruskal τ (Light and Margolin, 1971; Haberman, 1982; Agresti, 2002):

$$R_G(Y|X) = 1 - \frac{G(Y|X)}{G(Y)}, \tag{6}$$

the average fraction of total heterogeneity that is reduced by conditioning on X. Simple calculations lead to the following expression:

$$R_G(Y|X) = 1 - \frac{1 - \sum_{k=1}^{K} \sum_{j=1}^{J} p_{kj}^2 / p_k}{1 - \sum_{j=1}^{J} p_j^2}, \tag{7}$$

where K and J indicate the number of values taken by X and Y respectively.

Also in this case, the association measure can be derived for each sequence separately. For the toy sequences reported in Table 8.2, it is easy to obtain: $G(\mathbf{y}) = 0.62$, $G(\mathbf{y}|\mathbf{x}) = 0.41$, $R_G(\mathbf{y}|\mathbf{x}) = 0.34$, and $G(\mathbf{x}) = 0.58$, $G(\mathbf{x}|\mathbf{y}) = 0.39$, $R_G(\mathbf{x}|\mathbf{y}) = 0.32$. It can immediately be verified that the structure

Table 8.6 PSIN data: average of the individual $R_G(e|f)$ and $R_G(f|e)$ conditioned to gender and cohort (standard errors in parentheses)

$R_G(e|f)$	1961	1965	Whole sample
Females	0.44 (0.018)	0.38 (0.020)	0.42 (0.013)
Males	0.26 (0.017)	0.26 (0.018)	0.26 (0.012)
Whole sample	0.36 (0.013)	0.33 (0.014)	0.35 (0.010)

$R_G(f|e)$	1961	1965	Whole sample
Females	0.37 (0.015)	0.40 (0.019)	0.38 (0.012)
Males	0.21 (0.012)	0.28 (0.020)	0.24 (0.011)
Whole sample	0.29 (0.011)	0.34 (0.014)	0.31 (0.008)

of these numbers differs from that of the corresponding Shannon-based statistics, although $R_H(x|y) < R_H(y|x)$ and $R_G(x|y) < R_G(y|x)$.

In Table 8.6 we report the averages and the standard errors of the R_Gs obtained for all the individual pairs of f- and e-careers in the PSIN data conditioned to the combinations of gender and cohort.

Although R_G and R_H quantify different aspects of the data, comparing Table 8.3 with Table 8.6 reveals roughly the same structure of association coefficients.

Indeed, with R_G too, we detect appreciable differences in the strengths of association between men and women, but the differences between cohorts are not relevant, just as we saw with R_H. Clearly, the two measures behave differently, since they react differently to departure from independence. Even if we are here considering averages of the measures calculated for each individual separately, we can reasonably expect that R_G and R_H will mostly reveal similar patterns of association since $-p\log_2 p$, the building block of entropy, is monotone with $p(1-p)$, the brick of Gini's dispersion. However, R_H has the advantage of being well embedded in statistics and information theory (e.g., Kullback, 1959). Also, averaging over individuals in this case has a clear interpretation in the framework of loglinear models (Agresti, 2002). Of course, R_H and R_G are appropriate measures of association of life course sequences only if the underlying measures of variation are adequate. Although the Gini index has hardly been used (but see O'Rand, 1996) in life course research, entropy has often been used to characterize life course sequences. Good examples are in Billari (2001), Fussel (2005), Elzinga and Liefbroer (2007), and recently in Widmer and Ritschard (2009) and Morand and Toulemon (2009).

However, Elzinga (2010) argued that entropy and other dispersion measures cannot be optimal when focusing on sequences, since they are insensitive to order-differences. For example, for $x = aaabbb$ and $y = ababab$, we have that $H(x) = H(y)$ and $G(x) = G(y)$ while the different ordering of states in the two sequences may be quite significant and should influence the variation measure. A possible solution to this problem, as Elzinga argued, is to refer to the number of distinct *subsequences* as a basis for quantifying variation or complexity. This is considered in the next subsection.

Complexity Reduction: the Subsequence Excess

Following the ideas in Elzinga (2010), we introduce a coefficient of association based upon a measure of the variation depending on the number of distinct *subsequences*.

Given a T-long sequence $x = x_1 \ldots x_T$, defined over some alphabet $\mathcal{A}$, we say that a sequence u is a subsequence of x, denoted by $u \preceq x$, if all the states of u occur in x and in the same order. Note, however, that states which are contiguous in u need not be contiguous in x. For example, if $x = abac$, then a, aa, bc and abc are all subsequences of x. In particular, the empty sequence λ and the sequence x are also subsequences of x. Let $\phi(x)$ denote the number of distinct subsequences of x: $\phi(aaaa) = 5$ and $\phi(abcd) = 16$. Elzinga (2010) argued that $\phi(x)$ should be used as the basis to measure the variation or "complexity" of x. In general, for a T-long sequence, it is:

$$T + 1 \leq \phi(x) \leq 2^T, \tag{8}$$

but the upper boundary is not tight if the length $|x| = T$ exceeds the size $|\mathcal{A}|$ of the alphabet $\mathcal{A}$.[1] Therefore, Elzinga proposed to use

$$\log_2(T + 1) \leq C(x) = \log_2 \phi(x) \leq T \tag{9}$$

as a measure of the complexity of sequences (for details and the calculation of $\phi(x)$, refer to Elzinga et al., 2008; Elzinga, 2010).

When studying multiple life course domains, one often combines the characters of different alphabets to denote combinations of states from the different domains, as for example in Elzinga (2005), Aassve et al. (2007), or Pollock (2007). We call such a combination of sequences x and y the "mix" $\widehat{xy}$ of x and y : $\widehat{xy} = (x_1y_1)(x_2y_2)\ldots(x_Ty_T)$. $\widehat{xy}$ only exists if x and y have the same length, and $|\widehat{xy}| = |x| = |y|$. The alphabet underlying $\widehat{xy}$ consists of all the combinations of the elements taken from the alphabets used in the construction of x and y. For example, if x and y are constructed from the alphabets $\mathcal{A}_x = \{a, b, c\}$ and $\mathcal{A}_y = \{p, q, r\}$ respectively, the mix $\widehat{xy}$ would be constructed from the alphabet $\mathcal{A}_{xy} = \{ap, aq, ar, \ldots, cp, cq, cr\}$. Considering again the toy sequences used in the previous section, $x = (a/20; b/10; a/20; c/50)$ and $y = (p/30; r/10; q/20; r/40)$, one would obtain $\widehat{xy} = (ap/20; bp/10; ar/10; aq/10; cq/10; cr/40)$.

With these concepts and this notation, we return to quantifying the association between sequences. Within this framework perfect association between two sequences x and y occurs when one sequence is perfectly predictable from the other one and vice versa. Thus, any subsequence in x must be associated with one and only one subsequence in y, so that $\phi(x) = \phi(\widehat{xy}) = \phi(y)$. Note that for the considered toy sequences, some subsequences in x mix with more than one subsequence in y and vice versa. Hence, $\phi(\widehat{xy}) > \max\{\phi(x), \phi(y)\}$: x and y are thus not perfectly related. Instead, the two sequences $x = (a/20; b/10; a/20; c/40)$ and $y = (p/20; r/10; p/20; q/40)$, are perfectly related, since to each embedding of a particular subsequence in x one and only one particular subsequence in y is associated and vice versa. Since $\max[C(x), C(y)] \leq C(\widehat{xy})$, we propose

$$R_C(y|x) = \frac{C(y)}{C(\widehat{xy})} \tag{10}$$

as a measure of the proportional reduction of complexity that is attained by removing x from $\widehat{xy}$, an operation that is analogous to conditioning on x as in R_H or R_G.

When x is perfectly associated with y, no complexity will be added, yielding $C(y) = C(\widehat{xy})$ and hence $R_C(y|x) = 1$. Instead, if x and y are unrelated, $\widehat{xy}$ will turn out to be more complex than y, in which case $R_C(y|x)$ will be small.

To describe this measure in even more detail, remember that to calculate R_G or R_H we used a contingency table, defined on the alphabets $\mathcal{A}_x$ and $\mathcal{A}_y$, to derive marginal and conditional variation. We concluded that, for studying the association between sequences, this is not optimal since order effects are not accounted for. Now imagine that we construct a contingency table that is not defined just over the single elements of the alphabets but instead is constructed over the sets of all distinct subsequences of x and y, each row and each column representing a particular subsequence of x and y respectively. What we count with $\phi(\widehat{xy})$ is the number of non-empty cells in this table. There are a minimum amount of these: $\max\{\phi(x), \phi(y)\}$, and the excess of non-empty cells is caused by lack of association. So, it seems that we solved our problem—accounting for order-effects—by extending the contingency table based on the singletons, the elements of the alphabets, to a subsequence-contingency table based on all "tuples": we incorporated order into the contingency table. As an example in Table 8.7 we show a portion of the subsequence-contingency table for the sequences $x = abacba$ and $y = pqpqqq$ with $\phi(x) = 38$ and $\phi(y) = 22$.

One more "technical" issue merits attention. Note that since $|x| + 1 \leq \phi(x), R_C$ as defined in (10), has a positive lower bound being $\log_2(|x|+1) \leq C(x)$ (Elzinga et al., 2008). Therefore, we define

$$0 \leq R_C(y|x) = \frac{C(y) - \log_2(|x|+1)}{C(\widehat{xy}) - \log_2(|\widehat{xy}|+1)} \leq 1 \tag{11}$$

with the understanding that $|x| = |\widehat{xy}|$ by definition of a mix $\widehat{xy}$.

This definition of R_C was applied to calculating the association between all the individual pairs of e- and f-careers from the PSIN data. In Table 8.8

Table 8.7 Portion of the subsequence-contingency table for the sequences $x = abacba$ and $y = pqpqqq$

	p	q	pp	pq	qp	qq	ppp	...	pqp	...
a	1	1								
b		1								
c		1								
aa			1	1						
ab				1						
ac				1						
ba					1	1				
⋮										
aba									1	
⋮										

Table 8.8 PSIN data: average of the individual $R_C(e|f)$ and $R_C(f|e)$ conditioned to gender and cohort (standard errors in parentheses)

$R_C(e|f)$	1961	1965	Whole sample
Females	0.61 (0.014)	0.67 (0.015)	0.63 (0.011)
Males	0.56 (0.019)	0.67 (0.019)	0.61 (0.014)
Whole sample	0.59 (0.012)	0.67 (0.012)	0.62 (0.008)

$R_C(f|e)$	1961	1965	Whole sample
Females	0.55 (0.015)	0.47 (0.017)	0.52 (0.012)
Males	0.52 (0.019)	0.41 (0.019)	0.47 (0.014)
Whole sample	0.53 (0.012)	0.44 (0.013)	0.50 (0.009)

the averages of the R_Cs conditioned to the combinations of gender and cohort are reported, with their standard errors. The results differ substantially from those obtained with R_H and R_G. This is not so surprising, since R_C accounts for sequencing, while R_G and R_H focus on a quite different aspect, namely the state distribution within the sequences.

Using the subsequence approach, we observe relatively small differences in association between men and women. For both genders, high levels of association are observed. Also, we observe a reverse tendency of the coefficients relative to cohorts, as compared to those displayed in Tables 8.3 and 8.6. Over time, the conditional association of the e-career with the f-career increases, suggesting that events in the former become more strongly influenced by events in the latter. At the same time, the conditional association of the f-career with the e-career suggests that "progression" in family life is becoming less strongly dependent on developments in one's career.

Case-Bases Measures: A Perturbation Experiment

In this section we analyze the behavior of the case-based measures using simulated data. We start considering the four pairs of sequences in Figure 8.2. Note that the first two paired sequences, (x, y_a), are perfectly and symmetrically associated, since there is a one-to-one correspondence between their states and durations. In case (b) the last state in y_b is identical to the third visited state. As a consequence, while from x it is possible to perfectly predict y_b, the reverse does not hold, since to the same state in y_b two different states in x_b are associated. A similar situation also holds in cases when two adjacent states are merged into a single state or when one state is detailed into two different states. In case (c), y_c has the same number of states x but with different durations: the association between the two sequences should be lower. Lastly, in (d) y_d has been obtained by randomly permuting the states in y_a. In this situation, the association between x and y_d should be low in both directions. In Table 8.9, we report the values of the case-based measures for the pairs of sequences in Figure 8.2.

Table 8.9 Association between the pairs of sequences shown in Figure 8.2

Pair	$R_H(x\|y)$	$R_H(y\|x)$	$R_G(x\|y)$	$R_G(y\|x)$	$R_C(x\|y)$	$R_C(y\|x)$
(a)	1.000	1.000	1.000	1.000	1.000	1.000
(b)	0.755	1.000	0.648	1.000	1.000	0.991
(c)	0.837	0.831	0.774	0.761	0.756	0.760
(d)	0.029	0.029	0.025	0.025	0.185	0.971

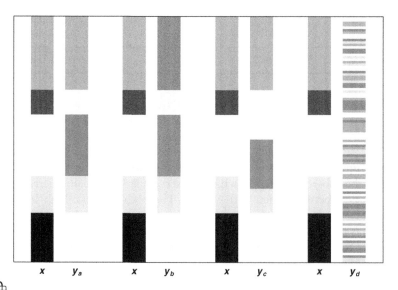

Figure 8.2 A synthetic career in the X-domain with four different synthetic careers in the Y-domain.

From the table, it is clear that R_H and R_G behave as expected. However, for R_C, the behavior in cases (b) and (d) is quite unexpected. Closer inspection reveals that $\phi(\mathbf{x}) = 2{,}978{,}976, \phi(\mathbf{y}_b) = 2{,}726{,}967$ and for the mix we find $\phi(\widehat{\mathbf{x}\mathbf{y}_b}) = \phi(\mathbf{y}_b)$. As for case (d), observe that the complexity of $\mathbf{y}_d$ is so high that the increase in complexity is beyond numerical precision when $\mathbf{y}_d$ is admixed to $\mathbf{x}$. Therefore, even if R_C is appealing from a theoretical point of view, its numerical behavior is not promising.

To compare the behavior of the case-based measures when gradually distorting perfect association, we generated five pairs of perfectly associated careers of 100 observations, and obtained five blocks of 20 identical sequences in the two domains by replicating them. The two panels in the first row of Figure 8.3 represent these $n = 100$ synthetic sequences. From the Y-sequences, we randomly selected a proportion π of the states from the $100T$ available and randomly permuted them. In the second and third row of Figure 8.3 we show the perturbed Y-sequences for some values of π to illustrate the amount of perturbation. The full report of this perturbation-experiment is given in Table 8.10 where the measures of the associations between the X-domain and the perturbed Y-domains are reported.

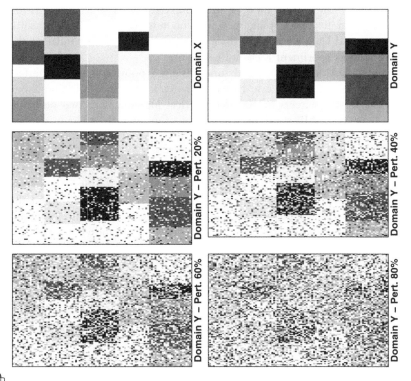

Figure 8.3 Two domains, X and Y. Y is subject to perturbations.

Table 8.10 Association between the X-domain and the perturbed Y. The amount of perturbation is determined by the parameter π, the proportion of random permutations of Y-states

π	$R_H(x|y)$	$R_H(y|x)$	$R_G(x|y)$	$R_G(y|x)$	$R_C(x|y)$	$R_C(y|x)$
0	1.0000	1.0000	1.0000	1.0000	1.0000	1.0000
0.1	0.8743	0.8067	0.8732	0.8038	0.4887	0.9925
0.2	0.7489	0.6521	0.7421	0.6323	0.3443	0.9900
0.3	0.6301	0.5259	0.6147	0.4879	0.2815	0.9917
0.4	0.5324	0.4297	0.5054	0.3727	0.2445	0.9909
0.5	0.4260	0.3334	0.3895	0.2633	0.2182	0.9907
0.6	0.3286	0.2524	0.2876	0.1782	0.2008	0.9910
0.7	0.2575	0.1939	0.2121	0.1194	0.1906	0.9911
0.8	0.1951	0.1456	0.1493	0.0760	0.1845	0.9908
0.9	0.1478	0.1097	0.1044	0.0490	0.1802	0.9912

Clearly, R_H and R_G behave similarly, even if R_G reacts more strongly to perturbation. As for R_C, although in a qualitative sense it behaves as it should, numerically it does not properly render the degree of association. Therefore, R_C should be rejected as a practical measure of case-based association.

Note that the significance of the observed values of the association measures can be tested using permutation tests (see e.g., Pesarin, 2001; Edington and Onghena, 2007) through a procedure that is similar to the perturbation-experiment reported here.

Before concluding it is worth underlining a point concerning perfect association as detected by case-base criteria. Refer to the SIPs in Figure 8.4, representing the careers of 100 hypothetical individuals in two domains. In the upper-left SIP, we show five groups of individual careers in domain X and within each group the careers are identical. In the upper-right panel, we present the careers of the same individuals, identically ordered, in domain Y. Clearly, the panels show that transitions in domain X and in domain Y are, always and for all individuals, simultaneous. So, *within* cases, association is perfect. This is perfectly reasonable, since as can be noted, the association between individual sequences in the two domains is *identical* for all the sample cases, and consequently the average of the individual measures describes in a satisfactory manner the (common) association. Also, note that knowing the career in one domain allows us to perfectly predict the career in the other domain.

The lower-left SIP is identical to the upper-left one. In the lower-right SIP, we show a permutation of the careers shown in the upper-right SIP. Now, the association between domains is still perfect at the individual level, since the transitions in one domain are simultaneous with the transitions in the other domain. Also, even in this case the average of the individual association indices provides good indications about the (common) underlying association. However, having observed a sequence in one domain does not provide us with any clue about the career in the

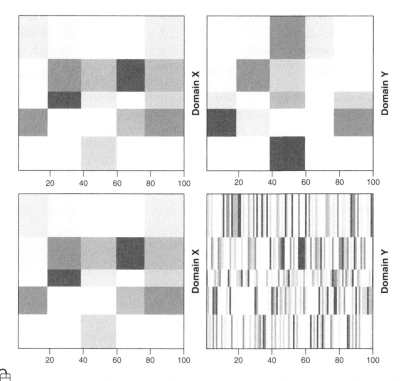

Figure 8.4 SIPs of two pairs of synthetic careers on domains X and Y. In both
pairs the timing of the transitions is identical. In the bottom pair, the
Y-sequences of the top pair have been permuted.

other domain. So, association as a measure of predictability may completely
fail if confined to case-based association.

Therefore, in the next section we consider additional types of association,
where the relation between domains is *modeled* by extending regression
concepts to the dissimilarity case.

Modeling the Relation Between Domains

From a statistical point of view, we are usually concerned with measures
of association properly quantifying if and to what extent information on a
"response" phenomenon can be improved based on an "explanatory" one.
This type of association can only be measured by focusing not on the
single individuals but, rather, on individuals having similar characteristics
with respect to the explanatory phenomenon. To better understand the
measures we will describe in a moment, it is worth better clarifying the
difference between global and local measures.

Consider a dependent variable Y explained using another variable X
through a model $Y = m(X) + \epsilon$, where $m(\cdot)$ is some regression function

and has to be estimated. A *global* approach assumes that the relation between Y and X can be well approximated by the same function for each X-value: therefore, for the *i*th X-value it is $y_i = m(x_i) + \epsilon_i$. When a linear model is used to describe the relation, it is $m(X) = a + bX$ and the strength of association can be measured through correlation.

In some cases the relation between Y and X is complex, so that the assumption of a linear relation or even of a monotone relation is not realistic and it is impossible or difficult to individuate a unique suitable function. *Smoothed* (or *local*) regression models are fitted to data by smoothing: $m(x_i) = m_i(x_i)$ is the value of a function fitted only to those observations "close" to x_i in a sense which will be detailed later. As is evident, this approach is more flexible, and allows the form of the relation to vary for each data point. In this section we describe both global and local approaches and discuss their characteristics.

Global Association Measures

As for the *global* approach, we here focus on linear or monotone relations. We say that sequences from a domain X are globally associated to sequences from a domain Y when: (i) cases that have sequences that are close or similar in one domain, have sequences that are close or similar in the other domain too; (ii) cases with very remote or dissimilar sequences in one domain, have sequences that are very remote or dissimilar in the other domain too.

Clearly, given cases with sequences in two domains, X and Y, we first have to establish distance or similarity between sequences. For the relation between distance and similarity, the reader is referred to Chen et al. (2009). Distances between sequences can be generated in many ways. In the social sciences, OM (see e.g., Abbott, 1995; Abbott and Forrest, 1986; Abbott and Hrychak, 1990; Anyadike-Danes and McVicar, 2010) or one of its variants (Halpin, 2010; Hollister, 2009; Lesnard, 2010) is used, but alternatives have also been successfully applied (Elzinga and Liefbroer, 2007; Bras et al., 2010). For a given pair of cases, say the *i*th and the *j*th, the distances (or dissimilarities) in the Y and the X domains, $d_Y(i,j)$ and $d_X(i,j)$ are obtained. All the pair-wise distances can be arranged into the two distance matrices $\mathbf{D}_X$ and $\mathbf{D}_Y$. As mentioned before, we are interested in measuring the extent of association between these two matrices or between their elements. Fortunately, several coefficients have been introduced with this aim (Abdi, 2007, 2010). Even if it is obvious, it is worth emphasizing from the very beginning that the association measures we are going to present *strongly depend upon the criterion chosen to measure the dissimilarities.*

A very intuitive index which measures linear association is the Mantel coefficient, R_M, introduced by Mantel (1967) and often used in ecology (see e.g., Legendre and Legendre, 1998; Manly, 1997). For two given matrices, $\mathbf{D}_X$ and $\mathbf{D}_Y$, R_M is defined as the correlation between their off-diagonal elements. Following the same logic, it is also possible to explore the existence of monotone relations between the dissimilarities over the two domains, avoiding the reference to a linear relation, which might be too strong an assumption when sequence-based dissimilarities are considered. This can be done by using rank-based correlation coefficients, for example the Kendall or the Spearman coefficient.

Another index specifically designed to measure the agreement between two dissimilarity matrices is the Escoufier coefficient R_V (see Escoufier, 1973; Robert and Escoufier, 1976). To introduce the coefficient, we start by considering the case when it is of interest to measure the dissimilarity between two *data* matrices, $\mathbf{X}$ and $\mathbf{Y}$, with n rows and P and Q columns respectively. For the sake of simplicity and without loss of generality, assume $\mathbf{X}$ and $\mathbf{Y}$ to be centered to have null centroids. Let $\mathbf{S}_X = \mathbf{X}^{\mathrm{T}}\mathbf{X}/n$ and $\mathbf{S}_Y = \mathbf{Y}^{\mathrm{T}}\mathbf{Y}/n$ denote the sample variance and covariance matrices of $\mathbf{X}$ and $\mathbf{Y}$ respectively, and $\mathbf{S}_{XY} = \mathbf{X}^{\mathrm{T}}\mathbf{Y}/n$ be the matrix whose elements are the sample covariances between the columns of $\mathbf{X}$ and those of $\mathbf{Y}$. Finally, let $\mathbf{C}_X = \mathbf{X}\mathbf{X}^{\mathrm{T}}$ and $\mathbf{C}_Y = \mathbf{Y}\mathbf{Y}^{\mathrm{T}}$ denote the matrices of cross-products of the two data matrices. The R_V coefficient is defined as:

$$0 \leq R_V(\mathbf{X}, \mathbf{Y}) = \frac{\mathrm{tr}(\mathbf{S}_{XY}\mathbf{S}_{YX})}{\sqrt{\mathrm{tr}(\mathbf{S}_X^2)\mathrm{tr}(\mathbf{S}_Y^2)}} \leq 1, \tag{12}$$

wherein tr denotes the trace, i.e. the sum of the diagonal elements of a square matrix. Note that geometrically, the R_V-coefficient is a generalized cosine between two vector spaces, i.e. it generalizes the familiar Pearson correlation coefficient for two variables.

To extend R_V to the case when dissimilarity matrices are considered, note first that simple calculations lead to the following alternative definition of R_V:

$$R_V(\mathbf{X}, \mathbf{Y}) = \frac{\mathrm{tr}(\mathbf{C}_X\mathbf{C}_Y)}{\sqrt{\mathrm{tr}(\mathbf{C}_X^2)\mathrm{tr}(\mathbf{C}_Y^2)}} \tag{13}$$

It has been shown (Mardia et al., 1979; Shawe–Taylor and Christianini, 2004) that if $\mathbf{D}_X$ and $\mathbf{D}_Y$ are Euclidean (i.e. resulting from inner products of vectors in a linear space), $\mathbf{C}_X = (-1/2)\mathbf{H}\mathbf{D}_X^2\mathbf{H}$, where $\mathbf{H} = \mathbf{I} - (1/n)\mathbf{1}\mathbf{1}^{\mathrm{T}}$, $\mathbf{I}$ is an $(n \times n)$ identity matrix, and $\mathbf{1}$ is an $(n \times 1)$ vector of "1"s (a similar result holds for $\mathbf{D}_Y$). Pre-multiplying $\mathbf{D}_X^2$ by $\mathbf{H}$ amounts to centering its columns and post-multiplying with $\mathbf{H}$ results in centering its rows. The R_V coefficient can thus be expressed in terms of the *doubly centered* distance matrices $\widetilde{\mathbf{D}}_X^2 = \mathbf{H}\mathbf{D}_X^2\mathbf{H}$ and $\widetilde{\mathbf{D}}_Y^2$:

$$R_V(\mathbf{D}_X, \mathbf{D}_Y) = \frac{\mathrm{tr}(\widetilde{\mathbf{D}}_X^2\widetilde{\mathbf{D}}_Y^2)}{\sqrt{\mathrm{tr}(\widetilde{\mathbf{D}}_X^2)^2\mathrm{tr}(\widetilde{\mathbf{D}}_X^2)^2}}. \tag{14}$$

The above derivations are valid only when the distances in $\mathbf{D}_X$ and $\mathbf{D}_Y$ are Euclidean, implying that the (unobserved) matrices $\mathbf{X}$ and $\mathbf{Y}$ consist of vector-representations of the sequences. Unfortunately, not all the sequence metrics meet this requirement. In particular, the OM-metric does not derive from a vector space. However, the subsequence-based metrics proposed by Elzinga (2003, 2005, 2010), by Wang and Lin (2007) or by Shawe-Taylor and Christianini (2004) derive from vector-representations. Hence, Escoufier's R_V could be used to express similarity between distance matrices as result of such metrics. Furthermore, note that $\widetilde{\mathbf{D}}_X^2 \widetilde{\mathbf{D}}_Y^2$ is a matrix whose elements are the inner products between the rows of $\mathbf{D}_X^2$ and the columns of $\widetilde{\mathbf{D}}_Y^2$. Since these matrices are both double-centered, the ith element on the main diagonal of $\widetilde{\mathbf{D}}_X^2 \widetilde{\mathbf{D}}_Y^2$ will be the *covariance* between the vector of the (squared) dissimilarities between the ith case and all the others in the X domain and the same vector in the Y domain. By adding these covariances for all the n cases, one obtains the numerator of R_V. Similarly, the quantities at the denominator represent the sum of the variances of the columns of the dissimilarity matrices in the two domains. R_V is thus related to the amount of *linear* association between the dissimilarities in the two domains individual (column) by individual.

The mentioned indices were calculated for PSIN data. The normalized distances between the e-careers and those between the f-careers were calculated using a method proposed by Elzinga (2005). In Table 8.11 we present the results obtained for R_M and R_V. Bootstrap was applied to estimate the standard error of the statistics. For the sake of completeness, the bootstrap estimates of the statistics are also reported in the table.

Note that R_M is quite small and that only for women is a weak association found. This is because linear association between dissimilarities is the most restrictive association structure. Also, it has to be stressed that

Table 8.11 $R_M(\mathbf{D}_e, \mathbf{D}_f)$ and $R_V(\mathbf{D}_e, \mathbf{D}_f)$, applied to the PSIN data and conditioned to gender and cohort (bootstrap estimates of the statistics and of their standard errors in parentheses)

$R_M(\mathbf{D}_e, \mathbf{D}_f)$	*1961*	*1965*	*Whole sample*
Females	0.18 (0.26; 0.03)	0.13 (0.25; 0.03)	0.15 (0.21; 0.02)
Males	0.06 (0.13; 0.03)	−0.05 (0.06; 0.03)	0.01 (0.06; 0.02)
Whole sample	0.09 (0.13; 0.02)	0.07 (0.14; 0.02)	0.08 (0.12; 0.02)

$R_V(\mathbf{D}_e, \mathbf{D}_f)$	*1961*	*1965*	*Whole sample*
Females	0.30 (0.37; 0.02)	0.26 (0.36; 0.02)	0.22 (0.28; 0.02)
Males	0.13 (0.21; 0.02)	0.17 (0.27; 0.03)	0.09 (0.15; 0.01)
Whole sample	0.14 (0.20; 0.02)	0.14 (0.21; 0.02)	0.11 (0.19; 0.02)

the particular distance metric chosen affects the value of the two indices. Actually, Elzinga's distance measure is based upon the number of common subsequences. Thus, if two individuals are characterized by similar "long term" trajectories, but one experiences short spells in a given state not visited by the other individual, the distance metric might turn out to be relevant too.

Also for these measures, permutation tests can be used to evaluate their significance. Usually (see Abdi, 2007, for a discussion), one dissimilarity matrix, say $\mathbf{D}_X$ is left unchanged, and a permuted dissimilarity matrix $\mathbf{D}_Y^\pi$ is obtained by randomly permuting the rows (and columns) of the original matrix $\mathbf{D}_Y$, so as to remove the possible association between dissimilarities. Association is then measured between $\mathbf{D}_X$ and $\mathbf{D}_Y^\pi$. This procedure can be repeated many times to generate the independence distribution, and the permutation p-value can be obtained as the proportion of permuted matrices with a level of association exceeding the observed value.

To evaluate whether this procedure is sensible, later we will evaluate how R_M and R_V react to deviations from the situation of perfect (global) association (see Figure 8.7 and Table 8.13 below).

As stated at the beginning of this section, global association is a very restrictive condition because of the monotonicity requirement on the distances. Such a requirement may not be realistic; for example, we expect to see women with very dissimilar family formation histories have the same career in the education/work domain. Therefore, it is interesting to study local, non-monotone association in the next section.

Local Association Measures

If there is association between domains, one can reasonably expect that cases close in one domain are close also in the other. However, non-monotone association does not imply the same relation holds for cases which are remote in one of the domains. Local association measures are based upon the "closeness principle," and aim at evaluating only whether X-similar sequences are also Y-similar. To apply this principle to distances and sequences, we (but see also Piccarreta, 2012) extend the ideas underlying smoothed models (see Härdle, 1992, for a review) to distances. Consider a model relating a dependent variable Y to another variable X, $Y = m(X) + \epsilon$, where $m(\cdot)$ has to be estimated. Assuming that cases similar with respect to X are also similar with respect to Y, the fit for the ith case, $m_i = m(x_i)$, can be estimated using a "local average" of the Y-values characterizing cases close to x_i. Of course, there are many definitions of "local." One possibility is to average over cases falling in a neighborhood around x_i. One popular approach to define the neighborhood of x_i consists of selecting the k nearest neighbors of x_i. Another obvious possibility is to fix a ball with radius, r, and only use the values in the ball, i.e. those that are closer than r to x_i.

To analyze the association between two domains Y and X using the information in the dissimilarity matrices $\mathbf{D}_X$ and $\mathbf{D}_Y$, we proceed as follows. For the ith case, we define the set N_i of its neighbors in domain X, using one of the two approaches described above. However, in general, it is not possible to determine the *average* of the Y-values for cases in N_i. Nonetheless, for any metric, it is possible to determine the *medoid* associated to N_i on domain Y. The medoid is defined as the most centrally located case in N_i, i.e. having the minimum distance from all other cases: $m_i = \arg\min_{j \in N_i} \{\sum_{i,j \in N_i} d_Y(i,j)\}$. The medoid is always a member of the set N_i, is a good representative of a set of cases (see e.g., Kauffman and Rousseeuw, 1990), and, unlike the mean, it only requires a proper metric. If Y is associated to X, then, according to the closeness principle, $d_Y(i, m_i)$ should be small since the objects in N_i are close to i in X.

Note that if one selects $k = n$, so that each N_i contains *all* the cases, then the "prediction" would coincide with the medoid of the Y-sequences, $\bar{m}$. This is the "prediction" one would obtain using only the marginal information about domain Y without taking domain X into account. Hence we propose (see also Piccarreta, 2007) to quantify association as:

$$R_P(\mathbf{D}_Y | \mathbf{D}_X) = 1 - \frac{\sum_i d_Y(i, m_i)}{\sum_i d_Y(i, \bar{m})}. \tag{15}$$

Observe that this index is the proportional reduction in variation when using the knowledge of X to "predict" Y, exactly as in Equation (1). Clearly, the index is asymmetric and the more "tight" the X-neighborhoods N_i are in domain Y, the more Y is associated to X.

So far, we have hardly discussed how to construct the neighborhood sets $N_i(k)$, the set of k nearest neighbors of case i, or $N_i(r)$, the set of points in a ball of radius r, centered at case i. Constructing such sets is, given a distance matrix, easy enough. The question is then what value of k or r to chose. Here we discuss some possibilities.

To optimize k, we advocate a "leave-one-out" cross-validation. Suppose we have a set $N_i(k)$ for some k and let m_i^* be the medoid of $(N_i(k) \setminus i)$, i.e. the medoid determined leaving out case i. Then the distance $d_Y(m_i^*, i)$ can be interpreted as the "prediction error" of $N_i(k)$ with respect to case i. For given k, we can calculate the total cross-validation error $CV(k) = \sum_i d_Y(m_i^*, i)$. Now k may be chosen such that $CV(k)$ is minimized. A similar procedure can be applied to determine r.

Clearly, the above procedure will generate one unique value of k for all i. A more flexible approach can be defined by combining the nearest neighbors and the enclosing-ball criterion. For example, given an optimal k, select $r_i = \max_{j \in N_i(k)} d_X(i,j)$: this will produce a case-specific volume r_i. Of course, k, could also be optimized *per case* through the cross-validation procedure as described above. For a full and detailed discussion of such methods, the reader is referred to Boj et al. (2010) or Friedman (1984).

Table 8.12 $R_P(\mathbf{D}_e|\mathbf{D}_f)$ and $R_P(\mathbf{D}_f|\mathbf{D}_e)$, Calculated for the PSIN data and conditioned on gender and cohort (bootstrap estimates of the statistics and of their standard error in parentheses)

$R_P(\mathbf{D}_e|\mathbf{D}_f)$	1961	1965	Whole sample
Females	0.29 (0.46; 0.04)	0.22 (0.42; 0.04)	0.26 (0.43; 0.03)
Males	0.23 (0.39; 0.04)	0.23 (0.39; 0.05)	0.21 (0.39; 0.03)
Whole sample	0.21 (0.41; 0.03)	0.21 (0.39; 0.03)	0.23 (0.39; 0.04)

$R_P(\mathbf{D}_f|\mathbf{D}_e)$	1961	1965	Whole sample
Females	0.37 (0.54; 0.04)	0.34 (0.52; 0.05)	0.41 (0.55; 0.03)
Males	0.27 (0.47; 0.05)	0.28 (0.47; 0.06)	0.32 (0.49; 0.03)
Whole sample	0.31 (0.50; 0.03)	0.29 (0.50; 0.04)	0.36 (0.51; 0.04)

To calculate R_P for PSIN data, we used again the distance matrices obtained using the approach in Elzinga (2005), and considered a combination of the nearest neighbors and the enclosing-ball criterion, by allowing both k and r to vary across cases. The results are reported in Table 8.12. Also in this case bootstrap was used to estimate the statistics' standard errors; for the sake of completeness the bootstrap estimates are also reported in the table.

Observe that, as already found with case-based measures, the association of the f-careers to the e-careers is stronger than the reverse.

Again, we find that this association is much stronger for women than for men. To see why, we refer to Figures 8.5 and 8.6 where we present the SIPs for the separate gender-groups. In the first figure, we report the SIPs of the trajectories observed in the e- and in the f-domain for women born in 1961. The careers in the two domains are both ordered according to the first MDS factor characterizing $\mathbf{D}_e$. In this way, the e-sequences have a reasonable ordering and it is easier to inspect which f-careers are associated to similar e-careers. Note that women working full time (mostly on the right side of the plot) tend to have fewer children, whereas women that work part-time or not at all, tend to have more children. Hence, women with similar e-careers are often characterized by similar f-careers.

In Figure 8.6, the SIPs are shown for men born in 1961. Here, the sequences are ordered according to the first MDS factor extracted by $\mathbf{D}_f$ so as to explore if and to what extent men with similar f-careers tend to have similar e-careers. Observe that the work careers for the men are all dominated by WF (working full time), irrespective of the composition of their family. This explains the low value of the R_P index. However, the association is not close to zero, since men who are single or cohabit are

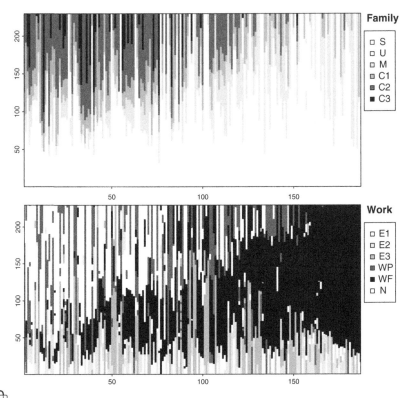

Figure 8.5 SIPs for the women born in 1961. Sequences are ordered according to the score on the first MDS factor extracted from $\mathbf{D}_e$.

those with the highest durations in the higher levels of education. So it seems as if these men decided to postpone parenthood.

A Permutation Experiment

For global and local measures it is important also to evaluate their behavior when gradually distorting perfect local association. With this aim, similarly to what was done in in Figure 8.3, we generated five pairs of perfectly associated careers of 100 observations, and obtained five blocks of 20 identical sequences in the two domains by replicating them. The two panels in the first row of Figure 8.7 represent these $n = 100$ synthetic sequences. From the Y-sequences, we randomly selected a proportion π of the $n = 100$ *cases* and randomly permuted them (please note that this is equivalent to permuting a sample of rows and columns of the distance matrix $\mathbf{D}_Y$), thus disrupting the perfect initial association between the two domains. In the second and third row of Figure 8.7 we show the perturbed Y-sequences

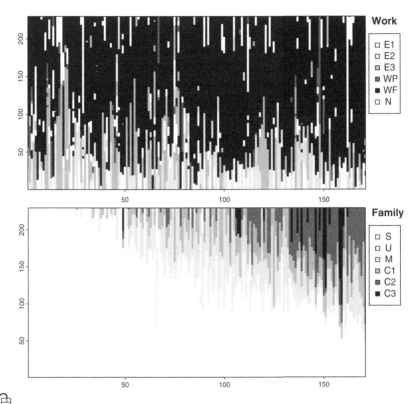

Figure 8.6 SIPs for men born in 1961. Sequences are ordered according to the score on the first MDS factor extracted from $\mathbf{D}_f$.

for some values of π to illustrate the amount of perturbation. Table 8.13 reports the complete results for for the global and the local association measures, R_M, R_V, and R_P. Note that the former measures assume values lower than 1 when no perturbation is applied to the data, since the original sequences are not globally perfectly associated (i.e., there is not a perfect linear association between dissimilarities). Nonetheless, we can observe that all the measures behave in the expected way, decreasing as the amount of perturbation increases. Permutation tests similar to those described for the global association indices can thus be used to verify the significance of local association.

Conclusion

In this chapter, we have explored three different kinds of association: case-based, local and global association. Case-based association relies on contingencies between states: conditioning on a state in one domain

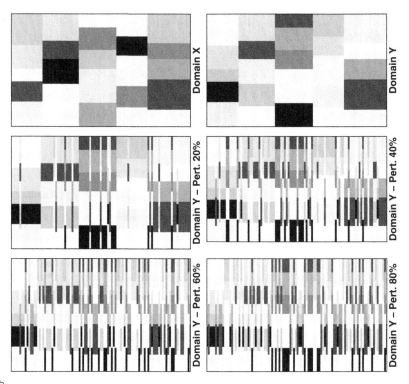

Figure 8.7 Two domains, X and Y. Y is subject to perturbations.

Table 8.13 Association between the X-domain and the perturbed Y. The amount of perturbation is determined by the parameter π; the proportion of cases permutated in the Y-domain

π	R_M	R_V	$R_P(x\|y)$	$R_P(y\|x)$
0	0.8724	0.9122	1.0000	1.0000
0.1	0.7844	0.8240	0.9500	0.9422
0.2	0.5622	0.6025	0.7973	0.7865
0.3	0.4497	0.4873	0.7196	0.6826
0.4	0.2845	0.3220	0.5439	0.5394
0.5	0.2177	0.2585	0.4620	0.4379
0.6	0.1100	0.1464	0.3502	0.3109
0.7	0.0622	0.0948	0.2131	0.2397
0.8	0.0669	0.1014	0.1937	0.2152
0.9	−0.0204	0.0190	0.0417	0.0318

reduces variation (i.e. increases predictability, of the states in the other sequence). Such an interpretation quite naturally leads us to use entropy and conditional entropy to quantify variation, and to use these to express association as the proportion of variance reduction due to conditioning. Hence, coefficients like R_H and R_G rely on contingencies of states. In the context of sequences, such coefficients suffer from the disadvantage that they are fully insensitive to the sequential nature of the data: different ordering of the same states does not affect their abundance, and hence does not affect (conditional) entropy and thus cannot influence the value of R_H or R_G. Thus, in a sequence-context, such measures are not very useful for association of sequences. Therefore, we focussed on the intuition that, when sequences are associated, subsequences in one domain must be contingent upon subsequences from the other domain. This in fact generalizes the notion of contingencies between simple states. Using complexity C instead of an entropy-like measure of variation thus emphasizes the sequential nature of the data. However, the numerical behavior of R_C turned out to be quite remote from our intuitions about strength of association. Hence we conclude that case-based association measures are not very useful in the context of mining for association between sequences.

Next we tried to use distance or dissimilarity data as the basis of quantifying association. We analyzed both global and local measures of association. We emphasized that these two types of measures detect different types of association. In particular, the value taken by the Mantel's and the Escoufier's coefficients are inadequate when the analysis of careers is taken into account, since the assumption of monotone relation between dissimilarities is a strong one, and one might reasonably expect a more complex (and more difficult to describe or to guess) relation between two domains. Nonetheless, on the basis of our simulations we noted that both the local and the global measures reasonably react to deviations from the situation of perfect association and can thus both be considered reliable criteria for quantifying the type of association they are focused on.

Notes

Part of this work was carried out in the Fall of 2007 when the first author was a visiting scientist at the Department of Social Research Methodology of the VU University of Amsterdam.

1 We use the notation $|\cdot|$ to denote both the size of a set—in which case the argument is in upper case—and the length of a (sub-)sequence—and then the argument is in lower case. This is quite standard in the literature on strings as in Crochemore et al. (2007).

References

Aassve, A., F. C. Billari, and R. Piccarreta (2007). "Strings of Adulthood: A Sequence Analysis of Young British Women's Work-Family Trajectories." *European Journal of Population* 23:369–388.

Abbott, A. (1995). "Sequence Analysis. New methods for Old Ideas." *Annual Review of Sociology* 21:93–113.

Abbott, A. and J. Forrest (1986). "Optimal Matching Methods for Historical Sequences." *Journal of Interdisciplinary History* 15:471–491.

Abbott, A. and A. Hrychak (1990). "Measuring Resemblance in Sequence Data: An Optimal Matching Analysis of Musicians' Careers." *American Journal of Sociology* 96:144–185.

Abdi, H. (2007). "RV Coefficient and Congruence Coefficient." In *Encyclopedia of Measurement and Statistics*, edited by Neil Salkind, pp. 849–853. Sage, Thousand Oaks (CA).

Abdi, H. (2010). "Congruence: Congruence coefficient, *RV* coefficient, and Mantel coefficient." In *Encyclopedia of Research Design*, edited by Neil Salkind, pp. 222–229. Sage, Thousand Oaks (CA).

Agresti, A. (2002). *Categorical Data Analysis*. Wiley Series in Probability and Statistics. Wiley Interscience, New York, second edition.

Anyadike-Danes, M. and D. McVicar (2010). "My Brilliant Career: Characterizing the Early Labor Market Trajectories of British Women From Generation X." *Sociological Methods & Research* 38:482–512.

Billari, F. (2001). "The Analyses of Early Life Courses: Complex Descriptions of the Transition to Adulthood." *Journal of Population Research* 18(2):119–142.

Blossfeld, H. P. and J. Huinink (1991). "Human Capital Investments or Norms of Role Transition? How Womens' Schooling and Career Affect the Process of Family Formation." *American Journal of Sociology* 97(1):143–168.

Boj, E., P. Delicado, and J. Fortiana (2010). "Distance-based Local Linear Regression for Functional Predictors." *Computational Statistics and Data Analysis* 54(2): 429–437.

Bras, H., A. C. Liefbroer, and C. H. Elzinga (2010). "Standardization of Pathways to Adulthood? An Analysis of Dutch Cohorts Born Between 1850 and 1900." *Demography* 47:1013–1034.

Chen, S., B. Ma, and K. Zhang (2009). "On the Similarity Metric and the Distance Metric." *Theoretical Computer Science* 410:2365–2376.

Cover, T. M. and J. A. Thomas (1991). *Elements of Information Theory*. Wiley Series in Telecommunications. Wiley, New York.

Crochemore, M., C. Hancart, and T. Lecroq (2007). *Algorithms on Strings*. Cambridge University Press, Cambridge, UK.

Edington, E. S. and P. Onghena (2007). *Randomization Tests*. Chapman & Hall, London.

Elzinga, C. H. (2003). "Sequence Similarity – A Non-Aligning Technique." *Sociological Methods & Research* 31:3–29.

Elzinga, C. H. (2005). "Combinatorial Representation of Token Sequences." *Journal of Classification* 22:87–118.

Elzinga, C. H. (2010). "Complexity in Categorical Time Series." *Sociological Methods & Research* 38:463–481.

Elzinga, C. H. and A. C. Liefbroer (2007). "De-Standardization and Differentiation of Family Life Trajectories of Young Adults: A Cross-National Comparison Using Sequence Analysis." *European Journal of Population* 23:225–250.

Elzinga, C., S. Rahmann, and H. Wang (2008). "Algorithms for Subsequence Combinatorics." *Theoretical Computer Science* 409:394–404.

Escoufier, Y. (1973). "Le traitement des variables vectorielles." *Biometrics* 29: 751–760.

Friedman, J. H. (1984). "A variable span smoother." Technical report, Dept. of Statistics, Stanford University.

Fussel, E. (2005). "Measuring the Early Adult Life Course in Mexico: An Application of the Entropy Index." In *The Structure of the Life Course: Standardized? Individualized? Differentiated? Advances in Life Course Research,* Volume 9, edited by R. Macmillan, pp. 91–122. Elsevier, Amsterdam.

Gabadinho, A., G. Ritschard, N. Müller, and M. Studer (2011). "Analyzing and visualizing state sequences in R with TraMineR." *Journal of Statistical Software* 40:1–37.

Gauthier, J.-A., E. D. Widmer, P. Bucher, and C. Notredame (2010). "Multichannel Sequence Analysis Applied to Social Science Data." *Sociological Methodology* 40:1–38.

Gini, C. (1912). *Variabilità e Mutabilità, contributo allo studio delle distribuzionie relazioni statistiche.* Cuppini, Bologna.

Gini, C. (1939). *Variabilità e Concentrazione,* volume 1 of *Memorie de metodologica statistica.* Giuffrè, Milano.

Guttman, L. (1941). "An Outline of the Theory of Prediction." In *The Prediction of Personal Adjustment,* edited by Paul Horst, pp. 261–262. Social Science Research Council, Bulletin 48, New York.

Haberman, S.J. (1982). "Analysis of Dispersion of Multinomial Responses." *Journal of the American Statistical Association* 77:568–580.

Halpin, B. (2010). "Optimal Matching Analysis and Life-Course Data: The Importance of Duration." *Sociological Methods & Research* 38:365–388.

Härdle, W. (1992). *Applied Nonparametric Regression.* Cambridge University Press, Cambridge.

Hollister, M. N. (2009). "Is Optimal Matching Sub-Optimal?" *Sociological Methods & Research* 38:235–264.

Kauffman, L. and P. J. Rousseeuw (1990). *Finding Groups in Data.* John Wiley and Sons, New York.

Kullback, S. (1959). *Information Theory and Statistics.* Wiley, New York.

Legendre, P. and L. Legendre (1998). *Numerical Ecology.* Elsevier Science, Amsterdam, 2nd edition.

Lesnard, L. (2010). "Setting Cost in Optimal Matching to Uncover Contemporaneous Socio-Temporal Patterns." *Sociological Methods & Research* 38:389–419.

Liefbroer, A. C. and M. Corijn (1999). "Who, What, Where and When? Specifying the Impact of Educational Attainment and Labor Force Participation on Family Formation." *European Journal of Population* 15(1):45–75.

Liefbroer, A. C. and C. H. Elzinga (2012). "Intergenerational Transmission of Behavioral Patterns: How Similar are Parents' and Children's Demographic Trajectories?" *Advances in Life Course Research* 17(1):1–10.

Liefbroer, A. C. and M. Kalmijn (1997). *Panel Study of Social Integration in The Netherlands 1987–1995 (PSIN8795) Codebook.* ICS Occasional Papers and Documents Series (ICS Code Books-30). Interuniversity Center for Social Science Theory and Methodology, Utrecht.

Light, R. J. and B. H. Margolin (1971). "An Analysis of Variance for Categorical Data." *Journal of the American Statistical Association* 66:534–544.

Manly, B. F. J. (1997). *Randomization, Bootstrap and Monte Carlo Methods in Biology.* Chapman and Hall, London, 2nd edition.

Mantel, N. (1967). "The Detection of Disease Clustering and a Generalized Regression Approach." *Cancer Research* 27:209–220.

Mardia, K. V., J. T. Kent, and J. M. Bibby (1979). *Multivariate Analysis*. Academic Press, London.

Mooi-Reci, I. (2008). *Unemployed and Scarred for Life. Longitudinal Analysis of How Unemploymentand Policy Changes Affect Re-Employment Careers and Wages in The Netherlands, 1980–2000*. Ph.D. thesis, VU University, Amsterdam.

Morand, E. and L. Toulemon (2009). "Analyse des séquences par optimal matching: le passage à l'âge adulte des femmes et des hommes en France." In *Xème Journées de Méthodologie Statistique de l'Insee, 23–25 mars 2009*.

O'Rand, A. M. (1996). "The Precious and the Precocious: Understanding Cumulative Disadvantage and Cumulative Advantage Over the Life Course." *The Gerontologist* 36:230–238.

Pesarin, F. (2001). *Multivariate Permutation Tests with Applications in Biostatistics*. John Wiley and Sons, Chichester.

Piccarreta, R. (2007). "Exploring the Relationships between Life Course Domains: An Approach Based on Sequence Analysis." Department of Decision Sciences, Dondena Centre for Social Dynamics, Bocconi University, Milan, Italy.

Picarreta, R. (2012). "Graphical and Smoothing Techniques for Sequence Analysis." *Sociological Methods & Reserach* 41(2):362–380.

Piccarreta, R. and O. Lior (2010). "Exploring Sequences: A Graphical Tool Based on Multi-Dimensional Scaling." *Journal of the Royal Statistical Society. Series A, Statistics in Society* 173:165–184.

Pollock, G. (2007). "Holistic Trajectories: A Study of Combined Employment, Housing and Family Careers by Using Multiple-Sequence Analysis." *Journal of the Royal Statistical Society. Series A* 170:167–183.

Robert, P. and Y. Escoufier (1976). "A Unifying Tool for Linear Multivariate Statistical Methods: The RV-coefficient." *Applied Statistics* 25:257–265.

Scherer, S. (2001). "Early Career Patterns: A Comparison of Great Britain and West Germany." *European Sociological Review* 17:119–144.

Shannon, C. E. (1948). "A Mathematical Theory of Communication." *The Bell System Technical Journal* 27:379–423, 623–656.

Shawe-Taylor, J. and N. Christianini (2004). *Kernel Methods for Pattern Recognition*. Cambridge University Press, Cambridge, UK.

Theil, H. (1972). *Statistical Decomposition Analysis*. North-Holland Publishing Company, Amsterdam.

Wang, H. and Z. Lin (2007). "A Novel Algorithm for Counting All Common Subsequences." In *2007 IEEE Conference on Granular Computing*, pp. 502–505. IEEE.

Widmer, E. D. and G. Ritschard (2009). "The De-Standardization of the Life Course: Are Men and Women Equal?" *Advances in Life Course Research* 14:29–39.

9 Exploratory Mining of Life Event Histories

Gilbert Ritschard, Reto Bürgin, and Matthias Studer

Introduction

This chapter explains how data-mining-based techniques can be used to discover interesting knowledge from sequences of life events, that is, to find out how people sequence important life events. We illustrate with data from the biographical survey conducted by the Swiss Household Panel in 2002. The focus is on the sequencing of events in the occupational life course and of events, such as starting a union and childbirth, that affect living arrangements. Addressed methods include finding of frequent sequential patterns, identification of discriminant subsequences, and clustering of event sequences.

Mining of sequentially organized data has been successfully exploited in many domains such as genetics, device control, and speech recognition as well as for automatic text, customer behavior, and web logs analysis. In the social sciences sequence exploration mainly focuses on state sequences, and consists typically of building typologies by means of the optimal matching approach (Abbott and Forrest, 1986). In contrast, we focus in this chapter on event sequences rather than state sequences. Let us clarify the difference between the two.

A state, such as being jobless, lasts the whole considered unit of time while an event, for example ending a job, occurs at a certain time point and has no duration. The event does not last, but provokes, possibly in conjunction with other events, a state change. In state sequences, the positions in the sequence reflect the duration since the beginning of the sequence, while in event sequences they just inform us about the number of precedent events. Therefore, while state sequences are particularly of interest for studying durations and timing in life courses, event sequences are especially useful when the concern is the sequencing (i.e., the order in which events occur). This does not mean that we cannot account for the time of occurrence of the events. However, since position does not convey time information, explicit time stamps are needed for that. Another important difference between states and events is that multiple events can occur at the same time point, while states are mutually exclusive. We cannot be in two different states at the same time. Furthermore, depending on the situation at the time the event occurs, the same event may characterize different transitions. For example, when living alone the event "having a child" provokes the transition from the state "Alone" to the state "Alone with a child" while for someone married it provokes the transition from state "Married" to the state "Married with a child." Although there is not always a clear univocal relationship between state sequences and event sequences, each of them is just an alternative way of looking at the same information about life trajectories. Event sequences need different tools than state sequences. For example, while state sequences are easily rendered with stacked segments of colored lines each representing the time spent in a given state, events cannot be rendered this way because they have no duration.

So, our aim in this chapter is to show what we can do with event sequences and the kind of results that we can expect from an event sequence analysis. While state sequences have received a lot of attention in the social sciences, especially since the popularization of the optimal-matching-based methods by Abbott in the late 1980s (Abbott and Forrest, 1986; Aisenbrey and Fasang, 2010), and there exist nowadays efficient pieces of software to explore such state sequences (e.g., Brzinsky-Fay et al., 2006; Gabadinho et al., 2011), event sequence analysis has received much less attention in the social sciences.

The approach followed in the few social science papers that consider event sequences (Abbott, 1983, 1991; Blockeel et al., 2001; Billari et al., 2006) mainly consists, with the noticeable exception of Blockeel et al. (2001), in looking at the frequencies of *a priori*-defined subsequences of interest. Here, we adopt a more holistic point of view and explore all subsequences that can be found in the data. We essentially place our approach in the line of the work developed by the data mining community (Agrawal and Srikant, 1995; Mannila et al., 1995; Bettini et al., 1996; Mannila et al., 1997; Zaki, 2001) as an extension of the mining of frequent—non-ordered—patterns. Behind these aspects, we also consider the measurement of pairwise dissimilarities between event sequences, which then gives access to any dissimilarity-based method.

Broadly, the addressed methods consist in finding the most frequent subsequences, i.e., the most common ways of sequencing life events, and then in finding out among them those that best discriminate between groups such as women and men for example. The measurement of pairwise dissimilarities between event sequences is also addressed and we show how such dissimilarities can be used for clustering event sequences. Regarding the event time-stamps, we show that, when available, they can be used to restrict the search for interesting subsequences through time constraints. Differences in timing can also explicitly be accounted for when computing dissimilarities between time-stamped event sequences.

It is worth mentioning that the approach proposed here differs from traditional event history analysis (Blossfeld et al., 2007; Yamaguchi, 1991), which is indeed survival analysis and as such focuses on the timing or hazard of only one specific event: death, marriage or a first job, for example. See Scott et al., Ghisletta, and Zhou et al. (chapters 14, 15, and 16) in this volume for applications using exploratory data mining approaches in a survival analysis context.

For the social sciences, the exploration of event sequences should permit us to answer questions such as:

- What is the most typical succession of family or professional life events?
- Are there standard ways of sequencing those events?
- What are the most typical events that occur after a given subsequence such as after leaving home and ending education?
- How is the sequencing of events related to covariates?
- Which event sequencings best discriminate groups such as men and women?

As already mentioned, our aim is to demonstrate the scope of sequential event pattern mining in social sciences. We exploit for that the features offered by the TraMineR package (Gabadinho et al., 2011) for event sequences[1] and illustrate with data on Swiss cohabitational and occupational life courses. TraMineR is a general toolbox for exploring state and event sequence data in the free open source graphical and statistical environment R (R Development Core Team, 2012). The name TraMineR stands for "Trajectory Miner in R."

The Data

We consider sequences derived from the biographical survey conducted in 2002 by the Swiss Household Panel.[2] We retain the 1,503 cases studied in Widmer and Ritschard (2009) with techniques for state sequences, but look at the events behind the state changes. The retained individuals were all aged 45 years or more at the survey time, i.e., they were born in 1957 or earlier, and the focus is on their life trajectories between 20 and 45 years, the data granularity being at the yearly level. Table 9.1 gives the alphabet of the cohabitational and occupational states analyzed in Widmer and Ritschard (2009) and Figure 9.1 shows the index plot—rendering of the individual sequences—of each of the two considered sets of state sequences. In those plots, each sequence is represented by a horizontal line colored according to the state at the successive positions. Color changes indicate state transitions and the length of each segment in the same color reflects the time spent in the corresponding state. In Figure 9.1 the sequences are sorted by states from the end to the beginning of the time frame.

The data were selected on the basis of fully informed cohabitational trajectories but incomplete sequences of occupational sequences were not excluded. There are even some cases for which we have no information

Table 9.1 Short and long state labels

Cohabitational		Occupational	
2P	Biological father and mother	Mi	Missing
1P	One biological parent	FT	Full-time
PP	One biological parent with her/his partner	PT	Part time
A	Alone	NB	Negative break
U	With partner	PB	Positive break
UC	Partner and biological child	AH	At home
UN	Partner and non-biological child	RE	Retired
C	Biological child and no partner	ED	Education
F	Friends		
O	Other		

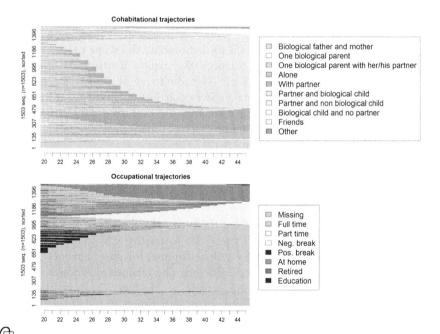

Figure 9.1 Cohabitational and occupational state sequences.

at all about occupation. These most probably correspond to people who never worked and stayed at home during the whole observed age interval, i.e., from 20 to 45 years old.

Cohabitational event sequences are derived from the cohabitational state sequences by specifying the events that cause the transition between any two states. The list of considered cohabitational events is given in Table 9.2 and the retained definition of the transitions in terms of those events is listed in Table 9.3. To give an example of how this latter table should be read,

Table 9.2 Cohabitational events

Short	Long label
2P	Start living with both parents
1P	Start living with one parent
PP	Start living with one parent and her/his partner
LH	Leaving home
A	Start living alone
U	Move in with partner
UE	Separation
C	Start living with a child
CL	Last child leaves home
O	Start other living arrangement

consider the list "LH,A" at the intersection of row 2P and column. This list of two events indicates that each time we observe a transition from the state 2P (leaving with both parents) to the state A (living alone) we assume that the concerned individual experienced the events LH ("leaving home") and A ("start living alone"). The terms on the diagonal of the table have a somewhat different meaning. They indicate the event that we assign when the state sequence starts in the corresponding row state. Those diagonal terms serve thus only at the beginning of the sequences. For example, the 2P in the first cell of the table means that we assign the event 2P at age 20 for all sequences starting in state "living with both parents" (2P).

For occupational trajectories, we define the events as the start of the spells in a same state. We do so, however, after having assimilated the states "missing" (Mi), for the reason mentioned above, and "retired" (RE), which is observed only four times, to the state "at home" (AH). We thus get six occupational events, starting one of the following: an at home spell, working full-time, working part time, a negative break, a positive break, and an education period, which we respectively denote as AH, FT, PT, NB, PB, and ED.

The categorical parallel coordinate plots (Bürgin et al., 2012) in Figure 9.2 show the diversity of event sequencing in the considered cohabitational and occupational, trajectories. The diversity is rendered by avoiding overlapping thanks to small displacements of the lines. The figure (see the book web page for a color version) also highlights with colors the most typical patterns, that is, here, those that are shared by at least 5% of the observed cases. In the graphic, sequences which form a prefix of another one, i.e., which are identical with the beginning of some other longer sequence, are merged into this longer sequence, and the plot renders the frequencies of events and embedded sequences with varying width of both the event points and the connecting segments between successive events. Let us look at the the lower panel of Figure 9.2 to clarify how the plots should be read. We first notice that the successive event squares of a given pattern

Table 9.3 Events associated to cohabitational state transitions

	2P	1P	PP	A	U	UC	UN	C	F	O
2P	"2P"	"1P"	"PP"	"LH,A"	"LH,U"	"LH,U,C"	"LH,U,C"	"LH,C"	"LH,A"	"LH,O"
1P	"2P"	"1P"	"PP"	"LH,A"	"LH,U"	"LH,U,C"	"LH,U,C"	"LH,C"	"LH,A"	"LH,O"
PP	"2P"	"1P"	"PP"	"LH,A"	"LH,U"	"LH,U,C"	"LH,U,C"	"LH,C"	"LH,A"	"LH,O"
A	"2P"	"1P"	"PP"	"A"	"U"	"U,C"	"U,C"	"C"	""	"O"
U	"2P"	"1P"	"PP"	"UE,A"	"U"	"C"	"C"	"C"	"UE,A"	"UE,O"
UC	"2P"	"1P"	"PP"	"UE,CL,A"	"CL"	"U,C"	"CL,C"	"UE"	"UE,CL,A"	"UE,CL,O"
UN	"2P"	"1P"	"PP"	"UE,CL,A"	"CL"	"C"	"U,C"	"UE,C"	"UE,CL,A"	"UE,CL,O"
C	"2P"	"1P"	"PP"	"CL,A"	"CL,U"	"U"	"CL,C"	"C"	"CL,A"	"CL,O"
F	"2P"	"1P"	"PP"	""	"U"	"U,C"	"U,C"	"C"	"A"	"O"
O	"2P"	"1P"	"PP"	"A"	"U"	"U,C"	"U,C"	"C"	"A"	"O"

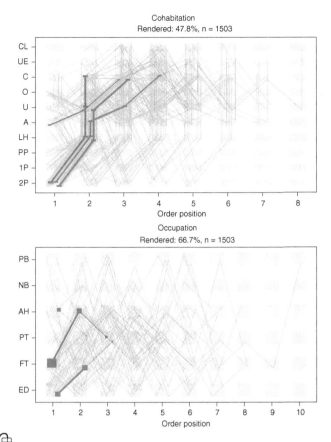

Figure 9.2 Event sequencing in cohabitational and occupational trajectories.

are always located at the same position in the light-grey zones drawn as background of the coordinate points. This facilitates tracking the successive events of a same sequence. Now, let us consider the green pattern. Since the green square at position 1 is clearly larger than the one at position 2, it follows that a large proportion of the observed people experienced only the first event, i.e., "start working full-time" (FT). Likewise, since the second green segment is thinner than the first one, it shows that only a fraction of those who start staying at home (AH) after having started to work full-time, return later to working part time (PT).

In the upper panel in Figure 9.2, we observe, for instance, that most of the considered individuals were living with their both parents (2P) at the beginning of the sequence, i.e., when they were 20 years old and that leaving home (LH) and moving in together with a partner (U) in the same year was very common in the 20th century Switzerland. It was also common to have a first child in the same year. More spaced events are quite frequent too, but the plot clearly shows that the standard order of the events is leaving home, starting a union, and then having a first child.

Regarding occupational trajectories, we see that the most frequent start points at 20 years old are education (ED) and working full-time (FT), and that it is very common to switch directly from education to full-time work and from full-time work to an at home stay (AH). We also observe that after switching from working full-time to an at home stay, it is quite common to return to working part time (PT).

Frequent Subsequences vs. Frequent Itemsets

The mining of frequent itemsets and association rules has been popularized in the 1990s with the work of Agrawal and Srikant (1994) and Agrawal et al. (1995) and their *Apriori* algorithm. The work was for instance intended to analyze "buying baskets," (i.e., to identify items that customers often buy together and rules such as "if a buyer buys a product A then he or she will most probably also buy B") or to the control of devices where the interest is in discovering symptoms that often occur together and announce a failure. Adopting a longitudinal perspective, the methods were then extended to account for the sequences of purchases or of symptom occurrences, and this is where the mining of frequent subsequences started.

The main idea of *Apriori* is to exploit the property that an itemset of size $k+1$ obtained by adding an item to a non-frequent itemset of size k cannot be frequent. It considers thus successively only itemsets of size $k+1$ derived from frequent itemsets of size k, which reduces considerably the number of itemsets to scan.

Mining typical event sequences is in some sense a specialized case of the mining of frequent itemsets. It is much more complex, however, and requires the user to specify time constraints and select a counting method. While there is general agreement about how to count occurrences of itemsets in the classical unordered framework, there is no such agreement for episodes (subsequences) that may occur more than once in the same sequence. The additional time dimension raises questions such as: What is the maximum time span, i.e., sequence length we want to analyze? Until which time gap should events be considered to occur together? Joshi et al. (2001) nicely present the various counting schemes and possibly useful time constraints. A discussion of those aspects in a social science perspective can be found in Ritschard et al. (2008) and we will recall some of the main points in the next section.

Efficient algorithms for extracting frequent subsequences have been proposed in the literature, among which the prominent ones are those of Bettini et al. (1996), Srikant and Agrawal (1996), Mannila et al. (1997) and Zaki (2001). The algorithm implemented in TraMineR is an adaptation of the prefix-tree-based search described in Masseglia (2002).

Mining Frequent Sequential Patterns

We show in this section how we can search for frequent subsequences in cohabitational and occupational event trajectories and the kind of knowledge we can gain from them. We also provide examples of the basic TraMineR commands used for extracting the frequent patterns.

Before turning to the frequent subsequences, it is worthwhile first specifying the terminology—from Studer et al. (2010)—we use, since there is no clear standard for it in the literature.

Terminology

Formally, as in Studer et al. (2010), we define a *transition* as a—non-ordered—set of events occurring at the same time point, namely the set of events that causes the state transition. For instance the joint occurrence of the two events "leaving home" and "moving in with a partner" results in the transition from the state "living with both parents" (2P) or possibly from "living with one parent" (1P) to the state "living with a partner" (U). A transition is also known in the data mining literature as a transaction, especially when the concern is the mining of customer behaviors.

An *event sequence* is then defined as an ordered list of transitions. We represent it as a succession of transitions separated by edges or arrows. Thus, in the sequence

$$(\text{LHome, Union}) \rightarrow (\text{Marriage}) \rightarrow (\text{Childbirth}) \qquad (1)$$

the first term (LHome, Union) is a transition defined by the two events LHome and Union, while the two next transitions each result from a single event. To account for the time gap between two consecutive transitions, we add *duration stamps* when necessary. With the following notations

$$\xrightarrow{22} (\text{LHome, Union}) \xrightarrow{3} (\text{Marriage}) \xrightarrow{2} (\text{Childbirth}) \xrightarrow{15}$$

we indicate that the considered individual gets married 3 years after moving in together with a partner and has a child 2 years after marriage. We also indicate that he or she leaves home to move in with a partner 22 years after the start of observation (birth), and that he or she remains under observation for 15 years after the birth of the child without experiencing any other event of interest.

An *event subsequence y* of an event sequence *x*, is an event sequence such that all its events also belong to *x* and occur in the same order as in *x*. For instance, (Lhome) → (Childbirth) is a subsequence of the example sequence (1) while (Lhome) → (Union) is not a subsequence of it.

The notion of state subsequence, as used for instance by Elzinga (2003, 2010), would more or less correspond to a subsequence of transitions.

Event subsequences are different in that each transition may contain more than one event and that the same event may be present in different transitions. For our data sets, event and state subsequences would be equivalent to the occupational trajectories for which the events indicate the start of each spell in a same state. They are, however, quite different for the cohabitational trajectories where we used the more complex transformation defined by Table 9.3.

Inputting Data to TraMineR

Data can be provided to TraMineR in the vertical time-stamped event form, i.e., with a different line for each event. The data would then appear as in Table 9.4.

In that example, the time stamp is the age at which the individual experiences the event and we can see that individual 101 both leaves the parental home and moves in with a partner at the age of 22.

We may also provide state sequences and select one of several methods for converting them into event sequences (see Gabadinho et al., 2009; Ritschard et al., 2009, for details). We adopt this solution for our illustrative data since they are organized as state sequences. For the cohabitational trajectories, we select the method consisting in associating to each state transition the events as specified by the transition definition, Table 9.3. Letting for instance seqs.coh be the state sequence object for the cohabitational trajectories and transition.coh.mat be the transition definition matrix for cohabitation trajectories given in Table 9.3, we derive the event sequence object from them with the following simple command

```
R> shpevt.coh <- seqecreate(seqs.coh,

          tevent=transition.coh.mat)
```

Table 9.5 shows how the first five event sequences lookout. Notice that each sequence has a start event corresponding to the element given on the corresponding diagonal element of the transition definition table.

For the occupational trajectories, we first recoded states Mi and RE as AH and then generated the event sequences with the option tevent="state," which assigns an event to the start of each spell

Table 9.4 Example of vertical time-stamped event data

ID	Time stamp	Event
101	22	LHome
101	22	Union
101	25	Marriage
101	27	Childbirth
102	18	Union
102	..	...

Table 9.5 First five cohabitational event sequences

	Sequence
1	$(2P) \xrightarrow{1} (LH,U) \xrightarrow{25}$
2	$(2P) \xrightarrow{2} (LH,U) \xrightarrow{6} (C) \xrightarrow{18}$
3	$(2P) \xrightarrow{1} (A,LH) \xrightarrow{5} (U) \xrightarrow{2} (C) \xrightarrow{18}$
4	$(2P) \xrightarrow{8} (LH,U) \xrightarrow{1} (C) \xrightarrow{17}$
5	$(A) \xrightarrow{6} (C,U) \xrightarrow{20}$

in the same state and names the event with the name of that state.

```
R> shpevt.occ <- seqecreate(seqs.occ,
         tevent="state")
```

Finding the Frequent Subsequences

As already mentioned above, a given subsequence may occur more than once in the same sequence and we have to chose a counting method to determine its frequency. In TraMineR we can chose from Joshi et al. (2001)'s six different counting methods. The first and perhaps most common one (COBJ) is to count the number of sequences that contain the subsequence of interest. The second one (CWIN) sets a sliding window size and counts in each sequence the number of windows that contain the subsequence and then adds up the counts. The third one (CMINWIN) proceeds similarly but uses in each sequence the smallest window size that contains the subsequence. The last two methods consist in counting the number of occurrences of the subsequence in each sequence, allowing for possible overlaps of the occurrences found in the fourth method (CDIST_O) and considering only non–overlapping occurrences in the fifth one (CDIST). In the last case, the solution may depend on the order in which we constitute the successively counted subsequences.

Most Frequent Cohabitational Subsequences

We now extract the cohabitational subsequences supported by at least 50 cases, which we obtain with the COBJ counting method. The TraMineR command is

```
R> cons <- seqeconstraint(countMethod='COBJ')
R> shp.fss.coh <- seqefsub(shpevt.coh,
         minSupport=50, + constraint=cons)
```

Table 9.6 Most frequent cohabitational subsequences with at least two events

	Subsequence	Support	Count	No. transitions	No. events
1	(2P) → (LH)	0.621	934	2	2
2	(2P) → (U)	0.582	874	2	2
3	(2P) → (C)	0.477	717	2	2
4	(LH,U)	0.454	682	1	2
5	(U) → (C)	0.429	645	2	2
6	(2P) → (LH,U)	0.392	589	2	3
7	(LH) → (C)	0.382	574	2	2
8	(A) → (U)	0.376	565	2	2
9	(2P) → (LH) → (C)	0.325	489	3	3
10	(C,U)	0.291	437	1	2

We get 85 subsequences. The most frequent ones contain a single event, which is not very instructive about event sequencing. Therefore, we display in Table 9.6 the 10 most frequent subsequences that contain at least two events.

We observe that the most frequent subsequences consist of only four out of the 10 considered events, namely: living with both parents, leaving home, starting a union, and having a first child. We learn from Table 9.6 that the most frequent and hence most typical event sequencing is to live with both parents and then leave home more than a year later. This succession of events is experienced by 62.1% of the 1,503 cases, while, after living with both parents when being 20 years old or more, 58.2% move in with a partner and 47.7% have a first child. People who follow the sixth most frequent subsequence (2P) → (LH,U) are also counted among those who experienced the most frequent two subsequences (2P) → (LH) and (2P) → (U). We thus deduce that there are $345 = 934 - 589$ individuals who left home without moving in with a partner at the same time. Other similar results can be derived from Table 9.6. For instance, if we look at the seventh and ninth most frequent subsequences, we can establish that 0.6% of the individuals in the sample started a union, had a first child more than a year afterwards, but did not at any time live with both parents after they were 20 years old.

We get very similar results by counting distinct occurrences (CDIST_O) of the subsequences rather than the number of sequences that include them. It leads to an increase of roughly 3 to 5% of the count support. This is not surprising, since most of the considered events occur only once in each sequence. It is more interesting to vary the maximal time span allowed for the subsequences. Setting the maximum time span as three years, we get only 29 subsequences with a support of 50 or more. The 10 most frequent are given in Table 9.7 and comparing this with Table 9.6 we can see some changes in the order of the subsequences. Leaving home to start a union, (LH,U), and having the first child when moving in with a partner, (C,U),

Table 9.7 Most frequent cohabitational subsequences with at least two events and a 3-year maximum time span

	Subsequence	Support	Count	No. transitions	No. events
1	(LH,U)	0.454	682	1	2
2	(C,U)	0.291	437	1	2
3	(2P) → (LH)	0.275	414	2	2
4	(U) → (C)	0.274	412	2	2
5	(A,LH)	0.244	367	1	2
6	(C,LH)	0.180	270	1	2
7	(C,LH,U)	0.175	263	1	3
8	(LH) → (C)	0.166	250	2	2
9	(A) → (U)	0.158	237	2	2
10	(2P) → (A)	0.148	223	2	2

are the most frequent patterns occurring within three years. There are also important drops in support. For example, only 412 cases out the 645 who experienced the sequence (U) → (C), had their first child within 3 years of starting to live with a partner.

Most Frequent Occupational Subsequences

The 10 most frequent occupational subsequences with more than one event are listed in Table 9.8. The sequencing of occupational trajectories clearly looks less standardized since the frequencies of the most frequent subsequences are about three times lower than those of the most frequent cohabitational trajectories. Remember that the events here are just the transitions between states and that we have assimilated the start of a retired (RE) or missing (Mi) spell to the start of an at home (AH) spell. There are no simultaneous events.

We learn from Table 9.8 that three transitions are experienced by more than 20% of the cases, namely starting education and later a full-time job

Table 9.8 Most frequent occupational subsequences with at least two transitions

	Subsequence	Support	Count	No. transitions	No. events
1	(ED) → (FT)	0.283	425	2	2
2	(FT) → (AH)	0.265	398	2	2
3	(FT) → (PT)	0.219	329	2	2
4	(AH) → (PT)	0.130	195	2	2
5	(ED) → (AH)	0.113	170	2	2
6	(ED) → (PT)	0.112	168	2	2
7	(FT) → (FT)	0.112	168	2	2
8	(FT) → (AH) → (PT)	0.105	158	3	3
9	(FT) → (ED)	0.073	109	2	2
10	(ED) → (FT) → (PT)	0.071	107	3	3

Table 9.9 Most frequent occupational subsequences with at least two transitions and a 3-year maximum time span

	Subsequence	Support	Count	No. transitions	No. events
1	(ED) → (FT)	0.185	288	2	2
2	(FT) → (AH)	0.067	100	2	2
3	(ED) → (AH)	0.042	73	2	2
4	(PT) → (FT)	0.036	56	2	2
5	(PT) → (AH)	0.034	53	2	2
6	(ED) → (PT)	0.031	52	2	2

(28.3%), working full-time and then staying at home (26.5%) and starting a full-time job and later working part time (21.9%). Among those who start an at home stay after having worked full-time, 60.3% (the 158 cases who experienced the eighth most frequent subsequence) return to working part time afterwards. These are, as we will see below, mainly women.

As for cohabitational trajectories, we get more or less the same most frequent subsequences with all counting methods. With CDIST_O, the counting supports increase by about 6 to 10%, i.e., a bit more than for cohabitational subsequences. This indicates that repeating events are slightly more frequent in occupational trajectories than in cohabitational ones. Let us look at subsequences with a three-year maximum time span. Table 9.9 reports the only six such subsequences that satisfy the minimum count-support of 50. Comparing with Table 9.8 we see that while 68% of those who started to work full-time after education, (ED) → (FT), did it within 3 years from the beginning of the education spell (which is most often also the beginning of the sequence), only 25% of those who stopped working full-time to stay at home, (FT) → (AH), did it within 3 years of starting to work full-time.

Merging Cohabitational and Occupational Channels

It is also interesting to combine both sets of events and to consider joint event sequences. Table 9.10 shows the most frequent subsequences of the combined cohabitational and occupational trajectories.

Unsurprisingly, in this list we find the most frequent cohabitational subsequences that are more than twice as frequent as the most frequent occupational subsequences. However, there are also some subsequences that combine cohabitational and occupational events. Among them, the two most frequent are moving in with a partner after having started full-time work and having the first child after having started full-time work. They are respectively shared by 69.5% and 58.3% of the cases. We also observe that it is common to leave home after having started to work full-time (55.5%), and even to leave home and move in with a partner in the same year after starting to work full-time (37.6%).

Table 9.10 Most frequent subsequences of combined cohabitational–occupational
sequences

	Subsequence	Support	Count	No. transitions	No. events
1	(FT) → (U)	0.695	1045	2	2
2	(2P) → (LH)	0.621	934	2	2
3	(FT) → (C)	0.583	876	2	2
4	(2P) → (U)	0.582	874	2	2
5	(FT) → (LH)	0.555	834	2	2
6	(2P) → (C)	0.477	717	2	2
7	(LH,U)	0.454	682	1	2
8	(U) → (C)	0.429	645	2	2
9	(2P) → (LH,U)	0.392	589	2	3
10	(LH) → (C)	0.382	574	2	2
11	(2P,FT)	0.378	568	1	2
12	(A) → (U)	0.376	565	2	2

In the previous exploration, we have been able to identify the standard ways of sequencing the life events of interest. The next step would be to study whether there are differences in those standards among socio-demographic groups such as those defined by gender or birth cohort. A possible solution would be to look for the most frequent subsequences separately in each group and then compare them. It is more efficient, however, to directly seek the most discriminant subsequences.

Discriminant Subsequences

When examining the most typical sequential patterns, questions naturally arise about their relationship with covariates such as sex or birth cohort. For instance, we may be interested in knowing which sequence pattern best characterizes women or youngest cohorts.

To answer those questions, we use the method proposed in Studer et al. (2010), which consists of measuring the strength of association of each subsequence with the considered covariate and then selecting the subsequences with the strongest association. The association can be measured with the Pearson independence Chi-square. We define for this a 0–1 presence indicator variable of the subsequence, cross tabulate it with the covariate, and compute the Pearson Chi-square of the resulting table. The tables are of the same size for all subsequences since they all cross tabulate a binary indicator variable with the same covariate. Hence, the Chi-square can be directly used for sorting the subsequences, the most discriminant one being the one with the highest Chi-square. We would get the same order with a normalized Chi-square such as Cramer's v or by sorting in increasing order of the associated p-values for the independence test.[3]

Differentiating Between Sexes

Below is the TraMineR command with which we obtain, in decreasing order of their discriminant power, the cohabitational subsequences that best distinguish between sexes. We limit the search to subsequences which discriminate sex at the 1% significance level.

```
R> shp.dss.coh <- seqecmpgroup(shp.fss.coh,
           group = seqssex, + pvalue.limit = 0.01)
```

The six most discriminating cohabitational subsequences are listed in Table 9.11 and the frequencies of all 13 subsequences that significantly discriminate for sex at the 1% level are plotted in Figure 9.3.

The colors used for the bars in Figure 9.3 (see the color version on the book's web site) indicate the sign and significance of the associated Pearson residual. This residual is the signed square root of the contribution to the Chi-square of the cell $(1, g)$ corresponding to the presence, 1, of the

Table 9.11 Cohabitational subsequences that best discriminate sex

	Subsequence	Chi-2	Support	Frequency men	Frequency women	Difference
1	(LH)	38.3	0.72	0.795	0.651	0.144
2	(2P) → (U)	22.4	0.58	0.642	0.521	0.122
3	(LH) → (U)	19.0	0.27	0.316	0.216	0.101
4	(LH) → (C)	18.3	0.38	0.436	0.328	0.109
5	(2P) → (LH)	18.3	0.62	0.676	0.567	0.108
6	(2P) → (A) → (U)	17.5	0.21	0.253	0.164	0.089

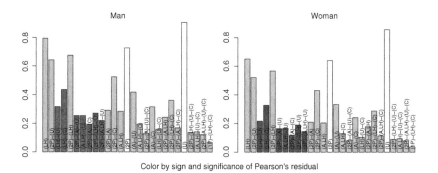

Figure 9.3 Cohabitational subsequences that discriminate sex at the 1% level.

subsequence in the group g

$$\text{Pearson residual} = \frac{(n_{1g} - e_{1g})}{\sqrt{e_{1g}}}$$

where n_{1g} is the observed count and e_{1g} the expected count under the independence assumption. At first glance, it may be surprising to get non-significant—small—Pearson residuals for discriminant subsequences. As can be seen in Figure 9.3, this happens for highly frequent subsequences such as those formed by the single event "starting to live with both parents" (2P) or "moving in with a partner" (U). The reason is that in such cases, the negation of a frequent subsequence, e.g., "never living with both parents" or "never moving in with a partner," will be small and the associated Pearson residual very large.

Although there are significant differences between men and women in the frequencies of some cohabitational subsequences, the differences are not very important and are probably due to timing effects that are hidden because we do not account for events occurring before the age of 20. For instance, the lower proportion of women that leave home between ages 20 and 45 can be explained by the higher proportion of women who leave home before their twentieth birthday.

Differences between men and women are much more important in the sequencing of occupational events. Table 9.12 lists the six most discriminant occupational subsequences while Figure 9.4 plots the frequencies of all subsequences which are significantly discriminant at the 0.1% level.

Interestingly, most of the occupational discriminant subsequences are quite frequent for women and rarely observed for men. The three exceptions are ED, ED $\rightarrow$ FT and FT $\rightarrow$ ED $\rightarrow$ FT, which indicate that more men than women start an education spell after they are 20 years old, that men more often start working full-time after education, and that men more often return to working full-time after an education break. All the other discriminant subsequences contain either the "at home," AH, or the "partial time," PT, event, which, in 20th century Switzerland, typically are events experienced by women. The results clearly demonstrate this fact.

Table 9.12 Occupational subsequences that best discriminate sex

	Subsequence	Chi-2	Support	Frequency men	Frequency women	Difference
1	(FT) $\rightarrow$ (AH)	322.7	0.26	0.060	0.470	−0.410
2	(AH)	317.5	0.41	0.181	0.634	−0.453
3	(PT)	269.7	0.28	0.088	0.469	−0.381
4	(FT) $\rightarrow$ (PT)	247.5	0.22	0.051	0.387	−0.337
5	(AH) $\rightarrow$ (PT)	195.5	0.13	0.008	0.252	−0.244
6	(FT) $\rightarrow$ (AH) $\rightarrow$ (PT)	161.5	0.11	0.004	0.206	−0.202

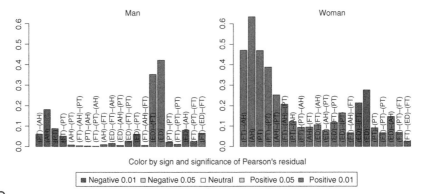

Figure 9.4 Occupational subsequences that discriminate sex at the 0.1% level.

As we did with frequent subsequences, it is interesting to look for discriminant subsequences of trajectories that combine the cohabitational and occupational events of each individual. The 10 most discriminant subsequences of the mixed sequences are listed in Table 9.13 and we can see that all of them contain one of the two events AH and PT which are typically experienced by women. In four cases, at least one of these two events is combined with "moving in with a partner" (U) and in one case with "first child" (C). The results clearly demonstrate that staying at home or switching to part time work some time after moving in with a partner is a typically female behavior. This is also true for switching to part time work some time after birth of the first child.

Differentiating among Birth Cohorts

Let us now look at differences in the event sequencing among birth cohorts. We consider three cohorts, namely people born in 1910–1924, 1925–1945,

Table 9.13 Mixed events: subsequences that best discriminate sex

	Subsequence	Chi-2	Support	Frequency men	Frequency women	Difference
1	(FT) → (AH)	322.7	0.26	0.060	0.470	−0.410
2	(AH)	317.5	0.41	0.181	0.634	−0.453
3	(PT)	269.7	0.28	0.088	0.469	−0.381
4	(U) → (PT)	260.4	0.20	0.036	0.373	−0.337
5	(FT) → (PT)	247.5	0.22	0.051	0.387	−0.337
6	(FT) → (U) → (AH)	228.2	0.16	0.016	0.302	−0.286
7	(U) → (AH)	226.0	0.20	0.041	0.350	−0.309
8	(AH) → (PT)	195.5	0.13	0.008	0.252	−0.244
9	(C) → (PT)	193.3	0.15	0.019	0.273	−0.254
10	(FT) → (U) → (PT)	192.7	0.16	0.027	0.289	−0.262

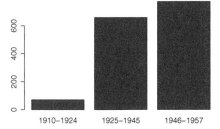

Figure 9.5 Birth cohort distribution.

and 1946–1957. Figure 9.5 shows the birth cohort distribution of our 1,503 cases.

We directly consider the sequences which combine cohabitational and occupational events. The subsequences which best discriminate among cohorts are listed in Table 9.14. Except for (U) → (C), all displayed discriminating subsequences comprise the "part time" (PT) event. In Figure 9.6, we see that there is a strong increase in the frequencies of those subsequences. We thus learn that, among the considered events, the emergence of part time working, and especially starting to work part time after a union or after the first childbirth, is the most prominent change which occurred over birth cohorts. The increasing frequencies of (U) → (C), is also a prominent evolution across cohorts. The latter result reflects the shift from frequent simultaneous U and C events to situations where the first childbirth occurs a year or more after moving in with a partner. This is confirmed by, for instance, the subsequence (2P) → (LH,U,C) which is the 22nd most discriminating one (result not displayed) and whose proportion falls from 27% in the oldest cohort to 12% in the youngest cohort.

Table 9.14 Mixed events: subsequences that best discriminate birth cohorts

	Subsequence	Chi-2	Support	1910–1925	1926–1945	1946–1957
1	(PT)	64.5	0.28	0.042	0.205	0.362
2	(U) → (PT)	63.0	0.20	0.014	0.135	0.281
3	(FT) → (PT)	56.1	0.22	0.014	0.156	0.291
4	(A) → (PT)	46.3	0.11	0.028	0.055	0.160
5	(A) → (U) → (PT)	39.4	0.08	0.014	0.039	0.125
6	(FT) → (U) → (PT)	38.5	0.16	0.000	0.114	0.210
7	(ED) → (PT)	36.8	0.11	0.028	0.065	0.159
8	(LH) → (PT)	35.9	0.15	0.014	0.109	0.204
9	(U) → (C)	34.2	0.43	0.239	0.370	0.497
10	(A,LH) → (PT)	34.0	0.05	0.000	0.020	0.084

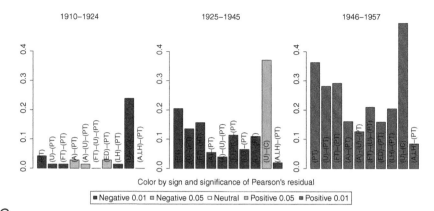

Figure 9.6 Mixed events: subsequences that best discriminate birth cohorts.

Clustering Event Sequences

Besides exploring the most frequent and most discriminant subsequences, it may also be of interest to run dissimilarity-based analyses—clustering or principal coordinate analysis, for instance—of event sequences as is typically done with state sequences (see for example Abbott and Tsay, 2000; Aisenbrey and Fasang, 2010). The only requirement for that is that the dissimilarity between time-stamped event sequences can be measured.

One possibility is to compute dissimilarities between event time stamped sequences with the OME distance, a variant of optimal-matching applicable to event sequences (Studer et al., 2010). OME is an edit distance like the optimal-matching distance between state sequences; i.e., OME is defined as the minimal cost of transforming one sequence into the other. We retain here the OME distance described in Studer et al. (2010), which extends a proposition by Moen (2000) to the case of possible simultaneous events. The transformation operations considered by OME are:

- the insertion/deletion of an event;
- a change in the time stamp of a given event.

Event dependent costs can be specified both for the insertion/deletion of an event as well as for a one-unit change in the time stamp of the event. The version implemented in TraMineR also allows us to choose between absolute or relative time alignment for the pairwise comparison of sequences.

As defined above, the distance mainly depends on the mismatches—it is the cost of transforming mismatches into matches—but does not account for the total number of events in the sequences and consequently does not account for the number of existing matches. However, two sequences distant by, for example, 2 look very dissimilar if they contain each only two or three events, while they are very similar if they each have hundreds of events. It is useful, therefore, to normalize the distance to account for the number of events in the sequence. We retain the following normalization:

$$d_{N,ome}(x, y) = \frac{2d_{ome}(x, y)}{\Omega(x) + \Omega(y) + d_{ome}(x, y)}$$

where $d_{ome}(x, y)$ is the OME dissimilarity between the time-stamped event sequences x and y, and $\Omega(x)$ the total cost for inserting all the events of x. Unlike the ratio $d_{ome}(x, y)/(\Omega(x) + \Omega(y))$ proposed by Moen (2000), the retained normalized distance $d_{N,ome}$ satisfies the triangle inequality.

Types of Cohabitational Trajectories

To illustrate, we compute the normalized OME dissimilarity matrix of our cohabitational trajectories using a constant indel cost of 1, a constant time change cost of 0.1, and absolute time for event alignment. We do not show the $1,503 \times 1,503$ dissimilarity matrix which is not of much interest per se, but use it to run a cluster analysis. We cluster the event sequences into five groups by partitioning them around medoids (Kaufman and Rousseeuw, 2005).[4] The resulting clusters are visualized in Figure 9.7. The top row renders the event sequencing, while the second row informs us about the timing of events by displaying the survival curves until the first occurrence of the four main events, namely "leaving home" (LH), "starting to live alone or with friends" (A), "moving in with a partner" (U) and "starting to live with a child" (C). The clusters are sorted in decreasing order of frequencies and are labeled with their medoid, i.e., the sequence with minimal sum of distances to the other sequences in the cluster.

Unsurprisingly, we get a different cluster for each of the four typical trajectories that we could observe in Figure 9.2. This confirms that the OME distance accounts for the sequencing of events. The survival curves show, however, that the clusters also differ in the timing of the events and in the final probability of experiencing the events. The fifth group corresponds to people who were living with their parents at 20 years old and did not

Table 9.15 Cohabitational trajectory types, distribution by sex

	Man	Woman	Overall
$(2P) \xrightarrow{2} (A,LH) \xrightarrow{5} (U) \xrightarrow{3} (C) \xrightarrow{16}$	0.298	0.216	0.257
$(2P) \xrightarrow{6} (C,LH,U) \xrightarrow{20}$	0.266	0.245	0.255
$(2P) \xrightarrow{4} (LH,U) \xrightarrow{4} (C) \xrightarrow{18}$	0.249	0.242	0.246
$(A) \xrightarrow{4} (U) \xrightarrow{3} (C) \xrightarrow{19}$	0.138	0.234	0.186
$(2P) \xrightarrow{26}$	0.049	0.063	0.056

Table 9.16 Cohabitational trajectory types, distribution by birth cohorts

	1910–1924	1925–1945	1946–1957	Overall
$(2P) \xrightarrow{2} (A,LH) \xrightarrow{5} (U) \xrightarrow{3} (C) \xrightarrow{16}$	0.183	0.235	0.282	0.257
$(2P) \xrightarrow{6} (C,LH,U) \xrightarrow{20}$	0.380	0.310	0.198	0.255
$(2P) \xrightarrow{4} (LH,U) \xrightarrow{4} (C) \xrightarrow{18}$	0.211	0.211	0.278	0.246
$(A) \xrightarrow{4} (U) \xrightarrow{3} (C) \xrightarrow{19}$	0.113	0.164	0.212	0.186
$(2P) \xrightarrow{26}$	0.113	0.080	0.030	0.056

experience any other events. Such trajectories are badly rendered by the event-sequence plots.

The first type of cohabitational trajectories (25.7%) consists in leaving home to live alone or with friends (about 5 years) before moving in with a partner, and (about 3 years) later having a first child. The three events LH, U, and C are spaced in time. Although this trajectory is representative of the first cluster, it does not mean that all individuals in the cluster followed exactly such a trajectory. It just means that the individuals in that cluster experienced a similar sequencing. The grayed lines in the top plot render the within-cluster diversity. We also learn from the survival curves that almost everyone in that cluster left home and experienced living alone or with a friend, that about 15% did not move in with a partner, and about 35% did not experience a childbirth.

In the second group (25.5%), the trajectories cluster around a situation where the individuals leave home, move in together with a partner, and have a first child simultaneously, i.e., all in the same year. This is confirmed by the overlapping of the three survival curves. Surprisingly, the union seems to precede leaving home for this second cluster. However, this is essentially a consequence of the retained coding, U serving as a start event for those who moved in with a partner before they were 20 years old, while we do not report the LH event of those persons.

In the third cluster (24.6%), we have trajectories of people who leave home to move in with a partner in the same year, and (about 4 years) later have their first child. The simultaneity of LH and U in this group is confirmed by the overlapping of the corresponding survival curves. About 20% of the people belonging to the cluster did not experience a childbirth.

The fourth cluster (18.6%) groups people mainly living alone when aged 20 years and who experienced spaced union start and childbirth events, the latter occurring about 4 years after moving in with the partner. A non-negligible proportion (30%) of the individuals in the cluster did not live alone, however. From the upper plot, we can see that these are trajectories with the O start event which correspond most probably to people who lived in an institution when aged 20.

The last cluster (5.6%) is formed by those who stay unmarried with their parents during the considered ages.

From the survival curves, we learn that while almost everybody in the second and third clusters moved in with a partner, a non-negligible proportion of the members of the first and fourth clusters did not experience living with a partner. The survival curves also reveal that while more than 80% of the cases experienced a childbirth in the second group, this proportion is lower in the other three clusters.

The relationship between the clusters and demographic covariates is informative. Cross tabulating cluster membership with sex, we see (Table 9.15) that both sexes are almost equally represented in the second and third group, while the spaced-life-events type (group 1) is clearly dominated by men. Although clusters 1 and 4 look dissimilar at first glance, a closer look reveals that the difference between them is essentially a matter of timing, union and childbirth occurring about 3 years earlier in group 4 than in group 1. Table 9.15 reveals that the "early spaced leaving home-union" cluster 4 is dominated by women while the "late spaced leaving home-union" group 1 mainly comprises men.

Cross tabulation with birth cohort (Table 9.16) reveals that the proportion of individuals following a trajectory of the third type $((2P) \xrightarrow{4} (LH,U) \xrightarrow{4} (C) \xrightarrow{18})$ remains stable, while the proportion of the second type $((2P) \xrightarrow{6} (C,LH,U) \xrightarrow{20})$ decreases from 38% among the older cohort to 20% for the younger cohort. In contrast, the proportion of the spaced-event types increases from 18% to 28% for cluster 1 (late spaced events), and from 11% to 21% for cluster 4 (early spaced events). The latter two types thus correspond to more modern ways of life organization. Finally, we notice that while staying with their parents during active life was quite common for the oldest cohort (11%), this way of life tends to disappear in the youngest cohort (3%).

Types of Occupational Trajectories

Partitioning around medoids from the OME distances between occupational event sequences, we identify the five clusters visualized in Figure 9.8.[5]

The five identified types are in the retained order of decreased frequencies: "Full-time" (39%), "At home after working full-time" (19%), "Short education" (18%), "Staying at home" (12%), and "Long education" (12%). From the survival curves, we see that all clusters also contain about 20% of individuals who experienced part time working, the higher percentage of part time (about 45%) being in the second cluster.

The correspondence with the trajectories highlighted in Figure 9.2 is, here, less evident than for the cohabitational trajectories. First, the most important cluster (39%) includes essentially only full-time trajectories, and

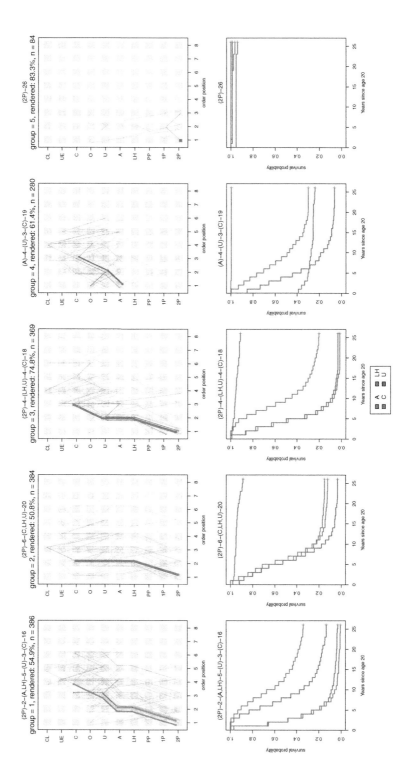

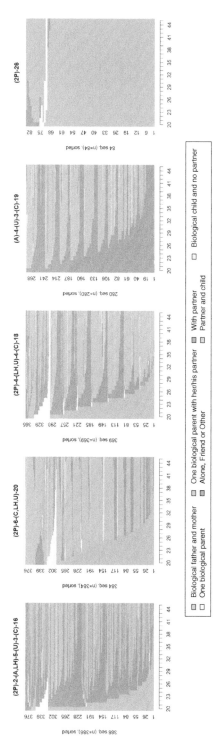

Figure 9.7 Cohabitational trajectories clustered from dissimilarities between event sequences.

Legend:
- Biological father and mother
- One biological parent
- One biological parent with her/his partner
- Alone, Friend or Other
- With partner
- Partner and child
- Biological child and no partner

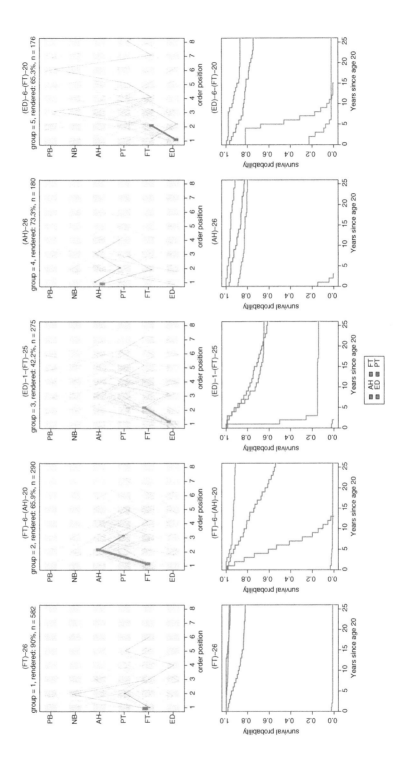

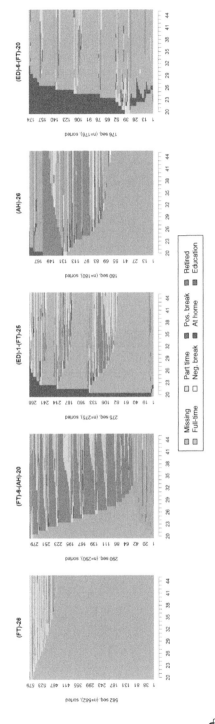

Figure 9.8 Occupational trajectories clustered from dissimilarities between event sequences.

the frequency of this full-time trajectory—with no other event than the starting event—is rendered by the green rectangle next to FT at position 1 in Figure 9.2 (see the color version on the book's web site). Likewise, the "Staying at home" trajectories are represented by the hardly visible colored square next to AH at position 1. This rendering emphasizes the trajectory much less than the heavy lines linking successive events. A second reason for the less clear correspondence between clusters and characteristic orderings is that two clusters, the third and the fifth ones, are characterized by the same sequence (ED) → (FT), the difference between the two being the timing of the events for which the plot in Figure 9.2 does not account.

Tables 9.17 and 9.18 provide insights on the cluster composition in terms of sex and birth cohort. They reveal that the "Full-time" and "Long education" trajectories are principally male ones, while the "At home after working full-time" type is typically female. The type "At home', also dominated by women, declines across birth cohorts in favor of the "Short education', (ED) $\xrightarrow{1}$ (FT), and "At home after working full-time', (FT) $\xrightarrow{6}$ (AH), types of trajectory.

Table 9.17 Occupational trajectory types, distribution by sex

	Man	*Woman*	*Overall*
(FT) $\xrightarrow{26}$	0.488	0.286	0.387
(FT) $\xrightarrow{6}$ (AH) $\xrightarrow{20}$	0.041	0.345	0.193
(ED) $\xrightarrow{1}$ (FT) $\xrightarrow{25}$	0.185	0.181	0.183
(AH) $\xrightarrow{26}$	0.100	0.140	0.120
(ED) $\xrightarrow{6}$ (FT) $\xrightarrow{20}$	0.186	0.048	0.117

Table 9.18 Occupational trajectory types, distribution by birth cohort

	1910–1924	*1925–1945*	*1946–1957*	*Overall*
(FT) $\xrightarrow{26}$	0.338	0.404	0.378	0.387
(FT) $\xrightarrow{6}$ (AH) $\xrightarrow{20}$	0.141	0.209	0.184	0.193
(ED) $\xrightarrow{1}$ (FT) $\xrightarrow{25}$	0.127	0.155	0.212	0.183
(AH) $\xrightarrow{26}$	0.239	0.135	0.096	0.120
(ED) $\xrightarrow{6}$ (FT) $\xrightarrow{20}$	0.155	0.097	0.131	0.117

Conclusion

Three methods for discovering useful knowledge from life event histories have been addressed: the mining of frequent sequential patterns, the identification of the most discriminant subsequences between given groups, and the definition of types of trajectories by means of unsupervised clustering. As illustrated by their application on a Swiss dataset on cohabitational and occupational trajectories, the three approaches provide complementary insights on the way Swiss people organized their life course during the 20th century and how it changed over time. The mining of frequent sequential patterns permits us to find out the overall most common characteristics of the analyzed sequences. The search for discriminant subsequences helps us to understand the salient distinctions between groups, such as between men or women or between successive birth cohorts. Finally, clustering and the resulting suggested typology serves to identify dominant types of trajectories.

The mining of frequent sequential patterns has received a lot of attention from the data mining community which proposed plenty of efficient algorithms, the main concern being scalability to very large datasets and the ability to handle time and content constraints. The mining of frequent patterns from life course data is quite new, however. The only such application that we know about is the study by Blockeel et al. (2001). One reason for this apparently low interest in the mining of frequent sequential patterns in social sciences is that the available tools remained barely accessible to non-computer scientists. We therefore put much effort into implementing simple commands—illustrated in the text—in our TraMineR R-package to assist the user in inputting the data in a suitable form and in running the mining process under various constraints. The tools for finding discriminant subsequences and computing OME distances between event sequences are unique features of TraMineR and are not currently available elsewhere.

It is worth mentioning that the clustering of event sequences as illustrated in the previous section, is just one possible use of the pairwise OME dissimilarities. Indeed, once we have the OME distances, we can run any dissimilarity-based analysis including, among others, self-organizing maps (Kohonen, 1997; Massoni et al., 2009), principal coordinate analysis (Gower, 1966), the search for representative sequences (Gabadinho et al., 2011), ANOVA-like discrepancy analysis and regression trees (Studer et al., 2011).

What did we learn from our exploration of Swiss life courses? As expected, the findings from frequent and discriminant subsequences essentially concern the sequencing of events. The main result is that while experiencing simultaneously in the same year leaving home, moving in with a partner, and the first childbirth predominated among the older cohorts (people born between 1910 and 1924), the norm tends towards more spaced events in younger cohorts. Differences between male and female cohabitational trajectories are not very important and essentially concern event timing. Regarding occupational events, however, there are very strong differences. The results clearly demonstrate that the 20th century is characterized by the emergence of part time working and that this phenomenon mainly concerns women. While staying at home remains quite common for women, we observed a slow shift across cohorts from cases where women stay at home after 20 years old to a model where women first work full-time during their early twenties and only later stop working to stay at home, generally after the first childbirth. The cluster analysis confirmed these results but complemented them by emphasizing important timing differences. For example, women most often stop education to start working full-time around 21 years old while it is more typical for men to do so around the age of 26.

Notes

This publication results from research work executed within the framework of the Swiss National Centre of Competence in Research LIVES, which is financed by the Swiss National Science Foundation. The authors are grateful to the Swiss National Science Foundation for its financial support. In addition the authors also thank Nicolas S. Müller for his participation in the coding of the TraMineR functions for event sequences.

1 See Studer et al. (2010) and the user's guide (Gabadinho et al., 2009) for details about TraMineR's event sequence features.
2 See www.swisspanel.ch
3 Although, as pointed out to us by Raffaella Piccarreta, directional measures such as the τ of Goodman and Kruskal (1954) or the uncertainty coefficient u of Theil (1970) would be better suited for this discrimination purpose, they are not yet implemented in TraMineR
4 The retained number of groups corresponds to an elbow from which quality measures such as the average silhouette (ASW = .24) and the discrepancy reduction ($R^2 = .39$) only grow slowly when we increase the number of clusters.
5 Again, the retained number of groups corresponds to an elbow in the evolution of quality measures such as the average silhouette (ASW = .39) and the discrepancy reduction ($R^2 = .55$).

References

Abbott, A. (1983). Sequences of social events: Concepts and methods for the analysis of order in social processes. *Historical Methods 16*(4), 129–147.
Abbott, A. (1991). The order of professionalization: An empirical analysis. *Work and Occupations 18*(4), 355–384.
Abbott, A. and J. Forrest (1986). Optimal matching methods for historical sequences. *Journal of Interdisciplinary History 16*, 471–494.
Abbott, A. and A. Tsay (2000). Sequence analysis and optimal matching methods in sociology, review and prospect. *Sociological Methods and Research 29*(1), 3–33 with discussion, pp. 34–76.

Agrawal, R. and R. Srikant (1994). Fast algorithm for mining association rules in large databases. In J. B. Bocca, M. Jarke, and C. Zaniolo (Eds.), *Proceedings 1994 International Conference on Very Large Data Base (VLDB'94), Santiago de Chile,* San-Mateo, pp. 487–499. New York: Morgan-Kaufman.

Agrawal, R. and R. Srikant (1995). Mining sequential patterns. In P. S. Yu and A. L. P. Chen (Eds.), *Proceedings of the International Conference on Data Engineering (ICDE), Taipei, Taiwan,* pp. 487–499. New York: IEEE Computer Society.

Agrawal, R., H. Mannila, R. Srikant, H. Toivonen, and A. I. Verkamo (1995). Fast discovery of association rules. In U. M. Fayyad, G. Piatetsky-Shapiro, P. Smyth, and R. Uthurusamy (Eds.), *Advances in Knowledge Discovery and Data Mining,* pp. 307–328. Menlo Park, CA: AAAI Press.

Aisenbrey, S. and A. E. Fasang (2010). New life for old ideas: The "second wave" of sequence analysis bringing the "course" back into the life course. *Sociological Methods and Research 38*(3), 430–462.

Bettini, C., X. S. Wang, and S. Jajodia (1996). Testing complex temporal relationships involving multiple granularities and its application to data mining (extended abstract). In *PODS '96: Proceedings of the fifteenth ACM SIGACT-SIGMOD-SIGART Symposium on Principles of Database Systems,* pp. 68–78. New York: ACM Press.

Billari, F. C., J. Fürnkranz, and A. Prskawetz (2006). Timing, sequencing, and quantum of life course events: A machine learning approach. *European Journal of Population 22*(1), 37–65.

Blockeel, H., J. Fürnkranz, A. Prskawetz, and F. Billari (2001). Detecting temporal change in event sequences: An application to demographic data. In L. De Raedt and A. Siebes (Eds.), *Principles of Data Mining and Knowledge Discovery: 5th European Conference, PKDD 2001,* Volume LNCS 2168, pp. 29–41. Freiburg in Brisgau: Springer.

Blossfeld, H.-P., K. Golsch, and G. Rohwer (2007). *Event History Analysis with Stata.* Mahwah NJ: Lawrence Erlbaum.

Brzinsky-Fay, C., U. Kohler, and M. Luniak (2006). Sequence analysis with Stata. *The Stata Journal 6*(4), 435–460.

Bürgin, R., G. Ritschard, and E. Rousseaux (2012). Visualisation de séquences d'événements. *Revue des Nouvelles Technologies de l'Information (RNTI), E-23,* 559–560.

Elzinga, C. H. (2003). Sequence similarity: A non-aligning technique. *Sociological Methods and Research 31,* 214–231.

Elzinga, C. H. (2010). Complexity of categorical time series. *Sociological Methods & Research 38*(3), 463–481.

Gabadinho, A., G. Ritschard, N. S. Müller, and M. Studer (2011). Analyzing and visualizing state sequences in R with TraMineR. *Journal of Statistical Software 40*(4), 1–37.

Gabadinho, A., G. Ritschard, M. Studer, and N. S. Müller (2009). Mining sequence data in R with the TraMineR package: A user's guide. Technical report, Department of Econometrics and Laboratory of Demography, University of Geneva, Geneva.

Gabadinho, A., G. Ritschard, M. Studer, and N. S. Müller (2011). Extracting and rendering representative sequences. In A. Fred, J. L. G. Dietz, K. Liu, and J. Filipe (Eds.), *Knowledge Discovery, Knowledge Engineering and Knowledge*

*Management,*Volume 128 of *Communications in Computer and Information Science (CCIS),* pp. 94–106. Springer-Verlag.

Goodman, L. A. and W. H. Kruskal (1954). Measures of association for cross classifications. *Journal of the American Statistical Association 49,* 732–764.

Gower, J. C. (1966). Some distance properties of latent root and vector methods used in multivariate analysis. *Biometrika 53*(3/4), 325–338.

Joshi, M. V., G. Karypis, and V. Kumar (2001). A universal formulation of sequential patterns. In *Proceedings of the KDD'2001 workshop on Temporal Data Mining, San Fransisco, August 2001.*

Kaufman, L. and P. J. Rousseeuw (2005). *Finding Groups in Data.* Hoboken: John Wiley & Sons.

Kohonen, T. (1997). *Self-Organizing Maps* (2nd edn.). Heidelberg: Springer.

Mannila, H., H. Toivonen, and A. I. Verkamo (1995). Discovering frequent episodes in sequences. In *Proceedings of the First International Conference on Knowledge Discovery and Data Mining (KDD-95), Montreal, Canada, August 20–21, 1995,* pp. 210–215. AAAI Press.

Mannila, H., H. Toivonen, and A. I. Verkamo (1997). Discovery of frequent episodes in event sequences. *Data Mining and Knowledge Discovery 1*(3), 259–289.

Masseglia, F. (2002). *Algorithmes et applications pour l'extraction de motifs séquentiels dans le domaine de la fouille de données: de l'incrémental au temps réel.* PhD thesis, Université de Versailles Saint-Quentin en Yvelines.

Massoni, S., M. Olteanu, and P. Rousset (2009). Career-path analysis using optimal matching and self-organizing maps. In *Advances in Self-Organizing Maps: 7th International Workshop, WSOM 2009, St. Augustine, FL, USA, June 8–10, 2009,* Volume 5629 of *Lecture Notes in Computer Science,* pp. 154–162. Berlin: Springer.

Moen, P. (2000). *Attribute, Event Sequence, and Event Type Similarity Notions for Data Mining.* PhD thesis, University of Helsinki.

R Development Core Team (2012). *R: A Language and Environment for Statistical Computing.* Vienna, Austria: R Foundation for Statistical Computing.

Ritschard, G., A. Gabadinho, N. S. Müller, and M. Studer (2008). Mining event histories: A social science perspective. *International Journal of Data Mining, Modelling and Management 1*(1), 68–90.

Ritschard, G., A. Gabadinho, M. Studer, and N. S. Müller (2009). Converting between various sequence representations. In Z. Ras and A. Dardzinska (Eds.), *Advances in Data Management,*Volume 223 of *Studies in Computational Intelligence,* pp. 155–175. Berlin: Springer-Verlag.

Srikant, R. and R. Agrawal (1996). Mining sequential patterns: Generalizations and performance improvements. In P. M. G. Apers, M. Bouzeghoub, and G. Gardarin (Eds.), *Advances in Database Technologies – 5th International Conference on Extending Database Technology (EDBT'96), Avignon, France,*Volume 1057, pp. 3–17. Berlin: Springer-Verlag.

Studer, M., N. S. Müller, G. Ritschard, and A. Gabadinho (2010). Classer, discriminer et visualiser des séquences d'événements. *Revue des nouvelles technologies de l'information RNTI E-19,* 37–48.

Studer, M., G. Ritschard, A. Gabadinho, and N. S. Müller (2011). Discrepancy analysis of state sequences. *Sociological Methods and Research 40*(3), 471–510.

Theil, H. (1970). On the estimation of relationships involving qualitative variables. *American Journal of Sociology 76,* 103–154.

Widmer, E. and G. Ritschard (2009). The de-standardization of the life course: Are men and women equal? *Advances in Life Course Research* 14(1–2), 28–39.

Yamaguchi, K. (1991). *Event History Analysis.* ASRM 28. Newbury Park and London: Sage.

Zaki, M. J. (2001). SPADE: An efficient algorithm for mining frequent sequences. *Machine Learning* 42(1/2), 31–60.

Part II

Applications

10 Clinical versus Statistical Prediction of Zygosity in Adult Twin Pairs

An Application of Classification Trees

Carol A. Prescott

Introduction

Studies of twins have been used to establish the importance of both genetic and environmental factors in the development of human traits and disease. A key requirement for twin studies is an accurate and efficient method for distinguishing identical (monozygotic, MZ) twin pairs from fraternal (dizygotic, DZ) pairs. This chapter compares three methods for combining self-report items used to classify pair zygosity: a clinical algorithm, logistic regression, and classification trees using the R program RPART (Therneau & Atkinson, 1997). These methods were applied to 16 self-report items on physical similarity, childhood environmental similarity, height and weight obtained from 554 adult twin pairs from the Virginia Adult Twin Study of Psychiatric and Substance Use Disorders (Kendler & Prescott 2006). The classification accuracy of each method was evaluated against zygosity obtained by genotyping. All three methods obtained high levels of accuracy, ranging from 90 to 93%. The three items contributing most to accurate classification were: the twins' report of how often as children they were mistaken for each other by strangers; whether they were "as alike as two peas in a pod;" and their opinion about whether they were identical. Items for environmental treatment, height and weight contributed little to the solutions, although the classification tree solution specific to females did use height and weight differences between twins. Overall, the results indicate the items commonly used in twin studies to assign zygosity give accurate assignment for the majority of twin pairs assessed as adults. If replicated, the algorithms based on regression trees may provide some additional accuracy for pairs that cannot be classified due to inconsistent responses to these items.

Twin studies have been instrumental in elucidating the basis of individual differences in risk for disease and in behavioral traits. Traditionally used to estimate "heritability," the genetic contribution to individual differences, twin designs are also useful for studying the effects of environmental influences, including prenatal effects, and family and social factors. The classical twin model apportions variation into proportions of genetic and environmental variance, but contemporary twin studies frequently employ measures of genetic and environmental influence as well. The focus of much current twin research is to understand the interplay of genotypes and environments (McGue et al., 2010; Visscher et al., 2008) and other processes underlying variation in gene function (e.g., epigenetics; Bell & Saffery, 2012; Visscher et al., 2008; Haque et al., 2009).

There are two different mechanisms that underlie twinning. Monozygotic (MZ) or "one-egg" twins occur from the splitting of a single zygote following fertilization of one egg by one sperm. Twins in an MZ pair are thus identical for inherited genetic variation. Dizygotic (DZ), "two-egg" twins occur when two eggs are fertilized by different sperm. Genetically, DZ twins are as similar as ordinary siblings, sharing (on average) 50% of their inherited genetic variation. (See Keith et al., 1995, for details on the biology of twinning.)

Traditional twin studies distinguish genetic and environmental sources of variation by comparing twin pair resemblance in MZ and DZ pairs. The two types of twin pairs are assumed to be equally similar for environmental exposures relevant to the development of the outcome being studied. Therefore, to the degree that resemblance among MZ pairs exceeds that of DZ pairs, genetic factors are implicated. If the two twin types are equally similar, this indicates the contribution of environmental factors shared by the twins. To the extent that identical twins are dissimilar, this implicates the role of environments and experiences that are unique to the individual (and any measurement error associated with the outcome).

Another application of twin pairs used commonly in epidemiology and molecular genetic studies is the cotwin-control design. These approaches employ pairs who have different outcomes and can be studied for differences in exposure to environmental risk factors or biological mechanisms. Prospective cotwin-control studies identify pairs that differ on a risk factor and follow them up to study similarities and differences in disease and behavioral outcomes. Some cotwin-control designs use just MZ pairs; others also include DZ pairs to evaluate how the same risk factor may be expressed with a different genotype.

Classifying Twin Pair Zygosity

A key requirement for all types of twin studies is having an accurate and efficient method for distinguishing MZ from DZ pairs. In the classical twin design, zygosity classification errors have the effect of reducing the observed MZ–DZ difference in pair resemblance, producing an underestimate of the impact of genetic variation. In cotwin-control designs and other methodologies that use differences within MZ pairs, zygosity errors will reduce the power to identify individual-specific environmental or biological processes, as these will be confounded with genetic differences.

Prior to the widespread use of hormonal fertility treatments, the twinning rate was about 1 in 80 births (i.e., 2.5% of individuals born). However, due to higher mortality among twins, a commonly used prevalence estimate is that 2% of adults over age 30 are twins. The MZ twinning rate is approximately constant, whereas the DZ rate varies across ethnicity and maternal age (Derom et al., 1987). In Caucasian populations, the MZ:DZ rate in unassisted pregnancies is about 1:2. Since half of DZ pairs are opposite-sex (male–female), this means the expected proportion of MZs among same-sex pairs is 50%. Hormonal treatments and fertilization procedures produce DZ twins, so the marked increase in multiple births since the mid-1980s is due primarily to the increase in DZ pairs.

Historically, zygosity typing of pairs relied on a combination of questionnaire responses and photographs, validated through typing of blood groups (e.g., Gottesman, 1963; Carter-Salzmann & Scarr, 1977). Zygosity can now be determined with essentially complete accuracy by typing a few genetic markers using DNA extracted from saliva or cheek cells (e.g., Becker et al., 1997). However, although genotyping is largely automated, the cost of typing a large number of pairs is not trivial. The supply and labor costs for the sample collection kit, sample processing, DNA extraction, and genotyping are currently about $100 per pair. Thus, it is highly desirable to have a means of identifying pair zygosity in advance rather than going to the expense of collecting and testing biological samples on pairs that may not be useful for the study design.

Why Don't We Just Ask Twins What Kind of Pair They Are?

Readers unfamiliar with twin research may wonder why zygosity typing requires more than a single self-report item. Many MZ pairs are virtually indistinguishable in appearance and their zygosity is obvious. But other MZ pairs are not as physically similar, and they become less similar over time due to differences in aging, weight gain, and cosmetic variation from hair styles and alteration of hair color and skin tone. It is also the case that many DZ pairs, like ordinary siblings, are quite similar in physical appearance, particularly in their youth and thus may be unsure of their zygosity.

In addition to pairs who are uncertain of their zygosity, many pairs believe they know their zygosity, but are wrong. In older birth cohorts, about 20% of pairs were mistaken about their zygosity (e.g., Nichols & Bilbro, 1966; Carter-Salzmann & Scarr, 1977). This was sometimes due to misinformation from physicians, who believed that twin pairs with a single placenta were always MZ. In fact, DZ pairs may have fused placentas and histological examination is needed to tell the difference. It is not uncommon for twins, when asked why they believe they are members of an MZ pair to say, "We look different but the doctor said we're identical." Consequently, algorithms developed to classify zygosity reply more on reports about physical appearance than on twins' opinions about their status.

Other sources of zygosity information that can be easily obtained are physical measures and photographs for rating facial similarity, and eye and hair color. Birthweight is not an accurate index of zygosity; there are several mechanisms by which MZ pairs may become quite discrepant in size due to overlapping vascular systems while in utero (Lopriore et al., 1995).

If zygosity classification errors were random, there would not be a great concern; such noise reduces study power but the tradeoff might be acceptable if the costs of obtaining additional pairs were less than the costs of genotyping. However, there are known biases in zygosity typing. For example, classifications based on height and weight or ratings of physical similarity based on photographs have yielded a bias for DZ pairs to be classified as MZ (Gottesman, 1963; Carter-Salzmann & Scarr, 1977). This is because classifying pairs as DZ based on physical dissimilarity is highly accurate whereas similar pairs may be MZ or DZ. For example, in a population based sample of U.S. twins born in the 1940s and assessed in high school, only 22% of same-sex pairs had dissimilar heights and weights and could be confidently classified as DZ (Prescott et al., 2011).

Study Aims

This study was designed to address three goals. The first was to use an exploratory approach to determine to what degree zygosity classification accuracy based on self-reported information about physical similarity could be improved over typically employed methods. Second was to explore whether other information commonly obtained from twins could improve classification over using just the standard items. Third was to explore whether there were any differences between female and male pairs in classification accuracy or in what variables were the best predictors of zygosity.

Method

Subjects

The data used for these analyses are from the Virginia Adult Twin Study of Psychiatric and Substance Use Disorders (VATSPSUD, Kendler & Prescott, 2006), a longitudinal clinical interview study of >9,000 twins. Pairs were originally identified through state birth certificates and then located by matching names and birthdates to department of motor vehicle records or other public sources. Pairs were born between 1940 and 1974 and the study was limited to pairs identified as "white" on birth certificates. The sample includes two parallel studies. The first was of twins from female–female

(FF) pairs. Later, twins from male–male (MM) and male–female pairs were included.

Cell samples for DNA extraction were requested from participants interviewed in wave 1 with MM pairs and wave 4 with FF pairs. These were obtained in person (for MM) or by mail (FF) by having participants rub four small cytology brushes gently against the inside of their cheeks. From among the pairs who provided samples, 269 FF and 296 MM pairs were selected for genotyping. These included 196 pairs randomly selected to validate the questionnaire zygosity classification. Additional pairs were genotyped at their request or because questionnaire responses were inconclusive or inconsistent between twins within a pair. Four of the genotyped pairs were excluded from the current analyses because of missing data: three twins refused to report their weights; another was missing one of the questionnaire items. The analyses reported here are based on 286 male pairs and 268 female pairs with complete questionnaire data.

DNA was genotyped using polymerase chain reaction (PCR) markers, with an average of 17.5 markers typed for the FF pairs and 11.5 for the MM pairs (Kendler & Prescott, 1999). The number tested was reduced for MM pairs as these were conducted later and it had been found with the FF pair analyses that fewer markers were needed to obtain a high level of precision.

Measures

Outcome

The outcome predicted in the analyses is zygosity classification (coded as $MZ = 0, DZ = 1$) obtained by genotyping. Pairs identical on all markers were classified as MZ, those with different genotypes on two or more markers were classified as DZ (Becker et al., 1997). Of the 554 pairs included in the current analyses, there are 141 FF and 141 MM pairs being classified as MZ and 127 FF and 145 MM pairs classified DZ by this method. The relatively similar distribution of pairs across zygosity and sex was not predefined, but reflects the expected proportion of 50% of same-sex twin pairs in this age range being MZ (Derom et al., 1987).

Predictors

A very large number of variables were available for inclusion in exploratory analyses. More than 5,000 variables have been collected from VATSPSUD participants and many of these differ in agreement between MZ and DZ twin pairs. However, most of these measures are behavioral phenotypes that could be *outcomes* of zygosity differences rather than determinants of it. Another consideration in measure selection was generalizability. With sufficient numbers of variables the zygosity groups could be classified perfectly, but such a tailored solution would not be expected to generalize to another sample.

Sixteen variables were selected for inclusion as possible predictors of zygosity. Five were from a standard set of questions by Nichols and Bilbro (1966) assessing the physical similarity of the twins during childhood (defined as up to age 12). Three of these items asked about the frequency with which, when they were children, the two twins were mistaken for each other by strangers, teachers, and parents (coded 1–3 for Often, Sometimes, Never). Another item asked if, as children, the twins were "as alike as two peas in a pod" (coded 1–3 for Yes, Not Sure, No). The fifth item asked the twin's opinion about the pair's zygosity (coded 1–4 for Definitely MZ, Probably MZ, Probably DZ, or Definitely DZ). These items are henceforth referred to as: Strangers, Teachers, Parents, Peas, and Opinion, respectively. The items were all coded in the direction of higher scores indicating less similarity and thus expected to predict being DZ.

We also included five items that are typically used to index the similarity of the twins' environment. Whether the twins were currently living together was coded as a binary variable (Cohabiting: 0=yes, 1=no). Four items on childhood rearing environment asked how often the twins shared a bedroom, dressed alike, shared playmates, and were in the same school classroom. These items were coded on a four-point scale (Usually, Often, Sometimes, Rarely).

We also included some demographic and physical variables. Sex was coded as 0 = female, 1 = male. Age was coded in years at the time of the data collection. We included the average height and weight for the pair and the absolute within-pair difference of height and of weight, calculated from self-reports.

These items were collected by interviews conducted in person for FF pairs (in 1987–89) and by telephone for MM pairs (in 1993–96). Twins within a pair were interviewed by two different interviewers who were unaware of the appearance or other characteristics of the co-twin.

Analysis Strategy

Four classification methods were compared against genotype-based zygosity: a simple cutting score, a clinical algorithm, a logistic regression model, and an exploratory approach using classification trees. Following the conventions from clinical decision making (e.g., Dawes, 1962), five measures of accuracy are reported for each method:

- Overall accuracy is defined as the proportion of individuals who are correctly placed by the method
- Sensitivity is the proportion of actual MZ pairs who are classified as MZ by the algorithm

- Positive predictive value (PPV) is the proportion of pairs classified as MZ by the algorithm who are actually MZ
- Specificity is the proportion of actual DZ pairs who are classified as DZ by the algorithm
- Negative predictive value (NPV) is the proportion of pairs classified as DZ by the algorithm who are actually DZ.

The proportion of pairs called MZ by each method is also reported. These six values are obviously redundant, as they are all based on a 2 × 2 classification table, but having the information in this form is useful for comparing across methods and for generalizing to other contexts with different base rates (e.g., Gottesman & Prescott, 1989).

Cutting Score on Unit-Weighted Scale

The four most commonly used items from the Nichols and Bilbro (1966) measure (Opinion, Peas, Strangers, and Teachers) were put into equivalent units (all 1–3) and summed. The Parents item was excluded as parents rarely confuse even their identical twin children so the absence of confusion is not evidence of MZ zygosity (although the presence of confusion is useful evidence). The possible range was 4 to 12. A score of 8 was selected *a priori* as the cutting score for classifying pairs as DZ. Scores of 8 and higher represented responding "Sometimes" or "Rarely" to all of the confusion items, giving a "Probable" or "Definite" DZ response to Opinion, and answering "Not Sure" or "No" to Peas. Pairs with scores below 8 were classified as MZ.

Clinical Method

The second method used a clinical algorithm based on the same items (Opinion, Peas, Strangers, Teachers, and Parents). A version of this algorithm used in the present case is depicted as a flowchart in Figure 10.1. This algorithm is based on the author's attempt to codify the clinical process, but it should be noted that often such a precise decision tree is not followed. Rather, an "expert rater" reviews the responses of both twins to these items along with any additional information, such as interviewer notes or photographs. For example, in the VATSPSUD FF sample, ratings of the pairs' similarity based on color photographs of the twins were combined with the questionnaire information to make the zygosity classifications (Kendler & Prescott, 2006). (Photograph ratings were not included in the current analyses because photos were not available for MM pairs). With pairs for whom the available information is inconclusive or contradictory between twins or items, the expert may seek a second opinion or employ biological validation.

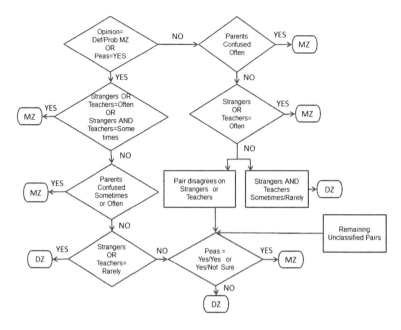

Figure 10.1 Flowchart of clinical algorithm for classifying zygosity in twin pairs.
Notes: Example of decision rules made by raters reviewing zygosity questionnaire responses for twin pairs. Diamond indicates decision point based on items. Classifications shown in MZ or DZ ovals. / indicates twins gave different answers; e.g., Def/Prob = one twin reported definitely the other twin reported probably.

The algorithm employed here attempts to adjust for two biases described previously—the tendency for twins to be mistaken about their zygosity and for some proportion of pairs called MZ based on physical similarity actually to be DZ. Although developed independently, it is similar to the process described by Goldsmith (1991) in which the algorithm first evaluates the hypothesis that a pair is DZ by rejecting that they are identical; if that cannot be rejected, the second part of the process is to evaluate the hypothesis the pair is MZ.

Logistic Regression

The third method compared is logistic regression. The analyses were conducted for the full sample combining across sexes and then separately within males and females. In the combined analyses, 16 predictors were entered in a model that estimated the probability of being DZ. Predictors were removed in a backward stepwise process based on highest p-level until all remaining predictors had a p-level below 0.10. This broad criterion was employed to allow a better chance for the final models in each method to be comparable. For the within-sex analyses, the same procedure was followed but starting with 15 predictors (all but sex).

Classification accuracy for the regression models was assessed as follows. The expected value of the logistic equation obtained from the final regression analysis was calculated for each subject. Pairs with a probability in the lower half ($<.50$) were assigned as MZ and those in the upper half ($>.50$) as DZ. These classifications were then compared against those obtained by genotyping. Although .50 differs from the base rate of MZ pairs in the sample, in general this would not be known and one would use the population expectation of 50%. Within the sexes, classification accuracy was examined in two ways: once using the regression equation estimates from the entire sample and then using estimates obtained from the sex-specific analysis.

Classification Trees

For our exploratory method, we used the Recursive Partitioning and Regression Trees package implemented through the R library (RPART, v 3.1, Therneau & Atkinson, 1997). As described elsewhere in this volume (e.g., Chapter 1 and 2), classification tree methods use an iterative process to identify a variable and cutpoint that produce the greatest classification accuracy against a criterion outcome. This first cutpoint results in two "branches" which are then evaluated separately to identify cutpoints on other variables that further improve the classification. The branching continues until a user-defined criterion is met. After the tree is "grown," a second step is "pruning," in which branches are cut back and the solution is cross-validated to determine the alpha level associated with the pruning criterion (see Therneau & Atkinson, 1997). The program output contains the number of cases that would be correctly classified at each step and as a result of the final splits (which produce "terminal nodes").

The analyses reported here used the "classification" option within RPART which maximizes purity of the nodes as measured by the Gini index (see Breiman et al., 1984). Further splitting on a particular branch would halt if the terminal node had fewer than 10 observations (i.e., RPART command: minsplit = 10). This is a somewhat arbitrary cutoff, but given the relatively large sample size and variable set, the intent was to seek a balance between not missing any novel findings versus capitalizing too much on features specific to the sample and producing solutions that would not generalize. Each solution was based on 10 cross-validations (i.e., xval = 10). Otherwise the default program options were used. As with the logistic regression models, the analyses were first conducted with the full sample and then separately for females and males.

Accuracy estimates were calculated for the full sample and within sex by using the classification rules implied by the grown tree. All individuals allocated to a terminal node were assigned to the zygosity group (MZ or

DZ) more common in that node. These classifications were then compared to those based on genotyping. As with the logistic regression analyses, classification accuracy was examined in two ways within the sexes: once using the algorithm developed on the entire sample and again using the algorithms resulting from the sex-specific analysis.

Results

Descriptive statistics for each of the predictor variables are given in Table 10.1 by sex and zygosity. As expected, responses to the standard items show marked differences between MZ and DZ groups, with the MZ pairs much more likely to classify themselves as Probably or Definitely MZ, to endorse being as alike as two peas in a pod, and to be confused by strangers and teachers. Consistent with results from studies of juvenile twins indicating that parents are quite accurate in distinguishing their MZ offspring (e.g., Rietveld et al., 2000; Heath et al., 2003), the MZ and DZ groups had only slight differences on the Parents item. The group averages

Table 10.1 Descriptive statistics for genotyped twin pairs by sex and zygosity

	Female pairs		Male pairs	
	MZ (n = 141)	DZ (n = 127)	MZ (n = 141)	DZ (n = 145)
	M (SD)	M (SD)	M (SD)	M (SD)
Standard items				
Twins' opinion[a]	2.0 (1.1)	3.7 (0.6)	1.7 (0.9)	3.6 (0.6)
Peas in a pod[b]	1.4 (0.7)	2.8 (0.5)	1.5 (0.7)	2.8 (0.4)
Confused by strangers[b]	1.2 (0.4)	2.3 (0.7)	1.1 (0.3)	2.3 (0.6)
Confused by teachers[b]	1.4 (0.5)	2.5 (0.5)	1.5 (0.5)	2.5 (0.5)
Confused by parents[b]	2.4 (0.5)	2.9 (0.2)	2.5 (0.5)	2.9 (0.2)
Sum4	5.7 (1.9)	10.4 (1.7)	5.6 (1.5)	10.3 (1.5)
Environmental similarity				
Dressed alike[a]	2.3 (0.7)	2.4 (0.7)	2.6 (0.8)	2.9 (0.7)
Shared room[a]	1.4 (0.6)	1.3 (0.5)	1.4 (0.6)	1.5 (0.7)
Shared classrooms[a]	2.6 (0.8)	2.9 (0.8)	2.8 (0.8)	2.9 (0.7)
Shared playmates[a]	1.4 (0.5)	1.8 (0.7)	1.7 (0.6)	2.0 (0.7)
Currently cohabiting	21%	12%	15%	9%
Biometric variables				
Age (years)	29.1 (7.3)	31.0 (7.9)	33.9 (9.0)	36.1 (8.4)
Mean height (in.)	64.5 (2.4)	65.0 (2.4)	70.4 (2.4)	70.3 (2.2)
Height difference (in.)	1.1 (1.3)	2.3 (1.7)	1.1 (1.5)	2.0 (1.9)
Mean weight (lb)	133.3 (23.1)	133.3 (23.1)	179.5 (30.5)	185.0 (24.2)
Weight difference (lb)	11.9 (14.1)	11.9 (14.1)	14.9 (14.9)	24.5 (19.3)

Values are based on ratings averaged across twins within a pair. Ordered responses, with [a] range of 1–4 (more to less similar); [b] range of 1–3 (more to less similar); Sum4 = Opinion, Peas, Stranger, and Teachers items equally weighted and summed (range 4–16).

on the unit-weighted sum of the first four items (all but Parents) amplify these groups' differences, with the MZ and DZ averages more than three SD apart.

In contrast, the average ratings on measures of childhood environmental similarity are about equal for MZs and DZs, particularly the items that reflect external control—being dressed alike, sharing a bedroom, and being in the same classrooms at school. There is slightly more similarity among MZ pairs than among DZ pairs for sharing playmates, which may reflect personal choice in peer selection. The proportion of pairs currently cohabiting as adults is much greater among MZ than DZ pairs (18% vs 10%) and is somewhat higher in females than males.

Due to sample differences in when these data were collected, the males on average were older (M = 35.0 years, SD = 8.7) than the females (M = 30.0, SD = 7.6), but they represent the same birth years (1940–74). The gender difference in cohabiting may reflect this age difference; some young adults may be cohabiting because they are still residing with their parents. MZ and DZ groups are very similar in height and weight and are representative of population averages (e.g., height of 5 feet 5 inches for women, 5 feet 10 inches for men). As expected, there are differences in within-pair variation: on average, MZ pairs differ in height by 1 inch compared to 2 inches for DZ pairs. The within-pair weight difference varied by gender; MZ and DZ female pairs were equally similar (average difference of 12 lb), but among males there was greater variation in DZ pairs (25 lb vs 15 lb in MZ).

Cutting Score Approach

Using the simple cutting score on the unit-weighted scale, 251 pairs were correctly classified as MZ and 251 were correctly classified as DZ, an overall accuracy of 90.6%. It was serendipitous that these numbers are equal. Although we picked a cutting score that was midway through the range of possible scores, there is no constraint that the data needed to be distributed symmetrically or that the classifications be similar for both groups. Overall, 49% of pairs were classified as MZ by this method. This represents a sensitivity of 89.0% (251 called MZ of the actual 282 MZ pairs), a PPV of 92.3% (251 of the 281 called MZ are actually MZ), specificity of 92.3% (251 called DZ of the actual 272 DZ pairs), and NPV of 89.0% (251 called DZ of the actual 273 DZ pairs). Table 10.2 gives these values and also the results when this cutting score is applied separately within FF and MM pairs. Overall, the algorithm obtained very similar accuracy for the two sexes.

Clinical Algorithm

Using the clinical algorithm shown in Figure 10.1, 253 pairs were correctly classified as MZ and 244 correctly classified as DZ, an overall accuracy of

Table 10.2 Classification accuracy by approach for 282 MZ and 272 DZ genotyped twin pairs

	Cutting score	Clinical algorithm	Logistic regression	Classification tree
All pairs				
Overall accuracy	90.6%	89.7%	91.3%	93.3%
MZ proportion	49.1%	50.7%	50.2%	50.0%
Correctly classified as MZ	$n = 251$	$n = 253$	$n = 256$	$n = 256$
Sensitivity	89.0%	89.7%	90.8%	90.8%
PPv	92.3%	90.0%	92.1%	92.4%
Correctly classified as DZ	$n = 251$	$n = 244$	$n = 250$	$n = 251$
Specificity	92.3%	89.7%	91.9%	92.3%
NPV	89.0%	89.4%	90.6%	90.6%

	Cutting score	Clinical algorithm	Logistic regression		Classification tree	
			Combined solution	Sex-specific	Combined solution	Sex-specific
Females pairs						
Overall accuracy	89.2%	89.2%	89.9%	86.9%	91.4%	90.7%
MZ proportion	49.3%	52.2%	52.2%	44.8%	50.0%	50.0%
Correctly classified as MZ	$n = 122$	$n = 126$	$n = 127$	$n = 113$	$n = 126$	$n = 128$
Sensitivity	86.5%	89.4%	90.1%	80.1%	89.4%	90.8%
PPV	92.4%	90.0%	90.7%	94.2%	94.0%	91.4%
Correctly classified as DZ	$n = 117$	$n = 113$	$n = 114$	$n = 120$	$n = 119$	$n = 115$
Specificity	92.1%	89.0%	89.8%	94.5%	93.7%	90.6%
NPV	86.0%	88.3%	89.1%	81.1%	88.8%	89.8%
Male pairs						
Overall accuracy	92.0%	90.2%	92.7%	92.3%	91.6%	92.0%
MZ proportion	49.0%	49.3%	48.3%	48.6%	50.0%	50.0%
Correctly classified as MZ	$n = 129$	$n = 127$	$n = 129$	$n = 129$	$n = 130$	$n = 125$
Sensitivity	91.5%	90.1%	91.5%	91.5%	92.2%	88.7%
PPV	92.1%	90.1%	93.5%	92.8%	90.9%	94.7%
Correctly classified as DZ	$n = 134$	$n = 131$	$n = 136$	$n = 135$	$n = 132$	$n = 138$
Specificity	92.4%	90.3%	93.8%	93.1%	91.0%	95.2%
NPV	91.8%	90.3%	91.9%	91.8%	92.3%	89.6%

Sensitivity, proportion of actual MZ pairs correctly identified; PPV = Positive predictive value, proportion of pairs classified MZ that are correctly classified; Specificity, proportion of actual DZ pairs correctly identified; NPV = Negative predictive value, proportion of pairs classified DZ that are correctly classified. n = number of pairs correctly classified as MZ or DZ
"Actual" classification based on genotyping: MZF = 141, DZF = 127, MZM = 141, DZM = 145.

89.7% (497/554). This represents a sensitivity of 89.7% (253/282), a PPV of 90.0% (253/281), specificity of 89.7% (244/272), and NPV of 89.4% (24/273). The algorithm had very similar performance when applied just to the female or male data. The pattern is very similar to that obtained for the Cutting Score approach.

Logistic Regression Analyses

Relative to the fit of a baseline model containing no predictors, the logistic regression model for both sexes combined and including all the predictors obtained a model likelihood ratio (LR) of 504 with 16 degrees of freedom (df). The backwards stepwise procedure resulted in a final model with six predictors that fit almost as well as the full model (LR = 500 vs baseline), implying that the other 10 variables did not independently contribute (in a linear way) to prediction of zygosity. The retained variables were (ordered by Wald chi-square value): Opinion, confused by Strangers, Peas in a pod, dressing alike, average height, and confused by Teachers. Table 10.3 gives the odds ratio (OR) associated with each variable in the direction of predicting DZ pairs. For example, the OR of 2.5 for the Opinion item indicates that for each higher point a pair obtained on this four-point question, the pair is estimated as 2.5 times more likely to be DZ than MZ. The magnitude of the ORs reflect the scaling of the original measures, so must be considered against the measurement units. For example, height is a significant predictor with an OR of 1.1, indicating a 10% increase in probability of being DZ associated with each additional inch of height.

The classification accuracy for this model is slightly better than observed for the clinical algorithm—but the difference is small, particularly considering that the regression result was obtained from predicting the known outcome whereas the clinical algorithm was not. The overall accuracy of the final regression model was 91.3% (506/554), with three more MZ pairs correctly classified and six more DZ pairs correctly classified than by the clinical algorithm. The largest difference in the other accuracy estimates is the PPV of 92%, reflecting that the logistic model classified four fewer individuals as MZ (277 vs 281) but a higher number of those classified MZ were actually MZ (256 vs 253). When the classification results from this model were examined within FF and MM pairs, it obtained slightly better classification for males than females (Table 10.2). This is expected given that a majority of pairs in the combined analysis were male.

We next ran the series of logistic regression models separately within sex. Among females, the full model obtained LR = 228 (df = 15), compared to LR = 223 for the reduced model with five variables. Higher scores on Opinion, Peas, Strangers, and height all predicted in the DZ direction. In the context of these other variables, sharing of a bedroom was in the opposite direction (significantly less frequent among MZs). These estimates (labeled female-specific in Table 10.2) resulted in somewhat

Table 10.3 Variables included in final models from different approaches

	Clinical algorithm variable importance	Logistic regression Odds ratio (95% CI)			RPART variable importance		
		All	Female	Male	All	Female	Male
Twins' opinion[a]	1	2.5 (1.8,3.6)	1.6 (1.0,2.5)	4.4 (2.4,8.1)	2	3	1
Peas in a pod[b]	3	2.8 (1.7,4.5)	3.3 (1.6,6.4)	4.8 (2.1,11.0)	3	1	2
Confused by strangers[b]	2	5.3 (2.4,11.6)	7.0 (3.1,15.6)	11.2 (4.2,30.1)	1	2	2
Confused by teachers[b]	2	2.1★ (0.9,4.8)	–	–	2	2	3
Confused by parents[b]	4	–	–	–	–	–	4
Dressed alike[a]	–	0.5 (0.3,0.8)	–	0.3 (0.1,0.7)	–	–	–
Shared room[a]	–	–	0.5 (0.2,0.9)	–	–	3X	–
Mean height (in.)	–	1.1 (1.0,1.2)	1.2★ (1.0,1.4)	–	–	–	–
Mean weight (in.)	–	–	–	–	–	4X	–
Weight difference (lb)	–	–	–	–	–	5X	–

All = male and female pairs combined. Variable importance based on order the variable appears in the branching (See Figures 10.1–10.4). Variables with the same value appeared at the same level of the tree. X = included in algorithm but its presence did not change ultimate classification of MZ or DZ. Separate results for female and male pairs based on sex-specific analyses (see Table 10.2). Variables other than height and weight have ordered responses, with 1 = more similar and higher values less similar: [a] range is 1–4; [b] range is 1–3 ★$p < .10$; all other estimates $p < .05$.

poorer overall accuracy (86.9% = 233/268) and a different classification pattern than obtained using the combined solution. There was higher PPV and specificity but lower sensitivity and NPV, reflecting fewer MZ but more DZ pairs correctly classified.

Among males, the fit of the full model was LR = 291 (df = 15) compared to LR = 286 for the reduced model, which had four variables. As with the combined sex and female analyses, higher scores on the Opinion, Peas, and Strangers items predicted being DZ. However, in the context of these other variables, more frequent dressing alike predicted being DZ rather than MZ. Among males, the accuracy values from the male-specific solution were almost identical to that of the combined sample solution. Comparing the ORs obtained from the female- and male-specific solutions (Table 10.3) shows that the usefulness of the items differs, with the Opinion, Peas, and Strangers items all stronger predictors among the MM than the FF pairs. This is not evident from examining the differences between MZ and DZ group average scores (Table 10.1), which are very similar in the two sexes. This underscores the importance of not relying on simple bivariate relations when devising complex decision rules.

Classification Tree Analyses

The result from the RPART classification tree analysis combined across sexes is depicted in Figure 10.2. In accordance with the other methods, the standard items are identified as contributing to classification of genotype-based zygosity. The resulting structure is rather simple. The first split occurs on the Stranger item between 227 pairs with both twins reporting they were confused by strangers Sometimes or Rarely, versus 327 pairs for which both twins reported Often (Often/Often) or one twin reported Sometimes and the cotwin reported Often (Sometimes/ Often). This cutoff on one item produces one group that is 93% DZ and another group that is 81% MZ. Among the mostly DZ group (shown on the left hand side of Figure 10.2) only one additional item improves the classification, and this had only a small effect: among eight pairs for which at least one twin responded Often to Teachers, five pairs were MZ.

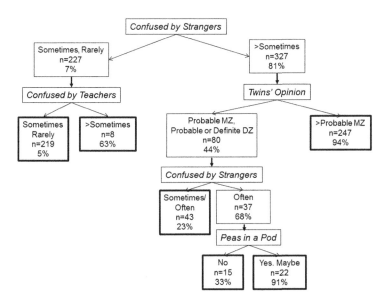

Figure 10.2 Results from classification tree analysis to assign zygosity in 554 male and female twin pairs.

Notes: Solution obtained using RPART with data combined across both sexes. Gray boxes indicate items. White boxes show item cutpoint, number of pairs included in category, and percent of those pairs who are actually MZ. > indicates one or both twins above specified level, e.g., >Probable MZ means both twins reported definite MZ or one twin reported definite MZ and the other reported probable MZ.

In the right branch, among pairs that were more than sometimes confused by strangers, the next split is on the Opinion variable. Among pairs where both twins said Definitely MZ or one reported Definitely MZ and the other Probably MZ, 94% are MZ and this is a terminal node. Among the remainder (i.e., both twins responded Probably MZ or some combination of Probably MZ, Probably DZ, or Definitely DZ), the branching continues by responses to Strangers. Since this variable was also used as a splitting criterion higher up, all that remains in this branch are pairs that responded Sometimes/Often (who are 77% DZ) and pairs that said Often/Often (who are 68% MZ). The 37 cases in the latter group are then split once more by Peas in a Pod, where pairs that agree on No are 67% DZ, whereas other options (all combinations of Yes and Not Sure) are 91% MZ. Unlike the logistic regression solution, none of the environmental similarity or biometric variables contributes to improved classification.

The overall accuracy of this solution is 93.3% (517/554), slightly better than that of the clinical algorithm (Table 10.2). The result among MZ pairs is virtually identical to that obtained with the logistic regression model; the RPART solution identifies the same proportion of actual MZs (256/282) but has a slightly better PPV because it incorrectly assigns five fewer DZ pairs as being MZ (i.e., PPV of 256/277 vs 256/282 in the logistic).

When this solution was applied within sex, the classification accuracy for males was similar to the other methods, but the results for females were more accurate than the other female-specific solutions, due to a larger proportion of the DZs pairs being correctly classified.

The final series of models were classification tree analyses conducted separately within sex. Figure 10.3 shows that the tree for females differs from that obtained with the combined sample. The four standard variables are represented early in the tree, but they appear in a different order, with Peas playing a more prominent role and Opinion important only for a subset of pairs. There are three variables included in the final solution for female pairs that do not appear in the combined tree. As with the female-specific logistic regression solution, Shared Room is important, again with the direction of sharing less often predictive of being MZ. Further along this branch (among pairs who reported as children they usually or always shared a room), the weight variables are predictive of zygosity, with almost all pairs with an average weight over 124 lb being DZ (72 of 73). Of those with average weight less than 124 lb, the within-pair difference in weight is predictive, with larger differences (>22 lb) more predictive of being MZ (50%) than smaller within-pair differences (0% MZ).

As shown in Table 10.2, the female-specific tree had overall accuracy of 90.7% (243/268). This and the other values are actually somewhat worse than those obtained with the combined solution, perhaps because the classification used a 50% base rate whereas the value for females was 53% MZ.

The solution among male pairs (Figure 10.4) contains the four standard items with the addition of confused by parents. The Parents item is relevant

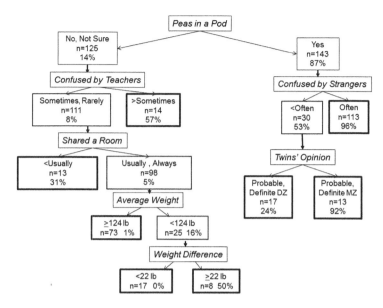

Figure 10.3 Results from classification tree analysis to assign zygosity in 268 female twin pairs.

Notes: Solution obtained using RPART. Gray boxes indicate items. White boxes show item cutpoint, number of pairs included in category, and percent of those pairs who are actually MZ. > indicates one or both twins above specified level, e.g., >Sometimes means both twins reported Often or one twin reported Often and the other reported Sometimes.

for only a small subgroup ($n = 35$) and even after splitting on this item, both groups have the majority of members classified as MZ. Thus, this split does not improve the binary classification accuracy of this model.

Overall accuracy for the male-specific tree was 92.0% (263/286), very similar to that obtained in the other male-specific analyses. However, the pattern of classification differed somewhat. This model had greater PPV and specificity, reflecting the higher proportion of DZ pairs correctly classified as such.

Additional Exploratory Analyses

In addition to the planned analyses, additional classification tree analyses were conducted to evaluate using within-pair disagreement and to identify sources of classification inaccuracy.

None of the algorithms makes explicit use of pair disagreement except for the clinical algorithm at one late step (Figure 10.1). However, using the pair mean combines pairs which agree completely and those which differ by two categories. For example, pairs for which both twins gave their opinion of their zygosity as Probably MZ are treated the same as pairs

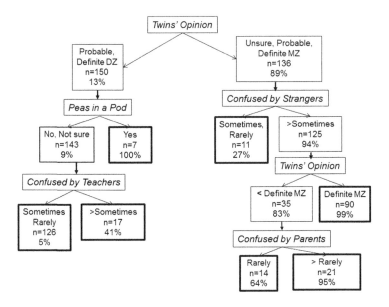

Figure 10.4 Results from classification tree analysis to assign zygosity in 286 male twin pairs.

Notes: Solution obtained using RPART. Gray boxes indicate items. White boxes show item cutpoint, number of pairs included in category, and percent of those pairs who are actually MZ. > indicates one or both twins above specified level, e.g., >Sometimes means both twins reported Often or one twin reported Often and the other reported Sometimes.

in which one twin says Definitely MZ and the other says Probably DZ. To evaluate whether explicitly including pair disagreement could improve classification, an additional analysis was run on the combined MM and FF dataset ($n = 554$ pairs) that included 25 predictors: the 16 original variables plus within-pair difference scores on the nine physical resemblance and shared environment items. Including this information provided little additional benefit. The solution was identical to that obtained previously (Figure 10.2) except that the 247 pairs in the far right terminal node, of which 94% were MZ, are further divided into those that differ on Opinion by 0 or 1 categories ($n = 229$, 96% MZ) and those that differed by more than 1 category ($n = 18$, 67% MZ).

Sources of Misclassification

As indicated in Table 10.2, the four classification methods gave very similar performance, obtaining about the same overall accuracy and classifying a similar number of pairs as MZ or DZ. However, it is not clear from these analyses as to what degree the methods are classifying the same pairs as MZ and DZ. The results of descriptive analyses to address this issue are shown in Figure 10.5. Of the 554 pairs, 504 (91%) were classified the same way

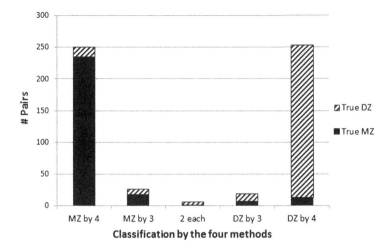

Figure 10.5 Agreement among methods on zygosity classifications.

Notes: There is high overlap across methods: more than 500 of the 554 pairs are assigned to the same category (MZ or DZ) by all methods and the majority of these are correctly classified into the categories determined by genotyping.

by all four methods. However, these included 29 pairs (13 MZ and 16 DZ) consistently placed in the wrong category by all four approaches. Another 44 pairs were classified the same way by three of the four methods: for 28 of these pairs the majority placement was correct, but for 16 pairs only one method had the correct zygosity category.

It is also of interest to understand what variables are associated with having the wrong classification. To investigate this, 17 variables were used to attempt to distinguish the pairs correctly classified by all four methods ($n = 475$) from pairs that were not ($n = 79$). The predictors included the original set of 16 plus genotype-based zygosity (coded as mz = 1, dz = 0). The resulting structure was quite complex. The Opinion, Stranger and Peas items formed the first few branches, after which were height, weight difference, height difference, age, and the Playmates items. It is likely that this solution overfits the data, as there were 18 branches and seven terminal nodes, none of which contained more than 15 pairs. However, some potentially useful results were that pairs who were misclassified tended to be younger, of average or shorter height, and differing in weight. In contrast, pairs who differed in weight but were older and taller tended to be classified correctly.

Discussion

The first goal of this study was to use an exploratory approach to determine to what degree classification accuracy could be improved over typically employed zygosity algorithms. The empirical approaches confirmed the importance of the traditionally employed items, including the clinical observation that the confusion by parents item is less informative than the others. However, as suggested by the classification tree in MM pairs (Figure 10.4), the Parent item may be useful when other information is inconclusive. Although absence of confusion in parents is not very informative, when parents are confused, this is strong evidence for pairs being MZ (Heath et al., 2003).

The clinical algorithm was found to perform nearly as well as the Logistic and Regression Tree approaches. This is impressive because the latter two methods were explicitly predicting a known outcome. Therefore, they would be expected to perform less well if the same equation were applied to a different dataset.

For the analyses reported here, the clinical method was translated into an explicit algorithm that could be evaluated by computer. In practice, protocols are often examined individually so that information such as pair disagreement or interviewer notes explaining a respondent's uncertainty or other remarks can be factored in. However, follow-up analyses indicated that pair disagreement did not add appreciably to accuracy, beyond using pair means for the physical similarity scores. The fact that something as simple as a unit-weighted sum of the items performed nearly as well as the other methods suggests that this individualized method is not needed for most pairs. In large samples of twins, it may be most efficient to employ a hybrid model in which a simple computer algorithm is used to assign most pairs and individual records are examined only for pairs for whom this method is inconclusive.

Follow-up analyses investigating the agreement among the four approaches found that not only did they achieve very similar levels of accuracy, they did so by classifying the same pairs in the same ways, including 475 pairs that were correctly assigned by all four methods. A small minority (45 pairs) were classified incorrectly by at least three of four methods. Although there were no strong findings, beyond that pairs who were misclassified had equivocal scores on the traditional zygosity items, there was some evidence that incorrectly assigned pairs tended to be younger, of average or shorter height, and differing in weight. If the within-pair differences in weight extended back to when the twins were children, it would explain why they would be less similar in physical appearance and would not be frequently confused despite being an MZ pair.

Utility of Other Items

A second goal for this study was to explore whether other information commonly obtained from twins could improve classification beyond using the standard items reflecting physical similarity and the twins' opinions.

In both the logistic regression and classification tree analyses conducted with female pairs, sharing a room was associated with being DZ. Sharing of a bedroom probably reflects family size. Because DZ pairs are likely to come from larger families with older siblings, they may have less opportunity to have separate bedrooms as young children than is the case for MZ pairs.

Physical attributes were somewhat useful in helping to classify the female pairs. In the classification tree analyses, adult weight predicted zygosity in a complex way for a small portion of the sample. Among the 98 pairs who frequently shared a room, almost all pairs with an average weight over 124 lb were DZ (72 of 73). Of those with average weight less than 124 lb, the within-pair difference in weight was predictive of zygosity, with larger differences (>22 lb) being more predictive of being MZ (50%) than having smaller within-pair differences (0%). Although this at first seems paradoxical, it must be considered in the context of the tree. In adulthood, a large within-pair difference in weight is indicative of dizygosity only for women who are thin. For pairs that are heavier, large differences in weight represent acquired weight rather than differences in body morphology expected to be more heritable.

The logistic regression identified height rather than weight, with a small effect of greater height predicting being DZ among female pairs. It has been noted in other twin registries that DZ pairs tend to be slightly larger. This may reflect the tendency for DZ twins and other offspring of older mothers (who have had prior pregnancies) to be heavier at birth.

Among the male pairs, in the context of the physical similarity variables, more frequent dressing alike predicted being DZ rather than MZ. This is a somewhat surprising finding and may be sample specific.

Overall, there were some hints that other information beyond the Nichols and Bilbro items might be useful, but the evidence was not consistent or particularly strong.

Sex Differences

The third study goal was to explore whether there were any differences between females and males in classification accuracy or in what variables were the best predictors of zygosity. There were small fluctuations across methods, but overall the algorithms resulted in similar classification accuracy. For both the logistic regression and classification tree methods, the accuracy obtained with the sex-specific solution was similar to that obtained when the estimates from the combined solution were applied. The combined and sex-specific logistic regression solutions pointed to the same variables with the same relative ordering of ORs: Strangers, Peas, and Opinion.

At first glance at Figures 10.3 and 10.4, the sexes seem to differ in the configural pattern of the classification tree solutions. However, the solutions use the same variables and approximately the same cut points. Given the substantial overlap among these items, the ordering is not meaningfully different. This is further underscored by the finding that the classification accuracy of the combined solution is not appreciably higher than for the sex-specific trees that have somewhat different structure.

As discussed previously, there were some differences between the sexes in terms of the additional variables included in the sex-specific classification trees. The main effect of weight in the FF solution may point to pairs that are highly discordant. Differences in weight were most predictive among pairs with a low average, but for a pair to have an average weight less than 124 lb, at least one of the twins must be rather light. This suggests using a different scaling of weight that characterizes the pair difference in terms of the proportion of the total weight rather than in absolute pounds. However, it should also be emphasized that the weight variables were informative for only a small proportion of pairs and so this feature of the solution may be sample specific and not generalizable to other contexts.

Limitations

Ideally, we would use data from an independent twin sample to evaluate the generalizability of these solutions. Correct classification obtained by the solutions obtained using exploratory methods would almost certainly be lower than we observed here. As our "clinical" algorithm achieved comparable accuracy to the exploratory solutions, we can speculate that this method would probably be superior in another sample.

These examples employed data from a heterogeneous sample of genotyped pairs. These included pairs who were randomly selected for genotyping along with pairs selected on the basis of uncertainty from their responses to the physical similarity and other zygosity items. Consequently, the proportion of pairs with ambiguous information is greater here than would be expected in a typical study. This means the classification rates obtained here underestimate the actual accuracy that would be expected. For example, the performance of the clinical algorithm in the randomly selected genotyped pairs in the VATSPSUD study was 95.5% (Kendler & Prescott, 1999).

The analyses used genotype-based zygosity as the gold standard and assumed it was measured without error. Although genotyping platforms are largely automated in procedures and calling genotypes, there is still room for human error—such as sample mislabeling when isolating the DNA twice or creating the sample plate, or data entry errors when the results are coded. The markers used for zygosity testing in VATSPSUD were PCR, which had some subjectivity as genotyping calls were based on quantitative differences judged by visual inspection rather than by machine-made assignments as is typical currently.

The twins included in these analyses were all Caucasian adults. Somewhat different solutions might have been obtained if the sample were more ethnically diverse or of younger age. In studies of juvenile twins, information to classify zygosity is typically obtained by questioning mothers about the physical similarity of their twins, so this might be expected to produce different patterns than obtained from an adult sample of twins. Adult twins have the additional information on how they have aged and become less similar, whereas mothers of juvenile twins are asked about these items without the perspective of hindsight.

Additional information used in some other studies was not employed here, including hair color, eye color, and blood group. Including these could have increased the accuracy of the algorithms, but they were not available for MM pairs.

Several other applications of zygosity estimation are beyond the scope of this chapter and were not dealt with here. The availability of computerized records on large population samples has led to attempts to create registries of twins who have just historical record data and are unavailable for questioning about their physical similarity. These methods use mixture distributions to estimate zygosity (e.g., Heath et al., 2003; Neale, 2003; Benyamin et al., 2006; Webbink et al., 2006) or other bootstrapping methods to evaluate biases associated with non-participation in volunteer registries (e.g., Prescott et al., 2005).

Conclusion

Over the past 50 years, many studies have been conducted to validate questionnaire measures of zygosity against blood typing or genetic markers. The majority have used linear regression approaches of main effects (e.g., Christiansen et al., 2003; Gottesman, 1963; Jackson et al., 2001; Magnus et al., 1983; Peeters et al., 1998; Reitveld et al., 2000). Others have calculated the joint probabilities of classification given combinations of traits and markers (e.g., Lykken, 1978), proposed decision trees (Goldsmith, 1991; Song et al., 2010), or used sensitivity analyses to evaluate the impact of different assumptions about zygosity (Prescott et al., 2011). As far as the author is aware, the results presented in this chapter represent the first example of classifying zygosity using a data mining strategy that permitted interactions and complex combinations of items, providing information about maximum classification accuracy.

Note

The Virginia Adult Twin Study of Psychiatric and Substance Use Disorders is directed by Kenneth Kendler at Virginia Commonwealth University (VCU) and has been funded by grants from the National Institutes of Health, the Keck Foundation, a gift from Mrs. Rachel Brown Banks and administrative funds from VCU. Richard Straub oversaw the genotyping of the twin pairs. Steven Aggen and Charles MacLean conducted the original zygosity analyses for the genotyped sample. Patsy Waring supervised interviewing for the FF study and Sarah Woltz for the MM study for which the zygosity information was obtained. Linda Corey was PI of the MidAtlantic Twin Registry and provided access to the twin sample. The author is grateful to an anonymous reviewer for helpful comments on an earlier version of this chapter and to the Borchard Foundation for providing support for this work.

References

Becker, A., Busjahn, A., Faulbabe,r H.-D., Bahring, S., Robertson, J., Schuster, H., and Luft F.C. (1997). Twin zygosity: automated determination with microsatellites. *Journal of Reproductive Medicine, 42,* 260–266.

Bell, J.T., and Saffery, R. (2012). The value of twins in epigenetic epidemiology. *International Journal of Epidemiology.* [doi: 10.1093/ije/dyr179]

Benyamin, B., Deary, I.J., and Visscher, P.M. (2006). Precision and bias of a normal finite mixture distribution model to analyze twin data when zygosity is unknown: Simulations and application to IQ phenotypes on a large sample of twin pairs. *Behavior Genetics, 36,* 935–946.

Breiman, L., Friedman, J.H., Olshen, R.A., and Stone, C.J. (1984). *Classification and Regression Trees.* Pacific Grove, CA: Wadsworth.

Carter-Saltzman, L., and Scarr, S. (1977). MZ or DZ? Only your blood grouping laboratory knows for sure. *Behavior Genetics, 7,* 273–280.

Christiansen, L., Frederiksen, H., Schousboe, K., Skythe, A., von Wurmb-Schwark, N., Christensen, K, and Kyvik, K. (2003). Age- and sex-differences in the validity of questionnaire-based zygosity in twins. *Twin Research, 6,* 275–278.

Dawes. R. (1962). A note on base rates and psychometric efficiency. *Journal of Consulting Psychology, 26,* 422–424.

Derom, C., Vlietinck, R., Derom, R., van Den Berghe, H., and Theiry, M. (1987). Increased monozygotic twinning rate after ovulation induction. *Lancet, 1,* 1236–1238.

Goldsmith, H.H. (1991). A zygosity questionnaire for young twins: A research note. *Behavior Genetics, 21,* 257–269.

Gottesman, I.I. (1963). Heritability of personality: A demonstration. *Psychological Monographs No. 572,* 77, 1–21.

Gottesman, I.I., and Prescott, C.A. (1989). Abuses of the MacAndrew MMPI Alcoholism Scale: A critical review. *Clinical Psychology Review, 9,* 223–242.

Haque, F.N., Gottesman, I.I., and Wong, A.H.C. (2009). Not really identical: Epigenetic differences in monozygotic twins and implications for twin studies in psychiatry. *American Journal of Medical Genetics, 151C,* 136–141.

Heath, A.C., Nyholt, D.R., Neuman, R., Madden, P.A.F., Bucholz, K.K., Todd, R.D., Nelson, E.C., Montgomery, G.W., and Martin, N.G. (2003). Zygosity diagnosis in the absence of genotypic data: An approach using latent class analysis. *Twin Research, 6,* 22–26.

Jackson, R.W., Snieder, H., Davis, H., and Treiber, F.A. (2001). Determination of twin zygosity: A comparison of DNA with various questionnaire indices. *Twin Research, 4,* 12–18.

Keith, L., Papiernek, E., Keith, D., and Luke, B. (Eds.), (1995). *Multiple Pregnancy.* New York: Parthenon Publishing Group.

Kendler, K.S., and Prescott, C.A. (1999). A population-based twin study of major depression in men and women. *Archives of General Psychiatry, 56,* 39–44.

Kendler, K.S., and Prescott, C.A. (2006) *Genes, Environment and Psychopathology: Understanding the Causes of Psychiatric and Substance Use Disorders.* New York: Guilford Publications.

Lopriore, E., Vandenbussche, F.P.H.A., Tiersma, E.S.M., de Beaufort, Z.J., and de Leeuw, J.P.H. (1995). Twin-to-twin transfusion syndrome: New perspectives. *Journal of Pediatrics, 127,* 675–680.

Lykken, D.T. (1978). The diagnosis of zygosity in twins. *Behavior Genetics, 8,* 437–473.

Magnus, P., Berg, K., and Nance, W. (1983). Predicting zygosity in Norwegian twins born 1915–1960. *Clinical Genetics, 24,* 103-112.

McGue, M., Osler, M., and Christensen, K. (2010). Causal inference and observational research: The utility of twins. *Perspectives in Psychological Science, 5,* 546–556.

Neale, MC (2003). A finite mixture distribution model for data collected from twins. *Twin Research and Human Genetics, 6,* 235–239.

Nichols, R.C., and Bilbro, W.C. (1966). The diagnosis of twin zygosity. *Acta Genetics et Statistica Medica, 16,* 265–275.

Peeters, H., van Gesterl, S., Vlietinck, R., Derom, C., and Derom, R. (1998). Validation of a telephone zygosity questionnaire. *Behavior Genetics, 28,* 159–163.

Prescott, C.A., Kuhn, J.W., and Pedersen, N.L. (2005) Twin pair resemblance for psychiatric hospitalization in the Swedish Twin Registry: A 32-year follow-up study of 29,602 twin pairs. *Behavior Genetics, 37,* 547–558

Prescott, C.A., McArdle, J.J., Lapham, S., and Plotts, C. (2011). Using Project Talent twin data to estimate the range of the components of variance of high-order cognition. *Behavior Genetics, 41,* 932.

Reitveld, M.J., van der Valk, J.C., Bongers, I.L., Stroet, T.M., Slagboom, P.E., and Boomsma, D.I. (2000). Zygosity diagnosis in young twins by parental report. *Twin Research, 3,* 134–141.

RPART version 3.1. R Partitioning and Regression Trees, http://127.0.0.1:10651/library/rpart/html/rpart.html

Song, Y.-M., Lee, D.-H., Lee, M.K., Lee, K., Lee, H.J., Hong, E.J., Han, B., and Sung, J. (2010). Validity of the zygosity questionnaire and characteristics of zygosity-misdiagnosed twin pairs in the Healthy Twin Study of Korea. *Twin Research and Human Genetics, 13,* 223–230.

Therneau, T.M., and Atkinson, E.J. (1997). *An introduction to recursive partitioning using the RPART routines.* Mayo Foundation.

Visscher, P.M., Will, W.G., and Wray, N.R. (2008). Heritability in the genomics era—Concepts and misconceptions. *Nature Reviews Genetics, 9,* 255–266.

Webbink, D., Roeleveld, J., and Visscher, P.M. (2006). Identification of twin pairs from large population-based samples. *Twin Research and Human Genetics, 9,* 496–500.

11 Dealing with Longitudinal Attrition Using Logistic Regression and Decision Tree Analyses

John J. McArdle

Introduction

This chapter is an application of several methods to solve the same problem—the improved understanding of, and correction for, non-random attrition in longitudinal studies. In this chapter, several forms of data analysis are described and used to deal with problems of longitudinal attrition or dropout. Then results from a specific application to real data from the Cognition and Aging in the USA (CogUSA) study are briefly presented. These results illustrate specific features of alternative data analysis models that can be applied to such a problem, including *Logistic Regression Models* (LRM) and *Decision Theory Analysis* (DTA). Due to its flexibility and accuracy, the new use of DTA was considered more successful than LRM here, but we know the use of these kinds of correction techniques is not likely to solve the entire problem of longitudinal attrition.

The Common Problems of Longitudinal Attrition and a Solution

In any behavioral study, there are persons who dropout of a study for many different reasons—dropout may be due to fatigue, lack of interest, high difficulty of tasks, or some combination of all of these. This is especially a problem in studies where people are asked to come back for additional visits—i.e., short or long term longitudinal studies (see McArdle, 2009). Our ability to handle persons who dropout in a random fashion has been a well-known problem for 100 years. In general, we do not have to do anything new when the dropout is random, but we do hope the resulting sample remains large enough (i.e., of sufficient power) to support the statistical tests (see King et al., 2001).

However, when people dropout in a non-random fashion it is possible our results can be misleading. That is, the results of complete cases can be very misleading. For example, mean scores may continue to rise because we have eliminated those who scored low. Or the variance of the outcomes may decrease because we have eliminated those who varied the most from the average. In a similar fashion, the correlation over time may change sign because we have selectively eliminated the full range of data. This all implies any final longitudinal model, even when fitted only to the complete cases, will give us a biased view of the true data generation (for examples, see McArdle & Hamagami, 1992; Zhang & Singer, 1999).

To correct for these problems using measured variables and missing at random assumptions it has been shown it is important to create a "selection model." To the degree we know the exact differences between the dropouts and non-dropouts, we can correct the non-dropout data and make more reliable inferences. But the creation of this kind of selection model is often very hard to do, mainly because the key variables and their relationships are not known. So we often simply conclude that either (a) the dropout was random, or (b) we do not really know why they did not come back. Any effort in this direction seems to satisfy most readers.

This is exactly the situation that cries out for some method of exploratory data mining analysis. That is, this is often a situation where there is very little prior theory available to distinguish between the resulting groups, other than the obvious ones of poor performance due to fatigue, so we do not really know why this event occurs (i.e., dropout at Time 2). We can always try to throw everything we have measured at this outcome and see if anything sticks, but we lose the basis of our probabilities, and we do not typically state this up front. At the same time, we know it is important to find observable reasons for attrition, including possible interactions and nonlinearity, mainly so that we can form a selection model to correct the available data we have measured (at Time 2) for the possibility of self-selection. Even if this approach does not work, it tells us that the attrition was unpredictable from the variables we measured. So, in this case, there is no doubt that some kind of exploratory analysis could be helpful. We consider Decision Tree Analysis (DTA) here.

Some Solutions for Selection Bias

The basic goal of any longitudinal analysis is to make an appropriate inference about all people at all occasions. We usually do this from data on people who were measured at a specific time, and we evaluate changes from the expected Time 1 to Time 2 means, standard deviations, and correlations. In any longitudinal study, we know that those individuals who in fact are measured twice (i.e., the Returners) share some characteristics in common with those who are measured only once (i.e., the Dropouts). But some differences can be assessed from the Time 1 data obtained on everyone.

A related problem in any study is the potential for bias in the parameters due to a particular selection of the people. Among several others, the work of Heckman (1976, 1979) is helpful here. The previous statistical models for this topic are well represented in the work of Heckman (1979; see Purhani, 2000) and we will only summarize a few key points here. The Heckman method was developed to address problems of self-selection among women participating in the labor force. This method makes it possible to assess whether selection bias is present, identify factors contributing to the selection bias, and to control for this bias in estimating the outcomes of interest. The Heckman method attempts to control for the effect of non-random selection by incorporating both the observed and unobserved factors that affect non-response. Evidently, not everyone is completely pleased about the use of this approach, but these critiques usually are fairly technical (see Muthén & Joreskog, 1983; Purhani, 2000).

In this application, we know that if we can distinguish between our two longitudinal groups (Returners and Dropouts) on some of the common characteristics we can use some kind of a statistical model to make an optimal prediction of who did or did not come back (i.e., the "selection" model). To the degree this selection model is an accurate statement of the group differences, we can create individual estimates of the "probability of Dropout" and we can use these individual estimates in various ways to correct for any "Selection Bias"—Heckman suggests using these estimates in further regressions, but here we simply create sample weights (see Stapleton, 2002) for those that did come back, and use these to adjust our statistical information (e.g., the Time 2 statistics)

The "Heckman" Method of Sample Bias Corrections

Heckman (1979) discussed bias from using non-random selected samples to estimate behavioral relationships as a form of "specification error." He also suggested a two-stage estimation method to correct the bias. In his original example, he recognized that people who work are selected non-randomly from the population, so estimating the determinants of wages from the subpopulation of persons who actually are selected to work may introduce bias. This correction involves a typical assumption of normality of the residuals, but it then provides a test for sample selection bias and a formula for a bias corrected model.

The generic specification model for this relationship was initially presented as a Probit regression of the form

$$Prob(D_n = 1 | Z_n) = \Phi(\zeta Z_n) \tag{1}$$

where D_n indicates dropout ($D_n = 1$ if the respondent is a dropout and $D_n = 0$ otherwise), Z_n is a vector of explanatory variables for the person, ζ is a vector of unknown group parameters, and Φ is the cumulative distribution function of the standard normal distribution, Estimation of the model yields results that can be used to predict this probability for each individual of either being a dropout or not. In the second stage, the researcher corrects for self-selection by incorporating a transformation of these predicted individual probabilities as an additional explanatory variable. The substantive equation may be specified in the usual multiple regression way as

$$Y[2]_n^* = \sum_{p=0}^{P} [\beta_p \, X_{p,n}] + e_n \qquad (2)$$

where $Y[2]_n^* =$ an outcome variable at Time 2 for everyone, so it is not observed if the person drops out, $X_p =$ a measured predictor variable which is constant over the group, and $\beta_p =$ an unknown regression coefficient (or weight) associated with the pth predictor. In this formulation we assume $X_{0,n} = 1$ so the first slope β_0 is actually an intercept term.

The conditional expectation of the score given the person has not dropped out in the substantive expression is then

$$E\{Y[2]_n | X_{p,n}, D = 1\} = \sum_{p=0}^{P} [\beta_p X_{p,\,n}] + E\{e_n | X_{p,n}, D = 1\}, \qquad (3a)$$

and, assuming joint normality of the error terms, we can simply write

$$E\{Y[2]_n | X_{p,n}, D = 1\} = \sum_{p=0}^{P} [\beta_p X_{p,n}] + \rho \sigma_e \lambda_n, \qquad (3b)$$

where ρ is the correlation between unobserved determinants of propensity to dropout, σ_e is the standard deviation of the residuals, $t\lambda$ is the "inverse Mills ratio" evaluated at the predictor score (i.e., ζZ_n). This model looks fairly complicated at first glance. But, essentially, a new predictor variable is created to represent the probability of this person either coming back or not coming back to the study. To the degree this new variable is correlated with the other variables we could see some changes in the basic substantive regression results—that is, they will be conditional on this sample selection because they will have taken this into account.

Many people have commented that this combined equation demonstrates Heckman's insight that sample selection can be viewed as a form of "omitted-variable bias" for a randomly selected sample. They also note that the dropout equation can be estimated by replacing the selection score (λ) with the Probit estimates from the first stage (i.e., $\zeta \, Z_n$), basically assuming they are predictive of Dropout, and then we include this estimate as an additional explanatory variable in any further linear estimation of the substantive equation. If the residual variance is greater than zero, the coefficient on the estimated values (of λ) can be zero. But this can only happen if the correlation between unobserved determinants of "propensity to Dropout "and the outcomes are also zero. This means that testing the null that the coefficient for selection (λ) is zero is equivalent to testing for sample selectivity. Of course, since all samples are likely to be selected in some way the changes in the other substantive coefficients are probably the most important.

There are some potential problems in using this approach (see Puhani, 2000). The model obtains formal identification from the normality assumption but an exclusion restriction is required to generate credible estimates—there must be at least one variable which appears with a non-zero coefficient in the selection equation but does not appear in the equation of interest (in econometrics this is termed an "instrument" (see Angrist & Krueger, 2001)—and this is not always available. The two-step estimator discussed above is known as a limited information maximum likelihood (LIML) estimator. In asymptotic theory and in finite samples, as demonstrated by Monte Carlo simulations, the full information maximum likelihood (FIML) estimator exhibits better statistical properties. However, the FIML estimator is more computationally difficult to implement (see Puhani, 2000; King et al., 2001). Also, since the covariance matrix generated by OLS estimation of the second stage is often inconsistent, the correct standard errors and other statistics can be generated from an asymptotic approximation or by resampling, such as through a bootstrap approach. It is worthwhile mentioning that Muthén & Joreskog (1983) presented a general likelihood expression for any model where the selection variable is used to select on the outcome variable directly and this essentially corrects for the person selection in a common factor process.

Sampling Corrections via Sample Weighting

Another important type of sample correction is also considered classical—sample *Inverse Probability Weighting* (IPW)—this is described by Kish (1995), Kalton and Flores-Cervantes (2003), Stapleton (2002), and Cole & Hernam (2008). The basic principle here is slightly different—some people (a) who are measured should have counted more than (b) other persons who are also measured, and this can be applied to any model. In most cases, this is because (a) are representatives of a group of people that are not as well represented as (b). In practice, a set of sample weights are created based on observable variables (typically using some form of regression) or other conditions and these are used to counterbalance what would otherwise be an under-weighting of some persons (a) and an over-weighting of other persons (b). Of course, this is precisely the same concept as that used in longitudinal attrition where we try to identify the degree to which each person we did in fact measure (at Time 2, $Y[2]$) is somewhat like (at Time 1, $Y[1]$) the people we did not measure (at Time $Y[2]$).

In any case, the sampling weights are then formed as a simple ratio of specific variables. If we had a special selection of males and females in a population, we might consider the sample adequate if we had the same proportion of males and females in the sample. This can be accomplished with sample weights in a simple way. Here we simply calculated the weight as the proportion of males we have measured ($P(m)$) compared to the proportion of males we hoped we had measured ($P(hm)$), by writing the

IPW for each person as

$$W_n = 1/\{P(m)_n/P(hm)_n\} * N. \tag{4}$$

The use of population weights, relative weights, or effective weights (Stapleton, 2002) is a choice that must be made by the researcher, and these are commonly used for different purposes. Also, in more complex sampling frames many different variables are used to create weights. In a standard example, the Time 2 data might be understood in terms of one variable or a linear regression of predictors (X_p) as

$$E\{Y[2]_n\} = \sum_{p=0}^{P} [\beta_p X_{p,n}] \tag{5}$$

which can be fitted in the standard way, or by assuming everyone has observed data (i.e., Missing At Random; see King et al., 2001). As an alternative, the new weights (W) are created for each person and these can be applied to the Time 2 parameters in a new observed variable regression for the persons as

$$E\{Y[2]_n W_n\} = \sum_{p=0}^{P} [\beta_p X_{p,n} W_n], \tag{6}$$

for all people we do have information about (e.g., Y[2] and X), and where the product variables (products created by weighted variables) are used in regression and we see added variation for all scores (see Stapleton, 2002). Obviously, if the weights are all equal to unity nothing in the regression changes. But if (a) the weight is higher than unity then the case is taken more seriously, and if (b) the weight is smaller than unity then the case does not count as much. Of course, it is also well known that the weights add variance to the predictions, so these kinds of weighted predictions are often less precise than un-weighted ones (see Kish, 1995; Stapleton, 2002). One thing that is not emphasized is that any weights are limited by their own accuracy in separating the groups apart.

Using Logistic Regression Modeling (LMR) Methods

The identification of predictor variables which can then be used to predict the binary dropout outcome is obviously crucial. We know there are many situations where we can predict a binary event using logistic regression methods (LMR; see Hosmer & Lemeshow, 2000; McArdle & Hamagami, 1994). The LMR is usually written as

$$ln\{R(g)/[1 - R(g)]\} = \sum_{p=0}^{P} [\beta_p X_{p,g}] \tag{7}$$

where for selected binary outcomes (0 = non-response or 1 = response), $R(g)$ = the observed rate of response in a group coded g (g = 1 to G), *ln* = the natural logarithm, $R(g) / [1 - R(g)]$ = the odds of responding, $X_{p,g}$ = a measured predictor variable which is constant over the group, and β_p = an unknown regression coefficient (or weight) associated with the *p*th predictor. In this formulation we assume $X_0 = 1$ so the first slope β_0 is actually an intercept term. In this case, the LRM is a linear regression prediction of the log of the odds of responding. This LRM can be fitted and extended in several different ways (e.g., see McArdle & Hamagami, 1994). Of some importance here, we can turn the expression into probabilistic form by calculating the inverse relationship and writing

$$E\{R(g)\} = e^{\{logit(g)\}}/1 - e^{\{logit(g)\}} \tag{8a}$$

where e = the exponential function (i.e., $e = 2.7183$), and

$$logit(g) = \sum_{p=0}^{P}(\beta p\ Xp)\} \tag{8b}$$

to obtain the expected response probability ($E\{R(g)\}$ is a direct function of the predictors. After doing this we can obtain a weight ($1/E\{R(g)\}$) for subsequent analyses for each person in each sub-group g defined by the pattern of predictors (X). That is, depending upon the scores on the predictors we will have a weight reflecting the probability of returning to the study. If the probability is high, because the person looks like all the others that returned, we would down-weight these data. However, if the probability of returning is low, we would up-weight these data because this person is like many other people who did not come back. Of course, the resulting weights will only be useful (and alter the previous model) if the original predictors are an accurate reflection of the log odds of responses.

This LRM approach was used in a longitudinal follow-up study of cognitive tests by McArdle et al. (2002). In this study, we tried to predict who came back from knowledge of a lot of prior demographics and cognitive test scores but no important linear predictors were found. We then assumed that the sample of longitudinal respondents was a valid sub-sample of everyone who was originally tested, but we also know that we made no effort to investigate measured nonlinearities or interactions.

Using Decision Tree Analysis (DTA) Methods

An historical view of *Decision Tree Analysis* (DTA) is presented in detail elsewhere (see Sonquist, 1970; Breiman et al., 1984; Strobl et al., 2009; Berk, 2009; McArdle, chapter 1, this volume). A few common features of these DTAs are:

1   DTAs are admittedly "explorations" of available data.
2   In most DTAs the outcomes are considered to be so critical that is does not seem to matter how we create the forecasts as long as they are "maximally accurate."

3   Some of the DTA data used have a totally unknown structure, and experimental manipulation is not a formal consideration.
4   DTAs are only one of many statistical tools that could have been used.

Some of the popularity of DTA comes from its easy to interpret dendrograms, or Tree structures, and the related Cartesian subplots. DTA programs are now widely available and very easy to use and interpret.

One of the more interesting interpretations of DTA comes from the regression interpretation of this form of data analysis. One source of this reorganization comes from Berk (2009, p.108) who suggested we could write the binary (or continuous) outcome (after the fact) in terms of "basis functions" where

$$f(X_n, Z_n) = \beta_0 + \beta_1[(I(X_n \leq \gamma(1)] + \beta_2[I(X_n > \gamma(1)$$
$$\& \ Z_n \leq \gamma(2)] + \beta_3[I(X_n > \gamma(1) \ \& \ Z_n > \gamma(2)], \tag{9}$$

where I = an indicator of the score for a person that depends on their individual scores on measured X and Z, and $\gamma(j)$ = some unknown cut-point on the score distribution of X or Z. The hard part of doing a DTA analysis is to know how to estimate which variables contribute to this expression, and on what specific cutting score. The previous approach is not used in optimization, but when considered after the fact, at least it allows us to put the DTA into the same framework as the previous logit regression. In this framework, DTA makes predictions from some unspecified combination of linear weights (β_k) and nonlinear cut-points $(\gamma(j))$. We can always count the number of parameters needed by considering the number of weights (K) and cut-points (J) required to produce the score expectation. In the example above (Eq. 9) we need to estimate four weights and three cut-points so seven parameters are estimated from the data, and these may not be all treated equally. Of course, a variety of other statistical rules could be applied to this kind of model.

The figures used in Sonquist (1970) and Morgan & Sonquist (1963; see McArdle, chapter 1, this volume, Figure 1.1) appear to represent the first application of DTA for a binary outcome, and their discussion is still worth considering in some detail. We can state that DTAs are based on the utterly simple ideas that we can deal with a multivariate prediction or classification by (a) using measured variables that separate only parts of the database at any time (i.e., partition), and (b) doing some procedure over and over again in subsets (i.e., recursively). The best "splits" or "partitions of the data" can be done using measured variables found using several different techniques (e.g., the Gini index). Differences here are minor. We then must determine when to stop model fitting—a tradeoff between "model complexity" (e.g., based on the degrees of freedom, *df*s) and "model misfit" (e.g., based on the model likelihood L^2; see Hastie et al., 2001; McDonald, 1999). Differences here could be major. Some form of cross-validation is often employed (but see Tibshirani & Tibshirani, 2009).

It is important to note that this DTA approach cannot be considered to produce a different probability for each terminal node. If the outcome variable is binary then the probability of the outcome can be determined by the simple proportion of persons in the node who still have the outcome. For example, if there are 100 people in a final node and 80 of these people came back to the study then the probability of dropout for this node is 20% (or 0.2).

Logistic Regression Compared to Decision Tree Analysis

There have been several published comparisons about the effectiveness and accuracy of LMR and DTA methods. Some of the most recent examples include those of Monahan et al. (2001), Lemon et al. (2003), Penny and Chesney (2006), and McArdle (2012, and chapter 1, this volume). The most basic result in terms of main effects is that the DTA should be very close to the LRM, especially in the overall probability differentiation. But the DTA allows many possible forms of nonlinearity of the predictors as well as interactions among the predictors, so it should be able to make more accurate predictions about the probability of the outcome for any group. This also means that when a nonlinear relation or interaction is found it can be placed back into the LRM for an improved prediction.

Comparisons to a related technique, Linear Discriminant Analysis, are also common (e.g., see Feldesman, 2002). But, as far as we know, there are no published studies that compare the use of LMR and DTA as alternative methods in the analysis of longitudinal attrition.

Methods

Participants

A recent experiment in cognitive adaptive testing has been carried out by the author and colleagues in the *Cognition and Aging in the USA* (CogUSA) study (see McArdle et al., 2005). CogUSA is a collaborative study of a large ($N > 1,500$) and representative sample of adults aged 50–90 from across the United States. In this and other collaborative studies (see McArdle et al., 2007) we have attempted to create new Adaptive Testing formatted measurements of several dimensions of adult cognitive functioning to evaluate the extent to which these tests could be used in the larger and ongoing *Health and Retirement Study* (HRS; Ofstedal et al., 2005). The focus in the current study is on measurements of the construct of both fluid intelligence (Gf) and quantitative reasoning (Gf-RQ) as presented in the Woodcock–Johnson (WJ III) Number Series (NS) test.

The current data set was collected in three sequential phases. Initially, a random digit dialing service (Genesys) was used to request survey data from adults over 50 years of age living in the 48 contiguous states of the United States. Of approximately 3,200 telephone numbers solicited,

$N = 1,514$ (50% response rate) agreed to our initial telephone testing. Of course, not everyone initially agreed to do this, and there is a pressing need for an analysis of this initial selection issue here, but since no variables were measured for the non-respondents, we cannot actually deal with this attrition problem any further.

In the second phase, a first individual telephone testing (T1) was based on a 12 carefully selected cognitive tests administered verbally via the telephone. The NS items were presented as a telephone administered *Computer Adaptive Tests Inventory* (CATI) with up to six items. The remaining short ability tests came from the HRS project. On average, the telephone interview took 38 minutes to complete, and each respondent was paid US$20 for their participation. Upon completion of the telephone interview, each person was asked if they were willing to participate in face-to-face (FTF) testing with similar tests within a period of 14 days (with time-lag randomization, as in McArdle & Woodcock, 1997), and they would be given US$60 to do so.

For our third phase, we retested everyone who agreed to do so. In this Face-To-Face (FTF) at time 2 (F2), only $n = 1,230$ (81% response rate) agreed to participate. Of course, this is a relatively high 81.2% response rate, especially for a repeated cognitive testing (see Ferrer et al., 2005), so we probably should be very pleased. However, there are many reasons to think this is not a random selection and that further corrections might be useful. During the second testing, F2 testing data was collected on a large set of the Woodcock–Johnson scales (WJ-R, WJ-III), including the NS, and other scales as well (e.g., the WASI; see McArdle et al., 2010 for details). During this phase of data collection the complete 47-item publication version of NS was administered (see McArdle, chapter XX, this volume) rather than the shorter CATI form. The average interview length during the second phase of face-to-face data collection was 181 minutes. Each respondent was paid US$60 for their participation in Phase II.

We also know that big problems can emerge when we use the F2 data but wish to make inferences to all T1 participants, and we certainly do wish to do this. In fact, we wish to make inferences about all the people we asked to participate at the beginning, but we realize this is far too difficult. So, a variety of alternative statistical models were used, were designed to isolate aspects of the selection mechanisms T1 → F2 so we could more effectively use the F2 (and subsequent T3) data.

Results

Initial Results

The current analysis focused on trying to distinguish those who came back ($n = 1,230$; 81.2%) from those who did not come back ($n = 284$; 18.8%) on various common (Time 1) characteristics. We have limited this to measured demographic and Time 1 characteristics. Since this analysis

Table 11.1 Initial results for the standard approaches to longitudinal attrition analyses

Both	Frequency	Percent	Cumulative Frequency	Cumulative Percent
T1 =0	284	18.76	284	18.76
T1F2=1	1230	81.24	1514	100.00

Analysis Variable : Years_ED (Unweighted)

Both	N Obs	N	Mean	Std Dev	Minimum	Maximum
0	284	284	13.8908451	2.4878618	3.0000000	17.0000000
1	1230	1230	14.2495935	2.3198957	1.0000000	17.0000000

was not actually driven by any major theoretical considerations, we simply started by seeing if there was a noticeable difference at Time 1 on any important characteristic. As it turned out, we immediately noticed that the self-reported level of *Educational Attainment* (EA) of those who came back was a bit higher (EA{Ret} = 14.2 ± 2.3 years) than or those who did not come back (EA{Drop} = 13.9 ± 2.5 years) (Table 11.1).

Similarly, we can plot the observed data on the NS test for Time 1 (Telephone, TEL, or T1) and Time 2 (Face-to-Face, FTF) in Figure 11.1. Perhaps the most striking feature of this diagram is the fact that some people who scored high at Time 1 (T1) did not do the same at Time Two (F2). Of course, there are also people who scored well at Time 2 but not at Time 1. The standard Pearson Product Moment Correlation for

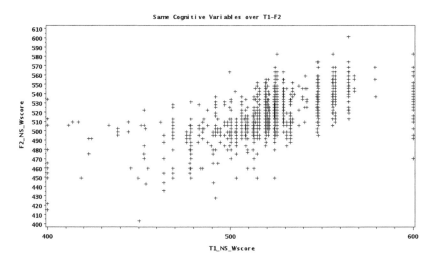

Figure 11.1 The empirical distribution of the number series CogUSA scores ($n = 1,223$; at T1 and F2).

the $n = 1{,}225$ persons who took both versions is $r[T1, F2] = .59$. This, of course, is a statistic we can examine closely as we look for the impacts of selection.

Logistic Regression (LRM) Results

In a first logistic regression model we predicted the outcome of coming back $(0 = \text{dropout, vs } 1 = \text{respondent})$ using five available demographic variables (Age, Education, Gender, and Ethnicity as Black or White) and we found a significant overall effect ($\chi^2 = 11$ on $df = 5, p < .05; \text{Max}R^2 = .01$) but no specific individual effects. The results are presented in Table 11.2a.

We then fitted a second logistic regression where we dropped one of the Ethnicities (White or Not) but added three two-way interactions with Education. Here we found a significant overall effect ($\chi^2 = 17$ on $df = 7$, $p < .05; \text{Max}R^2 = .02$), and now Education was considered notable ($\beta = 0.3$) along with Age ($\beta = 0.08$) and their interaction ($\beta = -0.005$). The new results are presented in Table 11.2b.

Since the initial demographics did not explain much of the outcome variability, we eliminated all interactions and ran another model with all cognitive variables measured at time one (i.e., telephone versions of nine scales). The results are presented in Table 11.2c. Here we did not find

Table 11.2a Model 1 results of the initial logistic regression approach to longitudinal attrition analyses

	R-Square	0.0100	Max-rescaled R-Square		0.0120

Testing Global Null Hypothesis: BETA=0

Test	Chi-Square	DF	Pr > ChiSq
Likelihood Ratio	11.3250	5	0.0453
Score	12.2114	5	0.0320
Wald	11.8839	5	0.0364

Analysis of Maximum Likelihood Estimates

Parameter	DF	Estimate	Standard Error	Wald Chi-Square	Pr > ChiSq
Intercept	1	-0.9828	0.7047	1.9452	0.1631
Years_AGE	1	0.00296	0.00630	0.2208	0.6384
Years_ED	1	-0.0530	0.0284	3.4835	0.0620
Eff_Gen	1	-0.0953	0.1341	0.5051	0.4772
black	1	0.6398	0.3855	2.7544	0.0970
white	1	0.0377	0.3143	0.0144	0.9045

Odds Ratio Estimates

Effect	Point Estimate	95% Wald Confidence Limits	
Years_AGE	1.003	0.991	1.015
Years_ED	0.948	0.897	1.003
Eff_Gen	0.909	0.699	1.182
black	1.896	0.891	4.037
white	1.038	0.561	1.923

Table 11.2b Model 2 results of the initial logistic regression approach to longitudinal attrition analyses

	R-Square	0.0113	Max-rescaled R-Square	0.0183

Testing Global Null Hypothesis: BETA=0

Test	Chi-Square	DF	Pr > ChiSq
Likelihood Ratio	17.1990	7	0.0162
Score	18.8170	7	0.0088
Wald	17.8080	7	0.0129

Analysis of Maximum Likelihood Estimates

Parameter	DF	Estimate	Standard Error	Wald Chi-Square	Pr > ChiSq
Intercept	1	-5.8068	2.3264	6.2304	0.0126
Years_ED	1	0.3013	0.1650	3.3350	0.0678
Years_AGE	1	0.0754	0.0342	4.8626	0.0274
Eff_Gen	1	0.7486	0.7985	0.8789	0.3485
black	1	0.2047	1.1564	0.0313	0.8595
Years_ED*Years_AGE	1	-0.00533	0.00247	4.6690	0.0307
Years_ED*Eff_Gen	1	-0.0609	0.0559	1.1885	0.2756
Years_ED*black	1	0.0291	0.0862	0.1136	0.7361

a significant overall effect ($\chi^2 = 22$ on $df = 14$, $p > .05$; $MaxR^2 = .02$), but we did find individual effects of Education ($\beta = -0.07$) along with a Self-Rating of Memory Scale (TICS_RM $\beta = -0.008$). Now it was easy to understand that those who came back for a second testing had higher Educational Attainment and had a higher Self-Rated Memory.

We did not go further using this LRM strategy because from these results alone we could begin to form a useful narrative about dropouts (i.e., these people were less educated and had a lower self-rating of their memory) and perhaps we could even learn to avoid future dropout problems. But we also came to realize this was decidedly an exploratory approach using confirmatory tools (LRM). This approach was not unlike those of many other researchers who select specific predictor variables for inclusion in an LRM but then highlight only the biggest impacts. Although this seems wrong, following the usual logic, we created sampling weights (see Table 11.3b) using these final predictions, and the weighted Education of those who came back was now only a bit higher (WE{Ret} = 14 ± 1.2 years) than those who did not come back (WE{Drop} = 13.5 ± 1.3 years). All we actually wanted to alter was the persons who did come back because they are the only ones for whom we have F2, and subsequent data.

Decision Tree Analysis Results

The previous logistic strategy is now commonly used to adjust the available data at Time 2 for the dropouts. But this is often done in the absence of a theory of dropout or with the inclusion of all possible interactions

Table 11.2c Model 3 results of the initial logistic regression approach to longitudinal attrition analyses

```
Testing Global Null Hypothesis: BETA=0
   Test                    Chi-Square       DF      Pr > ChiSq
   Likelihood Ratio          21.8156        14          0.0825
```

Analysis of Maximum Likelihood Estimates

Parameter	DF	Estimate	Standard Error	Wald Chi-Square	Pr > ChiSq
Intercept	1	0.3380	1.1272	0.0899	0.7643
Years_AGE	1	0.00401	0.00679	0.3490	0.5547
Years_ED	1	-0.0654	0.0307	4.5317	0.0333
Eff_Gen	1	-0.1119	0.1386	0.6511	0.4197
black	1	0.6686	0.3899	2.9399	0.0864
white	1	-0.00694	0.3187	0.0005	0.9826
T1_TIC_RM	1	-0.00812	0.00399	4.1489	0.0417
T1_TIC_RPM	1	-0.00332	0.00476	0.4873	0.4851
T1_Pcesd	1	0.000906	0.00332	0.0744	0.7850
T1_TIC_NA	1	0.0853	0.1561	0.2988	0.5846
T1_TIC_DA	1	-0.2607	0.1398	3.4787	0.0622
T1_TIC_BC	1	0.0658	0.1706	0.1489	0.6996
T1_TIC_S7	1	-0.0431	0.0536	0.6466	0.4213
T1_TIC_IR	1	0.0329	0.0625	0.2770	0.5987
T1_TIC_DR	1	-0.00332	0.0506	0.0043	0.9476

Odds Ratio Estimates

Effect	Point Estimate	95% Wald Confidence Limits	
Years_AGE	1.004	0.991	1.017
Years_ED	0.937	0.882	0.995
Eff_Gen	0.894	0.681	1.173
black	1.951	0.909	4.191
white	0.993	0.532	1.855
T1_TIC_RM	0.992	0.984	1.000
T1_TIC_RPM	0.997	0.987	1.006
T1_Pcesd	1.001	0.994	1.007
T1_TIC_NA	1.089	0.802	1.479
T1_TIC_DA	0.770	0.586	1.013
T1_TIC_BC	1.068	0.765	1.492
T1_TIC_S7	0.958	0.862	1.064
T1_TIC_IR	1.033	0.914	1.168

and nonlinearities. Basically, only main effects are fitted and we need to try to use these to make a story about dropouts. Unfortunately, we also know that the viability of the sample corrections is dependent on the accuracy of this prediction, so we really should include any measured variable at this point. In the absence of any further theory, we simply used standard methods of DTA to obtain a larger classification difference between our two groups.

The initial DTA was run using CART-PRO (Steinberg & Colla, 1999) and the results are presented in Figures 11.2(a) and (b). In Figure 11.2(a) we show the initial DTA structure of the separation of Returners/Dropouts using only the Demographic variables. Only six predictors were included, but all nonlinearities and interactions were examined, and the result (in Figures 11.2(c) and (d)) shows that 16 nodes (or groups of people) yielded a relative cost of RC=0.909 with an overall accuracy of ACC=61.9%

Table 11.3 Additional results for the adjusted weights approach to longitudinal attrition analyses

	Both	Frequency	Percent	Cumulative Frequency	Cumulative Percent
	T1 =0	284	18.76	284	18.76
	T1F2=1	1230	81.24	1514	100.00

Analysis Variable : Years_ED (Unweighted)

Both	N Obs	N	Mean	Std Dev	Minimum	Maximum
0	284	284	13.8908451	2.4878618	3.0000000	17.0000000
1	1230	1230	14.2495935	2.3198957	1.0000000	17.0000000

Analysis Variable : Years_ED (Weighted Logit3)

Both	N Obs	N	Mean	Std Dev	Minimum	Maximum
0	284	283	13.5351807	1.3227358	3.0000000	17.0000000
1	1230	1218	13.9709048	1.1545143	1.0000000	17.0000000

Analysis Variable : Years_ED (Weighted CART2)

Both	N Obs	N	Mean	Std Dev	Minimum	Maximum
0	284	284	13.9154230	1.4485225	3.0000000	17.0000000
1	1230	1230	13.9428634	0.8828511	1.0000000	17.0000000

(or 66% for Dropouts, and 58% for Returners; see Figure 11.2(b)). The relative importance of each measured variable in this prediction is listed in Figure 11.2(e), and this shows years of Education is most important (listed as 100%), and this is followed by years of Age (78%), and all other variables are used much less frequently (less than 25%).

In Figure 11.3(a) we show the results of a prediction of the same outcome for the same data adding all the Demographic and all the Cognitive variables measured at Time 1. This model included 17 predictors (see Figure 11.3(a)) and all possible nonlinearity and interactions reducing the uncertainty even further (see Figures 11.3(b)–(e)). The final model used here has 50 nodes (sub-groups or groups), we now have a relative cost of RC=0.977 with an ACC=73.3% (or 82% for Dropouts, and 65% for Returners; see Figure 11.3(d)).

The final analysis was convincing enough that we tried to figure out how to interpret what it actually meant. First off, we realized we now had a system that did a much better good job at predicting dropouts (at 82% correct) and we basically wanted to know why. To understand this prediction system even better, we pruned the tree even further. As seen in Figure 11.3(c), we found this prediction was largely due to Education, and then within the lower group of those (Education < 16.5) there seemed to be evidence that lower rated Memory (TICS_RM < 30) was indicative of higher probability of Dropout (Prob = 30.8 %), and among those with

(a)
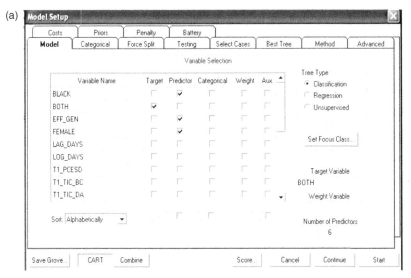

(b)
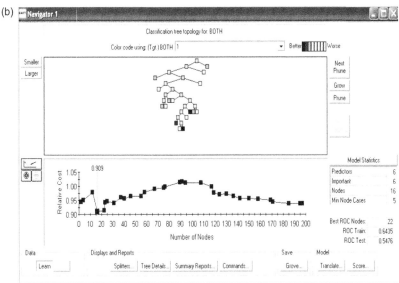

Figure 11.2(a–b) (a) Input for CART-PRO model 1—Demos → Dropout. (b) Initial DTA from CART-PRO model 1.

(c)

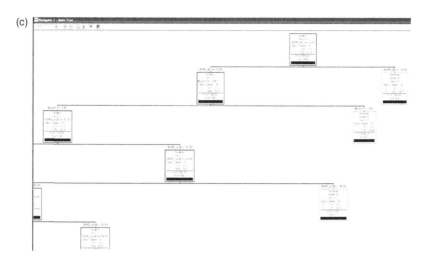

(d)

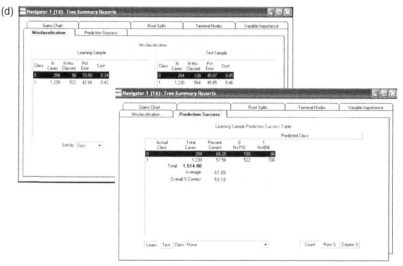

Figure 11.2(c–d) (c) Details of tree from CART-PRO model 1.
(d) Classification from CART-PRO model 1.

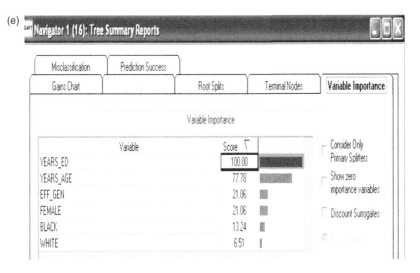

Figure 11.2(e) VI from CART-PRO model 1.

higher rater Memory (TICS_RM> 30) we had a higher dropout among Black participants (31.1%).

We also created sampling weights for members of all 50 groups by simply assigning each person the average probability of Dropout within the cell. This is displayed in Figure 11.4. We created sampling weights using these DTA predictions, and the weighted Education of those who came back (WE{Ret}=13.9 ± 1.5 years) was now very similar to that of those who did not come back (WE{Drop}=13.9 ± 0.9 years) (Table 11.3).

Comparison of Results

We first compared these individual weights to those created by the previous logistic regression model (see Figure 11.5). Not so surprisingly, the DTA weights (P_CART2) were not identical to the logit regression weights (P_LOGIT3), but the correlation among these weights for individuals (of $r=0.26$) was surprisingly low and the plots of the scores show the weights can be dramatically different. Perhaps this suggests that in the DTA something new and different was being found.

When the summary statistics were weighted we found slightly closer results. Table 11.4 is a listing of the basic summary statistics for all $n = 1,228$ persons measured on the Number Series at T1 and F2. We should note that there were two persons who did not have any valid information, and NS scores below 400 or above 600 were considered as "missing information" (see King et al., 2005; McArdle, 1994; see McArdle & Cattell, 1994). With this in mind, the first column lists the summary statistics (MLE and T-value) (a)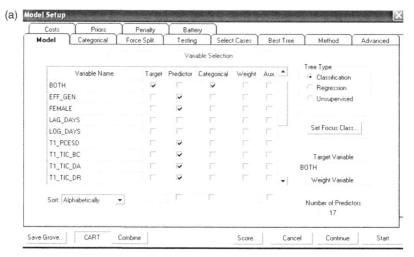

(b)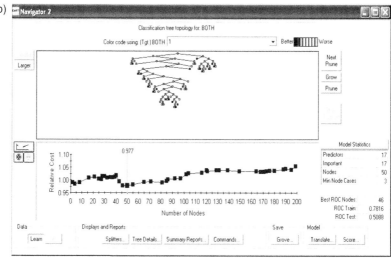

Figure 11.3(a–b) (a) Input to CART PRO model 2—Demos+Cogs → Dropout.
(b) Initial DTA from CART model 2.

for the two means, two variances, and one covariance (or [correlation] $r[T1,F2]$). Here it is clear that the Telephone to Face-To-Face Number Series test performance has an estimated positive correlation ($r[T1,F2] =$ .604 (27.)). The second column shows the same results when the Logit3 Weights are used, and here the means decrease, the variance increases, and the estimated correlation is higher but less precise ($r[T1,F2] = .613$ (25.).) The third column gives the same statistics for the CART2 model weights, and now the estimate means are smaller, the variances are up and

(c)
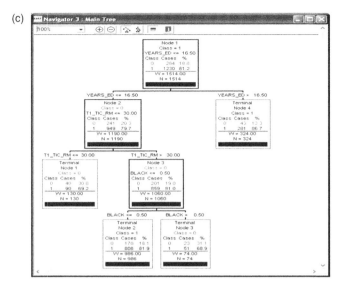

(d)
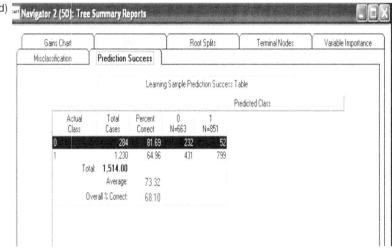

Figure 11.3(c–d) (c) "Pruned" CART-PRO model 2. (d) Classification from CART-PRO model 2.

down, and the estimated correlation is highest and most precise ($r[\text{T},\text{F}] = .674$ (29.)).

The model to be considered is one where the Time 1 Demographics (e.g., Age, Education, Gender, Ethnicity, and Time-Lag) are use to describe the T1_NS data, and then these Demographics plus the T1_NS are used to predict the F2_NS. The object of this analysis was to see if there was anything unique in the T1_NS CATI to warrant its use instead of the

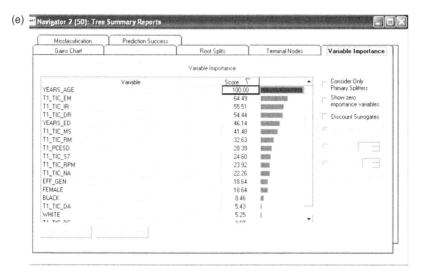

Figure 11.3(e) VI CART model 2 results.

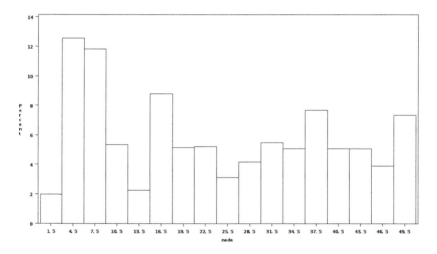

Figure 11.4 Histogram of assigned nodes (1–50) from full CART2.

longer F1_NS Test. The results for three alternative weight schemes are presented in Table 11.5.

The first model (*M*0) is not weighted and shows the prediction of the original T1_NS from the five demographics and the biggest effects here are positive differences due to Education ($\alpha = 4.65$ W points per year) and negative differences due to Age ($\alpha = -0.68$ W points per year). Given this,

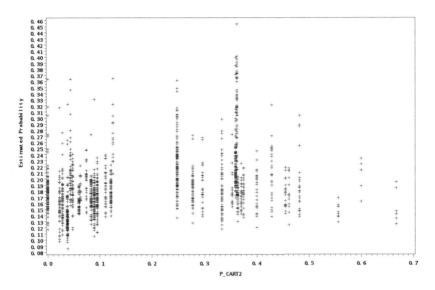

Figure 11.5 Comparison of alternative dropout probabilities ($r = 0.26$).

Table 11.4 Summary statistics for the adjusted weights approach to longitudinal attrition analyses of the key correlation of number series scores ($n = 1,228$; Mplus 6.0; see Appendix Table 11.A4)

Statistic	M0: Un-Weighted	M1: Logit3 Weights	M2: CART2 Weights
Mean T1_NS	516.0 (569.)	514.0 (536.)	512.2 (466.)
(Var T2_NS)	1007. (19.)	1029. (18.)	769.0. (15.)
Mean F2_NS	519.1 (764.)	516.8 (674.)	516.3 (496.)
(Var F2_NS)	566.1 (19.)	614.0 (17.)	674.2 (14.)
Covariance T1 & F2	456.5 (16.)	487.5 (15.)	485.2 (13.)
[Correlation T1 & F2]	[.604] (27.)	[.613] (25.)	[.674] (29.)

Females and Whites also have higher scores, but the Time-Lag variation makes no difference at all. The explained variance of T1_NS is $R^2 = .22$. The second equation in this model, where F2_NS is the outcome, shows that T1_NS is an independent positive predictor ($\beta = 0.30$ per W unit), with the rest of the predictors showing the same pattern of impacts, now independent of T1_NS. The explained variance of F2_NS is $R^2 = .50$.

The second model ($M1$) is Logit3 weighted and shows much the same effects. Here the prediction of the original T1_NS from the five demographics has the same positive differences due to Education ($\alpha = 4.55$ W points per year) and negative differences due to Age ($\alpha = -0.65$ W

Table 11.5 Summary statistics for the adjusted weights approach to longitudinal attrition analyses of the key regression for number series scores ($n = 1,228$; Mplus 6.0; see Appendix Table 11.A5)

Statistic	M0: Un-Weighted	M1: Logit3 Weights	M2: CART2 Weights
T1_NS Equation			
α Intercept T1	484.0 (61.)	482.3 (58.)	499.3 (52.)
α Age (Years)	−0.68 (9.)	−0.65 (8.)	−0.71 (9.)
α Education (Years)	4.65 (13.)	4.55 (11.)	3.30 (7.)
α Time-Lag	−0.04 (1.2)	−0.03 (1.0)	−0.04 (1.2)
α Gender (Effect)	−5.04 (3.1)	−4.97 (3.0)	−6.07 (3.2)
α Ethnicity (White)	11.16 (3.8)	12.07 (4.2)	15.84 (4.9)
Residual Variance T1	785.2 (48.)	791.9 (16.)	567.8 (14.)
Explained Variance T1	[.22] (10.)	[.23] (9)	[.26](8.)
F2_NS Equation			
β Intercept F2	346.0 (28.)	340.4 (24.)	271.1 (15.)
β T1_NS (prior)	0.30 (13.)	0.31 (11.)	0.43 (12.)
β Age (Years)	−0.50 (10.)	−0.51 (10.)	−0.44 (7.)
β Education (Years)	2.87 (12.)	2.88 (11.)	3.24 (8.)
β Time-Lag	−0.03 (1.6)	−0.03 (1.3)	−0.04 (1.3)
β Gender (Effect)	−3.92 (3.9)	−3.63 (3.3)	−2.97 (2.2)
β Ethnicity (White)	9.67 (5.)	11.22 (6.)	9.78 (4.)
Residual Variance F2	281.1 (18.)	295.6 (16.)	285.8 (12.)
Explained Variance F2	[.50] (24.)	[.50] (22.)	[.57] (22.)

points per year), with Females and Whites having higher scores, and with a Time-Lag variation that makes no difference at all. The explained variance of T1_NS is $R^2 = .23$. The second equation in this model, where F2_NS is the outcome, shows that T1_NS is an independent positive predictor ($\beta = 0.31$ per W unit), with the rest of the predictors showing the same pattern of impacts, now independent of T1_NS, but possibly with less accuracy. The explained variance of F2_NS is also $R^2 = .50$.

The third model (M2) is CART2 weighted and still shows much the same effects. Here the prediction of the original T1_NS from the five demographics has smaller but still positive differences due to Education ($\alpha = 3.30$ W points per year) and slightly larger negative differences due to Age ($\alpha = -0.71$ W points per year), with Females and Whites having even higher scores, but with a Time-Lag variation that makes no difference at all. The explained variance of T1_NS is $R^2 = .26$. The second equation in this model, where F2_NS is the outcome, shows that T1_NS is a larger independent positive predictor ($\beta = 0.43$ per W unit), with the rest of the predictors showing the same pattern of impacts, now independent of T1_NS, but possibly with less accuracy. The explained variance of F2_NS is also raised slightly to $R^2 = .57$.

Thus, the weighting used impacts all parameters, but only slightly. The model basically describes the same effects no matter what we do, although the CART2 weights seem to produce the most precision.

Conclusion

The examples used here were designed to illustrate a broad range of corrections for any longitudinal application. In our first analysis, using LRM, we indicated the "probability of non-response" (or conversely, staying in) in longitudinal data. Even after we put everything we had measured into the linear model, we still could only account for about 2% of the outcome variance. From this model, we created relative sampling weights that could be used for a statistical correction. In the next example, we used DTA to do the same thing, and with DTA we accounted for almost 18% of the outcome variance. In DTA we did not need to do anything more than specify the outcome, the measured variables to be used, and the criterion for stopping the tree building. In comparison to the first analysis, this was much easier to do. This is much more precise so we expect different results when we apply these weights to the data, but this does not really happen. In both cases, these analyses were used directly, to create optimal contrasts in future analyses, or we might say were useful as a baseline against which to judge the success of any previous analysis.

Why do we expect DTA was more helpful than LRM? It is clear that DTA is typically most useful when there are many predictors and we do not know which ones to consider first. This was certainly the case here. But DTA is also useful when there are many possible nonlinearities or interactions and we do not know which ones were the most crucial. This could have been done in LRM but it was thought to be so tedious that it was not even attempted. Of course, this mostly only happens when the model assumptions are not met due to nonlinearity, or outliers, or even combinations of both.

If the correct selection model was linear and additive and all assumptions were met, then the LRM would naturally be best. This means that DTA is both easy and hard to use. The easy parts come because DTA is an automated stepwise procedure that considers most nonlinearity and all interactions. DTA does not need assistance with scaling and scoring and variable redundancy is allowed (we can simply put all IVs in). DTA also does a lot of numerical searching, but it is very fast. But the hard parts come when we realize we need to know the key outcome variable (which we do here), and we need to judge how many nodes should be used to fit the current data (which we do not), and still have a repeatable model (which we may or may not have).

This leaves us with several other hard questions to answer. One of our most important is simply stated—"Is this CogUSA sample of persons an adequate representation of the population?" As confessed initially, this is not really a question we can answer because we did not measure anything on the people that initially refused to participate. We can say a bit more about the transition from the telephone testing (T1) to the face-to-face testing (F2) but we might only be 18% accurate because dropout may be due to unmeasured variables. In any case, the sampling weights we created from both LRM and DTA are now a part of the CogUSA database (available from the ICPSR website) and perhaps only future research with CogUSA or other samples will tell us if either of these alternative weights are useful.

Note

I thank the National Institute on Aging (Grant # AG-07137-22) and the APA Science Directorate for funds to support this ongoing research. Portions of this paper were also presented at the Department of Psychology, University of California at Davis, May 2010. I also thank my many colleagues at the University of Southern California, including Drs. Kelly Kadlec, Dr. Kelly Peters, and Ms. Ellen Walters, for helping me organize the CogUSA data plans. Similarly, I thank many members of the Institute of Social Research (ISR), at the University of Michigan, including Dr. Gwen Fisher, Dr. Brooke Helppie-McFall, Dr. Willard Rodgers, Ms. Zoanne Blackburn, and Ms. Halimah Hassan, for carrying out the unusual CogUSA data collection plan. I am also especially grateful to Dr. Gilbert Ritschard, for pointing out where the Decision Tree Analysis approach can be useful and where it can fail.

References

Angrist, J. D. & Krueger, A. B. (2001). Instrumental variables and the search for identification: From supply and demand to natural experiments. *Journal of Economic Perspectives, 15*(4), 69–85.

Berk, R. A. (2009). *Statistical Learning from a Regression Perspective.* New York: Springer.

Breiman, L., Friedman, J., Olshen, R., & Stone, C. (1984). *Classification and Regression Trees.* Pacific Grove, CA: Wadsworth and Brooks/Cole.

Cole, S. R. & Hernan, M. A. (2008). Constructing inverse probability weights for marginal structural models. *American Journal of Epidemiology, 168*(6), 656–664.

Feldesman, M. R. (2002). Classification trees as an alternative to linear discriminant analysis. *American Journal of Psychical Anthropology, 119*, 257–275.

Ferrer, E., Salthouse, T. A., McArdle, J. J., Stewart, W. F., & Schwartz, B. (2005). Multivariate modeling of age and retest effects in longitudinal studies of cognitive abilities. *Psychology and Aging, 20*(3), 412–442.

Hastie, T., Tibshirani, R., & Freidman, J. (2001). *The Elements of Statistical Learning: Data Mining, Inference, and Prediction.* New York: Springer.

Heckman, J. J. (1976). The common structure of statistical models of truncation, sample selection and limited dependent variables and a simple estimator for such models. In *Annals of Economic and Social Measurement, Volume 5, number 4* (pp. 475–492). NBER.

Heckman, J. J. (1979). Sample selection bias as a specification error. *Econometrica, 47*(1), 153–161.

Hosmer, D. W. & Lemeshow, S. (2000) *Applied Logistic Regression*. New York: John Wiley & Sons.

Kalton, G. & Flores-Cervantes, I. (2003). Weighting methods. *Journal of Official Statistics, 19*, 81–97.

King, D. W., King, L. A., Bachrach, P. S., & McArdle, J. J. (2001). Contemporary approaches to missing data: The glass is really half full. *PTSD Quarterly, 12*(2), 1–7.

Kish, L. (1995). Methods for design effects. *Journal of Official Statistics, 11*, 55–77.

Lemon, S. C., Roy, J., Clark, M. A., Friedmann, P. D., & Rakowski, W. (2003). Classification and regression tree analysis in public health: Methodological review and comparison with logistic regression. *Annals of Behavioral Medicine, 26*(3), 172–181.

McArdle, J. J. (2009). Latent variable modeling of longitudinal data. *Annual Review of Psychology, 60*, 577–605.

McArdle, J. J. (2010). Ethical Issues in factor analysis. In A. Panter (Ed.), *Statistics Through an Ethical Lens*. Washington, DC: APA Press.

McArdle, J. J. (2012). Exploratory data mining using CART in the behavioral sciences. In H. Cooper & A, Panter (Eds.), *Handbook of Methodology in the Behavioral Sciences* (Chapter 20). Washington, DC: APA Books.

McArdle, J. J. & Cattell, R. B. (1994). Structural equation models of factorial invariance in parallel proportional profiles and oblique confactor problems. *Multivariate Behavioral Research, 29*(1), 63–113.

McArdle, J. J. & Hamagami, F. (1994). Logit and multilevel logit modeling studies of college graduation for 1984–85 freshman student-athletes. *The Journal of the American Statistical Association, 89* (427), 1107–1123.

McArdle, J. J. & Woodcock, J. R. (1997). Expanding test-retest designs to include developmental time-lag components. *Psychological Methods, 2*(4), 403–435.

McArdle, J. J., Ferrer-Caja, E., Hamagami, F., & Woodcock, R.W. (2002). Comparative longitudinal multilevel structural analyses of the growth and decline of multiple intellectual abilities over the life-span. *Developmental Psychology, 38*(1), 115–142.

McArdle, J. J., Fisher, G. G. & Kadlec, K. M. (2007). Latent Variable Analysis of Age Trends in Tests of Cognitive Ability in the Health and Retirement Survey, 1992–2004. *Psychology and Aging, 22*(3), 525–545.

McArdle, J. J., Fisher, G. G., Rodgers, W., & Kadlec, K. M., (2005). The *Cognition in the USA (CogUSA) Study*. Unpublished Manuscript, Department of Psychology, University of Southern California, Los Angeles, CA.

McArdle, J. J., Fisher, G. G., Rodgers, W., Kadlec, K. M., & Helppie-McFall, B. (2010). Notes on a New *Cognition in the USA (CogUSA) Study*. Unpublished Manuscript, Department of Psychology, University of Southern California, Los Angeles, CA.

McDonald, R. P. (1999). *Test Theory: A Unified Treatment*. Mahwah: Erlbaum.

Monahan, J., Steadman, H. J., Silver, E., Appelbaum, P. S., Robbins, P. C., Mulvey, E. P., Roth, L., Grisso, T., & Banks, S. (2001). *Rethinking Risk Assessment: The MacArthur Study of Mental Disorder and Violence*. Oxford: Oxford University Press.

Morgan, J. N. & Sonquist, J. A. (1963). Problems in the analysis of survey data: And a proposal. *Journal of the American Statistical Association, 58*, 415–434.

Muthén, B. & Joreskog, K. (1983). Selectivity problems in quasi-experimental studies. *Evaluation Review, 7*, 139–174.

Ofstedal, M. B., Fisher, G. G., & Herzog, A. R. (2005). *Documentation of cognitive functioning measures in the Health and Retirement Study* (HRS/AHEAD Documentation Report DR-006). Ann Arbor: University of Michigan.

Penny, K., & Chesney, T. (2006). A comparison of data mining methods and logistic regression to determine factors associated with death following injury. In S. Zani, A. Cerioli, M. Riani, & M. Vichi (Eds.), *Data Analysis, Classification and the Forward Search* (pp. 417–423). Heidelberg: Springer-Verlag.

Puhani, P. A. (2000). The Heckman correction for sample selection and its critique. *Journal of Economic Surveys, 14*(1), 53–68.

Sonquist, J. (1970). *Multivariate Model Building*. Ann Arbor, MI: Institute for Social Research.

Stapleton, L. M. (2002). The incorporation of sample weights into multilevel structural equation models. *Structural Equation Modeling, 9*(4), 475–502.

Steinberg, D. and Colla, P. (1999). *CART: An Introduction*. San Diego: Salford Systems.

Strobl, C., Malley, J., & Tutz, G. (2009). An introduction to recursive partitioning: Rationale, application and characteristics of classification and regression trees, bagging and random forests. *Psychological Methods, 14*(4), 323–348.

Su, X., & Tsai, C. (2005). Tree-augmented Cox proportional hazards models. *Biostatistics, 6*(3), 486–499.

Swets, J. A., Dawes, R. M., & Monahan, J. (2000). Psychological science can improve diagnostic decisions. *Psychological Science in the Public Interest, 1*(1), 1–26.

Tibshirani, R. J. & Tibshirani, R. (2009). A bias correction for the minimum error rate in cross-validation. *The Annals of Applied Statistics, 3*(2), 822–829.

Zhang, H. & Singer, B. (1999). *Recursive Partitioning in the Health Sciences*. New York: Springer.

Appendix

Appendix Table 11.A1 A SAS program for logistic regression analysis of attrition

```
TITLE2 'Data Description';
PROC FREQ;
     TABLE Both;
RUN;

TITLE2 'Initial Model';
PROC LOGISTIC;
     MODEL Both = Years_Age Years_Ed Eff_Gen Black White /RSQ;
     RUN;

TITLE2 'Model with Interactions';
PROC LOGISTIC;
     MODEL Both = Years_Ed Years_Age Eff_Gen Black Years_Ed*Years_Age
Years_Ed*Eff_Gen Years_Ed*Black /RSQ;
RUN;

TITLE2' "More Advanced Models';
PROC LOGISTIC;
     MODEL Both = Years_Age Years_Ed Eff_Gen Black White
     T1_TIC_RM T1_TIC_RPM T1_Pcesd
     T1_TIC_NA T1_TIC_DA T1_TIC_BC T1_TIC_S7 T1_TIC_IR T1_TIC_DR /RSQ;
     OUTPUT p=p_logit3 OUT=temp1;
RUN;
```

Appendix Table 11.A2 Weighting the data with SAS based on the logistic selection model (for details, see Stapleton, 2002).

```
DATA temp2; SET temp1; Weight=P_logit3; Lambda3= 1501 / 1218;
W_logit3=Weight*lambda3; RUN;

PROC CORR NOPROB; VAR p_logit3 weight w_logit3; RUN;

PROC MEANS DATA=temp1; CLASS Both; VAR Years_ED; WEIGHT w_logit3; RUN;

PROC SURVEYMEANS DATA=temp1; CLASS Both; VAR Years_ED; WEIGHT w_logit3; RUN;

PROC SURVEYMEANS DATA=temp1 (WHERE=(BOTH=0)); VAR Years_ED; WEIGHT w_logit3;
RUN;
PROC SURVEYMEANS DATA=temp1 (WHERE=(BOTH=1)); VAR Years_ED; WEIGHT w_logit3;
RUN;
```

Appendix Table 11.A3 A selected portion of the SAS input based on output from the CART PRO V6 program.

```
/******************************************************************
 * The following SAS-compatible code was automatically generated
 * by the TRANSLATE feature in the Salford Systems CART(tm)
 * program, version: 6.2.0.160
 ******************************************************************/
MODELBEGIN:
NODE1:
if YEARS_ED gt .z then do;
    if YEARS_ED <= 16.500000 then goto NODE2;
    else goto NODE39;
    end;
  else goto NODE2;
NODE2:
if T1_TIC_RM gt .z then do;
    if T1_TIC_RM <= 30.000000 then goto NODE3;
    else goto NODE5;
    end;
  else goto NODE5;
NODE3:
if T1_TIC_DR gt .z then do;
    if T1_TIC_DR <= 3.500000 then goto NODE4;
    else goto TNODE3;
    end;
else if T1_TIC_EM gt .z then do;
    if T1_TIC_EM <= 37.500000 then goto NODE4;
    else goto TNODE3;
    end;
else if T1_TIC_IR gt .z then do;
    if T1_TIC_IR <= 3.500000 then goto NODE4;
    else goto TNODE3;
    end;
else if YEARS_AGE gt .z then do;
    if YEARS_AGE <= 81.000000 then goto TNODE3;
    else goto NODE4;
    end;
else if T1_TIC_S7 gt .z then do;
    if T1_TIC_S7 <= 0.500000 then goto NODE4;
    else goto TNODE3;
    end;
else if T1_TIC_NA gt .z then do;
    if T1_TIC_NA <= 2.500000 then goto NODE4;
    else goto TNODE3;
    end;
  else goto TNODE3;
```

Appendix Table 11.A4 A selected Mplus input for the weighted correlation (see Table 4).

```
TITLE:  Correlation Run of CogUSA scores (McArdle, 2012)

DATA:       FILE = 'Z:/BANK/COGUSA/DATA_2012/CogUSA_selection.dat' ;

VARIABLE: NAMES=
            Both Eff_Gen Lag_days Age EDUC White
            T1_NS_WSCORE F1_NS_WSCORE P_logit3 P_CART2 NS_Wdiff
            T1_NS F2_NS;

            USEVAR = T1_NS F2_NS;

            USEOBS=(BOTH==1);

            WEIGHT = P_LOGIT3;

            MISSING=.;

ANALYSIS:  TYPE=MISSING H1; ESTIMATOR=MLR;

MODEL:     T1_NS WITH F2_NS;

OUTPUT:    PATTERNS SAMPSTAT RESIDUAL STANDARDIZED TECH4;
```

Appendix Table 11.A5 A selected Mplus input for the weighted regression (see Table 5).

```
TITLE:  Regression Run of CogUSA scores (McArdle, 2012)

DATA:       FILE = 'Z:/BANK/COGUSA/DATA_2012/CogUSA_selection.dat' ;

VARIABLE: NAMES=
            Both Eff_Gen Lag_days Age EDUC White
            T1_NS_WSCORE F1_NS_WSCORE P_logit3 P_CART2 NS_Wdiff
            T1_NS F2_NS;

            USEVAR = Eff_Gen Lag_days Age EDUC White T1_NS F2_NS;

            USEOBS = (BOTH == 1);

            WEIGHT = P_CART2;

            MISSING=.;

ANALYSIS:  TYPE=MISSING H1; ESTIMATOR=MLR;

MODEL:     F2_NS ON T1_NS Eff_Gen Lag_days Age EDUC White;

           T1_NS ON Eff_Gen Lag_days Age EDUC White;

           Lag_Days ON Eff_Gen Age EDUC White;

           EDUC ON Eff_Gen Age White;

! just to bring in all data

           Eff_Gen ON Age White;

           Age ON White;

OUTPUT:    PATTERNS SAMPSTAT RESIDUAL STANDARDIZED TECH4;
```

---

12 Adaptive Testing of the Number Series Test Using Standard Approaches and a New Decision Tree Analysis Approach

John J. McArdle

Introduction

In this chapter the author attempts to demonstrate the utility of a *Decision Tree Analysis* (DTA) approach to *Computerized Adaptive Testing* (CAT). The basic psychometric premise used here is that if the overall (or total) score for any individual comes from a full set of items (I), the behavior on a smaller number of items ($i < I$) can be used to mimic the overall tests score with lower but substantial accuracy. If so, then these specific items may be administered instead of the complete set and this could save valuable administration time. In essence, once the test is constructed, the DTA approach is not theory bound and neither is the CAT approach. The reduction in accuracy could always be compared to the time gained by not having to administer irrelevant items. This principle is illustrated with data from a specific test, 47 items from the published *Number Series* (NS) test (from the *Cognition and Aging in the USA* (CogUSA) experiment), where all individuals ($N > 1,200$) were effectively administered all 47 items. Standard CAT strategies are compared to the DTA approach and we conclude that the DTA is much more accurate with a scale reliability of $\rho^2 = 0.85$ with only 4–7 items administered. Other costs versus benefits type experiments are suggested.

Psychometrics is often defined as the sampling of measures or scales or items designed to measure psychological constructs. Although this history is often overlooked (see McDonald, 1999), most multivariate measurement concepts are evaluated by using specific items or specific scales. In theory, the common factor model applied to item and scale measurement (McDonald, 1985; McArdle et al., 2001, 2007) is a useful starting point. The theoretical model of measurement used is also important when considering item scoring, the presentation of data, and the creation of more time-efficient shorter tests. More time-efficient shorter tests will be the focus of this chapter.

Several practical and problems have emerged when attempting to repeatedly measure adults on cognitive tasks—"cognitive attrition," "cognitive fatigue," and "practice effects" (e.g., McArdle & Woodcock, 1997). If the tests are too long then otherwise cooperative people may not volunteer to repeat them a second time. Tests based on a shorter number of items can minimize some of these problems. One must first start by counting the total number of required items in a particular measurement scale. The use of shortened tests also makes practical sense because, if it is not possible to administer the required number of items in the available person testing time, one probably cannot gather the needed data. The biggest problem when using shortened tests with fewer items is decreased reliability of the measurement scale (see Guilford, 1956). The benefits (shorter testing time) versus the costs (decreased reliability) can be conceptualized as a key experimental design issue (Fisher, 1940; McArdle, 1994; see McArdle & Cattell, 1994). The purpose of the current chapter is to address, via the presentation of practical and theoretical methods, the benefit–cost trade-off of shortened tests of psychological constructs.

The Number Series Test

The *Number Series* (NS) test is attributed to T. G. Thurstone (1962) as a test included in the *Primary Mental Abilities* (PMA) battery (e.g., see Schaie, 2005). When using the PMA-NS, a person is shown a series of numbers (e.g., 6, 11, 15, 18, 20, __) and is asked to identify the number that would correctly continue the series. In the PMA-NS a person is provided a maximum time of 4.5 minutes to solve 20 NS items. Research has demonstrated that performance on the PMA-NS test does not relate to number facility or fluency (as in the different PMA-N test). On the other hand, Number Fluency (N) is a narrow cognitive ability in Carroll's (1993) meta-analytic summary of the extant factor analysis literature. Instead, and because of the puzzle-like demands of the numerical tasks, the PMA-NS scores are considered good indicators of inductive (Gf-IN) or quantitative reasoning (Gf-RQ), under the broad ability domain of *fluid reasoning* (Gf).

A contemporary variation of the PMA-NS test is included in the *Woodcock–Johnson Psycho-Educational* battery (WJ-III; Woodcock, McGrew, & Mather, 2007). Although conceptually similar to the PMA-NS, the WJ III Number Series tests includes different items and a different response format (e.g., 2, 4, 6, __, 10). The WJ-II Number Series test includes a total of 47 items ordered by increasing item difficulty (see Woodcock, 1990; McArdle, et al., 2001). The WJ III Number Series is interpreted as an indicator of the narrow Cattell-Horn-Carroll (CHC) ability of quantitative reasoning (RQ) which is subsumed under the broad domain of *fluid reasoning* (Gf; after Cattell, 1941, 1998; Horn, 1972, 1988, 1991, 1998; also see Carroll, 1993, 1998).

Methods for Constructing Shortened Tests

A number of different approaches have been proposed for developing short forms of longer tests. Under classical true score theory (Allen & Yen,

2002; Novick, 1966) each individual test item is assumed to be empirically related to the latent (underlying) true score as per the same mathematical model. Thus, this classical true score assumption implies that in a test where the relations (i.e., correlations) between the latent true score and the items are higher, the measurement scale will produce a higher internal consistency reliability. It is not surprising that the number of items used in the construction of a scale (test) can be used as a first estimate of the overall internal consistency reliability of the scale. If a test is "shortened" by some simple method, such as administering only half of the items, then we are guaranteed to lose some reliability because more items of the same construct typically produce higher internal consistency reliability. The well-known Spearman–Brown prophecy formula can be considered an index of the loss of reliability when less than a full set of items is used (see Guilford, 1956, p. 453; McArdle & Woodcock, 1997). Although we expect to lose information in shorter tests, they certainly take less time to administer.

If another basic model of measurement (i.e, the Rasch model; Rasch, 1960) is assumed to be correct, the number of individual item responses (i_n) required to indirectly measure any latent construct for any specific person can determined in a more efficient fashion. In particular, Rasch-based *Adaptive Testing* (AT) techniques (see Baker & Kim, 2004; Embretson & Reise, 2000) can be used to devise unique subsets of items (from the larger total set of items) that are "tailored" to the latent ability of each person. AT methods are well known and have been applied in a wide variety of testing environments (see Baker & Kim, 2004). The first requirement for the simplest form of AT is that a single common latent factor model represents a good fit to the observed item data (see McDonald, 1999). In this approach, all items in a scale or test are required to measure only one underlying ability dimension (the uni-dimensionality assumption).

Any AT approach typically involves: (a) the administration of one or more Rasch-scaled (calibrated) items from the test, (b) the immediate estimation of the person's likely total ability score for the test based on the person's performance on the item(s) administered to this point, (c) followed again by the selection of another item nearby the previous estimate of the person's latent ability, and (d) the iterative repetition of this process until a specified stopping criterion is met. Since this repetitive iterative administration and estimation process can be difficult to track for an examiner, computer programs are often needed to create, administer, and score *Computer Adaptive Tests* or *Computer Administered Tests* (CATs). In theory and practice, the most notable feature of an adaptive testing procedure, even one based on a fixed number of items or time limits, is that the specific selection and number of items administered is not necessarily the same for each person. CAT focuses the testing on the particular subset of items, from the complete test item pool, that are closest to the person's level of ability, thus resulting in the administration of more person-centered item selection and administration.

Finally, it is possible to employ other computer-based data-analytic tools to produce person-centered AT approaches. As outlined in the remainder of this chapter, *Decision Tree Analysis* (DTA; Sonquist, 1970; Breiman et al., 1984; Berk, 2009; McArdle, chapter 1, this volume) using regression type data analytic methods will be described and demonstrated as an alternative method for creating AT or shorter tests. DTA procedures are not used in typical AT construction, so they will be explained in greater detail below.

The Current Study

The Woodcock–Johnson (WJ) III tests (Woodcock, 1990; McArdle & Woodcock, 1998) are ideal for the current research project as they were initially created using the Rasch model as the basis for item calibration and evaluation. Although the WJ III tests have basal and ceiling rules, they have not been created in a fully adaptive test form. Several approaches for adapting specific WJ III tests will be described here, followed by analysis of adult population data gathered with adapted versions of the WJ III Number Series (NS) test. In the next section we describe the data collection methods used in a recent empirical study of NS. This is followed by a description of the mathematical basis of the short form and AT strategies described here, including CART. Results from the application of each CAT strategy to the NS data are then presented. Finally, the Discussion will cover issues of implementation of the different procedures as well as potential problems.

Methods

Persons

A recent experiment in cognitive adaptive testing has been carried out by the author and colleagues in the *Cognition and Aging in the USA* (CogUSA) study (McArdle, Fisher Rodgers & Kadlec, 2005). CogUSA is a collaborative study of a large ($N > 1,200$) and representative sample of adults aged 50–90 from across the United States. In this and other collaborative studies (see McArdle, Fisher & Kadlec, 2007) we have attempted to create new AT-formatted measurements of several dimensions of adult cognitive functioning to evaluate the extent to which these tests could be used in the larger and ongoing *Health and Retirement Study* (HRS; Freedman, Aykan & Martin, 2002; Futz, Ofstedal, Hertzog & Wallace, 2003; Ofstedal, Fisher, & Herzog, 2005). The focus in the current study is on measurements of the construct of both fluid intelligence (Gf) and quantitative reasoning (Gf-RQ) as presented in the something similar to the Woodcock–Johnson (WJ III) Number Series (NS) test.

The current data set was collected in two phases. First, a random digit dialing service (named Genesys) was used to request survey data from adults over 50 years of age living in the 48 contiguous United States. Of approximately 3,200 telephone numbers solicited, $N = 1,514$ (50% response rate) agreed to initial testing. The telephone testing was based on a battery of 12 cognitive tests administered verbally via the telephone. In this first step, the NS items were presented as a telephone administered CAT (with up to six items). The remaining short ability tests came from the HRS project. On average, the telephone interview took 38 minutes to complete. Each respondent was paid US$20 for his or her participation.

Upon completion of the telephone interview, each person was asked if they were willing to participate in face-to-face testing using similar tests within a period of 14 days (with time-lag randomization, as in McArdle & Woodcock, 1997, 1998). In this second phase of data collection, $n = 1,230$ (81% response rate) agreed to participate. The dropout of these participants has been studied elsewhere (see below, and McArdle, chapter 11, this volume). During the second phase, face-to-face testing data was collected on a large set of the Woodcock–Johnson scales (WJ-R, WJ-III), including the NS. During this phase of data collection the complete 47-item publication version of NS was administered. The average interview length during the second phase of face-to-face data collection was 181 minutes. Each respondent was then paid US$60 for their participation in Phase II.

The number of dropouts from the first-phase telephone testing to the second phase face-to-face interviewing was $n=284$ (19%). It was determined that the "telephone only" persons differed slightly from persons that also participated in the face-to-face assessments. In order to account for these group differences, an initial selection model was developed (after Heckman, 1979). Logistic regression was used to predict participation (0 or 1) from all known demographics and telephone measures, and the logistic pseudo-explained variance was relatively small (max $< 2\%$). We also applied nonlinear (i.e., DTA) procedures to this respondent sample difference problem, with the resulting prediction equation accounting for over 18% of the variation between the two groups (for details, see McArdle, Fisher, Rodgers & Kadlec, 2005; or McArdle, chapter 11, this volume). We used these selection bias models to develop person sampling weights (viz., relative weights as per Stapleton, 2002) that were then applied to the persons that participated in both phases of data collection, in order to insure more representative sample analyses.

Available Number Series Data

At the NS item level, some items were found to be very easy (i.e., answered correctly by almost everyone; items 1–10 had a 98% success rate), whereas other items were more difficult (items 29–47 had less than a 50% success

Table 12.1 Summary statistics for CogUSA Face-to-Face complete case sample
($N = 1,223$)

1a: Univariate

Variable	Mean (sd)		Minimum	Maximum
Years of age	64.6	(11.3)	38	96
Year of education	13.9	(2.2)	1	17
Females (1)	51.3%	..	0	1
White (1)	86.0%	..	0	1
Dyad (1)	67.7%	..	0	1
Wscore47	516.3	(24.9)	415.2	588.0

1b: Bivariate Correlations

	Years of age	Years of education	Females	White	Dyad	Wscore47
Years of age	1.00					
Year of education	−0.25	1.00				
Females (1)	−0.07	−0.09	1.00			
White (1)	0.02	0.12	−0.03	1.00		
Dyad (1)	−0.26	0.10	−0.20	0.15	1.00	
Wscore47	−0.41	0.52	−0.13	0.27	0.27	1.00

rate), and some were found to be very hard (items 42, 44, 46, and 45 had less than a 10% success rate). Interestingly, the previously most difficult item (item 47) had a relatively high success rate (28%) in this sample and this could be due to the item or the people. Correlations among NS items are discussed in the following section.

Weighted summary descriptive statistics for the persons who completed the face-to-face testing are presented in Table 12.1. The sample persons, on average, were approximately 65 years old, had 13 years of education, were approximately equally split between females (51%) and males (49%), were predominately white (86%), and approximately 2/3 (68%) were living as dyads. A WJ III raw score-to-W-score scoring procedure was used to assign each person a NS W-ability score on the $I = 47$ item test. As summarized in Table 12.1, the sample had a mean NS W-score of 516 (SD = 25). The NS W-score distribution is displayed in Figure 12.1, and represents a fairly symmetric and nearly normal distribution of NS W-abilities.

Analytic Strategy—Masking Some Available Data

The general approach used in this study was to evaluate the impact on the NS ability score distribution when a majority of items were eliminated

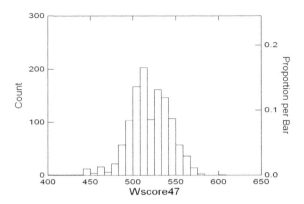

Figure 12.1 The empirical distribution of the W-scores for Face-to-Face number series task in the CogUSA sample ($N = 1,223$).

from the total test. The rationale for this elimination strategy was based on the fact that, even if we followed the prescribed WJ III NS standard basal and ceiling rules, it would take too much time to administer the complete NS test as one part of a much larger survey data collection protocol (e.g., the HRS). The criterion used to evaluate the impact of eliminating items was the magnitude of the correlation of the total 47-item NS W-score (NS47) with all other NS W-score estimates based on the administration of fewer items. Given the known relation between test length and test reliability, we would expect some loss of reliability and validity for several of the adaptive forms of the NS test compared to full length NS. To evaluate the trade-off between fewer items (less testing time) and subsequent decreases in reliability and validity, we employed three different analytic techniques.

The overall strategy employed was the "masking data" approach created by Bell (1954) and used by McArdle (1994). From the data collection procedures previously described we know each person's full-length (47-item test version) NS W-score (WFULL47). However, we recognize that the respondent data collection burden for the complete NS test is not possible. Thus, we use the item level data as a starting point to answer the questions—"What if we did not administer all of the full-length NS test? What score "would have been" a person's full-length NS W-score if they had taken even fewer items using different forms of AT strategies? The goal of each analysis is identical—given a smaller number of items ($i < I$) to estimate the full-length total W-scores, how accurately can we reproduce the total full-length WFULL47. All techniques listed below can be compared directly using indices based on standard correlation and regression methods.

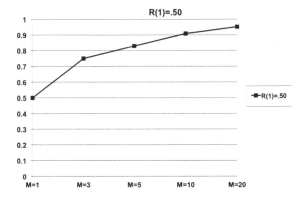

Figure 12.2 Expected score reliability from the Spearman–Brown (1910) "prophecy" formula based on test length (*I*) and inter-item reliability (r_{ic}^2).

Method 1: Creating Classical Short Forms

Considering Classical Short Forms

In this CogUSA sample, the 47-item NS test was observed to have a Chronbach's alpha (α) of 0.925. The highest item correlation with the total score was $r = 0.65$ (item 39). Four other items had similarly high correlations (items 37, 47, 36, and 33). However, low item-to-total score correlations were observed as well, including three below $r = 0.20$ (items 1, 2, 3) and 10 below $r = 0.30$. The average inter-item correlation for the NS test was approximately $r = 0.50$.

Results can be based on the classic Spearman and Brown (both in 1910) equations (see Zimmerman & Williams, 1966) to estimate the reliability of a test as a function of the number of items and the average inter-item correlation. For clarity here, this classical expression has been calculated for five different item set sizes ($n = 1, 3, 5, 10, 20$). These varying levels of inter-item correlation ($R(i) = 0.50$) are plotted as a single curve in Figure 12.2.

This curve suggests that if we have a test where each item has an inter-item correlation of $r(i) = 0.50$, we can create a test with different internal consistency reliabilities based on the total number of items: $R_{ic}(1) = 0.50; R_{ic}(3) = 0.75; R_{ic}(5) = 0.825; R_{ic}(10) = 0.90; R_{ic}(20) = 0.95$. In this calculation all items must have exactly the same correlation with each other. Given this assumption, and depending upon the desired level of reliability we seek, we can use the information contained in the curve in Figure 12.2 to create short form tests based on any specific number of items. Each short form test can, in turn, be evaluated as a function of time taken to administer a specific item set. In addition, inspection of the curve in Figure 12.2 demonstrates that considerable useful information is obtained in short form item sets ranging from five to 10 items. Conversely, the curve in Figure 12.2 also suggests that if we have a test with 10 items ($I = 10$, $R_{ic}(10) = 0.95$) and we eliminate half the items to produce a five-item test ($i = 5$), there will inevitably be a reduction in the accuracy of the individual W-score estimate (to about $R_{ic}(5) = 0.825$). And perhaps this is would not be a terrible loss.

In the Classical Short Form calculation described above, we can start with items that meet the prior conditions. That is, we can ignore the classical model of equal true score loadings and only isolate the items that are the best surrogates for the longer full length NS test. In this case the obtained reliability will be directly related to the average correlation between the items. We hope that we can start with a big enough item pool so enough items remain available.

In the first analysis here we consider the scores that result from taking only i specific items in a specific adaptive order from the full $I = 47$ item pool. In this case, new W-scores were calculated for four different sets of items ($i = 4$, 5, 6, and 7) and two different approaches to short form item selection—either a (1) fixed interval (FIXED) of items spread equally across the scale, or (2) random interval (RANDOM) selection of items for each person.

The empirical results obtained from creating the 10 different short forms (five item sets by two methods) is displayed in Figure 12.3. In Figure 12.3 the X-axis represents the number of items selected (i.e., 3 to 7). The Y-axis represents the resulting correlation of the new W-score with the W-score estimated from the full length NS test. The use of a fixed set of items (a common short form procedure), has a correlation with the full length W-score scale of $r = 0.50$ with $i = 3$, which increases monotonically to $r = 0.85$ when using $i = 7$. Of course, the reported correlations in Figure 12.2 do not represent the reliability of the new W-score, because the original 47 item test has a limited reliability ($\alpha = 0.925$). The product of these two reliability indices must be used—Although it is not emphasized here, the reliability of a FIXED seven-item scale would be only 0.67 (i.e., 0.85*085*0.93). However, the inspection of these results leads to the conclusion that the use of a fixed set of items does not require the administration of all 47 items. As expected, short form tests will result in a quantifiable loss of reliability.

The use of a random set of items procedure (RANDOM) sometimes works well as a basic design principle, especially when prior knowledge is not available (see McArdle & Woodcock, 1997). However, inspection of the lower curve in Figure 12.3 indicates that this procedure yields much less accuracy for any number of selected items (i.e., $r = 0.50$ with $i = 5$). For this reason, the RANDOM short form procedure cannot be recommended in the current design context.

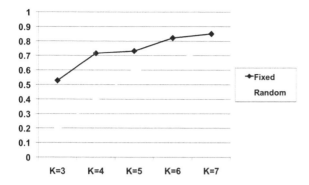

Figure 12.3 Empirical correlations of different fixed and random short forms ($N = 1,223$).

Method 2: Creating Classical Adaptive Tests

Creating Adaptive Tests

The general theme of individualized *adaptive testing* (AT) has followed directly from the same logic (see Embretson & Reise, 2000; Wainer, 2000). Adaptive tests are "abbreviated" but *not* "shortened" forms of classic tests, and they are usually based on some form of *item response theory* (IRT) model (see the next section). There are numerous methods for creating adaptive tests. Regardless of specific AT approach, the most important design decisions include: (a) the set of initial items, (b) the selection of the next items in a sequence, and (c) the stopping rule(s). In IRT, a reasonably accurate estimate of the person's ability can be based on far less than the full set of complete items (e.g., $i_n = 0.2 * I$). In practice, the benefits accrue because the person is asked fewer questions (resulting in shorter testing times) while maintaining adequate accuracy and reliability. Furthermore, if there is a concern about practice effects from repeated administration of the same items (see McArdle & Woodcock, 1997; Ferrer et al., 2005), IRT-based AT designs can insure that the repetition of any item in a second testing does not occur (i.e., no item-specific practice).

The Rasch Model of Item Measurement

In a very basic form of measurement we assume that some attribute or construct (c) score exists for each individual ($n = 1$ to N), and a set of items ($i = 1$ to I) with different levels of calibrated item difficulty (δ_i) can be found. We may assume any person's responses to these items follow the simple expression

$$f\{\pi_{in}\} = c_n - \delta_i \tag{1}$$

where π_{in} = the probability of person n responding correctly to item i is some undefined function (f) of the score on the person's score on the underlying construct (c_n, more typically termed "theta") compared to the difficulty of the item (δ_i). (Note: Greek letters are only used here when parameters are estimated.) The simple idea that the probability of a person's correct response (π) to any item is directly related to the distance of the individual's construct score compared to each item's difficulty is a fundamental form of the models in *item response theory* (IRT; see Rasch, 1960).

In the elegant IRT theory, only one construct (c) is measured with the items (assumption of uni-dimensionality), and each item is related to this common factor in exactly the same way (i.e., no difference in the factor loadings; see McDonald, 1999). In practice, the function (f) is typically a logistic (or probit) function, and there may be a constant and multiplier in the fitted equation. More complex models can be created for more complex responses (see Masters, 1982; Baker & Kim, 2004; Wilson, 2005).

Also important here is the notion of a *standard error of measurement of the construct* ($se\{c\}$), which is typically defined (for binary or dichotomous items) in the Rasch model as

$$se\{c\} = 1 / \left(\sum_{i=1}^{I} (\pi_{in} \cdot 1 - \pi_{in}) \right)^{1/2} \tag{2}$$

where the accuracy with which a person is measured is both a function of the specific items selected (i) and their relevant probability for the specific individual. If many items are measured at many levels of difficulty then the resulting $se\{c\}$ will be very small. The standard errors of the sum of the observed response scores (r_n) can be calculated as

$$se\{r\}_n = 1 / \left(\sum_{i=1}^{I} (r_{in} \cdot 1 - r_{in}) \right)^{1/2} \tag{3}$$

so the relation between this standard error of measurement of the score and the standard error of the observed scores gives us an IRT estimate of the average test reliability as

$$\rho^2 = \sum_{n=1}^{N} (se\{c\}_n / se\{r\}_n)^2 / (N - 2). \tag{4}$$

We will not use this last formula directly, but we will consider the concepts it implies.

Adaptive Test Strategies

A typical IRT-based adaptive testing strategy starts with a full set of I items which have been previously administered and have known calibrated or scaled difficulty levels (i.e., $\delta_i =$ known). The following general sequence is then followed: (a) administer a first set of items somewhere near the middle of the difficulty scale for that person, (b) as soon as a person gets one right and one wrong, create an *estimate* of the (c_n) score for that person, and (c) then select the next item so it is as close to that difficulty level (δ_i) as possible, (d) re-estimate the person's (c_n) score, (e) determine if the obtained score meets a specified stopping criterion, and if not, (f) continue with steps (c) through (e) in an iterative fashion until the specified stopping criterion is achieved. Although it is never clear when to stop, possible "stopping rules" might be at a specified time limit, or at i items, or when the $se\{c\}$ is less than some pre-specified value (for further details, see Baker & Kim, 2004).

In early work on the iterative IRT AT topic, Lord (1955, 1970) showed that the *maximum likelihood estimate* (MLE) for a person's ability level (c_n) can be created from the prior item responses using an iterative calculation schema (see Baker & Kim, 2004) based on the number of items administered up to some point (t). A convenient computational formula for this MLE of ability is:

$$c[t+1]_n = c[t]_n + \sum \{r_{in} - \pi_i[t]\} / \sum \{\pi_i[t]1 - \pi_i[t]\}, \tag{5}$$

where $r_{in} =$ the person's response (0 or 1) to a prior item, $\pi_i[t]$ is the probability of a correct response to item i under the given item characteristic curve model up to time t. This procedure is repeated over and over again (i.e., iterated) as more responses are obtained for each person.

One implication of the computational MLE formula (5) is that one cannot actually calculate the MLE of the person's ability level until he or she has answered at least one correct and one incorrect item. If a person answers all items the same way (either all correct or all incorrect) the MLE formula is not calculable. A second implication of the MLE formula is that the choice of an optimal item is never perfectly defined—matching the estimated ability level with the difficulty of the item is optimal only if the model is completely correct.

Complexities of Item Selection

The above described IRT-based AT approaches may be started at any point on the scale, a design consideration that has been the focus of considerable past research (see Baker & Kim, 2004). In IRT, the optimal selection of the initial items would be different for persons of different ability levels, but these ability levels are not known in advance of the testing. If we have no *a priori* information about a person, then the initial selection of items is typically near the middle of the scale, or tends towards easier items. However, if we have any *a priori* measurement information regarding a person (i.e., past test scores, demographics, etc.) this information can used these to provide a more accurate and efficient test starting point.

IRT-based AT stopping rules can be complex. As alluded to previously, testing may discontinue when a predetermined number of i items has been administered, as per a specified plan. Alternatively, testing can stop if an *a priori* testing time limit has been reached (i.e., five minutes) and as much information as possible is desired. As a result, it is possible that different persons will have been administered different numbers of items (and produce different $se\{c\}$ estimates). Finally, a stopping rule can be proscribed based on scale accuracy (e.g., when $se\{c\} < 1$). The use of accuracy as a final stopping rule for the AT will result in persons being administered different numbers of items, but with similar estimated score accuracies.

In sum, there are a number of unknown factors that may lead to the final accuracy of estimation using any of these approaches, including: (1) the size of the original pool of items (I), (2) the size of the number of items chosen ($i < I$), (3) the spread of item difficulties across the scale (hopefully a fairly uniform spread), and (4) the stopping rule applied. Each of these design considerations can make a significant difference to the success or failure of an AT strategy.

Adaptive Test Results

The first AT analyses employed a simple *Half AT* strategy (termed HAT) based on the selection of items either "halfway up or halfway down" the prior difficulty scale. For example, in the HAT procedure persons are started with the item in the exact middle of the full-length WJ III NS test (ITEM24), and if this item was responded to correctly by the persons, an examination of their responses to a harder item, exactly half way up the difficulty scale (ITEM36) occurred. If this first item was incorrect (ITEM24=0) the procedure next examined the person's response to an easier item, exactly halfway down the difficulty scale (ITEM12). The halfway procedure was used to select each subsequent item until the requisite number of items had been administered (e.g., 4–7), at which time a final W-score was created based only on the HAT administered items.

An alternative set of analyses involved the examination of a *Fully adaptive test* strategy (termed FAT) that was based on the selection of items that are fully determined by the prior responses and the known difficulty scale. For example, in the FAT procedure, persons were started at the item in the exact middle (ITEM24), and if this item was responded to correctly (ITEM24=1), an examination of the person's responses to a harder item occurred, using the halfway principle (ITEM36). If this first item was incorrect (ITEM24=0) an examination of the person's response to an easier item occurred, again using the halfway principle (ITEM12). This halfway strategy was repeated until the person had at least one item right and one item wrong. When this criterion was met, the next item was selected to be as close to the estimated W-score ability level as possible (by MLE,

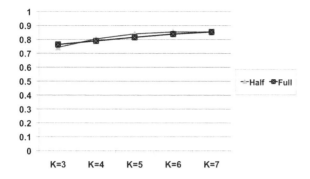

Figure 12.4 Empirical correlations of the half and fully adaptive tests
($N = 1,223$).

see below). At the end of the FAT procedure each person was assigned
a W-score based only on the unmasked $i_n = 6$ items, items that were not
guaranteed to be the same six items for each person.

The results presented in Figure 12.4 show the Pearson correlations of
the full length NS ($I = 47$) W-score with the W-scores estimated using
from $i = 3$ to 7 items under both HAT and FAT schemes. It is useful to
recall that the RANDOM or FIXED selected items approached $r > 0.7$
of the full-length score until $i = 6$ or 7 items had been administered. As
the curves in Figure 12.4 demonstrate, the HAT and FAT strategies display
reasonably high correlations after only $i_n > 3$ items are administered. As
displayed in Figure 12.4, correlations of the HAT with the full length
WFULL 47-item NS range between $r = 0.73$ ($i = 3$) to $r = 0.85$ ($i = 7$).
For FAT with the full length 47-item NS W-score the correlation ranges
between $r = 0.77$ ($i = 3$) to $r = 0.85$ ($i = 7$). Of course, such results
will vary with the content and the construct being measured, but it is
evident that the full length ($I = 47$) NS test is only needed if the goal
is to make precise statements about each *individual*. Only $i_n = 6$ adaptive
items, using either HAT or FAT strategies, are necessary for *group* level
accuracy using the IRT AT NS scale. These surprisingly good results are
due in large part to the fact that the same items are not used for each
person tested.

Method 3: Decision Tree Analysis

The typical uses of multiple regression prediction allow for many alternatives (e.g., Hierarchical, Stepwise, Best Subsets) to accomplish the analytic
goal of dealing with multiple predictors of a dependent variable. As
demonstrated throughout this book, the *Decision Tree Analysis* (termed DTA
here) approach is a relatively new form of "computer-assisted" data analysis

that has been formally developed over the past three decades (e.g., McArdle, chapter 1, this volume; Sonquist, 1970; Brieman et al., 1984; Zhang & Singer, 1999; Hastie, Tibshirani, & Freidman, 2001; Berk, 2009). Some of DTA's popularity comes from its visual–graphic topographic Trees and Cartesian subplots. DTA programs, such as the *Classification and Regression Tree* (or CART) algorithm are now widely available and easy to use and interpret (e.g., see www.SalfordSystems.com, www.XLMINER.com, etc.).

DTA calculations are based on the simple idea that multivariate prediction/classification can be dealt with by (a) using only parts of the database at any time (partition, segmentation), and (b) then completing a procedure over and over again in subsets (i.e., recursively). In a DTA analysis the first step is to define one dependent variable (DV) and a set of multiple independent variables (IVs). The DV and IVs are typically in binary form but may be measured using either categorical or continuous scaling. DTA begins by searching for the best predictor of the DV among all IVs, considering all "ordered splits" of IVs. At each step, the data is split (partitioned) into two parts according to the optimal predefined splitting rule (i.e., the maximum Point Biserial correlation). Next the search strategy is reapplied on each part of the data. This is repeated over and over (recursively) until a final split is not warranted or a "stopping criterion" has been reached. Clearly the repetitive recursive engine that drives DTA methods would not be possible without modern day computers (Breiman, 2001; McArdle, chapter 1, this volume).

The use of DTA procedures for the creation of an AT does not seem to have been attempted before, but it is an approach with the potential to solve many of the problems raised above in more traditional AT approaches. In a DTA-AT approach to the current problem, the first item selected (from the total item pool) would be the item most highly correlated with the total NS W-score (Point Biserial R). This DTA first-item selection procedure is an objective means to select an initial item for the test; however, the first item selected is sample dependent. In a similar manner, a second item selected for administration is chosen for those persons who incorrectly answered the first item (the left branch in a DTA topographic tree) and a second item is chosen for those persons who corrected answered the first item (the right branch of the DTA tree). It is important to note that there is no DTA requirement for the second item to be the same within each partitioning of the overall data. The previous procedure is applied recursively until a designated stopping point is reached (to be discussed below) and where each node in the recursive tree represents a particular pattern of selected items. The final tree nodes effectively isolate subgroups of participants who are most similar to one another and different from the persons in different final nodes. The DTA-AT procedure does not guarantee that each individual will require as many items to be administered to reach the same reliability as any other individual. The use of the DTA-AT procedure should allow for the prediction the total NS W-score with maximum accuracy. DTA-AT maximum accuracy can then be compared to the reliabilities of the resulting HAT or FAT.

Decision Tree Analysis Results

In this DTA-AT approach the same set of data and starting points were used as in the previous analyses. However, here only DTA procedures were used to select optimal items pools. This approach is initiated by predicting the full-length 47-item NS test (NS47). It is important to note that the DTA algorithm automatically selects the best starting item by examining every item's relation to the outcome or DV to be predicted (the WFULL47 NS score). Some versions of DTA automatically create stopping rules, however these are not required to be of the same length for different starting patterns. The empirical behavior of DTA-AT is impacted by the distribution of abilities in the sample, a distribution that can be adjusted by sampling weights (see Stapleton, 2002; McArdle, chapter 1, this volume). In the current analyses the distribution of ability sampling weighting was employed.

The empirical results from the DTA analysis were calculated using standard CART software (CART PRO 6.0; see Steinberg & Colla, 1999). The results are presented in several tables and figures. Figure 12.5 presents the complete tree selected using $I = 47$ NS items and relative sampling weights for $n = 1,223$ participants. In Figure 12.5 each dark box represents a node. More nodes could have been chosen but the DTA tree depicted in Figure 12.5 was determined by the CART algorithm to provide the best tradeoff between relative errors of prediction (0.153) and the complexity of the model (29 terminal nodes). The final DTA model in Figure 12.5 was also internally cross-validated 10 times (an automatic option of V6.0 of the CART program used). Additional detail for the same DTA tree is presented in Figure 12.6 (for numerical details, see the Appendix). This is included here to illustrate the detailed output provided by DTA models as well as their obvious complexity.

The information portrayed in the visual-graphic DTA figures is rewritten in tabular form in Table 12.2. The information included in Table 12.2 documents the suggested "routes" to be taken within each tree node in terms of specific items and responses. To assist in the interpretation of this tabular tree description, blocks of persons with same behavior on the first three items are separated. The overall tree starts with the responses of all persons on ITEM39—nodes 1–21 are used if I39=0 (incorrect) and the nodes 22–29 are used when I39=1 (correct). For clarity, these have been subdivided into blocks of nodes based on a common set of responses to the first three items. In Block 1 all persons are administered ITEM39, then ITEM21, and then ITEM16. In Node 1 the individuals' responses to ITEM6 is examined and it is observed that they all failed the item (I6=0). There are only six persons in the dataset who have this response set, and no

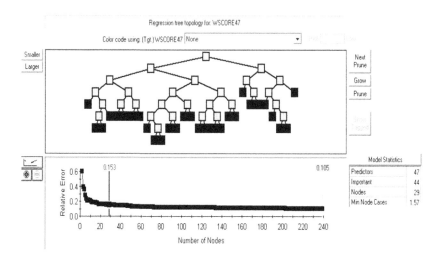

Figure 12.5 Empirical behavior of the regression tree approach ($N = 1{,}223$).

other items need to be administered to the persons that characterize this DTA tree node. That is, after failing the above mentioned four NS items these $n = 6$ persons have a mean W-score of 417 (SD=10). To create a best guess of this set of persons' full length NS W-score (NS47), all six persons are assigned a DTA-based W-score (WDTA4) of 417. Furthermore, the SD=10 is akin to the SEM of these specific WDTA4 W-scores.

The persons in the remaining three nodes in the first block (for details, see Figure 12.6 or Appendix table) responded correctly to ITEM6 (I6=1), and were thus designated to be administered ITEM24, which if correct (I24=1), invoked a stopping rule at $i = 5$ and an estimated W-score of WDTA5=483 (SD=5, $n = 4$). The remaining two nodes represent persons who failed ITEM24 (I24=0), which resulted in the examination of their performance on ITEM1and the specification of two additional groups—If we observes I6=0, thenWDTA6=415 (SD=0, $n = 2$) or if I6=1, WDTA6=483 (SD=5, $n = 4$).

The examination of the other five blocks of nodes is similar, but obviously includes more persons. There are several patterns that account for a substantially larger number of sample persons—Node 12 has a common response pattern (I39=0, I21=0, I28=0, I26=0, I25=1, and I34=0) with WDTA6=504 (SD=6, $n = 153$); Node 16 has a common response pattern (I39=0, I21=0, I28=0, I47=0, I34=0, and I29=0) with WDTA06=512 (SD=7, n=107); Node 14 has a common response pattern (I39=0, I21=0, I28=0, I26=1, and I43=0) with WDTA05=511 (SD=7, $n = 100$). In this result, there are only two nodes where seven items are required—Node

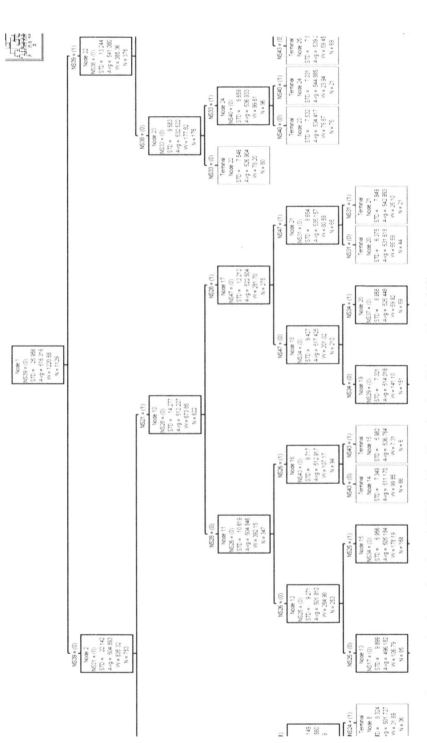

Figure 12.6 More details on the selected AT Tree ($J = 29$ nodes; $I = 47$ items; $N = 1,223$).

10 (WDTA07=496, SD=6, $n = 91$) and node 11 (WDTA07=510, SD=5, $n = 11$).

Inspection of the complete DTA results indicates that not every person is administered the same number of items. This seems reasonable because if the pattern of responses is clear, very few items need to be administered. However, if the item response pattern is inconsistent in the presence of no prior experience with the NS scale (e.g., answering an easier item incorrectly but a harder one correctly), then more items must be administered for increased precision of NS W-score estimation. A count of the number of persons who would have taken any specific number of items results in the conclusion that some persons only required the administration of three items (10%; $n = 124$ in two nodes), some required four items (14%; $n = 165$ in four nodes), others required five items (31%; $n = 378$ in 10 nodes), most people here required six items ($n = 37\%$; $n = 453$ in 13 nodes), and some required seven items (8%; $n = 102$ in two nodes).

DTA analysis also provides information regarding the *Variable Importance* (VI) of each predictor in the complete DTA tree. VI here is an indicator of what each item contributes to the overall WDTA07 score, and the diminishing returns of each successive additional item. Figure 12.7 summarizes the relative importance of each NS item in the complete DTA tree. A brief review of Figure 12.7 indicates that ITEM47 is the most important item and is assigned a baseline importance value of 100%. Indeed, this was earlier considered a questionable item because it was supposedly the hardest item in previous research, but here about 28% of the people got it right, so it was a discriminating item. In any case, next in importance are

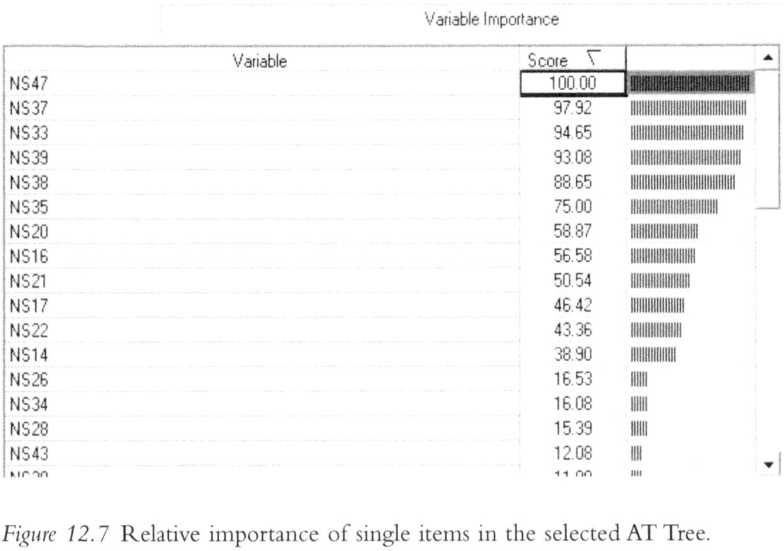

Variable Importance

Variable	Score ▽	
NS47	100.00	
NS37	97.92	
NS33	94.65	
NS39	93.08	
NS38	88.65	
NS35	75.00	
NS20	58.87	
NS16	56.58	
NS21	50.54	
NS17	46.42	
NS22	43.36	
NS14	38.90	
NS26	16.53	
NS34	16.08	
NS28	15.39	
NS43	12.08	
NS39	11.00	

Figure 12.7 Relative importance of single items in the selected AT Tree.

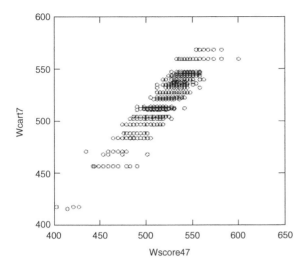

Figure 12.8 Relative W-scores estimated using all 47 items vs CART 7 ($r = 0.96$).

ITEM37 (98%), ITEM33 (95%), and ITEM39 (93%). The reason ITEM39, the most predictive item used in Step 1, was not also the most important is that it was only used once in the big subdivision of nodes. Items 38, 35, 20, 16, 21, 17, 22, and 14 are assigned relative importance values at or above 39%. Most informatively, the information summarized in Figure 12.7 indicates that only approximately half the NS items would be needed in a starting pool to capture all the variation in NS abilities observed in this CogUSA sample.

In total, the final DTA nodes (29) represent subgroups of persons with common response patterns, and some subgroups are small and others large. The overall weighted average of the WDTA07 (termed such because it includes up to seven items for a few persons) is 516.4 with an SD=7.1. The actual full-length WFULL47 compared to the predicted WDTA07 is displayed in Figure 12.8. If these data are indicative of a broader population, the WDTA07 to WFULL47 correlation is an astounding Pearson $r = 0.96$. Although an extremely high correlation, $r = 0.96$ is not a perfect correlation, and the overall scale reliability is probably much less at $\rho^2 = 0.85$ (e.g., 0.97*0.97*0.93) reflecting some unnecessary noise at different levels of the new DTA-AT NS scores. While the new W-scores are decidedly less accurate at the level of the individual person, there is a clear benefit in reduced testing time (i.e., time used to administer 4–7 items instead of 47!). This gain in efficiency could be a driving force in the use of fewer items per person.

Table 12.2 Statistical comparison of predictive validity of five demographics on alternative number series scores

Predictor	WFULL47	(t-value)	WDTA07	(t-value)	WDELTA40	(t-value)
Intercept	513	(388)	513	(390)	0.6	(1.0)
Age	−6.5	(12)	−5.4	(10)	−1.1	(5)
Education	17.2	(18)	16.2	(17)	1.0	(2.5)
Gender	−4.6	(4.2)	−5	(4.5)	0.4	(0.8)
Ethnicity	11.9	(6.7)	4.6	(5.5)	2.3	(3.0)
Dyad	4.6	(3.8)	4.6	(3.8)	0.0	(0.1)
Prob. F (5, 2129)	$p < .0001$		$p < .0001$		$p < .0001$	
Explained variance	0.38		0.34		0.03	

Notes: All variables are rescaled. Age $=$ (Years_Age(n)-50)/10; Education $=$ (Years_Educ(n)-12)/4; Gender, Ethnicity and Dyad are all $-1/2$ or $+1/2$.

Comparisons of Predictive Validity

Although faster NS testing times are impressive, shorter duration is a necessary but not sufficient indicator of the success of AT approaches. In a final analysis, the loss of predictive ability vis-à-vis shorter testing was evaluated with the short-form-derived estimated W-scores. Table 12.2 presents the results of the prediction of the full-length NS W-score (WFULL47) from five demographic indicators—in order of importance, Education (years), Age (years), Gender, Ethnicity, and Dyad (all effect coded). As summarized in Table 12.3, the weighted results display a collective $R^2 = 38\%$ of explained variance with strong independent contributions for each variable. Additionally, higher scores are associated with persons who are more educated, younger, white, female, and living in dyads. When this exact regression model is applied to the DTA-AT-based estimated W-scores (WDTA07), the $R^2 = 34\%$, so there is not much loss of explained variance.

To test for any demographic difference between the results, the discrepancy between the two individual estimates of scores (WDELTA40 = WFULL47-WDTA07) was calculated and fitted to the regression model (i.e., the weighted difference was the outcome or DV for the same set of IV predictors). This final discrepancy score model, literally based on 40 additional items, accounted for only $R^2 = 3\%$ of explained variance from the demographic variables.

Although this model showed minimal discrepancies, it was statistically significant here, so this suggested potential biases as a function of age and ethnicity. To further explore these potential bias problems in detail we could certainly use other DTA techniques (see McArdle, chapter 1 this volume; Ritschard, chapter 2, this volume). But instead we now present a

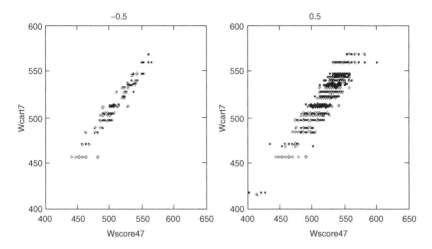

Figure 12.9 Potential bias for majority white participants (+0.5) in CART 7 W-scores.

relatively simple analysis where the available data was simply split into two groups—non-whites and whites. The non-white and white WDTA07 and WFULL47 relations are plotted in Figure 12.9. The information presented in Figure 12.9 suggests the estimation of WDTA07 scores for the non-whites (the left panel in Figure 12.9) is excellent when compared to the WFULL47, but the estimated WDTA07 scores for the white group may have some unintended noise. Obviously, this should be investigated in more detail, and perhaps other variables can be used to improve our understanding of this potential bias. This type of multi-group analysis is hardly ever done, but it should be.

Discussion

There is an increasing literature on the theory and practice of creating short forms and adaptive psychological tests. The purpose of this chapter was to demonstrate how different concepts of short form and adaptive testing strategies can be evaluated at an empirical level. Any of the testing strategies described can be used on any measurement scale as long as item-level data is available for the scale. However, even in the presence of empirical evaluation data, an optimal adaptive item selection rule may not necessarily be clear since the underlying model is never completely correct for an available test. Thus, in the current study we used known Rasch-estimated W-scores from a full test as the target guide. Various empirical methods were employed to estimate the known total target score using five different forms of adaptive strategy:

1   *Fixed Items* (FIX)—A set of i items representing an equal spread across the total W-score scale were selected. This mimics a classical "short form." The empirical behavior of the FIX item strategy was fairly good, but it was concluded that the administration of more items produces considerable benefits.

2   *Random Items* (RAN)—A set of i items that were randomly selected across the total W-score scale. This approach might be suitable in situations where the prior scaling is not available and the user would like a range of items to be administered. Of the five different strategies used here, the RAN approach was determined to be the worst strategy of all.

3   *Half-Adaptive Items* (HAT)—A set of i_n items were selected based on prior person responses. If the person's prior response was correct, the next item selected was "halfway up" the W-score scale. If the person scored incorrectly, the next item selected was "halfway down" the W-score scale. Not all persons received the same items, but the stopping rule was identical across all persons. The empirical behavior of the HAT procedure was remarkably good, even though it does have some known theoretical flaws (i.e., initial items define the limits). On a practical basis, the HAT strategy was actually easy to program and implement, and this may be an important consideration.

4   *Fully Adaptive Items* (FAT)—A set of i_n items were selected which were optimal based on the persons' correct responses to the item model (see Baker & Kim, 2004). Once again, not all persons received the same set of items, but the stopping rule was identical. The empirical behavior of the FAT strategy was very good, but it did not reach its peak efficiency until $i_n = 6$–7 items were administered. This could be true because we cannot start FAT until at least one item is answered correctly and one item is answered incorrectly.

5   *DTA Adaptive Items* (DTA-AT)—The DTA algorithm selected a set of i_n items as described above. The DTA approach clearly exceeded all other approaches, which it was supposed to do, although many different types of person response patterns were identified. The internal cross-validation techniques led to a clear separation of person item response patterns. The DTA approach produced an $r = 0.96$ with the total W-scale score, and required between $i_n = 3$ and 7 items depending upon the consistency of the response patterns. The small sizes of the DTA tree nodes were internally cross-validated, so they are expected to reflect important response patterns of persons when examining other distributions of abilities.

Of course, very little is known about how this DTA procedure would operate with a new sample of people. Since the creation of special item response patterns with very few persons can be time consuming, the expected generalization of the identified response patterns requires cross-validation in a new sample. It is worthwhile mentioning that the DTA approach described here is very similar to the Digit Recognition example presented by Breiman et al. (1984, pp. 43–49). In the approach used here we first assume we have a group of persons (*N*) who have been administered all items (*I*). In DTA we use all individual item scores (0 or 1) as predictors of the continuous total score actually obtained, so this is a regression tree (i.e., not a classification tree). As far as we can tell, the DTA approach described here has not previously been used to create an AT.

Dealing with Incomplete Data

Linking, equating, and calibrating refer to a series of statistical methods for comparing scores from tests (scales, measures, etc.) that do not contain the same exact set of items, but are designed or engineered to measure the same underlying construct. Classical test theory approaches to linking (e.g., mean equating, equi-percentile equating) require strong, and often un-testable, assumptions. Also, restrictive equating designs (e.g., random group design, single group design) are necessary for these approaches and these methods do not generalize when equating more than two tests. The introduction of *item response theory* (IRT) methods led to improved linking techniques (e.g., common–item equating, common-person equating) as IRT models have built-in linking mechanisms (McDonald, 1999; Embretson & Reise, 2000) due to the inherent manner in which IRT methods are used to handle incomplete data.

This IRT model based approach has led to extensive research on incomplete data problems at the level of test items. Linking, equating, and calibrating refer to a series of statistical methods for comparing scores from tests (scales, measures, etc.) that do not contain the same exact set of measurements (items), but are presumed to measure the same underlying construct. Summaries of recent work on this topic are in Dorans, Pommerich and Holland (2007) and McArdle, Grimm, Hamagami, Bowles, & Meredith (2009). In addition, a recommended overview of linking scores is provided by Dorans (2007) who examines the general assumptions of different data collection designs, and provides explicit definitions of equating, calibrating, and linking. Dorans (2007) also provides a compelling example of the importance of adequate linking using multiple health outcome instruments, and how an individual's health may be misunderstood if alternative tests presumed to measure the same construct fail to do so.

Longitudinal studies present an additional dimension (e.g., time/age) that must be considered when linking item level data (see, McArdle et al., 2009; McArdle & Grimm, 2010). Linking is particularly important in longitudinal studies since researchers are most interested in studying change in people via the use of underlying scales that measure a constant construct across time.

However, longitudinal study item linking is often overlooked or assumed since the identical test is often administered at the different data collection stages in longitudinal studies. However, design constraints in longitudinal studies often make it impossible to administer the same test at each stage of data collection (see McArdle et al., 2009; McArdle & Grimm, 2010). Potential reasons that preclude administration of the same scale at each stage of longitudinal data collection include age appropriateness of tests, improved and revised editions of tests becoming available mid-project, and prior poor experiences with previously administered scales.

Nevertheless, analyzing item level data in any study, longitudinal or otherwise, should become the norm in applied research because of its many benefits. This being stated, there are limitations to analyzing item level data, some of which researchers and publishers may be able to overcome. The benefits of using item level AT data in longitudinal studies include the reduction of practice effects, checks and tests of item drift and differential item functioning across time/age, more precise estimates of ability, and information regarding the relative magnitudes of measurement error within and across time. One way to reduce practice effects is to administer a test based on new items at each successive occasion. This would reduce item-specific retest effects, which can bias any pattern of systematic changes in any ability. This information can be used in the estimation of statistical models to provide more precise estimates of important model parameters.

The drawbacks in using item level adaptive test data in longitudinal research stem from sample size restrictions and the limited number of available user-friendly software programs for combining item response models with higher-order statistical models when examining applied research questions. Item response models often require the estimation of many item and person ability parameters, which are poorly estimated with small and non-representative samples (the types of samples that are often found in psychological and longitudinal research). One possible solution, not without potential copyright and proprietary issues (in the case of commercially published tests), is for researchers and test authors and/or publishers to publish the final item parameter estimates calibrated on large and more representative samples. Indeed, this could be a good task for the biggest publishing companies. In this approach, the known "standard" item parameters could be fixed to the known values in new research studies and the whole IRT approach would be more useful in applied research studies.

Cost–Benefit Analysis

An important question that we tried to face here was how we could judge the success or failure of any specific AT strategy (see McArdle, 2010). In the current study we used a simple indicator of internal replication, the correlation between the AT-derived score with the full length total scale score. This reduces the current cost–benefit analysis to a comparison of the lower reliability of the true score (cost) versus the obvious gains in testing time (benefit). But it is still not clear where the stopping point should be.

The first key analysis is the empirical relationship between the estimated W-scale score of the person using one of the AT strategies and the true score of the person when the full set of items is administered (c_n). In most cases we would not really know the person's true score, but if available, or if we had an estimate based on a larger number of items, the goal would be for the score estimated with fewer items (i) to be as close as possible to the score estimated with more items ($I > i$). It follows that any analytic techniques based on the discrepancy between these scores, possibly framed as either a correlation or a regression, can provide an indicator of the cost of using any AT procedure.

The second key feature is the number of items administered and, rather than focusing on an optimal design ratio (i.e. R^2/i), we describe results obtained from allowing different numbers of items to be administered. Also, the external validation may prove to be more important (see Table 12.3 and Figure 12.9). This DTA-AT approach proved better than several other seemingly sensible methods for adaptive testing (e.g., FAT). And perhaps this DTA-AT should not be considered one of the alternative methods at all, but it just be considered as the optimal empirical standard against which to judge all other techniques. These issues obviously require further investigation with real data.

Note

This research was supported by NIH Grant # AG-07137-22. In addition, I thank Dick Woodcock for his sage advice and free access to important data sources, and Kevin McGrew for excellent editorial suggestions. Portions of this chapter were originally included in F. Schrank (Ed.), *The Woodcock–Munoz Research Foundation Journal*, Volume 1 (2009) and in McArdle (2010).

References

Allen, M. J. & Yen, W. M. (2002). *Introduction to measurement theory*. Long Grove, IL: Waveland Press.

Baker, F. B. & Kim, S.-H. (2004). *Item Response Theory parameter estimation techniques*, 2nd Edn. New York: Marcel Dekker.

Bell, R. Q. (1954). An experimental test of the accelerated longitudinal approach. *Child Development, 25*, 281–286.

Belson, W. A. (1959). Matching and prediction on the principle of biological classification. *Applied Statistics, 8*(2), 65–75.

Berk, R. (2009). *Statistical learning from a regression perspective*. Springer: New York.

Blau, D. M. & Gilleskie, D. B. (1998). *A dynamic structural model of health insurance and retirement*. University of North Carolina-Chapel Hill, Dept. of Economics.

Breiman, L. (2001). Statistical modeling: The two cultures. *Statistical Science, 16*(3), 199–231.

Breiman, L., Friedman, J. H., Olshen, R. A., & Stone, C. J. (1984). *Classification and regression trees.* Pacific Grove, CA: Wadsworth.

Carroll, J. B. (1993). Human Cognitive Abilities: A Survey of Factor Analytic Studies. Cambridge, Eng.: Cambridge University Press.

Carroll, J. B. (1998). Human Cognitive Abilities: A Critique (pps. 5–24) In McArdle, J. J. & Woodcock, R. W. (Eds) *Human abilities in theory and practice.* Mahwah, NJ: Erlbaum.

Cattell, R. B. (1941). Some theoretical issues in adult intelligence testing. *Psychological Bulletin, 38,* 592.

Cattell, R. B. (1998). Where is intelligence? Some answers form the triadic theory. In J. J. McArdle & R. W. Woodcock (Eds.), *Human cognitive abilities in theory and practice* (pp. 29–38). Mahwah, NJ: Lawrence Erlbaum Associates.

Dorans, N. J. (2007). Linking scores from multiple health outcome instruments. *Quality of Life Research, 16,* 85–94.

Dorans, N. J., Pommerich, M., & Holland, P. W. (2007). *Linking and aligning scores and scales.* New York: Springer.

Embretson, S. E. & Reise, S. P. (2000). *Item Response Theory for psychologists.* Mahwah, NJ: Erlbaum.

Ferrer, E., Salthouse, T. A., McArdle, J. J., Stewart, W. F., & Schwartz, B. (2005). Multivariate modeling of age and retest effects in longitudinal studies of cognitive abilities. *Psychology and Aging, 20*(3), 412–42.

Fisher, R. A. (1940). An examination of the different possible solutions of a problem in incomplete blocks. *Annals of Eugenics, 10,* 52–75.

Freedman, V. A., Aykan, H., Martin, L. G. (2002). Another look at aggregate changes in severe cognitive impairment: Further investigation into the cumulative effects of three survey design issues. *The Journals of Gerontology: Social Sciences, 57B*(2), S126–131.

Fultz, N. H., Ofstedal, M. B., Herzog, A. R., & Wallace, R. B. (2003). Additive and interactive effects of comorbid physical and mental conditions on functional health. *Journal of Aging and Health, 15*(3), 465–481.

Guildford, J. P. (1956). *Fundamental statistics in psychology and education.* New York: McGraw-Hill.

Hastie, T., Tibshirani, R., & Freidman, J. (2001). *The elements of statistical learning: data mining, inference, and prediction.* New York: Springer.

Horn, J. L. (1972). State, trait, and change dimensions of intelligence. *The British Journal of Mathematical and Statistical Psychology, 42*(2), 159–185.

Horn, J. L. (1988). Thinking about human abilities. In J. R. Nesselroade (Ed.), *Handbook of multivariate psychology* (pp. 645–685). New York: Academic Press.

Horn, J. L. (1991). Measurement of intellectual capabilities: A review of theory. Chapter 7 in McGrew, K. S., Werder, J. K., & Woodcock, R. W. (Eds.), *Woodcock–Johnson Technical Manual* (pp. 197–246). Allen, TX: DLM Teaching Resources.

Horn, J. L. (1998). A basis for research on age differences in cognitive abilities. In J. J. McArdle & R. Woodcock (Eds.), *Human cognitive abilities in theory and practice* (pp. 25–28). Chicago: Riverside.

Lord, F. M. (1955). Estimation of parameters from incomplete data. *Journal of the American Statistical Association, 50,* 870–876.

Lord, F. M. (1970). Some test theory for tailored testing. In W. H. Holtzman (Ed.), *Computer-assisted instruction, testing, and guidance* (pp. 139–183). New York: Harper & Row.

Masters, G. N. (1982). A Rasch model for partial credit scoring. *Psychometrika, 47*, 149–174.

McArdle, J. J. (1994). Structural factor analysis experiments with incomplete data. *Multivariate Behavioral Research, 29*(4), 409–454.

McArdle, J. J. (2010). What life-span data do we really need? In W. F. Overton (Ed.), *Biology, cognition and methods across the life-span.* Volume 1 of R. Lerner (Ed.), *The handbook of life-span development* (pp. 36–55). Hoboken, NJ: Wiley.

McArdle, J. J., Ferrer-Caja, E., Hamagami, F., & Woodcock, R. W. (2001). Comparative longitudinal structural analyses of the growth and decline of multiple intellectual abilities over the life span. *Developmental Psychology, 38*, 115–142.

McArdle, J. J., Fisher, G. G., & Kadlec, K. M. (2007). Latent Variable Analysis of Age Trends in Tests of Cognitive Ability in the Health and Retirement Survey, 1992–2004. *Psychology and Aging, 22*(3), 525–545.

McArdle, J. J., Fisher, G. G., Rodgers, W., & Kadlec, K. M., (2005). The *Cognition in the USA (CogUSA) Study.* Unpublished Manuscript, Department of Psychology, University of Southern California, Los Angeles, CA.

McArdle, J. J. & Grimm, K. J. (2010). An empirical example of change analysis by linking longitudinal item response data from multiple tests. In A.A. von Davier (Ed.), *Statistical models for test equating, scaling, and linking, statistics for social and behavioral sciences* (pp. 71–88). New York, Springer.

McArdle, J. J., Grimm, K., Hamagami, F., Bowles, R., & Meredith, W. (2009). Modeling life-span growth curves of cognition using longitudinal data with multiple samples and changing scales of measurement. *Psychological Methods, 14*(2), 126–149.

McArdle, J. J. & Woodcock, J. R. (1997). Expanding test-rest designs to include developmental time-lag components. *Psychological Methods, 2*(4), 403–435.

McArdle, J. J. & Woodcock, R. W. (Eds.), (1998). *Human cognitive abilities in theory and practice.* Mahwah, NJ: Lawrence Erlbaum Associates.

McDonald, R. P. (1999). *Test theory: A unified treatment.* Mahwah, NJ: Lawrence Erlbaum Associates.

McDonald, R. P. (1985). *Factor analysis and related methods.* Hillsdale, NJ: Erlbaum.

McGrew, K. S., Schrank, F., & Woodcock, R. W. (2007). *Technical manual. Woodcock–Johnson III normative update.* Rolling Meadows, IL: Riverside Pub. Co.

Novick, M. R. (1966). The axioms and principal results of classical test theory. *Journal of Mathematical Psychology, 3*(1), 1–18.

Ofstedal, M. B., Fisher, G. G., & Herzog, A. R. (2005). *Documentation of cognitive functioning measures in the Health and Retirement Study.* University of Michigan Institute for Social Research; HRS Documentation Report Series DR-006.

Rasch, G. (1960). *Probabilistic models for some intelligence and attainment tests.* Chicago: UC Press.

Ritschard, G. (2007). *CHIAD.* Unpublished MS, Dept. Econometrics, University of Geneva. Geneva, Switzerland.

Schaie, K. W. (2005). *Developmental influences on adult intelligence: The Seattle Longitudinal Study.* New York: Oxford University Press.

Sonquist, J. A. (1970). *Multivariate model building: The validation of a search strategy.* Ann Arbor, MI: Institute for Social Research, UM.

Steinberg, D. & Colla, P. (1999). *CART: An Introduction.* San Diego: Salford Systems.

Stapleton, L. M. (2002). The incorporation of sample weights into structural equation modeling. *Structural Equation Modeling, 9*(4), 475–502.

Thurstone, T. G. (1962). *PMA (primary mental abilities)*. Chicago, IL: Science Research Associates.

Wainer, H. (Ed.), (2000). *Computerized adaptive testing: A primer*, 2nd Ed. Mahwah, NJ: Erlbaum.

Wechsler, D. (1981). *Wechsler Adult Intelligence Scale – revised manual*. San Antonio, TX: The Psychological Corporation.

Wilson, M. (2005). *Constructing measures: An item response approach*. Mahwah, NJ: Erlbaum.

Woodcock, R. W. (1990). Theoretical foundations of the WJ-R measures of cognitive ability. *Journal of Psycho-Educational Assessment, 8*(23), 1–25.

Woodcock, R. W., McGrew, K. S., & Mather, N. (2001, 2007). *The Woodcock-Johnson Tests of Psycho-Educational Skills, III*. Chicago: Riverside Publishing Company.

Zhang, H. & Singer, B. (1999). *Recursive partitioning in the health sciences*. New York: Springer.

Zimmerman, D. W. & Williams, R. H. (1966). Generalization of the Spearman-Brwon Formula for test reliability: The case of non-independence of true scores and error scores. *British Journal of Mathematical and Statistical Psychology, 19*(2), 271–274.

Appendix

Table 12.A1 Twenty-nine nodes from CART evaluation of the number series total scores ($N = 1,223$)

Node	Item 1	Item 2	Item 3	Item 4	Item 5	Item 6	Item 7	W_Mean	W_Sd	Size	wtd sum	wtd SD
1	39=0	21=0	16=0	6=0		1=0		417	10	6	2502	60
2	39=0	21=0	16=0	6=1	24=0	1=1		415	0	2	830	0
3	39=0	21=0	16=0	6=1	24=0			456	9	50	22800	450
4	39=0	21=0	16=0	6=1	24=1			483	5	4	1932	20
5	39=0	21=0	16=1	22=0	18=0			470	8	20	9400	160
6	39=0	21=0	16=1	22=0	18=1			483	10	23	11109	230
7	39=0	21=0	16=1	22=1	24=0			489	6	25	12225	150
8	39=0	21=0	16=1	22=1	24=1			502	8	32	16064	256
9	39=0	21=1	28=0	26=0	25=0	17=0		467	12	5	2335	60
10	39=0	21=1	28=0	26=0	25=0	17=1	32=0	496	6	91	45136	546
11	39=0	21=1	28=0	26=0	25=0	17=1	32=1	510	5	11	5610	55
12	39=0	21=1	28=0	26=0	25=1	34=0		504	6	153	77112	918
13	39=0	21=1	28=0	26=0	25=1	34=1		513	6	25	12825	150
14	39=0	21=1	28=0	26=1	43=0			511	7	100	51100	700
15	39=0	21=1	28=0	26=1	43=1			537	6	7	3759	42

(continued)

Table 12.A1 (Continued)

Node	Item 1	Item 2	Item 3	Item 4	Item 5	Item 6	Item 7	W_Mean	W_Sd	Size	wtd sum	wtd SD
16	39=0	21=1	28=1	47=0	34=0	29=0		512	7	107	54784	749
17	39=0	21=1	28=1	47=0	34=0	29=1		521	5	34	17714	170
18	39=0	21=1	28=1	47=0	34=1	37=0		522	7	45	23490	315
19	39=0	21=1	28=1	47=0	34=1	37=1		536	5	15	8040	75
20	39=0	21=1	28=1	47=1	31=0			532	6	56	29792	336
21	39=0	21=1	28=1	47=1	31=1			543	8	25	13575	200
22	39=1	38=0	33=0	40=0				527	8	78	41106	624
23	39=1	38=0	33=1	40=0				534	8	76	40584	608
24	39=1	38=0	33=1	40=1				545	7	24	13080	168
25	39=1	38=1	46=0	43=0	45=0			539	8	59	31801	472
26	39=1	38=1	46=0	43=1	45=1			547	7	86	47042	602
27	39=1	38=1	46=0	43=1	45=1	41=0		546	3	6	3276	18
28	39=1	38=1	46=0	43=1	45=1	41=1		568	7	11	6248	77
29	39=1	38=1	46=1					559	10	46	25714	460

Table 12.A2 Twenty-nine nodes from CART prediction of the number series total scores from all items ($I = 47$; $N = 1223$; $R[pred] = 0.9612$)

Block	Node	Item 1	Item 2	Item 3	Item 4	Item 5	Item 6	Item 7	W_Mean	W_Sd	Sample Size
1	1	39=0	21=0	16=0	6=0				417	10	6
	2	39=0	21=0	16=0	6=1		1=0		415	0	2
	3	39=0	21=0	16=0	6=1	24=0	1=1		456	9	50
	4	39=0	21=0	16=0	6=1	24=0			483	5	4
						24=1					
2	5	39=0	21=0	16=1	22=0	18=0			470	8	20
	6	39=0	21=0	16=1	22=0	18=1			483	10	23
	7	39=0	21=0	16=1	22=1	24=0			489	6	25
	8	39=0	21=0	16=1	22=1	24=1			502	8	32
3	9	39=0	21=1	28=0	26=0	25=0	17=0		467	12	5
	10	39=0	21=1	28=0	26=0	25=0	17=1		496	6	91
	11	39=0	21=1	28=0	26=0	25=0	17=1		510	5	11
	12	39=0	21=1	28=0	26=0	25=1	34=0	32=0	504	6	153
	13	39=0	21=1	28=0	26=0	25=1	34=1	32=1	513	6	25
	14	39=0	21=1	28=0	26=1	43=0			511	7	100
	15	39=0	21=1	28=0	26=1	43=1			537	6	7

(continued)

Table 12.A2 (Continued)

Block	Node	Item 1	Item 2	Item 3	Item 4	Item 5	Item 6	Item 7	W_Mean	W_Sd	Sample Size
4	16	39=0	21=1	28=1	47=0	34=0	29=0		512	7	107
	17	39=0	21=1	28=1	47=0	34=0	29=1		521	5	34
	18	39=0	21=1	28=1	47=0	34=1	37=0		522	7	45
	19	39=0	21=1	28=1	47=0	34=1	37=1		536	5	15
	20	39=0	21=1	28=1	47=1	31=0			532	6	56
	21	39=0	21=1	28=1	47=1	31=1			543	8	25
5	22	39=1	38=0	33=0	40=0				527	8	78
	23	39=1	38=0	33=1	40=1				534	8	76
	24	39=1	38=0	33=1					545	7	24
6	25	39=1	38=1	46=0	43=0	45=0			539	8	59
	26	39=1	38=1	46=0	43=1	45=1			547	7	86
	27	39=1	38=1	46=0	43=1	45=1	41=0		546	3	6
	28	39=1	38=1	46=0	43=1	45=1	41=1		568	7	11
	29	39=1	38=1	46=1					559	10	46

13 Using Exploratory Data Mining to Identify Academic Risk among College Student-Athletes in the United States

Thomas S. Paskus

Introduction

In this chapter we use the techniques of Exploratory Data Mining (EDM) to identify "academic risk" among college students. Many of the specific analyses described below used data from the original Academic Performance Study (APS) of Division I student-athletes collected by the National Collegiate Athletic Association (NCAA). We present two examples of how we can define academic risk, but each example offers slightly different analytic challenges. For example, there is a very limited set of variables to craft a national initial eligibility standard. Nevertheless, there are many options to consider for combining scores and weighting the variables with utility weights that are defined to set rational cut-scores. Standard modeling approaches handle some aspects of the problem well (variable weighting), but others are challenging (evaluating the accuracy of selection when rule type, cut-scores, and utilities vary simultaneously). Non-standard exploratory approaches were then designed to address these challenges in a sufficiently nuanced manner given the high stakes to both the prospective student-athletes and the colleges. Although EDM techniques such as Decision Tree Analysis (DTA) can be adapted for these purposes, there is room for additional method development in dealing with this sort of problem.

Across most sports, the typical developmental pathway for elite athletes in the United States is atypical of what occurs in the rest of the world. Aspiring American athletes pursue their sporting dreams through competition in teams that are embedded within their secondary schools and compete interscholastically. Those athletes who excel may then have the opportunity to participate in a college-sponsored team and a few may move on from college to professional athletics opportunities.[1]

The embedding of athletics within the educational system has fostered substantial debate that is remarkably unchanged in topic and tenor since the 19th century (see Crowley, 2006). At the university level, concerns have often centered around participant safety and well-being, development of a tightly-regulated system to ensure competitive balance, enforcement of rules and standards for ethical behavior among athletes and administrators, inclusion of women and racial/ethnic minorities, and the financing of the athletics enterprise. However, the primary question within American college athletics has over time involved the "commitment to amateurism and the connection between education and athletics in which education is the principal partner" (Crowley, 2006, p. 11).

Since 1906, college athletics competition in the United States has been governed primarily by the National Collegiate Athletic Association (NCAA), a membership organization comprised of over 1,000 four-year colleges and universities that promote intercollegiate athletics as an integral part of the college educational experience. The NCAA's policies are set by college presidents, athletics administrators, faculty and student-athletes working in tandem with a national office staff that manages the day-to-day functions of the organization.[2] Although there is often a perception that colleges are not fully invested in educating their athletes, the policies and academic outcomes of the past several decades indicate that this is generally not the case. Since 1986 within the NCAA's Division I, a subset of 340 institutions that compete athletically at the highest collegiate level,[3] progressively stricter academic standards have been enforced that have resulted in graduation rates for student-athletes that are higher than among full-time non-athlete students at the same schools.[4]

Currently, these Division I academic standards include minimum high school or junior college academic criteria for receipt of an athletics scholarship and even higher standards to be able to compete for the team as a first-year student.[5] Student-athletes must then stay on track to earn a bachelor's degree within five years of entering any college or they are declared ineligible to compete. Since 2003, Division I teams have also been held to aggregate academic performance standards as measured by the Academic Progress Rate (APR), a real-time composite of student-athlete academic progress and retention, which has been shown to predict eventual student-athlete graduation rates. Teams not meeting a specified minimum APR are declared ineligible for NCAA championship competitions, are required to replace some athletics training with academic activities, and may face more severe sanctions if academic performance does not improve.

Given such strong embedding of educational expectations within the American developmental athletics system (especially since 1986 in major college sports), it should not be surprising that the membership of the NCAA currently has a great deal of interest in the development of models for predicting student-athlete academic success. Beginning with the Academic Performance Study (APS) in Divisions I and II during the late 1980s and early 1990s, the NCAA's research department[6] has collected high school and college transcript data and attempted to use it to inform national academic policy decisions. Today, these academic transcripts are captured for over 100,000 prospective and enrolling student-athletes each year. In conjunction with additional national survey data from student-athletes about their academic experiences, the NCAA's collection of data is quite substantial.

Such large datasets allow us the freedom to pair conventional statistical analysis/modeling with EDM techniques to better understand what variables, benchmarks, and academic trajectories relate to positive academic/educational outcomes for NCAA student-athletes. Examples of how we have used the combination of traditional and EDM methods to inform NCAA academic policy decisions are detailed here for both student-athletes entering college directly from high school (Example 1) and for others who may initially fail to qualify academically out of high school to play NCAA Division I sports but enter these universities later after attending a junior/two-year college (Example 2). These descriptions are not intended to be highly technical; see previous work by Paskus (1997) and Strobl, Malley, and Tutz (2009) for more thorough technical treatments of the analytic methods employed. Rather, I wish to highlight the application and potential advantages of EDM in a high-stakes national policy setting.

Example 1: Setting Academic Preparation Expectations for High School Student-Athletes

The NCAA does not dictate admissions decisions for individual member schools but it does want student-athletes to be fully prepared to succeed academically in college and eventually graduate. Therefore, since 1986, Division I of the NCAA has set strict academic preparation criteria for incoming student-athletes who wish to engage in competition immediately. These so-called initial eligibility (IE) standards have historically attempted to equate high school academic records across a diverse population of schools and individuals by using high school grades and ACT or SAT scores[7] to distinguish academically prepared student-athletes from those who are not prepared to immediately handle the rigors of college academics. Although the earliest initial eligibility standards may not have been set based on rigorous national research studies (see historical discussion in Petr & McArdle, 2012), subsequent research using traditional sorts of analyses (e.g., linear and logistic regression) indicated several base findings:

1   High school grades and ACT/SAT scores are independently related to various academic outcomes including first-year grades in college and eventual graduation (McArdle & Hamagami, 1994; McArdle, Paskus & Boker, 2013; Petr & McArdle, 2012).
2   High school grades are better predictors when based on core courses in areas such as English, math, and science (Petr & McArdle, 2012).
3   There is little evidence of prediction bias as a function of group differences (e.g., race/ethnicity), but selection bias (how cut-scores on high school grades or ACT/SAT scores differentially impact various groups) may exist (McArdle, 1998).

These analyses pointed NCAA policymakers toward variables (ACT/SAT scores and high school GPA in core academic classes) and even relative weights (core GPA should be weighted equal to or higher than test scores in setting minimum academic standards for student-athletes) that can be used to predict academic success. However, they are less useful in determining how to optimally combine the variables for selection purposes within a high-profile and high-stakes national context. That is, should minimums be set on high school core GPA (HSCGPA) and ACT/SAT (ACTSAT) separately, on the linear combination of the two, or in some other non-linear fashion? Simultaneously, what cut-scores on the variables/composite are most appropriate as indicating levels at which the risk of academic failure is sufficiently high to warrant the prospective student-athlete taking remedial actions before being able to compete (for example, a transition year without athletics competition at the four-year school or a diversion to a two-year college)?

Our efforts in developing exploratory methods to best answer these two questions (How do we optimally combine these variables? How do we best set cut-off scores?) have been a work in progress over the past 15 years and are detailed below.

Data Used to Study NCAA Academic Standards

Many of the specific analyses described below used data from the original NCAA Academic Performance Study (APS) of Division I student-athletes that began in the late 1980s. Division I schools were assigned via stratified random sampling to follow one of five sequential incoming classes of student-athletes and to record college academic transcript data on each member of the group annually for a maximum of six years. High school academic variables (e.g., grades and ACT/SAT scores) and personal demographics were also recorded. Study participation by colleges was voluntary, but high (70 to 96% within the five cohorts). Data from the 1984 and 1985-entering cohorts ($n=3,412$ student-athletes from 82 colleges) were most widely studied by the NCAA using the methods described below because they were uncensored by the imposition of national IE standards that originated in 1986 and may have led some prospective student-athletes with lower academic profiles not to enter Division I colleges. APS study methods and data are described in detail elsewhere (see McArdle & Hamagami, 1994; Paskus, 1997).

Reference will also be made to analysis of contemporary data from the NCAA's on-going Academic Performance Census (APC).[8] The APC data collection is a full census of high school academic transcripts for all recruits/prospective student-athletes (centralized academic certification of entering student-athletes has occurred since 1994) and term-by-term college academic outcomes for all who eventually take part in Division I athletics (academic reporting by member colleges required by the NCAA since 2003). Longitudinal records are created for students with both high school and college academic data. In total, high school data are obtained on about 75,000 to 125,000 students each year, about 25,000 of whom go on to Division I universities. Given that IE rules have been in effect in some form since 1986, HSCGPA and ACTSAT may be censored at the low end to varying degrees. That is, some prospective student-athletes who knew they would not meet the prescribed Division I IE standard may not have submitted their high school record for certification or attended a Division I school if they knew they could not compete during their freshman year. However, over the past decade, the APC data have provided NCAA policymakers with a powerful tool for examining ways to improve IE standards. The APC data and related NCAA data collections are detailed by Petr and Paskus (2009) and elsewhere (e.g., Scott, Paskus, Miranda, Petr, & McArdle, 2009; Petr & McArdle, 2012).

Graphical Depiction of the IE Research Issue

As illustrated by the joint ACTSAT–HSCGPA distribution from the APS study in Figure 13.1, combining high school academic indicators and setting a cut-score(s) for selection will never cleanly separate future

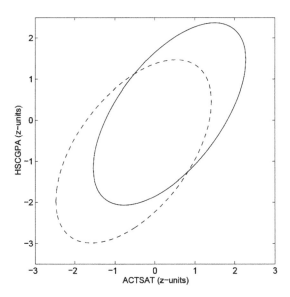

Figure 13.1 Distribution of high school core course grades (HSCGPA) and ACT/SAT score (ACTSAT) for eventual college graduates and non-graduates (from NCAA APS study).
Note: Solid = 95% confidence ellipsoid around joint distribution for eventual college graduates; Dashed = 95% confidence ellipsoid for eventual non-graduates.

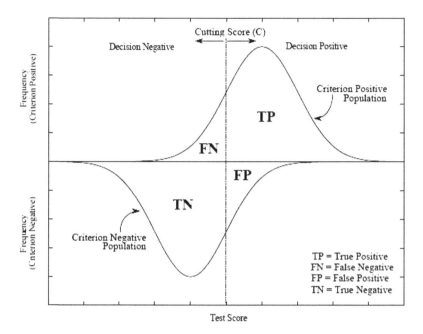

Figure 13.2 Decision analysis representation for a dichotomous outcome (Example of results of cutting on single variable = Test score). (From Paskus (1998), p. 284.)

academic successes from failures. These indicators predict college academic outcomes well in aggregate (e.g., reduce model misfit in a logistic regression prediction of graduation), but are certainly not error-free in a selection context. Consider Figure 13.2, in which college academic success is defined dichotomously[9] (criterion positive distribution can be thought of as representing eventual graduates, criterion negative as non-graduates) and a cut-score is imposed on a single variable (test score). Some student-athletes who could have succeeded academically will be denied competition eligibility (false negatives) while others will be granted eligibility to compete but will not meet academic expectations later in college (false positives). The task for NCAA researchers has been to help find appropriate selection lines to balance false negatives and false positives in a way that satisfies a diverse membership of colleges that have different academic missions and dissimilar tolerances for false negative and false positive decision outcomes.

Figure 13.3 displays a small sample of alternative selection rules that could be employed for combining HSCGPA and ACTSAT in determining initial eligibility. In Figure 13.3a, selection is based on a linear combination

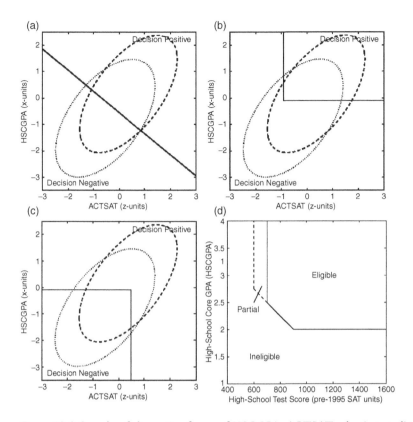

Figure 13.3 Sample of alternative forms of HSCGPA–ACTSAT selection studied.

of high school core GPA and test score (typically called the "sliding scale" within the NCAA). Selection could be optimized by varying the weighting of the two components (changing the slope of the line) or by shifting the line (setting a different composite cut-score) for a particular weighting of components. Figure 13.3b shows what is commonly referred to as a conjunctive or logical "AND" rule. With such a standard, student-athletes would be judged based on meeting a minimum HSCGPA *and* a minimum ACTSAT score. Either cut-score could be varied, effectively changing the relative weighting between the two components. A disjunctive or logical "OR" rule is displayed in Figure 13.3c. Meeting either a HSCGPA standard *or* an ACTSAT standard would establish eligibility. In addition to these three categories of selection rules, the components could be combined in many other complex fashions, including the one shown in Figure 13.3d, a rule implemented in Division I in 1996 despite research documenting concerns with such a selection (see Petr & McArdle, 2012). Figure 13.4 shows the

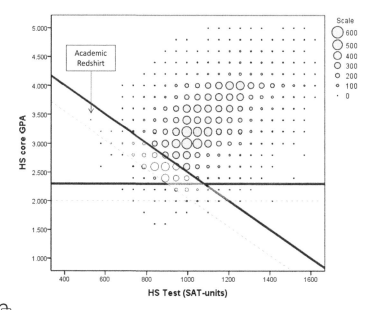

Figure 13.4 New division I initial eligibility standard approved for student-athletes entering college beginning in 2016.

selection rule employed in Division I since 2003 (everyone with scores in the shaded band and above can compete immediately) and a modification to the standard that is slated to take effect in 2016 (first-year competition only if HSCGPA and ACTSAT are above the shaded region).[10]

Using Traditional Modeling Techniques to Find Optimal IE Rules

A question that arises is to what degree can traditional forms of statistical analysis (for example, linear or logistic regression) help us choose variable combinatorial strategies and cut-scores for minimizing false positives/ negatives when we are dealing with a very limited decision space? That is, by the nature of this particular research problem, there are few predictor variables that can be considered; in a national setting across a wide range of diverse high schools and colleges, ACTSAT and HSCGPA (along with course content) tend to be the only common measures of academic preparation.

This question has been addressed in previous work by NCAA researchers (see Paskus, 1997, especially Tables 5–10 and supporting text), but continues to be a methodological focus for us. Generally, traditional modeling techniques are useful in determining optimal predictor weights when the selection rule is a linear combination of the two variables. Within the context of Figure 13.3a, our results of HSCGPA being two to three times more predictive than ACTSAT suggest that flattening the slope of the diagonal line would be optimal. However, the R-square difference between the optimal linear combination and an equally weighted combination is small and essentially disappears when a floor on HSCGPA (favored by NCAA policymakers) is added.

The traditional techniques do not definitively answer questions about the optimal cut-score on that linear combination or whether the linear composite will outperform an optimal conjunctive or disjunctive rule (Figures 13.3b and 13.3c). Programs such as TableCurve 3D (Systat Software, San Jose, CA) allow one to compare R-square values for linear prediction models versus those for complex non-linear combinations of predictors. For example, when used with the APS data, we could see that R-square values were not substantially improved upon when any of the non-linear models available in the program's library were applied (Paskus, 1997). Again, questions about which variable combinations in conjunction with specific cuts on the variables/composites lead to a minimization of false positives and false negatives are not directly answerable.

Using EDM Techniques to Find Optimal IE Rules

First, consider the quantification of selection rule quality. Diagnostic accuracy can be measured in a number of different ways, but we often convey it to NCAA policymakers using something like overall decision accuracy (ODA) when the outcome of interest is dichotomous. ODA simply represents the percentage of correct diagnoses (true positives and true negatives) out of the total number of cases judged or

$$ODA(\%) = 100 * \frac{(TP + TN)}{(TP + FN + TN + FP)}.$$

It is important, especially in studies where the event of interest has a low baserate, to couple the reporting of observed ODA with information on how it compares to the ODA for a selection rule that declares everyone "decision positive" or one that deems everyone "decision negative" (see Figure 13.2). Essentially the worst ODA you can have in a diagnostic setting is the higher of the criterion baserate (as a percentage) or one minus the baserate, provided false positive and false negative errors are not valued differently (an assumption that will be discussed shortly). An ODA of 91% sounds impressive but it would not be if only 10% of the sample was in the criterion positive distribution (you would obtain an ODA of 90% if you declared everyone decision negative). Similarly, our policymakers want to know how well a selection rule works within each distribution.

For example, what proportion of eventual graduates are correctly identified by HSCGPA and ACTSAT? This corresponds with examining a selection rule's sensitivity (ability to correctly classify criterion positive cases) and specificity (correct identification of criterion negative events). In summary, a single number representing selection rule quality will rarely suffice, but packaging a few can give a fairly comprehensive picture.

The inability to map traditional statistical techniques onto the full breadth of the IE research problem has led us to adopt, adapt, and develop other techniques to characterize decision accuracy as a function of alternative variable combinations and cut-scores. One method adopted that is now a standard component of many current statistical software packages is Classification and Regression Tree Analysis (CART). CART is used here to describe any of a family of iterative methods for sequentially selecting variables and cut-scores to isolate sets of observations with similar outcomes (see Breiman, Friedman, Olshen & Stone, 1984; Strobl et al., 2009). Described in its typical application within Example 2 of this chapter (taking a complex multivariate decision space and sequentially examining conditional predictor relationships), our adaptation of CART for studying initial eligibility rules has involved attempting to employ it to partition the low-dimensional IE decision space. Indeed, CART can be used to choose between linear, conjunctive, and other simple non-linear combinations of HSCGPA and ACTSAT as a function of ODA and similar outcomes (Paskus, 1997).

But, our analyses raised questions about whether CART's sequential algorithms necessarily hit upon a globally optimal decision rule or just a local ODA maximum. For example, do you necessarily find the best conjunctive rule if you first optimize a cut on either HSCGPA or ACTSAT and then optimize a cut on the other predictor as opposed to examining the ODA of every joint HSCGPA–ACTSAT conjunctive rule prior to establishing a cut-line on either? Figure 13.5 shows the result of a technique developed by the author to compare exhaustive sequential and simultaneous searches for an optimal conjunctive rule in the APS data (programmed using the MATLAB software). This plot can be read like a contour map for displaying mountainous terrain. Each point on the grid represents the vertex of a conjunctive selection rule as in Figure 13.3b—only student-athletes with HSCGPA above and ACTSAT to the right of the vertex would be academically eligible by that rule. Points along connected lines in Figure 13.5 correspond with particular values of ODA, with the innermost contours having the highest ODA. This analysis, along with others, demonstrated that sequential classification algorithms (many forms of CART are sequential) may not always hit upon global maxima for decision accuracy. In this example, while the global maximum was observed for the conjunctive rule defined by lines that intersect at z-units of approximately [−0.75, −0.75], the optimal sequential selection chose the rule defined by [−0.90, −0.10].

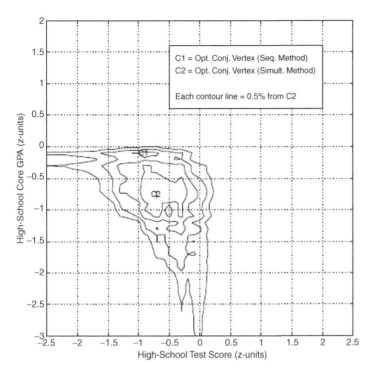

Figure 13.5 Finding an optimal conjunctive rule vertex in the prediction of graduation (FP and FN outcomes weighted equally).

In a similar fashion, families of univariate (HSCGPA or ACTSAT alone), linear (both equal and unequal combinations of HSCGPA and ACTSAT), and disjunctive rules have been examined for a variety of college academic outcomes available in our data (e.g., graduation, grading benchmarks). The global maxima for each of the bivariate selection methods are generally quite similar within each outcome although linear models tend to have the highest ODAs and may require less combinatorial precision. That is, conjunctive and disjunctive rules appear sensitive to the relative weighting of our predictors while linear composites require less precision in component weighting. Indeed, as others discovered long ago (e.g., Wackwitz & Horn, 1971; Einhorn & Hogarth, 1975; Wainer, 1976), when linear composites are used as selection lines, slope differences in the lines matter only at the outer edges of the decision space where few cases tend to exist. Where data points are dense in the middle of the bivariate predictor space, even lines with disparate slopes will cut the joint distribution in similar ways. Movements of conjunctive and disjunctive vertices would typically occur within the heart of the distribution of cases and thus change impacts to a larger degree.

Interestingly, the NCAA initial eligibility rule in place at the time these investigations began (Figure 13.3d) did not exhibit a comparable ODA to the best rules found, nor has any subsequent NCAA IE rule. Although this seems highly problematic at first blush, it led us to an important concept— the assumption that false positives and false negatives are considered equally problematic by relevant decision makers is often inappropriate within a selection context. In this case, have NCAA policymakers implicitly decided over time that identification of members of certain criterion groups (e.g., potential graduates) is more important? This speaks to a concept known as utility weighting, which is well-described by Cronbach and Gleser (1965), Ben-Shakhar, Kiderman, and Beller (1996), Gross and Su (1975), and many others. Utilities are essentially weights that represent actual or perceived values associated with each decision outcome. For example, if one has strong concerns about falsely excluding potential graduates from competition but is less concerned with potentially admitting non-graduates, one could weight FNs higher by a multiplicative factor in comparison to FPs in the calculation of ODA and similar selection outcome measures.

The most mathematically parsimonious way to quantify these decision utilities was first noted by Rorer, Hoffman and Hsieh (1966) and is detailed by the author in previous work (Paskus, 1997). Classification optimization is not dictated uniquely by the four decision cells in Figure 13.2, but rather by the ratio of the within-criterion group utility differences:

$$UtilityRatio = U = \frac{(U_{TP} - U_{FN})}{(U_{TN} - U_{FP})}$$

For a given selection problem, the impact of a decision maker possessing a particular utility ratio rather than accepting equal relative weights for all decision outcomes is the same as if one calculated ODA after re-scaling one of the criterion distributions using the utility ratio as a baserate multiplier (each case weighted by U). This re-weighting necessitates an alternative cut-score and perhaps even an alternative choice of variable combination strategy (Paskus, 1997) to optimize ODA.

In this form, perceived utilities can be operationalized easily as data weights, and methods such as linear/logistic regression and CART can be employed with weighted data to optimize selection under the desired utilities. Given the methodological problems already discussed, we have tinkered with more exploratory methods for examining utility-weighted selection optimizations. Figure 13.6 shows optimization curves for a single academic predictor (HSCGPA) using unit-weighted data (Figure 13.6a; the standard assumption of equal outcome utilities) and data re-weighted to represent an emphasis on identifying academically successful college student-athletes (Figure 13.6b; $U = 2$). As you might expect, when the emphasis shifts to concern about the correct identification of students who

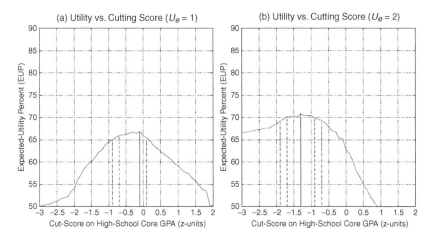

Figure 13.6 Expected utility at various Cut-scores for varying outcome utility ratios $U = (UTP-UFN)/(UTN-UFP)$.

will be academically successful ($U > 1$), the optimal cut-score decreases to lower the number of FN identifications.

As shown in Figure 13.7, employing iterative programming to obtain the optimization curve for a full range of utility ratios and plotting information on the resulting peaks of each curve (each curve's peak is represented topographically as a vertical slice along the utility ratio on the x-axis), we discovered an interesting monotonic relationship between utility ratio and optimal predictor cut-score. Essentially, as the perceived importance of the criterion positive group (U) increases, the cut-score should decrease to optimize utility-weighted ODA. Further, those ODA values at each curve's peak can be plotted for each selection rule and utility ratio to produce a full map of how optimal cut-scores of various combinatorial rules compare within any given utility ratio (vertical slices in Figure 13.8). Only slight global maximum differences are seen in Figure 13.8 between optimal conjunctive and optimal disjunctive rules across the range of utility possibilities. Note also how optimal prediction for any rule improves only marginally over baserate prediction when large utility weights are applied to either criterion distribution (vertical slices at left and right ends of the plot). Significantly, we also see that the NCAA selection from Figure 13.3d (labeled as "16" here) and a previous conjunctive NCAA standard ("48") are both optimized at a utility ratio of approximately 2.0. Assuming those NCAA rules were designed to be efficient, we can work backwards from this curve to infer the perceived utilities of the policymakers involved in the development of the NCAA's 1980s/1990s standards—avoid undue harm to eventual graduates.

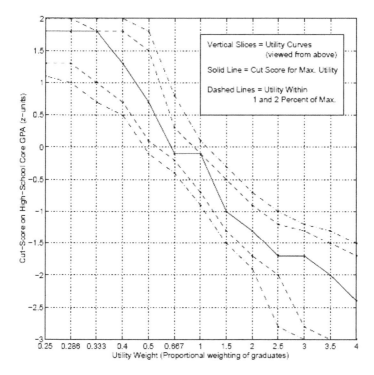

Figure 13.7 Optimal cutting score on HSCGPA as a function of U_e.

An NCAA membership review was conducted throughout 2010 and 2011 and led the NCAA's Division I Board of Directors to agree to further strengthen IE standards in 2016 (as shown in Figure 13.4). This review relied heavily on research (see Paskus, 2012, for a summary) and on discussions about utilities. However, we have found over time that non-statisticians struggle with the specifics of defining numeric utility weights. So, rather than presenting them in the form described above, utilities were generally discussed with our decision makers in terms of academic risk levels. Contour maps were produced that showed predicted first-year GPA or probability of graduation and the impacts of superimposing various rules over the contours. The 2016 standard will set a cut-score for receipt of aid and practice (bottom of the shaded region in Figure 13.4) that predicts roughly a 2.10 or 2.20 first-year GPA, while the cut-score for competition (top of the shaded area) predicts a first-year GPA of about 2.60. The decision makers deemed these appropriate standards as they relate to eventual graduation, and chose a rule (equally weighted linear combination with increased HSCGPA floor) that best followed the desired contour with maximum possible efficiency and fairness. Both traditional and EDM analyses contributed to the eventual decisions made.[11]

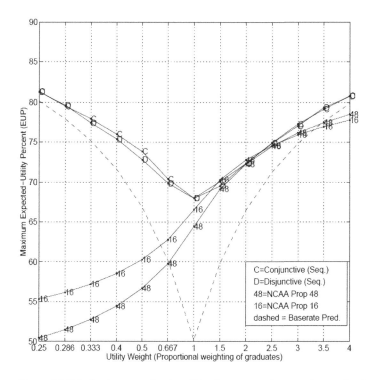

Figure 13.8 Inferring perceived utilities by comparing previous selection rules to optimal selection rules.

Example 2: Transitioning Academically from a Two-Year College to a Four-Year College

In evaluating the academic outcomes of Division I teams using the APR metric described earlier, it became apparent that student-athletes transferring from two-year colleges into Division I are currently experiencing academic success at a much lower level than student-athletes entering directly from high school or those transferring in from another four-year college[12] (see Table 13.1 and Paskus, 2012). Whereas many non–athlete students spend time at a two-year school on their way to a four-year degree because of the relative affordability of two-year colleges, student-athletes often transition through two-year schools as an alternative pathway when they fail to meet the NCAA's IE standards out of high school. Currently, any student-athlete earning a two-year degree can come to and compete immediately for a Division I school regardless of that student's high school academic record. Two-year transfers (typically called "2–4 transfers" in NCAA parlance) presently make up only about 5% of the

Table 13.1 Academic outcomes among transfer students in division I

	Non-transfers only	4-year transfers	2-year transfers
Average high school core GPA	3.36	3.25	3.00
Average high school test score	1070	1027	949
APR (2008–9)	971	949	926
APR 0-for-2s (2008–9 academic year)	2.2%	3.8%	5.5%
APR exhausted eligibility 0-for-2s (2008–9)	2.4%	4.3%	7.6%
Graduation success rate (2002 cohort)	80%	77%	65%

Division I student-athlete population, but 15–20% of the population in the high-profile sports of men's basketball and baseball.

Concerned about the poor academic outcomes for two-year college transfers, the NCAA's member colleges asked for a study of the relationship between high school/two-year college academic preparation and subsequent outcomes at four-year schools. Could we predict 2–4 transfer successes/failures and perhaps craft national academic legislation to enhance overall success rates? As we began analysis using traditional statistical methods, another question arose—were they the best tools for identifying 2–4 transfer risk profiles?

Data Used to Study Student–Athlete 2–4 Transfers

The NCAA's APC data collection was supplemented by two-year college transcript data submitted by member colleges for all football, baseball, and men's/women's basketball 2–4 transfers who entered Division I in the fall of 2008. These sports were chosen because the vast majority of all NCAA 2–4 transfers occur in these sports (and it allowed us to limit the data procurement/entry burden on the colleges in the study). In total, approximately 1,500 transcripts were collected and linked with prior high school records and subsequent first-year outcomes at the receiving four-year colleges. Two-thirds of 2–4 transfers in baseball had met IE standards out of high school, but that proportion was less than one-half in women's basketball and one-third in men's basketball/football. Outside of baseball, 2–4 transfers in these sports were more likely to be from a racial/ethnic minority group and more than one-third earned credits from multiple two-year colleges. Academic outcomes studied were generally limited to first-year performances at the four-year school including GPA, credits earned, end-of-year retention, and end-of-year academic standing.

Table 13.2 Univariate correlations with first-year GPA in division I for 2–4 transfers in 2008–9

	Univariate correlations	
	r	r
2-year credits earned/transferred	−.02	.02
English credits earned/transferred	−.03	.06
Math credits earned/transferred	.06	.17*
Science credits earned/transferred	.14*	.16*
PE credits earned/transferred	−.08*	−.10*
Overall GPA/transfer GPA	.50*	.39*
Number two-year colleges	−.11*	
Terms at last two-year college	.02	
HSCGPA	.44*	
HS Test Score	.38*	
HS Core Units	.14*	

Note: * indicates univariate correlation significantly different from zero at p<.01 level.

Using Traditional Modeling Techniques to Predict Success of 2–4 Transfers

Univariate correlations between selected predictors and first-year college GPA are displayed in Table 13.2. In the top half of the table, correlations with the outcome measure are displayed separately for credits/GPA earned at the two-year school (left column) and credits/GPA that transferred to the four-year school (right column). GPA at the two-year school and high school academic performance were substantially related to this and other first-year outcome variables. Overall credits accumulated at the two-year college and credits in English language/literature classes were not related to this outcome, although this is likely a result of the absence of variability due to pre-existing NCAA and two-year college expectations in those areas. Other predictors were examined but none were independently predictive of our outcomes once controlling for the variables in Table 13.2.

One of the numerous regression models examined is displayed in Table 13.3. Across outcome variables, the results were very similar. First, GPA from the two-year college was the strongest predictor of first-year success. Although HSCGPA, ACTSAT, and other high school variables were significant univariate predictors of success, they were not needed in our models once two-year college performance was assessed. Second, credits in physical education (PE) activity classes (for example, classes that primarily involve time spent learning/playing a sport, which are common electives at many two-year colleges) was a reliable negative indicator of academic success. First-year GPAs at four-year colleges tend to be overpredicted for 2–4 transfers to the degree their two-year college GPA was built upon physical education classes. Third, having earned credits in science classes was

Table 13.3 Regression prediction of first-year GPA in division I for two-year college transfers in 2008–9

Outcome=GPA1	Linear regression	
	Beta wt	p < .01
Associate degree earned		ns
Two-year credits transferred		ns
English credits transferred		ns
Math credits transferred		ns
Science credits transferred	.082	*
PE credits transferred	−.065	*
Transfer GPA	.500	*
Attended multiple 2-year colleges		ns
HS IE Status		ns

Note: Model R-square = .31.

independently predictive of four-year college success. Grouped together, English, math, and science (EMS) classes portend later academic success while physical education classes do not. Fourth, there is no direct benefit to knowing whether a student-athlete actually earned the two-year college degree once we have information about the classes they took and the grades they earned.

Similar to the initial eligibility problem described in Example 1, figuring out how to use these variables to set national eligibility expectations for 2–4 transfers does not follow directly from the results of models like the one shown in Table 13.3.

Application of EDM to Studying 2–4 Transfers

In contrast to setting national benchmarks on HSCGPA and ACTSAT for academic eligibility out of high school, NCAA committee members charged with crafting new standards for 2–4 transfers entertained a much larger set of possible predictors. As such, the sorts of highly iterative EDM methods developed specifically for our IE analyses were less tenable in this research. CART and similar iterative tree procedures were applied here along the lines of their typical use described earlier—making sequential choices about variables and cut-scores that identify subsets of cases with similar academic outcomes. Under standard variations in these models, the outcomes can be either discrete (often referred to as classification tree analysis) or continuous (regression tree analysis).

A typical classification tree (one of many examined during the research process) constructed using the SPSS Decision Trees add-on (IBM, 2010) is displayed in Figure 13.9. Only student-athletes who were IE non-qualifiers out of high school (thus needing to stay at the two-year school through completion of an associate's degree) were included in this particular analysis. The outcome, APR "0-for-2" status, was a dummy indicator coded=1

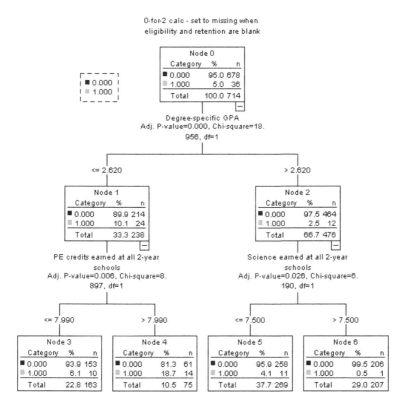

Figure 13.9 Sample CART model—prediction of 0-for-2 status after first year in division I.

for those student-athletes who dropped out of a four-year school while academically ineligible. The box at the top of the tree labeled "Node 0" describes the initial sample—out of 714 two-year transfers in this sample, 5% failed out of school in their first year at the four-year destination. After searching for a variable and cut-score to best differentiate the two outcome categories, the program chose GPA at the two-year school and simultaneously set a diagnostic cut-score of about 2.60. Among students with a GPA below 2.60 the failure rate was 10% vs. only 2.5% for those above 2.60. The program then determined it could further isolate academic successes/failures; the low-GPA group was subdivided using the number of PE credits earned while the high-GPA group was split based on science credits earned. No further tree growth improved outcome identification in this example. Various mathematical algorithms can be applied to develop these trees, set growth stopping criteria and even "trim" the obtained tree, but are not described here—see Breiman et al. (1984) and Strobl et al. (2009) for an introduction to these topics.

The final tree, characterized by the "terminal" nodes at the bottom of the figure, provides an interesting perspective on the data that is not captured in standard regression analyses. Two nodes in particular, node 4 and node 6, show how highly homogeneous outcome groups can be identified with just two variables/cuts each. Node 4, defined based on a GPA below 2.60 and 8 or more PE activity credits earned, contains a group of students with a 19% aggregate first-year failure rate. Node 6, containing those students with a GPA above 2.6 and 7.5 science credits or more, counted only one failure among 207 students with that profile. Nodes 3 and 5 reflect similar outcome proportions as for the sample on the whole and might benefit from additional study (Monahan and colleagues, 2001, present an interesting methodological approach involving pooling and reanalyzing these cases). In essence, CART operated here like a regression analysis with a propensity toward creating complex conditional splits/statistical interactions.

Figure 13.10 shows a regression tree with a continuous outcome, first-year GPA at the four-year college. This example tree was also allowed to grow more extensively than the previous one resulting in more variable splits and terminal nodes. Within-node homogeneity is described by the predicted first-year GPA for the cases in each node. Members of node 5

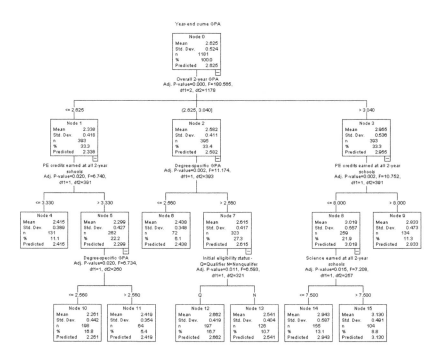

Figure 13.10 Sample CART model—prediction of first-year GPA in division I.

(two-year GPA below 2.60, more than 3.3 PE credits) had a low predicted GPA that was slightly reduced by also factoring in another variation of two-year GPA (node 10). Highest predicted GPA was seen in node 15 among student-athletes with two-year GPA above 3.04, fewer than 8 PE credits and more than 7.5 science credits. In all, these results were rather similar to those of the previous tree despite the change in dependent variable and the increased tree complexity.

Partitioning procedures currently available in the R computing environment (R Core Team, 2010) confirmed the basic results of SPSS decision trees. Figure 13.11 displays an example tree that used the PARTY recursive partitioning toolbox (Hothorn, Hornik, & Zeileis, 2006), in which low two-year GPA characterized poor subsequent performance and high two-year GPA/science credits accumulated was associated with positive performance. The RANDOM FOREST (Liaw & Wiener, 2002) package in R was also applied to these data with similar results, additionally providing statistical measures of variable importance (in our case, highlighting the significance of PE credits in predicting later success). See an excellent paper by Strobl and her colleagues (2009) for additional information on using these recursive techniques generally and within the R framework.

As we saw in our study of IE selection rules, the standard analytic assumption is that all classification mistakes are perceived as equally costly. Neither the SPSS Decision Trees module nor any of the R procedures described earlier allow for the explicit application of outcome utility weights. However, as described previously, conceptualizing the utility ratio as a baserate multiplier and treating it as a data weighting factor accomplishes the task. Weighting the criterion positive distribution (eventual academic failures or APR 0-for-2s) by a factor of 3.0 and running the R program's PARTY procedure produced the tree structure shown in Figure 13.12. Interestingly, selection variables and cut-scores did not change appreciably versus the trees obtained with the original data. However, this model did point to a failure risk defined by not only low two-year GPA and high PE credits but also low credit attainment in math.

After reviewing all of our research, including summary descriptions of these recursive EDM methods, the NCAA's leaders decided upon new 2–4 transfer rules to be able to compete immediately in athletics upon entering a Division I college. These included a 2.50 two-year GPA minimum, addition of a science course requirement to English and math ones that had recently been added, and a strict cap on the number of PE activity courses that may count toward one's two-year GPA and overall credit requirements. Based on the data reviewed and in line with observed historical utilities for IE policymaking, these cut-scores on each variable generally indicate implicit utilities leaning slightly away from the creation of false negatives (e.g., avoid denying eligibility to students who could eventually graduate).

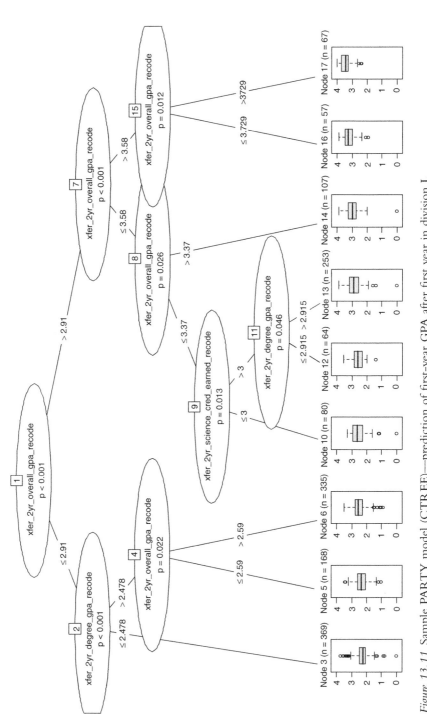

Figure 13.11 Sample PARTY model (CTREE)—prediction of first-year GPA after first year in division I.

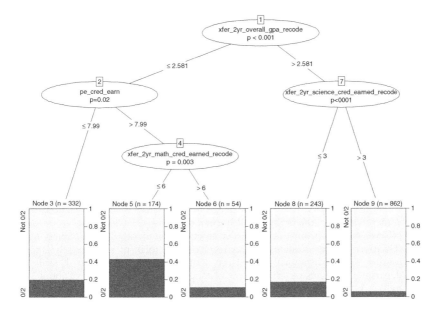

Figure 13.12 Sample PARTY model (CTREE)—prediction of 0-for-2 status after first year in division I (0/2 WT=3).

Conclusion

These two examples presented slightly different analytic challenges. In the first, we had a very limited variable space in which to craft a national initial eligibility standard but many options to consider for combining/weighting the variables and setting cut-scores. Standard modeling approaches handle some aspects of the problem well (variable weighting), but others are challenging (evaluating the accuracy of selection when rule type, cut-scores and utilities vary simultaneously). Non-standard exploratory approaches had to be designed on the fly to address these challenges in a sufficiently nuanced manner given the high stakes both for prospective student-athletes and colleges. Although EDM techniques such as CART can be adapted for these purposes, there is room for additional method development in dealing with this sort of problem.

In the second example, policymakers were open to considering many different variables, necessitating methods that could accurately resolve potential multivariate confusion. Although one could construct a logistic regression risk probability based on a couple of dozen transcript variables, within our national policy setting, selection rules have to be simple, transparent, and effective. EDM methods such as CART and its recursive partitioning offspring are well-suited for finding the pockets of highest academic risk in a student population. Questions about result stability within various samples, creation of ensemble results across variations in algorithm options, and routine application of utility weights to selection problems are being addressed to various degrees within the field and will hopefully lead to future programming enhancements.

Notes

1 In the highest-profile sports (those with the most professional opportunities), roughly 3–7% of high school participants go on to compete in college. In most of these sports, only about 1% of college participants then move on to national or professional teams (NCAA, 2012).

2 The NCAA staff includes our small research group that conducts national studies on various issues in college athletics in order to enhance the NCAA's ability to make data-driven policy decisions and to fully evaluate student-athlete academic, social, athletic, and health outcomes.

3 NCAA schools group into three competitive divisions. At the highest level (Division I) many student-athletes receive some form of athletics scholarship toward their educational and basic living expenses. Division I athletics department budgets vary from a low of roughly 3 million to a high of about 135 million US dollars for the more competitive programs (Fulks, 2012). Division II (limited athletic scholarships) and III (no athletic scholarships) colleges typically compete on a more regional basis with much smaller athletics budgets ranging from <1 to about 19 million US dollars (Fulks, 2012).

4 Presently 65% of Division I student-athletes graduate within six years from their initial college of entry versus 63% of non-athlete students at the same colleges (NCAA, 2011). When accounting for transfer to other four-year colleges, it is believed that more than 80% of student-athletes are earning a degree during that time (a parallel rate among members of the general student body is not currently calculated on a national basis).

5 Each division's membership independently sets its athletic and academic policies. In terms of academics, Division II uses similar initial eligibility and progress-toward-degree standards as Division I but with different academic benchmarks. Division II does not use a penalty system for team academic performance. In Division III, there are no national academic standards; all schools autonomously set academic entrance requirements and performance expectations.

6 The NCAA research program was launched in the 1980s by Ursula Walsh, who brought in Todd Petr (current NCAA managing director of research) and Jack McArdle to develop the APS study and other novel national research initiatives on student-athlete academic success.

7 Student-athletes can satisfy their IE standardized test requirements with an appropriate score on either of these national examinations. For ease of display and discussion in this chapter, scores on these exams are shown on a common scale (either z-scores or SAT-units).

8 The Academic Performance Census is part of a broader academic initiative called the Academic Performance Program (APP). As such, the academic outcomes data is sometimes referenced interchangeably as the APC or APP study.

9 A dichotomous outcome is presented throughout Example 1, but these methods can be applied with other types of outcome variables.

10 The 2016 modifications also include a stipulation that core academic classes areas such as English, math, and science be taken in an educationally appropriate sequence and timing.

11 In May 2013, NCAA leaders revisited the research and decided to temper the proposed increase in the sliding scale cut-score planned for 2016, primarily due to concerns about false negative outcomes among students from less advantaged academic backgrounds. This served as an interesting real-world example of how utility weights can change over time.

12 In Table 13.1, APR is the NCAA's real-time measure of term-by-term academic performance. It is calculated as a function of the proportion of student-athletes persisting in good academic standing each term. It is measured on a 0–1000 scale; teams with scores below a certain level (900 currently, 930 in future years) face sanctions. APR "0-for-2" percentages describe how many student-athletes fail out of school each year. Graduation success rate (GSR) is a national graduation rate that builds transfer movement into the formula.

References

Ben-Shakhar, G., Kiderman, I., & Beller, M. (1996). Comparing the utility of two procedures for admitting students to liberal arts: An application of decision-theoretic models. *Educational and Psychological Measurement*, 56, 90–107.

Breiman, L., Friedman, J.H., Olshen, R.A., & Stone, C.J. (1984). *Classification and Regression Trees*. Belmont, CA: Wadsworth International Group.

Cronbach, L.J., & Gleser, G.C. (1965). *Psychological Tests and Personnel Decisions*. Urbana, IL: University of Illinois Press.

Crowley, J.N. (2006). *In the Arena: The NCAA's First Century*. Indianapolis, IN: National Collegiate Athletic Association.

Einhorn, H.J., & Hogarth, R.M. (1975). Unit weighting schemes for decision making. *Organizational Behavior and Human Performance*, 13, 171–192.

Fulks, D.L. (2012). *Revenues and expenses 2004–2011: NCAA Division I intercollegiate athletics programs report*. Indianapolis, IN: National Collegiate Athletics Association.

Gross, A.L., & Su, W. (1975). Defining a "fair" or "unbiased" selection model: A question of utilities. *Journal of Applied Psychology*, 60, 345–351.

Hothorn, T., Hornik, K., & Zeileis, A. (2006). Unbiased recursive partitioning: A conditional inference framework. *Journal of Computational and Graphical Statistics*, 15(3), 651–674.

IBM (2010). *IBM SPSS Statistics for Windows, Version 19.0*. Armonk, NY: IBM Corporation.

Liaw, A., & Wiener, M. (2002). Classification and regression by randomForest. *R News,* 2(3), 18–22.

McArdle, J.J. (1998). Contemporary statistical models for examining test bias. In J.J. McArdle & R.W. Woodcock (Eds.), *Human Cognitive Abilities in Theory and Practice* (pp. 157–195). Mahwah, NJ: Lawrence Erlbaum Associates.

McArdle, J.J., & Hamagami, F. (1994). Logit and multilevel logit modeling of college graduation for 1984–1985 freshman student-athletes. *Journal of the American Statistical Association*, 89, 1107–1123.

McArdle, J.J., Paskus, T.S., & Boker, S.M. (2013). A multilevel multivariate analysis of academic performances in college based on NCAA student-athletes. *Multivariate Behavioral Research*, 48(1), 57–95.

Monahan, J., Steadman, H.J., Silver, E., Appelbaum, P.S., Robbins, P.C., Mulvey, E.P., Roth, L.H., Grisso, T., & Banks, S. (2001). *Rethinking Risk Assessment: The MacArthur Study of Mental Disorder and Violence*. New York: Oxford University Press.

NCAA (2011). Trends in graduation-success rates and federal graduation rates at NCAA Division I institutions. http://www.ncaa.org/wps/wcm/connect/e9eb8a0048d2623fb424ffb1fe52de76/GSR+and+Fed+Trends+2011+-+Final+10_20_11.pdf?MOD=AJPERES&CACHEID=e9eb8a0048d2623fb424ffb1fe52de76.

NCAA (2012). Estimated probability of competing in athletics beyond the high school interscholastic level. http://www.ncaa.org/wps/wcm/connect/public/test/issues/recruiting/probability+of+going+pro.

Paskus, T.S. (1997). Alternative prediction, selection and utility models for identifying academically successful college student-athletes (Unpublished doctoral dissertation). University of Virginia, Charlottesville, VA.

Paskus, T.S. (1998). Decision validity methods applied to cognitive assessments of brain disorder. In J.J. McArdle & R.W. Woodcock (Eds.), *Human Cognitive Abilities in Theory and Practice* (pp. 283–295). Mahwah, NJ: Lawrence Erlbaum Associates.

Paskus, T.S. (2012). A summary and commentary on the quantitative results of current NCAA academic reforms. *Journal of Intercollegiate Sport*, 5, 41–53.

Petr, T.A., & McArdle, J.J. (2012). Academic research and reform: A history of the empirical basis for NCAA academic policy. *Journal of Intercollegiate Sport*, 5, 27–40.

Petr, T.A., & Paskus, T.S. (2009). The collection and use of academic outcomes data by the NCAA. *New Directions for Institutional Research*, 144, 77–92.

R Core Team (2010). *R: A Language and Environment for Statistical Computing.* Vienna, Austria: R Foundation for Statistical Computing.

Rorer, L.G., Hoffman, P.J., & Hsieh, K. (1966). Utilities as base-rate multipliers in the determination of optimum cutting scores for the discrimination of groups of unequal size and variance. *Journal of Abnormal Psychology*, 50, 364–368.

Scott, B.M., Paskus, T.S., Miranda, M., Petr, T.A., & McArdle, J.J. (2008). In-season vs. out-of-season academic performance of college student-athletes. *Journal of Intercollegiate Sport*, 1, 202–226.

Strobl, C., Malley, J., & Tutz, G. (2009). An introduction to recursive partitioning: Rationale, application, and characteristics of classification and regression trees, bagging, and random forests. *Psychological Methods*, 14(4), 323–348.

Wackwitz, J.H., & Horn, J.L. (1971). On obtaining the best estimates of factor scores within an ideal simple structure. *Multivariate Behavioral Research*, 6, 389–408.

Wainer, H. (1976). Estimating coefficients in linear models: It don't make no nevermind. *Psychological Bulletin*, 83, 213–217.

14 Understanding Global Perceptions of Stress in Adulthood through Tree-Based Exploratory Data Mining

Stacey B. Scott, Brenda R. Whitehead,
Cindy S. Bergeman, and Lindsay Pitzer

Introduction

A wealth of research has linked perceptions of stress to both psychological well-being and physical health. What is less well understood are the circumstances that contribute to perceptions that life is overwhelming, uncontrollable, and considered stressful. Although theoretical accounts describe complex roles, experiences, and demands inherent to adult life, these life events and chronic stressors are rarely examined in concert. Furthermore, the diversity that exists among middle-aged and older adults suggests that concentrating on a single predictor or pathway may be short-sighted and limits understanding of the myriad ways that life is experienced. Exploratory data mining, though the use of regression tree and random forests analysis, offers a novel approach to this research problem. Three studies, assessing predictors of perceived stress in middle and later life, are presented here using these methods. Generally speaking, regression trees can offer an idiographic perspective by highlighting complex interactions resulting in multiple routes to similar levels of stress and how individuals with comparable scores on a given predictor may report very different levels of global perceptions of stress based on their other life circumstances. Random forests, in contrast, typically offer more nomothetic information by examining many trees and providing indices of predictor importance that can direct future studies. Together, these techniques offer the opportunity to inform life stress theory, direct intervention and prevention efforts, and guide future work.

Stress is very challenging for researchers to define and understand because it is a highly subjective experience that differs across individuals (Monroe, 2008). What is known, however, is that experiencing major life events or chronic strains, or even having the perception that life is stressful, has important implications for health and well-being in middle and later life (Thoits, 2010). Unfortunately, the current state of the literature is disconnected and provides little direction for developing hypotheses for how these stressors and strains may interact and contribute to global perceptions of stress. Exploratory Data Mining (EDM) techniques, therefore, are especially well-suited for investigations of this type. By using EDM approaches, such as regression trees and random forests, investigators can leverage flexibility and data-sensitivity in order to uncover effects describing characteristics of the available sample that can be used to develop predictions for future samples and to refine theory to address interrelationships among predictors.

In traditional regression models the primary objective is "to identify an *average* set of conditions associated with a given outcome" (Gruenewald, Mroczek, Ryff, & Singer, 2008, p. 332). A key advantage of tree-based methods, in contrast, is that they allow for *different* weightings and contributions of life events and strains for different individuals. A goal in the tree-based approach is to find a hierarchy of predictors that best describes differences in the outcome. To accomplish this aim, the analysis identifies variables that divide the sample on the outcome and the level at which the selected predictor best divides the sample. In so doing, it is possible to produce a more nuanced description of the contributors to different degrees of the outcome (Gruenewald et al., 2008). Recursive partitioning, of which regression trees and random forests are examples, is not yet common in developmental research, but examples include: (1) identifying cumulative and compensatory effects of academic and forms of social competence in predicting depressive symptoms in children (Seroczynski, Cole, & Maxwell, 1997); (2) pathways to positive and negative affect in adulthood and later life (Gruenewald et al., 2008); (3) interactions among protective factors for well-being in old age (Wallace, Bergeman, & Maxwell, 2002); and (4) combinations of biomarkers predicting mortality in later life (Gruenewald, Seeman, Ryff, Karlamangla, & Singer, 2006).

In this chapter, we use regression tree and random forests techniques to identify and assess the predictors of global perceptions of stress in middle age and later life and to understand the combinations of stressors that contribute to similarities and differences in individuals' perceptions of being stressed. Prior to our introduction of the sample and more detail on the techniques themselves, it is important to point out the technical and substantive advantages of using tree-based methods. From a technical standpoint, there are many good reasons to use tree-based methods (for an overview see Berk, 2008). Tree analyses are ideally suited for exploratory research—in cases, such as the applications below, in which there are many possible predictors, but prior research has not examined them together, thus providing little guidance regarding their additive or interactive effects. In traditional regression analyses, researchers must often choose from the available predictors of interest, and only examine interactions between a subset of them because of power constraints. By not forcing the researcher to choose among predictors *a priori*, the researcher introduces less bias at the outset of tree analysis. Regression trees are non-parametric with respect to the covariates and thus do not assume a model about the predictors involved. Regression trees are unaffected by monotonic transformations of the predictors (Wallace et al., 2002), therefore, rescaling and normalization of the covariates does not change the results. Finally, trees are robust to outliers in the covariates.

From a substantive standpoint, there are also many benefits from using tree-based analysis. Tree analyses can provide valuable information about nomothetic (group) and idiographic (individual) influences on traits of interest. *Nomothetic* approaches identify the major agents involved in understanding etiology by providing crucial information about the average or most probable influences at the group level. An *idiographic* pathways approach builds on this, but shifts attention to the interactions by which outcomes are attained, and highlights differences between individuals in these routes and eventual outcomes. This second approach is vital for understanding diversity in process and outcome. Individual regression trees embody both idiographic and nomothetic representations of data. As we describe in more detail below, the root node in a given tree provides a description of the entire sample (e.g., nomothetic information). The terminal nodes at the bottom of a regression tree reveal individual differences (e.g., idiographic information) in the traits of interest. In principle, a very large and complex tree with many predictors could split the sample until each terminal node describes a single person, an extreme level of idiographic detail. Examining different levels of a particular tree allows the investigator to shift focus between the two extremes of nomothetic and idiographic information for a sample. Random forests can further shift to a more nomothetic perspective by examining many different trees made from variations of the sample and predictors. Random forests aid in understanding *which* domain of risk is most important, whereas regression trees can also help us to understand *for whom* certain circumstances are particularly important.

Perceptions of Stress across the Adult Life Span

Aging is often thought of as a process that occurs in the same way across individuals, with "normal" aging frequently identified with the presence of disease and age-related physiological changes. The reality is, however, that individuals do not necessarily experience aging in the same way (Bergeman, 1997; Dannefer & Sell, 1988); people age differently, and investigations into the dynamic interactions among physical, psychological, and social processes that contribute to these different patterns and outcomes of aging are key to advancing understanding in this area. Interestingly, older adults tend to be more different from one another in terms of health, psychological functioning, and aspects of social interaction than are younger adults (Nelson & Dannefer, 1992). This has led to the idea that members of a given age cohort "fan out" as they age, becoming increasingly dissimilar from peers on any given characteristic (Baltes, 1979). Two broad concepts in developmental psychology, equifinality and multifinality, help to explain how the contextual experiences that shape people's lives create these differences in life trajectories (von Bertalanffy, 1969). *Equifinality*, refers to how different experiences in everyday life (e.g., social isolation and financial difficulties) may relate to a similar outcome (i.e., decreased feelings of well-being). For example, two individuals may find their lives very stressful, but for very different reasons—one may have recently suffered a job loss whereas the other may be experiencing ongoing difficulty with a teenage child. *Multifinality*, on the other hand, refers to how one etiologic factor can be associated with any of several outcomes, depending on person or context (e.g., a person who is extroverted may be upset by a lack of social interaction, but another individual, who is more introverted, may be satisfied with few social exchanges). Thus, there may be multiple contributors to the perception of stress within any individual and combinations of exposures to events, pressures, hassles or trauma may differ between individuals who perceive equal amounts of stress.

The idea that feelings of being stressed are precipitated by many different types of stressors that occur in a variety of different domains and that dynamically interact with one another to create unique stress experiences and reactions, is not novel. Theorists have used a constellation of terms—including acute, chronic, ambient, and role—to describe different types of stressors that can occur within a myriad of life domains, including those relating to physical health, finances, social relationships, neighborhood experiences, and work (Pearlin & Skaff, 1996; Turner, Wheaton, & Lloyd, 1995; Wheaton, 1997). Empirical studies of stress, however, tend to focus on a single domain (e.g., bereavement, work strain) or simply count up the number of events or hassles that have occurred without attending to the synergy among them (Miller, 1993). In fact, theoretical notions of stress overload define a particular experience as stressful, at least in part, because of concomitant demands (Almeida & Horn, 2004) and it has been speculated that the dynamic nature of a confluence of stressors may be more noxious than their independent or additive effects (Monroe & Simons, 1991). Thus, it is important that the interrelationships among different stressors be considered in any study investigating how stress manifests and operates within people's lives (Pearlin & Skaff, 1996; Wheaton, 1983).

Global Perceived Stress

Individuals' overall perceptions of how stressful life is may be an informative link between individual domain-specific stressors and negative physical and mental health outcomes. Global perceived stress, as assessed by the

Perceived Stress Scale (Cohen, Kamarck, & Mermelstein, 1983), may be individuals' aggregations of recent life events, ongoing strains, and lack of resources that together result in people appraising their current life situation as uncontrollable or overwhelming. Perceived stress has been found to predict depression symptomatology (Cohen et al., 1983), but despite being highly correlated with depression measures, it independently predicts physical health symptoms (Cohen & Williamson, 1988; Cohen et al., 1983). Global perceived stress also predicts other physical outcomes including changes in smoking rate (Cohen et al., 1983), susceptibility to the common cold (Cohen, Tyrell, & Smith, 1993), decreased gray matter volume in the hippocampus (Gianaros, Jennings, Sheu, Greer, Kuller, & Matthews, 2007), and biological correlates of longevity (e.g., telomere length, oxidative stress; Epel et al., 2004). This subjective assessment of current overall life stress, as indexed by global perceived stress, has important implications for how individuals respond to the challenges of daily life, as well. Individuals who report high levels of current global perceived stress are more affected by the experience of daily stress (van Eck, Nicolson, & Berkhof, 1998; Stawski, Sliwinski, Almeida, & Smyth, 2008). As measured by the Cohen scale, perceived stress assessments typically use a recall period of the last month. Therefore, although it may change slowly, perceived stress from socio-ecological conditions may be more of a prolonged state than an unchanging characteristic of a person. For example, Sliwinski and colleagues (Sliwinski, Almeida, Smyth, & Stawski, 2009) found substantial within-person variability in global perceived stress across biannual assessments, which moderated within-person variability in emotional responses to daily stress. Specifically, individuals exhibited larger effects of daily stress on negative affect during periods when they reported higher global levels of stress, compared to other periods when their perceived stress was lower.

In sum, global perceptions of stress are related to a variety of outcomes. Although it is theorized to provide aggregate information from multiple different domains of current life stress, few studies have examined the connection between the experience of specific individual stressors and the possible intermediary index represented by global perceptions of stress. To the extent that different combinations of stressful circumstances are related to similar levels of global perceived stress, this diversity in inputs has important implications for both theory and intervention.

Stressful Events and Conditions: Life Events and Chronic Stressors

One of the primary considerations when conceptualizing stress and its effects is to identify whether stress stems from the occurrence of *life events*, which tend to be acute and time-restricted occurrences, such as an injury or the death of a loved one or from more insidious *chronic stressors*, which are persistent problems that tend to lack any foreseeable resolution, such as a conflicted marital relationship or discontent at work (Wheaton, 1983). These two sources of life stress can also be intertwined. That is, life events can lead to the occurrence of chronic difficulties—the death of a spouse necessitates an adjustment to living alone and divorce may result in the chronic stress associated with single-parenthood. On the other hand, the ongoing strain associated with providing care to an ill spouse may end in conjugal loss, a major life event. Although life events and chronic strains may be related, they must be considered independently because reactions to, and experiences of, these events and difficulties may be different across people and circumstances (Brown & Harris, 1989; Wheaton, 1994, 1997). Each is briefly described below.

Life Events

Stress researchers have long recognized the detrimental effects of experiencing major life events, such as the loss of a spouse or a job, on physical and mental health (Palmore, Cleveland, Nowlin, Ramm, & Siegler, 1979; Holmes & Rahe, 1967; Neugarten, 1979). Adults not only nominate these experiences, but others such as personal illness, death of a loved one, and a loved one's illness as being especially difficult to manage (Hardy, Concato, & Gill, 2002; Seematter-Bagnoud, Karmaniola, Santos-Eggimann, 2008). Studies examining the impact of major life events have shown effects for depression and anxiety (De Beurs, Beekman, Geerlings, van Dyck, van Tilburg, 2001; Kessing, Agerbo, & Mortensen, 2003), disease precursors and symptoms (Golden-Kreutz et al., 2005; Mitsonis, Potagas, Zervas, & Sfagos, 2009; Rafanelli et al., 2005; Rosengren et al., 2004; Wigers, 1996), and mortality (Clémence, Karmaniola, Green, & Spini, 2007). Because major events often change a person's status or circumstances, the life stress they produce can be particularly pernicious.

Chronic Stressors

Chronic stressors may, by their very ubiquity, be particularly important to an individual's general perception of how unmanageable, overwhelming, and stressful life is at a particular point in time. These ongoing difficulties have been found to independently predict the onset of major depression (Brown & Harris, 1978, 1989), although not to the extent that is seen with severe life events. It is clear, however, that they play a contributing role in psychological adjustment, especially for those with recurrent depressive episodes (Monroe, Slavich, Torres, & Gotlieb, 2007), and to detrimental health outcomes (McEwen, 1988). Although a major life event can be devastating at its peak level of stress, the direct effects of acute stressors decline over time (Wheaton, 1994). Compare this with the nature of a chronic stressor, which does not necessarily have a perceptible beginning or a predictable end, and which may not even be perceived by the individual until he or she begins to experience the physical and psychological "wear and tear" that can result from a constant state of stress activation (McEwen & Seeman, 1999; Wheaton, 1994). We highlight previous research in this area below.

Chronic stressors are generally categorized into two types: (1) *role stressors*, which result from the fulfillment of certain family, interpersonal, or occupational roles, and (2) *ambient stressors*, which stem from the external environment (Pearlin & Skaff, 1996; Wheaton, 1997). Although the majority of the chronic stressors we consider here fall in to the role category, as they either directly (family, friend, and work role stressors) or indirectly (financial and health stressors) impact the degree to which a given individual feels overwhelmed in the context of a given role, we additionally consider those in the *ambient* category, such as neighborhood issues (e.g., deteriorating conditions or safety concerns). We also examine loneliness as a form of chronic stress due to social isolation, given its connections with stress appraisals (Hawkley, Burleson, Berntson, & Cacioppo, 2003) and morbidity and mortality (see House, Landis, and Umberson, 1988).

Role Stressors: Parenting, Work, Caregiving

Across the lifespan, adults juggle multiple roles and responsibilities that tend to change in nature across adult development. For example, those in midlife may be dealing with the chronic stress associated with the demands of parenting roles, spousal roles, occupational roles, social roles, and even caregiving roles if their parents' health begins to deteriorate, whereas adults in later life may experience caregiving role stress when a spouse becomes ill, stress related to retirement or volunteer roles, and demands related to parent- and grandparenthood.

The day-in, day-out nature of family roles, along with their near-universal presence, makes them a key contributor to the experience of chronic strain. The potentially explosive combination of differing role expectations and emotional attachment in the context of primary role identities contributes to the presence of stress-inducing conflict that often occurs within close family relationships (Pearlin & Skaff, 1996). Experiencing conflict with loved ones— particularly conflict that is repetitive and unresolved—is likely to have a substantial impact on the degree to which adults report feeling "stressed."

Studies examining work stress have identified both environmental and personality factors that contribute to the extent to which chronic problems at work contribute to overall levels of stress. Here, *effort–reward imbalance*—when individuals feel that their work is not being rewarded or acknowledged appropriately—can contribute to dissatisfaction with work and ultimately the experience of stress (Buddeberg-Fischer, Stamm, Buddeberg, & Klaghofer, 2010). Additionally, uncertainty about the stability of one's employment situation is also a substantial source of occupational stress (Scott-Marshall & Tompa, 2011). Research has indicated that chronic work-related stress contributes to stress responses associated with negative health indicators and outcomes, including heightened cortisol response (Schlotz, Hellhammer, Schulz, & Stone, 2004) and increased allostatic load (Bellingrath, Weigl, & Kudielka, 2009).

Caregiving roles are inherently stressful, as they not only demand time and instrumental aid, but also bring on the psychological and emotional struggles that come with observing the deterioration of a loved one. A large literature has established the negative impact of caregiving stress on well-being in both midlife and older adults (Mioshi, Bristow, Cook, & Hodges, 2009; Saban, Sherwood, DeVon, & Hynes, 2010). Pearlin and colleagues have suggested that one stress-inducing aspect of caregiving is that this role is typically not anticipated by the caregiver, and becomes an all-consuming "unexpected career" (Aneshensel, Pearlin, Mullan, Zarit, & Whitlatch, 1995; Pearlin & Skaff, 1996). The long-term nature of the caregiving role, which tends to become increasingly demanding as the health of the care receiver deteriorates, also contributes to the high levels of distress reported by caregivers.

Health

The presence of chronic illness or multiple health concerns is stressful for both midlife and older adults (Piazza, Charles, & Almeida, 2007; Whitbourne, 2001), but the fact that the presence and severity of chronic health problems increases with age means that health concerns are likely to have a greater contribution to perceptions of stress for adults in later life. Studies examining the stressful effects of specific chronic diseases support this idea: older adults with rheumatoid arthritis who report more severe or more frequent symptoms also report elevated levels of stress (Tak, Hong, & Kenneday, 2007), and patients with fibromyalgia report more perceived stress when they report more severe symptoms and/or higher levels of functional limitations resulting from those symptoms (Murray, Daniels, & Murray, 2006).

Finances

The nature of financial-related stressors tends to be different for adults in mid- and later life. Midlife adults, who are typically at the height of their careers and therefore financially stable, are more likely to experience stress related to multiple monetary demands, such as home mortgages, college tuition, retirement savings, and medical or assisted living care for aging parents (Antonucci, Akiyama, & Merline, 2001; Easterlin, 2006). Adults in later life, however, are more likely to face financial stress stemming from the limitations of a fixed income. The effects of financial stressors—and the extent to which they contribute to an individual's perceptions of stress—tend to be magnified in times of general financial hardship such as that stemming from the recent economic recession (Meltzer, Bebbington, Brugha, Jenkins, McManus, & Stansfeld, 2010).

Neighborhood

Stress related to the condition and safety of one's neighborhood fall into the ambient chronic stressor category mentioned above, as it stems from the external environment in which the individual is embedded. When an individual's neighborhood is not suitable for his or her needs (e.g., poor access to grocery, religious, educational, and/or medical facilities) or is unsafe (e.g., rising crime, unoccupied or untended residences), it can contribute to overall levels of perceived stress (Pearlin & Skaff, 1996; Yen, Michael, & Perdue, 2009). This may be an especially potent factor for older adults, who may be "stuck" in their homes for financial reasons—such as a fixed income or the inability to sell a property during difficult economic times—or who may be less able to relocate due to poor health or lack of support.

Loneliness

Loneliness, or social isolation, is associated with higher levels of perceived stress across adulthood (Aanes, Middlemark, & Hetland, 2010; Cacioppo, Hawkley, & Berntson, 2003), and has also been linked with health processes and outcomes that stem from extended stress exposure including heightened levels of cortisol (Doane & Adam, 2010; Steptoe, Owen, Kunz-Ebrecht, & Brydan, 2004), decreased immune function (Glaser, Kiecolt-Glaser, Speicher, Holliday, 1985; Kiecolt-Glaser, Garner, Speicher, Penn, Holliday, & Glaser, 1984), and increased morbidity and mortality risk (Berkman, Leo-Summers, & Horwitz, 1992; House et al., 1988; Penninx, van Tilburg, Kriegsman, Deeg, Boeke, & van Eijk, 1997). Although detrimental for all age groups, social isolation may be especially associated with later life, as physical limitations, widowhood, and smaller social networks are increasingly common with age. Particularly relevant to global perceptions of stress, Hawkley et al. (2003) found that compared to non-lonely college student peers, lonely students appraise their daily activities as more demanding and evaluate themselves as less able to meet these demands. These daily activities (e.g., work, chores, errands, health care, socializing) as measured in the Hawkley study tap many of the chronic stressor domains above, thus interactions between loneliness and other domains may be particularly informative in understanding how individuals appraise life stress.

Utility of Exploratory Data Mining for Understanding Stress

Applications of Tree-Based EDM to the Study of Stress in Middle and Later Adulthood

Design and Sample

The Notre Dame Study of Health & Well-Being (NDHWB) is broadly aimed at investigating risk, and resilience processes in middle-aged and older adults. NDHWB data were used to test hypotheses pertaining to how different combinations of contextual features (e.g., social isolation, neighborhood quality, health problems, age discrimination, financial concerns, and recent life events) contributed to overall feelings of stress in mid- (Scott, Bergeman, Whitehead, & Pitzer, 2012) and later life (Scott, Jackson, & Bergeman, 2011) samples.

The full sample of NDHWB participants ($N = 783$; $n = 338$ in the older cohort and $n = 445$ in the midlife cohort) were recruited from lists of adults in the northern Indiana area, which were provided by a market research firm and compiled based on multiple information sources including census data and the Survey of Residential Households. Participants received the survey packets in the mail, which they then completed and returned in a postage-paid return envelope. Participants received gift cards for an establishment of their choice for their participation (US$20.00 a year for the yearly questionnaire); informed consent was obtained prior to participation in each step of the project.

The data summarized in the first two studies (Study 1 and Study 2) came from different waves of the longitudinal study, prior to the harmonization of measures across samples. The Midlife sample included 410 adults ranging in age from 37 to 64 years ($M = 52.12$ years, $SD = 6.18$ years), and the Later Life sample comprised 282 adults aged 54 to 91 ($M = 68.79$ years, $SD = 5.10$ years). Detailed demographic information and full descriptions of the measures are provided in the original manuscripts (Scott et al., 2011; Scott et al., 2012) or are available upon request. A third study (Study 3) uses the Study 2 dataset for the Midlife participants and a data from a later wave with consistent measures for the Later Life participants. Study 3 demographic information is provided in Table 14.1. The mean age in the combined sample is 58.8 years (range 37–80, $SD = 10.1$).

Analysis

In our initial work with these data, we analyzed the Later Life (Study 1; Scott et al., 2011) and the Midlife (Study 2; Scott et al., 2012) samples separately due to differences in the available variables and because we wanted to focus on the possibly distinct influences in these developmental

Table 14.1 Study 3 demographics

	N	%
Total *N*	666	—
Gender		
Female	398	59.8
Male	268	40.2
Marital Status (missing = 7)		
Married	340	51.6
Divorced	153	23.2
Single	84	12.8
Widowed	71	10.7
Separated	11	1.7
Race		
Caucasian	572	85.9
Black/African American	59	8.8
Hispanic/Latin American	14	2.1
Asian/Pacific Islander	3	.5
Native American/Aleutian Islander	6	.9
Other	2	1.8
Education		
Grade school (grades 1–6)	2	.3
Middle school (grades 7–9)	14	2.1
High school (grades 10–12)	209	31.5
Vocational education	50	7.5
Some college classes	163	24.5
College degree	129	19.4
Post college professional degree	42	6.3
Graduate, medical, or law degree	56	8.4
Annual household income (missing = 13)		
Less than $7,500	27	4.2
$7,500 to $14,999	66	10.1
$15,000 to $24,999	92	14.1
$25,000 to $39,999	151	23.1
$40,000 to $74,999	200	30.6
$75,000 to $99,999	62	9.5
$100,000 or more	55	8.4

Note. The number of people missing on each variable is indicated by the digits in parentheses.

periods. We first summarize these results and then present the findings from a new analysis of the combined data (Study 3). As the NDHWB is a longitudinal study, with harmonized measures across the two subsamples at later waves, we can now examine whether the structure and variables identified in the more theoretically rich set of predictors from the Midlife sample are also useful in an older sample. Furthermore, we can assess the extent to which the interactions we found depend on age. Global perceived stress was the outcome variable in all studies and was treated as a continuous variable. Therefore, the analyses involve regression trees rather than classification analyses, which is another application of these methods. In Study 1, all of the predictors (life events, chronic health problems, somatic health problems, finances, neighborhood, loneliness, age discrimination experiences) were treated as continuous values. In Studies 2 and 3, categorical (e.g., work status, caregiver status) and continuous (e.g., life events, chronic health problems, somatic health problems, finances, neighborhood, loneliness, relationship chronic stress, family chronic stress, parenting chronic stress, role conflict, time pressure chronic stress, work chronic stress, income chronic stress, discrimination experiences of any kind) predictors were used. We also included age in the Study 3 analyses. In each of these applications, all of the listed predictors were given to the tree algorithm for the resulting tree and forests results; only those identified in the selected tree are displayed in the tree figure, the permutation values for all predictors are included in the variable importance figure from the forests results for Study 3.

RECURSIVE PARTITIONING APPROACHES TO EDM

Both regression trees and random forests involve recursive partitioning of the data—that is, repeatedly splitting the data in order that observations with similar values on the outcome are grouped together (Strobl, Malley, & Tutz, 2009). Choosing and using a tree algorithm involves three primary aspects: a node splitting rule, a stopping rule, and assignment to a terminal node. First, depending on the algorithm chosen, a node splitting rule is selected. The regression tree analysis begins with a *root node*, which contains all participants in the dataset. Using this node splitting rule, the software compares all possible levels of all predictor variables to determine an optimal predictor and its cutoff point, which is selected such that it minimizes the within-group variance on the outcome, global perceived stress. A binary split is made at this point, dividing the sample into two groups, called *child nodes*. This search for predictor and cutpoint is repeated for each of the child nodes. Second, the researcher sets limits regarding when to stop splitting. One example of a stopping rule is to set a minimum number of cases (subjects) in a terminal node; this is done to allow a large enough number of observations in each node to reliably estimate a model in the terminal node. A side effect of this constraint is that it reduces model-complexity by not allowing the tree to become overgrown (e.g., the extreme situation in which each person is represented by a node). In our examples the minimal terminal node size was set to 10. Third, based on these decisions and the resulting splits, observations are assigned to terminal nodes. These terminal nodes group observations into classes, values, or models of the outcome variable.

The results of these analyses are typically displayed in an inverted tree figure. In these graphic depictions, the splits resemble branches. At the end of these branches are *terminal nodes*, which represent subsamples of individuals who cannot reliably be further divided based on the constraints

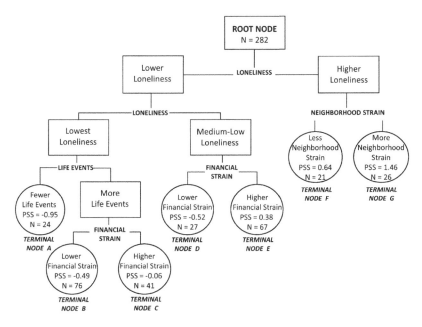

Figure 14.1 Study 1 perceived stress regression tree for later life sample.
Notes: Root and child nodes are displayed in boxes; terminal nodes are displayed in circles. Splitting variables are indicated at the intersection of branching lines. PSS: Mean Perceived Stress Scale for node. See Scott et al. (2011) and for more detail on measures and results.

that the investigator applied to the analysis. In figures, we use the label Node 1 for the root node, numbered rectangles for the child nodes, and lettered circles for the terminal nodes. The cutpoints for each splitting variable are displayed inside the boxes in Figure 14.3, to streamline our summaries of the Study 1 and 2 results these are omitted from those figures (Figures 14.1 and 14.2). Following Gruenewald et al. (2008), we refer to the series of interactions producing a given terminal node as a pathway and detail these combinations of risk and protective influences in the results section. We used the commercial software program CART Pro Version 6.0 (Steinberg & Colla, 1997) to conduct the regression tree analysis; regression trees can also be produced using free programs using R (R Development Core Team, 2012) packages such as *rpart* and *party*.

Split sample approaches (e.g., test set error and pruning) can be applied to help identify an informative tree structure. There are several ways to test the degree to which a selected tree is over fit, or is overly sensitive to, the idiosyncrasies of a particular dataset. In large samples, it is possible to split the data and grow a tree on part of the sample (often referred to as the *learning sample*) and test the error and cost on the other portion (called

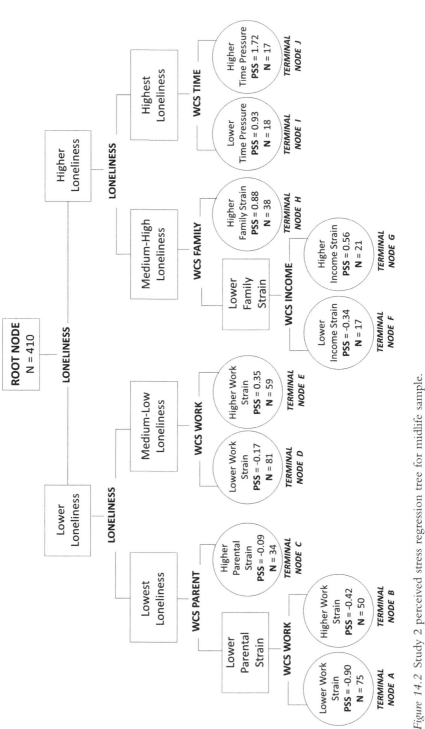

Figure 14.2 Study 2 perceived stress regression tree for midlife sample.

Notes: Root and child nodes are displayed in boxes; terminal nodes are displayed in circles. Splitting variables are indicated at the intersection of branching lines. PSS: Mean Perceived Stress Scale for node, WCS PARENT: parent role, WCS WORK: work role, WCS FAMILY: family role, WCS INCOME: financial strain, WCS TIME: time/pressure. See Scott et al. (2012) for more details on measures and results.

the *test sample*). Using CART Pro, the analysis begins by growing a very large tree. This initial process avoids missing key splits that could occur if the procedure was stopped too early, which cannot be known *a priori*. These large trees, however, have the drawback of being difficult to interpret and unlikely to replicate. To limit these complex and unreliable structures, the software *prunes* (e.g., removes splits that may capitalize on chance relationships) the large tree. Splits that do not improve the accuracy in cross-validation are pruned by CART Pro. Other tree algorithms, including some of the trees by R and most forests analysis, use p values to select variables and identify stopping points and so do not have to grow overly large trees and prune back (Strobl, Malley, & Tutz, 2009). The goal is to identify a tree that balances the idiographic (e.g., profiling the characteristics of individuals) with nomothetic (e.g., finding a set of variables likely to replicate to the larger population) goals.

A generalization of split sample approaches is to use cross-validation to compare the performance of several trees. These methods can be used in studies with smaller samples (Hastie, Tibshirani, & Friedman, 2001) in which the sample is too small to split into a learning and test sample. In the studies here, we selected 10-fold cross-validation in which the sample was randomly divided into 10 equal subsamples. In each iteration, the program grows a tree on nine of the subsamples and uses the remaining 10% to function as a pseudo-test sample. The sum of squared error for the tree is calculated. This comparison is repeated 10 times so that all the subsamples serve as both learning and pseudo-test samples. This provides an estimate of the generalizability of the tree (for a more detailed description of cross-validation, see Merkle & Shaffer, 2011).

The cross-validation procedures employed above reflect one method we used to balance our idiographic goal of describing the unique predictors and interactions in the sample with our interests in also identifying consistent predictors that could be useful in future work and with other samples. Indeed, Strobl and colleagues caution researchers about the sensitivity and instability of individual tree structures, stating that "the entire tree structure could be altered if the first splitting variable, or only the first cut point, was chosen differently because of a small change in the … data" (Strobl, Malley, & Tutz, 2009, p. 330). Thus, it is important to balance the idiographic approach with techniques that inform companion goals at the nomothetic level.

Random forests analysis is another recursive partitioning technique that leverages this variability in individual trees and aggregates over many structures to find predictors that are consistently important (Breiman, 2001). Random forests analysis involves varying the sample as well as the predictors across a large number of trees. One part of this is called *bagging*, or bootstrap aggregating. In bagging, a set of models (i.e., trees) are produced using bootstrapped samples of the data, then the predictions of these models are aggregated into a single prediction. In random forests, these models are weighted by their performance on the *out-of-bag* sample, the data that was not chosen in each bootstrap. In this way, random forests samples from the current data many times and then uses the remaining data as a test sample as a way of estimating how well the models would generalize to other samples. Another way a random forests analysis introduces variability among trees is by sampling the predictors with replacement. This sampling of predictors allows the procedure to identify other predictors and interactions that may have been missed when a particularly strong predictor is always in the analysis.

Random forests analysis involves the production of many, many trees — in the examples presented here we used 500 to 1,000 trees for each random forests analysis—as such, it is not possible to visually sort through them and determine the useful predictors across all the results. Instead, random forests analysis produces indices of variable importance (Breiman, 2001). In our applications, we used the *cforests* procedure in the *party* package in R (Strobl, Boulesteix, Zeileis, & Hothorn, 2007; Strobl, Boulesteix, Kneib, Augustin, & Zeileis, 2008). We specified a constraint of at least 10 participants per terminal node, and the default setting for bagging (subsampling .632*n), number of trees (ntree = 500 to 1,000), and sampling from the predictors (mtry = 5). We varied the starting seed because of our relatively small sample sizes.

Permutation importance is the key result of random forests analysis. Random permutation involves randomly shuffling or reordering the values of a predictor, such as neighborhood strain, so that the original relationship between neighborhood and the outcome, perceived stress, would be broken. If in the original data, neighborhood stress was related to global perceived stress, then using this shuffled version along with the non-permuted other predictors would result in significantly worse prediction. Variables that are poor predictors of perceived stress, in contrast, would only show a small random decrease—or sometimes even a small increase—after permutation. Rather than the permutation values themselves, it is the relative ranking of variables in terms of their permutation importance that is informative (Strobl, Hothorn, & Zeileis, 2009). We report conditional permutation importance, because we have correlated predictors (Strobl, Hothorn, & Zeileis, 2009).

MISSING DATA AND DATA TRANSFORMATIONS

Listwise deletion can be used for missing data in cases for which fewer than 5% of the total observations would be excluded (Berk, 2008). Other options include imputation prior to analysis as well as coding categorical predictors as missing (Hastie et al., 2009). The CART Pro software, as well as some of the R packages, has the option to manage missing data by using *surrogate* predictor variables as splitters. Surrogates are variables that behave similarly to the preferred (but unavailable) predictor in determining outcomes such as node size, composition, and which cases are assigned to left and right child nodes (Steinberg & Golovyna, 2006). They are used as splitting variables in situations for which the primary splitter is missing. Although surrogates can be useful, we chose not to use them for our purposes, because our primary interest was in determining the relationships between specific contextual risk variables and perceived stress in these samples. That is, for our descriptive purposes—particularly in the Study 1 application in which we further profiled the individuals in terminal nodes with in-depth interviews—the particular predictors that were involved, rather than their next-closest substitute, which may be from an entirely different theoretical stress domain, were key. In addition, Berk (2008) cautions against using surrogates as it is unclear at what amount of missingness the substitution will bias the results. Nevertheless, surrogates are an option in most tree-based analysis including CART and R, and provided that the missingness in the data meets the missing at random assumptions, the results should be unbiased and will increase the power to detect effects.

In the examples that follow, we limited our analyses to complete cases. Of the 297 participants in the Later Life sample (Study 1), 15 participants (approximately 5% of the sample overall) were missing at least one of the variables of interest. Of the 439 Midlife participants from the NDHWB who provided perceived stress data for Study 2, 29 (approximately 7%) were missing at least one of the predictors and were excluded from the dataset. In Study 3, the data from 291 subjects from Wave 3 (collected from Spring 2008–Spring 2009) of the Later Life Cohort and 439 subjects from Wave 1 (collected from Fall 2007–Fall 2008) of the Midlife Cohorts were pooled and standardized; 666 of the available 730 subjects had complete data (91% complete). We standardized the outcome and predictors in each study to increase the interpretability of the cutpoints and to characterize the samples in the terminal nodes.

Study 1: Later Life

Regression Tree Results

In Study 1 (Scott et al., 2011), we examined global perceived stress among a later life sample, based on the interactions between contextual stressors across many life domains (i.e., life events, neighborhood, finances, ageism, loneliness, chronic health, and somatic health). Figure 14.1 displays a summarized version of the selected regression tree. This tree includes six pairs of branches and seven terminal nodes. Each of the terminal nodes serves to describe relationships between contextual risk variables and overall perceived stress for members of the sample. As is depicted in the figure, four splitters were identified in this tree: loneliness, financial strain, neighborhood, and life events, to differentiate levels of perceived stress in this sample. Loneliness appears twice as a splitter—first, splitting the root node, which separates those who report very high loneliness (branch to right) from the rest of the sample. From this it is evident that older adults in the high loneliness pathway can further be described and separated by their level of neighborhood strain. Second, those with less extreme levels of loneliness (branch to left), can further be separated in terms of their social isolation, resulting in different profiles for different groups of people. For instance, for those who report moderate-to-low levels of loneliness, financial strain further explains the differences in overall level of perceived stress. For those who report very low levels of loneliness recent life events further subdivides this group, differentiating those who report extremely low perceived stress (Terminal Node A, which is almost a full standard deviation below the sample average) from those who report more low-to-average levels of perceived stress.

Older adults who reported the lowest loneliness and the fewest life events scores (Terminal Node A), reported perceived stress scores that are nearly one standard deviation below the average person in the study (one standard deviation reflects almost six points on the perceived stress scale in this sample). Although the older adults in this terminal node had similarly low levels of loneliness to those in the branch to the right, the latter reported more life events and their perceived stress could be further differentiated by their financial strain. That is, older adults in Terminal Node B had average or lower levels of financial strain and reported perceived stress about one-half a standard deviation below the sample average, whereas adults in Terminal Node C reported higher levels of financial stain, but had scores at about the average of perceived stress.

Continuing across the figure, there are pathways captured by more moderate levels of loneliness, which interact with very different contextual variables. Individuals in Terminal Node D have similar levels of perceived stress to those in Terminal Node B (i.e., one-half a standard deviation below the sample average), but the interaction between low-to-medium levels of loneliness and very low financial strain helps to differentiate Terminal Node D from Terminal Node B. Furthermore, those in Terminal Node E have higher levels of financial strain, and thus higher levels of perceived stress (.40 standard deviations above the sample average).

Moving to the pathways on the far right of the figure, the upper right node encapsulates individuals who report very high levels of loneliness. Here, neighborhood strain distinguished levels of perceived stress. For example, individuals in Terminal Node F report high levels of loneliness, but low-to-average neighborhood strain (i.e., average perceived stress in this group is .60 standard deviations above the average). Individuals in Terminal Node G, however, reported higher levels of neighborhood strain and thus reported perceived stress 1.5 standard deviations above the average. Indeed, these people seem to be at risk for feeling that life is overwhelming and unpredictable.

Random Forests Results

Although the tree in Figure 14.1 tells a compelling story, trees can be sensitive to unique characteristics of the sample. The random forests analysis, described in detail in Scott et al. (2011) provided good support for the variables selected as splitters, but also provides evidence that cautions researchers against interpreting a single tree. For example, loneliness was the strongest predictor and permuting loneliness decreases its predictive utility. Other stronger predictors include financial strain, neighborhood strain, and ageism. Because of their importance, all would likely replicate as predictors of perceived stress in other samples. Chronic and somatic health problems, on the other hand, were poor predictors—negative permutation importance values indicate that randomly shuffling responses on these health variables actually increases prediction on perceived stress. Life events had a small permutation value, suggesting that the life events splits in the tree should be interpreted cautiously, as they may be the product of random variation in the dataset. Depending on the sample and other predictors in the analysis, life events may not appear as an important predictor.

Study 2: Midlife Sample

Regression Tree Results

Building on the findings in the Later Life sample, recursive partitioning methods were also useful in examining interactions among stress contexts (e.g., social roles—work, caregiving, relationship, parent, role restriction, time/pressure, income, family; somatic health, chronic health, life events, financial strains, and neighborhood) in a midlife sample, and to underscore the role of loneliness in how middle-aged adults appraise the extent to which their life is stressful (Scott et al., 2012). A summarized version of the tree from this study is depicted in Figure 14.2, more detailed information is provided in the original report. The examination of this tree (especially in light of the results of EDM analyses of the Later Life sample in Study 1) emphasizes the fundamental importance that loneliness plays in how people perceive the stress in their lives. As is noted, individuals in the left branch have below average levels of stress compared with those in right branch. Within those with lower loneliness, loneliness further distinguishes those with lower (labeled lowest loneliness) from those with more average levels (labeled medium-low loneliness) of perceived stress, and those with more moderate (labeled medium-high loneliness) from those with very high levels of perceived stress (labeled highest loneliness).

Each of these loneliness groupings of perceived stress are further distinguished by interactions with other variables. For instance, among those reporting the least loneliness, higher parental role demands distinguish those in Node C from those in Node A, and from those in Node B, whose work role strains can be used to separate their moderately low global perceived stress (for whom perceived stress is about one-half a standard deviation below average) from those with perceived stress nearly one standard deviation below the sample average (Node A). Among the participants with medium-low levels of loneliness, work strain was also informative in separating those with below average levels of perceived stress (Node D) from those with average to moderate levels (Node E).

In the higher loneliness branches of the tree, different predictors were useful in describing perceptions of life stress. Among those with medium-high loneliness, family role strains separate those with relatively high levels (Node H) from those with more moderate perceptions of life stress (labeled lower family strain). Within this node, there is still substantial diversity in levels of perceived stress. Income demands help to partition those who, despite their loneliness, have relatively low levels of perceived stress (Node F) from those with above average perceptions of life stress (Node G). Those in the highest loneliness branch of the tree report very high levels of perceived stress overall, over one standard deviation from the average person in the sample. The interaction between this high level of loneliness and time pressure make it possible to distinguish very high levels of perceived stress (Node I) from extreme levels of appraised life stress (Node J).

Random Forests Results

Results of the random forests provided additional information on predictors of perceived stress in the Midlife sample. First, randomly shuffling the values of loneliness substantially diminishes the tree's prediction accuracy. Financial concerns, work, parent role, and time/pressure were relatively similar in conditional permutation importance, not as high as loneliness, but consistently the next highest permutation values. Finally, there is evidence for a small but consistent effect of permuting somatic health complaints, life events, neighborhood strain, work, role restriction, income, and family role. As in the Later Life sample, the predictors identified across many trees using random forests are not completely the same as those in the single selected tree that we examined in more detail above.

Study 3: Combined Sample Analysis

In Study 3, we supplemented the data from Study 2 with data from a wave in which the Later Life sample had completed the same measures. This allowed us to examine a larger set of contextual stress predictors than in

Study 1 as well as to explore the possibility of age interactions among the variables examined in Study 2.

Regression Tree

A regression tree on the combined sample is displayed in Figure 14.3. The CART software nominated a tree with loneliness splitting the sample twice and producing only three terminal nodes. Although this tree was nominated by the program, a number of other trees fell within one standard error of this tree and so could be equally good at describing the data but may be more informative to our interests. From our work in Study 1 and 2, we already had an idea that loneliness was a predictor involved in many interactions, as indicated by its appearance near the top of the tree figures, and that it was an influential predictor across many trees, as indicated by the random forests analysis. Therefore, we examined other trees that CART produced and selected the tree that had the lowest cross-validated relative error (e.g., 0.71 compared to 0.74 in the CART-nominated tree). This tree has 14 terminal nodes, ranging from 14 to 157 participants in the terminal nodes. It is complex, but not the most complex of the available choices. Loneliness, time pressure, work, family, role inoccupancy, finances, somatic health, life events, and age appear as predictors in the selected tree.

The root node at the top of Figure 14.33 contains the entire combined sample ($N = 666$). The predictors and outcome (i.e., global perceived stress as measured by the Perceived Stress Scale) have been standardized so that the cutoff values and node means can be interpreted relative to the sample averages. Loneliness was the first splitter selected by the tree. We begin by following the paths to the left, which can be followed to most of the lowest perceived stress scores, then return to the root node and discuss the interactions involved in the branches to the right.

Those in Node 2, with loneliness just under half (.42) a standard deviation and below the mean (i.e., moderate and low levels of loneliness), report perceived stress that is −.31 standard deviations below the sample average. This node represents a large portion of the sample ($N = 472$) and can be further subdivided by loneliness. Those with very low loneliness (−.43 standard deviations; Node 3) report levels of perceived stress that are about a half a standard deviation below the sample average. These individuals are further divided by their reports of time pressure, as assessed by the Wheaton Chronic Stress subscale. Those reporting low time pressure form Terminal Node A which has the lowest perceived stress (−.77 standard deviations below sample mean). Those with higher time pressure form Node 4 can be subdivided by their recent life events. Individuals reporting relatively few life events in the last year (Terminal Node B) report perceived stress about one half a standard deviation below the sample mean, whereas those with more life events (Terminal Node C) report levels of perceived stress near the sample average.

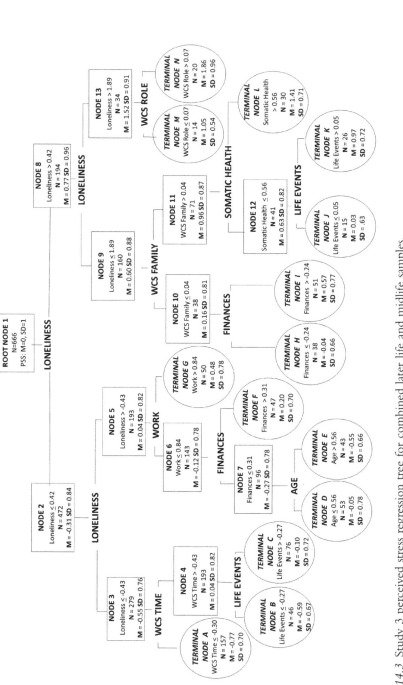

Figure 14.3 Study 3 perceived stress regression tree for combined later life and midlife samples.

Notes: Root and child nodes are displayed in boxes; terminal nodes are displayed in circles. Splitting variables are indicated at the intersection of branching lines. For each node, cut-point on the splitting variable, number of participants in that node, and perceived stress mean and standard deviation are reported. PSS: Mean Perceived Stress Scale for node; WCS FAMILY: family role, WCS INCOME: financial strain, WCS TIME: time/pressure, WCS ROLE: role inoccupancy/restriction. All variables listed in Figure 14.4 were included as predictors; those selected by the algorithm are displayed in the tree.

Node 5 represents those participants who reported low and moderate levels of loneliness (between −.43 and .42 standard deviations from the sample average, formed by the repeated splits on loneliness). They can be further differentiated by their reports of work stress. Those with high levels of work stress (.84 standard deviations above average) form Terminal Node G and report levels of perceived stress that are nearly one half a standard deviation above the sample mean. Those with lower levels of work stress (Node 6) report lower levels of perceived stress and are further divided by their reports of finances. Again, those with higher levels of concern about finances go to the right and form Terminal Node F, with higher levels of perceived stress than those in Node 7. The Node 7 participants with lower levels of financial concerns can further be split by age. Age operates in the opposite direction to the other predictors here—younger persons from Node 7 (formed by the interactions between loneliness, work, and finances) report average levels of perceived stress (Terminal Node D) whereas older persons (Terminal Node E) report perceived stress about one half a standard deviation below the sample average.

Returning to the root node, we now look at the branches to the right. Those with higher levels of loneliness (Node 8) can further be subdivided using loneliness. Individuals with moderate to very high loneliness (between .42 and 1.89 standard deviations above the mean, Node 9) report levels of perceived stress .60 standard deviations above the average. They can be further differentiated by their family role stress, as assessed by the Wheaton Chronic Stress subscale. Those with average and lower family concerns (Node 10) are split by their reports of finances. Individuals in Terminal Node H report relatively low levels of financial concerns and perceived stress scores around the sample mean, whereas those in Terminal Node I report higher financial concerns and perceived stress .57 standard deviations above average. Returning to the interactions between loneliness and family role stress, those in Node 11 with above-average concerns regarding their family role can be divided by their reports of somatic health problems. Those reporting more somatic health complaints comprise Terminal Node L and report perceived stress levels that are quite high (1.41 standard deviations above the sample mean). On the other hand, those with lower levels of somatic health problems (Node 12) can be split into groups with very different levels of global perceived stress: Terminal Node J, who experienced average or fewer life events and report average levels of perceived stress, and Terminal Node K who report more life events and perceived stress levels nearly one standard deviation above average. Finally, those reporting the highest levels of loneliness (Node 13) can be divided by their reports of role inoccupancy and restriction. Those in Terminal Node M report average or lower levels of role inoccupancy and restriction and global perceived stress scores about one standard deviation above the sample mean. Those in Terminal Node N, in contrast, report more role inoccupancy and restriction concerns and also the highest levels of perceived stress in the sample, 1.86 standard deviations above average.

Random Forests

Loneliness again emerged as the strongest predictor across variations of the sample and predictors (see Figure 14.4). Life events and finances were the next strongest predictors by their conditional permutation importance values, followed by work stress. Time pressure and somatic health complaints

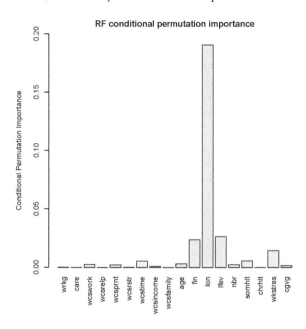

Figure 14.4 Study 3 random forest conditional variable importance results for combined later life and midlife samples.

Notes: Outcome variable: perceived stress scale, Cohen et al. (1983). Wrkg: work status (0: not working, 1: working), categorical. Care: caregiver status (0: not a caregiver, 1–7: caregiver type), categorical. Fin: financial strain, items from National Study of Midlife in the United States (MIDUS; Brim et al., 2007); CHlth: chronic health problems, Belloc et al. (1971); SHlth: somatic health problems, Belloc et al. (1971); Lonl: loneliness, UCLA Loneliness scale, Russell et al. (1980); LifeEv: life events, Elders Life Stress Inventory, Aldwin (1990); Nbr: neighborhood strain, items from Keyes (1998) and Ryff, Magee, Kling, and Wing (1999); Discr: lifetime discrimination experiences, based on Williams, Yu, Jackson, and Anderson (1997); Work: Knox work strain, Knox, Theorell, Svensson, and Waller (1989); CgvBur: caregiver burden, Multidimensional Caregiver Burden Inventory, Novak and Guest (1989). The following predictors are subscales from the Wheaton Chronic Stress Scale (Wheaton, 1991): wcswork: Wheaton Chronic Stress work strain, wcsrelp: Wheaton Chronic Stress relationship problems, wcsprnt: Wheaton Chronic Stress relationship problems subscale, wcsrstr: Wheaton Chronic Stress role restriction, wcstime: Wheaton Chronic Stress time pressure, wcsincome: Wheaton Chronic Stress financial problems, wcsfamily: Wheaton Chronic Stress family role.

ranked next. Caregiving, neighborhood, age, work (as measured by the Wheaton subscale), and parent role also emerged as predictors, but exert a relatively small amount of influence.

Discussion

Regression Trees: an Idiographic Depiction of Contributors to Global Perceptions of Stress in Later Life and Middle Age

Our interests in Studies 1 and 2 were in the specific interactions and variables that described the relationships between contextual life stressors and global perceptions of stress during these life periods. Those results prompted us to examine how a diverse set of predictors may interact with age in a combined sample of middle-aged and older adults. We focus our discussion here on the results regression tree and random forests analysis from Study 3. One way to consider the results of the regression tree analysis is to consider cumulative and compensatory pathways to perceptions of stress (Seroczynski et al., 1997; Wallace et al., 2002). *Cumulative pathways* represent a compounding of risk or, alternatively, protection through the interactions of multiple variables; whereas *compensatory pathways* indicate different ways in which being of low risk on one variable may offset high risk status on another. For example, the splits that lead to Terminal Nodes A, B, L, and N (Figure 14.3) are good examples of cumulative pathways, in which the experience of each of the risk variables results in splits that further differentiate the least (Nodes A and B) and the most (Nodes L and N) stressed groups. For instance, the highest stressed individuals report compounding very high levels of loneliness and role restriction/inoccupancy (Node N) and high loneliness, family role stress, and somatic health problems (Node L). On the least-stressed side, low loneliness or high social integration and support coupled with low time pressure (Node A) or low loneliness and fewer major life events (Node B) related to low levels of perceived stress. Generally speaking, it is probably not surprising that greater severity of stressors accompanies higher perceptions of being stressed and vice versa. Of greater interest are the compensatory pathways that reside in the center nodes of Figure 14.3. One example of this is subjects in Terminal Node H. Although subjects reported high loneliness relative to the rest of the sample, there may have been something protective about their financial situations that resulted in average levels of global perceptions of stress, but ones that were much lower when compared with Node I, who had levels of perceived stress that were more than a half a standard deviation above the mean. In other words, having below average financial stress may have helped to offset the pervasive influence of loneliness on perceived stress.

Results of the regression tree analysis also provide interesting examples of equifinality, or multiple paths to a single outcome, particularly among

Nodes D, H, and J. In this example, perceptions of stress are at the sample average, but the contexts of their lives contribute differently to individuals' feelings of stress. Intervention at the level of social isolation, for example, would benefit individuals in Nodes H and J more so than those in Node D, who appraise their social opportunities as adequate. Although individuals in Nodes H and J report high levels of loneliness, low family and financial stress (Node H) and low somatic health concerns or major life events may help to compensate for the lack of social integration—a good example of how individuals with similar levels of stress have different predictive risks. If we had relied on a simpler model of additive effects or chosen only to examine loneliness as a predictor given the strength of its correlation with global stress appraisal, we might have missed the possible offsetting influences of experiencing relatively low family demands and income concerns or low somatic health problems and few major life events in the context of social isolation.

Multifinality is exemplified by the myriad perceived stress outcomes that are associated with specific life stressors. The best example is loneliness. For some individuals, being socially isolated relates to high levels of stress (e.g., individuals in Node N have perceived stress levels that are almost two standard deviations above the mean) whereas for others, (Terminal Node J), there is little or no effect. Thus, for any given stressor identified, similar levels of exposure, may lead to different levels of influence on the outcome of interest. Using EDM, it is possible to assess combinations of stressors and to evaluate their cumulative and compensatory pathways. This approach helps researchers to better understand why particular levels of a stressor differentially relate to perceptions of stress and how to better harmonize intervention and prevention strategies.

Random Forests: Nomothetic Understanding of Global Perceptions of Stress

The more idiographic information provided by the individual trees selected in Figures 14.1, 14.2, and 14.3 can be useful in reminding researchers and theorists of the complex nature of individuals' lives. Rarely does an individual experience only one form of stress or strain at a given time, instead the trees give examples of the multiple demands that adults face and how the confluence of these may contribute to a person's general assessment of how stressful their life is. This complexity, however, does not provide clear information about which predictors may be particularly important across persons (see the limitations discussed above). To understand the life conditions that are tied to how generally stressful individuals perceive their lives to be, while still considering these complex interactions with other predictors, many varied trees must be examined.

In all of the studies highlighted here, random forests analysis underscored the importance of loneliness/social isolation. Much research has shown that social dynamics have complex relationships with stress, serving as both a source of the stress and as a coping or resilience mechanism. It is clear from random forests analyses, however, that the lack of social interactions is especially detrimental, and is related to elevated perceptions of stress. In comparison to the other variables included, loneliness was the most influential predictor for perceived stress in all three studies. Financial strain was also identified as a key predictor in the forests results for all the studies. In the Midlife-only sample, role strains were also addressed and several of these (e.g., work, parenting, time pressure) were also influential in predicting global perceptions of stress. With the inclusion of a broader age range in Study 3, time pressure and work remained as important predictors and life events also emerged.

Age Effects

One area of interest in combining the samples was to assess age interactions that importantly influenced perceptions of stress in mid- and later life. Interestingly, when age was allowed to enter the regression tree analysis, only one split based on age was identified for the selected tree. Furthermore, the random forests analysis did not identify age as a particularly important contributor to perceived stress. Another approach to understanding how age relates to perceptions of stress is to look at node membership. Not surprising, Node E was entirely comprised of older adults. Beyond that, however, several other nodes had high percentages of individuals from the older cohort (Node A = 62.4%; Node B =73.9%; Node F = 63.4%, and Node J = 60%). These findings suggest that although age per se did not show high importance as a splitter, older adults tended to have similar patterns of chronic stress and life events that contribute to their perceptions of stress, and, generally speaking, older adults were more likely to be in terminal nodes that represented average to below average levels of stress. Other factors (e.g., gender, ethnicity/race, education, socioeconomic status) may show similar patterns. Thus, another benefit of regression trees is that it is possible to study the qualitative or descriptive characteristics of individuals in the various nodes. For the interested reader, Study 1 (Scott, et al., 2011) included excerpts of "life stress" interviews to further elucidate differences associated with node membership.

Conclusion

The response to our use of EDM techniques for applied social science research has been overwhelmingly positive. Briefly, we comment on our experiences presenting these EDM techniques and results to substantive audiences in hopes that this may be useful to other applied researchers interested in using these techniques. Audiences at national conferences focused on adult development as well as exploratory (i.e., primarily qualitative) methods have been encouraging regarding our use of these techniques. Particularly for developmental researchers, the branching tree is a familiar metaphor for the possible pathways to adaptive and maladaptive outcomes (see Sroufe, 1997). The graphical presentation of the regression tree results, then, is extremely attractive and intuitive to many researchers working from this framework. Our interactions with qualitative researchers suggests that they were particularly excited about the possibilities for analyzing in a more purposefully exploratory way the survey data which they often collect with their more open-ended interviews and other traditionally-qualitative data.

In response to this enthusiasm, we want to not only highlight the benefits of this approach, but to emphasize its limitations as well. First, and particularly to those interested in development, the individual tree figures may look like causal flow-charts and seemingly fit theoretical depictions of risk and resilience processes. Our applications of them, however, use cross-sectional data that cannot speak to cause or even temporal precedence. Thus, we have attempted to be cautious in our language and describe the members of particular nodes as people with similar levels of global perceived stress who have certain combinations of strains rather than the presence of chronic strains in domain X1 and X2 leading to someone perceiving their life generally as particularly stressful. Future work using EDM with longitudinal data could examine how specific series of life experiences may be related to a later well-being outcome or the ideas of cumulative stress. These could be explored using techniques such as those outlined in this volume (chapter 9) examining sequences of life events (i.e., leaving parents' home, cohabiting with partner, living alone) as related to individual difference factors such as gender and cohort (Widmer & Ritschard, 2009). Second, we chose to use a combination of regression trees and random forests. This allowed us to let the selected trees help us to entertain the theoretical ideas of pathways, equifinality, and multifinality and stretch our thinking on how these multiple—and dynamic, though not assessed as such in these studies— stressors might be examined in future studies. By incorporating random forests we were able to further emphasize the limitations of individual regression trees and gain additional information about variable importance, which our audiences often assumed was available in the selected regression trees by simply finding the first splitter. Third, in presentations and manuscripts, we have attempted to make clear a tenet familiar to qualitative researchers: exploratory research does not mean unsystematic research. Indeed, the more exploratory the approach, the stronger the need for documenting the decisions that produced the results. The commercial software is extremely easy to use in comparison to a more script-driven interface; both, however, involve decisions that set the constraints underlying the analysis. Finally, although we acknowledge that our samples are rather small in comparison to the typical tree-based analysis applications, using techniques such as cross-validation and also considering random forests results can preclude findings that capitalize on chance characteristics of the sample.

With these cautions in mind, the results of these studies add valuable perspectives on perceptions of stress in adulthood. Perceived stress is a commonly used predictor, but little research has examined the role of contextual factors in global perceptions of stress. In all three studies, the regression trees give examples of the intricate combinations of life strains that are related to individual differences in global perceptions of stress. The selected trees also point out different patterns of interactions and predictors related to similar levels of the outcome. Further, the random forests results indicate that although loneliness is a powerful predictor in both studies, other life strains matter as well. Preventive programs that focused on a certain threshold on a single predictor, such as loneliness, could miss those participants who have high levels of perceived stress related to their other life domains or who compensate for life stress in one domain with low stress in others. On the other hand, attempting to target individuals for intervention based solely on their levels of their global levels of perceived stress or lack thereof, may be ineffective if it is presumed that a single set of life conditions contributed to this state. These points are important for intervention and prevention research to improve quality of life by examining individuals' multiple burdens and resources. Further, they direct stress theory and research to incorporate and address this complexity.

References

Aanes, M. M., Middlemark, M. B., & Hetland, J. (2010). Interpersonal stress and poor health: The mediating role of loneliness. *European Psychologist, 15*(1), 3–11.

Aldwin, C. M. (1990). The Elders Life Stress Inventory: Egocentric and non-egocentric stress. In A. M. P. Stephens, J. H. Crowther, S. E. Hobfoll, & D. L. Tennenbaum (Eds.), *Stress and coping late life families* (pp. 49–69). New York: Hemisphere.

Almeida, D. M., & Horn, M. C. (2004). Is daily life more stressful during middle adulthood? In D. Almeida (Ed.), *How healthy are we? A national study of well-being at midlife* (pp. 425–451). Chicago, IL: University of Chicago Press.

Aneshensel, C. S., Pearlin, L. I., Mullan, J. T., Zarit, S. H., & Whitlatch, C. J. (1995). *Profiles in caregiving: The unexpected career*. San Diego, CA: Academic Press, Inc.

Antonucci, T. C., Akiyama, H., & Merline, A. (2001). Dynamics of social relationships in midlife. In M. E. Lachman (Ed.), *Handbook of midlife development* (pp. 571–598). Hoboken, NJ: John Wiley & Sons, Inc.

Baltes, P. B. (1979). Life-span developmental psychology: Some converging observations on history and theory. In P. B. Baltes & O. G. Brim, Jr. (Eds.), *Life-span development and behavior.* (Vol. 2, pp. 256–279). New York: Academic Press.

Bellingrath, S., Weigl, T., & Kudielka, B. M. (2009). Chronic work stress and exhaustion is associated with higher allostatic load in female school teachers. *Stress, 12*(1), 37–48.

Belloc, N. B., Breslow, L., & Hochstim, J. R. (1971). Measurement of physical health in a general populations survey. *American Journal of Epidemiology*, *93*, 328–336.

Bergeman, C. S. (1997). *Aging: Genetic and environmental influences.* Thousand Oaks, CA: Sage.

Berk, R. A. (2008). *Statistical learning from a regression perspective.* Springer: New York.

Berkman, L. F., Leo-Summers, L., & Horwitz, R. I. (1992). Emotional support and survival after myocardial infarction. *Annals of Internal Medicine*, *117*, 1003–1009.

Breiman, L. (2001). Random forests. *Machine Learning*, *45*, 5–32.

Brim, O. G., Baltes, P. B., Bumpass, L. L., Cleary, P. D., Featherman, D. L., Hazzard, W. R., … Shweder, R. A. (2007). *National Survey of Midlife Development in the United States (MIDUS), 1995–1996.* Ann Arbor, MI: Inter-University Consortium for Political and Social Research.

Brown, G. W., & Harris, T. O. (1978). *Social origins of depression: A study of psychiatric disorder in women.* New York: Free Press.

Brown, G. W., & Harris, T. O. (1989). *Life events and illness.* New York: Guilford Free Press.

Buddeberg-Fischer, B., Stamm, M., Buddeberg, C., & Klaghofer, R. (2010). Chronic stress experience in young physicians: Impact of person- and workplace-related factors. *International Archives of Occupational and Environmental Health*, *83*, 373–379.

Cacioppo, J. T., Hawkley, L. C., & Berntson, G. G. (2003). The anatomy of loneliness. *Current Directions in Psychological Science*, *12*(3), 71–74.

Clémence, A., Karmaniola, A., Green, E. G. T., & Spini, D. (2007). Disturbing life events and wellbeing after 80 years of age: A longitudinal comparison of survivors and the deceased over five years. *Aging & Society*, *27*, 195–213.

Cohen, S., Kamarck, T., & Mermelstein, R. (1983). A global measure of perceived stress. *Journal of Health and Social Behavior*, *24*, 385–396.

Cohen, S., Tyrrell, D. A., & Smith, A. P. (1993). Negative life events, perceived stress, negative affect, and the susceptibility to the common cold. *Journal of Personality and Social Psychology*, *64*, 131–140.

Cohen, S., & Williamson, G. M. (1991). Stress and infectious disease in humans. *Psychological Bulletin*, *109*, 5-24.

Dannefer, D. & Sell, R. R. (1988). Age structure, the life course and "aged heterogeneity": Prospects for research and theory. *Comprehensive Gerontology–B*, *2*, 1–10.

De Beurs, E., Beekman, A., Geerlings, S., Deeg, D., van Dyck, R., & van Tilburg, W. (2001). On becoming depressed or anxious in late life: Similar vulnerability factors but different effects of stressful life events. *The British Journal of Psychiatry*, *179*, 426–431.

Doane, L. D., & Adam, E. K. (2010). Loneliness and cortisol: Momentary, day-to-day, and trait associations. *Psychoneuroendocrinology*, *35*, 430–441.

Easterlin, R. A. (2006). Life cycle happiness and its sources: Intersections of psychology, economics, and demography. *Journal of Economic Psychology*, *27*, 463–482.

Epel, E. S., Blackburn, E. H., Lin, J., Dhabhar, F. S., Adler, N. E., Morrow, J. D., & Cawthon, R. M. (2004). Accelerated telomere shortening in response to life stress. *Proceedings of the National Academy of Sciences*, *101*, 17312–17315.

Gianaros, P. J., Jennings, J. R., Sheu, L. K., Geer, P. J., Kuller, L. H., & Matthewes, K. A. (2007). Prospective reports of chronic life stress predict decreased grey matter volume in the hippocampus. *Neuroimage, 35*, 795–803.

Glaser, R., Kiecolt-Glaser, J. K., Speicher, C. E., & Holliday, J. E. (1985). Stress, loneliness, and changes in herpes virus latency. *Journal of Behavioral Medicine, 8*(3), 249–260.

Golden-Kreutz, D. M., Thornton, L. M., Gregorio, S. W.-D., Frierson, G. M., Jim, H. S., Carpenter, K. M., Shelby, R. A., & Anderson, B. L. (2005). Traumatic stress, perceived global stress, and life events: Prospectively predicting quality of life in breast cancer patients. *Health Psychology, 24*, 288–296.

Gruenewald, T. L., Mroczek, D. K., Ryff, C. D., & Singer, B. H. (2008). Diverse pathways to positive and negative affect in adulthood and later life: An integrative approach using recursive partitioning. *Developmental Psychology, 44*(2), 330–343.

Gruenewald, T. L., Seeman, T. E., Ryff, C. D., Karlamangla, A. S., & Singer, B. H. (2006). Combinations of biomarkers predictive of later life mortality. *Proceedings of the National Academy of Sciences of the United States of America, 103*(38), 14158–14163. doi:10.1073/pnas.0606215103

Hardy, S. E., Concato, J., Gill, T. M. (2002). Resilience of community-dwelling older persons. *Journal of the American Geriatrics Society, 52*, 257–262.

Hastie, T., Tibshirani, R., & Friedman, J. (2001). *Elements of statistical learning: Data mining, inference, and prediction.* Springer: New York.

Hawkley, L. C., Burleson, M. H., Berntson, G. G., & Cacioppo, J. T. (2003). Loneliness in everyday life: Cardiovascular activity, psychosocial context, and health behaviors. *Journal of Personality and Social Psychology, 85*, 105–120. doi:10.1037/0022-3514.85.1.105

Holmes, T. H., & Rahe, R. H. (1967). The Social Readjustment Rating Scale. *Journal of Psychosomatic Research, 11*, 213–218.

House, J. S., Landis, K. R., & Umberson, D. (1988). Social relationships and health. *Science, 241*, 540–545.

Kessing, L. V., Agerbo, E., & Mortensen, P. B. (2003). Does the impact of major stressful life events on the risk for developing depression change throughout life? *Psychological Medicine, 33*, 1177–1184.

Keyes, C. L. M (1998). Social well-being. *Social Psychology Quarterly, 61*(2), 121–140. doi:10.2307/2787065

Kiecolt-Glaser, J. K., Garner, W., Speicher, C., Penn, G. M., Holliday, J., & Glaser, R. (1984). Psychosocial modifiers of immunocompetence in medical students. *Psychosomatic Medicine, 46*(1), 7–14.

Knox, S. S., Theorell, T., Svensson, J. C., & Waller, D. (1985). The relation of social support and working environment to medical variables associated with elevated blood pressure in young males: A structural model. *Social Science & Medicine, 21*(5), 525–531. doi:10.1016/0277-9536(85)90036-X

McEwen, B. S. (1988). Stress, adaptation, and disease: Allostasis and allostatic load. *Annals of the New York Academy of Sciences, 840*, 33–44.

McEwen, B. S., & Seeman, T. (1999). Protective and damaging effects of mediators of stress: Elaborating and testing the concepts of allostasis and allostatic load. *Annals of the New York Academy of Sciences, 896*, 30–47.

Meltzer, H., Bebbington, P., Brugha, T., Jenkins, R., McManus, S., & Stansfeld, S. (2010). Job insecurity, socio-economic circumstances and depression. *Psychological Medicine, 40*, 1401–1407.

Merkle, E. C., & Shaffer, V. A. (2011). Binary recursive partitioning: Background, methods, and application to psychology. *British Journal of Mathematical and Statistical Psychology, 64*, 161–181.

Miller, T. W. (1993). The assessment of stressful life events. In L. Goldberger and S. Breznitz (Eds.), *Handbook of stress: Theoretical and clinical aspects* (pp. 161–173). New York: Free Press.

Mioshi, E., Bristow, M., Cook, R., & Hodges, J. R. (2009). Factors underlying caregiver stress in frontotemporal dementia and Alzheimer's disease. *Dementia and Geriatric Cognitive Disorders, 27*, 76–81.

Mitsonis, C. I., Potagas, C., Zervas, I., & Sfagos, K. (2009). The effects of stressful life events on the course of multiple sclerosis: A review. *International Journal of Neuroscience, 3*, 315–335.

Monroe, S. M. (2008). Modern approaches to conceptualizing and measuring human life stress. *Annual Review of Clinical Psychology, 4*, 33–52.

Monroe, S. M., & Simons, A. D. (1991). Diathesis-stress theories in the context of life stress research: Implications for the depressive disorders. *Psychological Bulletin, 110*, 406–425.

Monroe, S. M., Slavich, G. M., Torres, L. D., & Gotlib, I. H. (2007). Severe life events predict specific patterns of change in cognitive biases in major depression. *Psychological Medicine, 37*, 863–871.

Murray, T. L., Daniels, M. H., & Murray, C. E. (2006). Differentiation of self, perceived stress, and symptom severity among patients with fibromyalgia syndrome. *Families, Systems, & Health, 24*(2), 147–159.

Nelson, E. A., & Dannefer, D. (1992). Aged heterogeneity: Fact or fiction? The fate of diversity in gerontological research. *Gerontologist, 32*, 17–23.

Neugarten, B. L. (1979). Time, age, and the life cycle. *The American Journal of Psychiatry, 136*, 887–894.

Novak, M., & Guest, C. (1989). Application of a multidimensional caregiver burden inventory. *Gerontologist, 29*, 798–803.

Palmore, E., Cleveland, W. P., Nowlin, J. B., Ramm, D., & Siegler, I. C. (1979). Stress and adaptation in later life. *Journal of Gerontology, 34*, 841–851.

Pearlin, L. I., & Skaff, M. M. (1996). Stressors and adaptation in late life. In M. Gatz (Ed.), *Emerging issues in mental health and aging* (pp. 97–123). Washington, D.C.: American Psychological Association.

Penninx, B. W., van Tilburg, T., Kriegsman, D. M., Deeg, D. J., Boeke, A. J., & van Eijk, J. (1997). Effects of social support and personal coping resources on mortality in older age: The longitudinal aging study Amsterdam. *American Journal of Epidemiology, 146*(6), 510–519.

Piazza, J. R., Charles, S. T., & Almeida, D. M. (2007). Living with chronic health conditions: Age differences in affective well-being. *Journals of Gerontology: Psychological Sciences, 62*, P313–P321.

R Development Core Team (2012). *R: A Language and Environment for Statistical Computing.* Vienna, Austria: R Foundation for Statistical Computing.

Rafanelli, C., Roncuzzi, R., Milaneschi, Y., Tomba, E., Colistro, M.C., Pancaldi, L. G., & Di Pasquale, G. (2005). Stressful life events, depression and demoralization as risk factors for acute coronary heart disease. *Psychotherapy and Psychosomatics, 74*, 179–184.

Rosengren, A., Hawken, S., Ounpuu, S., Sliwa, K., Zubaid, M., Almahmeed, W. A., Blacket, K. N., Sitthi-amorn, C., Sato, H., & Yusuf, S. (2004). Association of psychosocial risk factors with risk of acute myocardial infarction in 11119 cases and 13648 controls from 52 countries (the INTERHEART study): Case-control study. *The Lancet, 364,* 11–17.

Russell, D., Peplau, L. A., & Curtona, C. (1980). The revised UCLA loneliness scale: Concurrent and discriminant validity evidence. *Journal of Personality and Social Psychology, 39*(3), 472–480.

Ryff, C. D., Magee, W. J., Kling, K. C., & Wing, E. H. (1999). Forging macro–micro linkages in the study of psychological well-being. In C. D. Ryff & V. W. Marshall (Eds.), *The self and society in aging process* (pp. 247–278). New York: Springer-Verlag.

Saban, K. L., Sherwood, P. R., DeVon, H. A., & Hynes, D. M. (2010). Measures of psychological stress and physical health in family caregivers of stroke survivors: A literature review. *Journal of Neuroscience Nursing, 42,* 128–138.

Schlotz, W., Hellhammer, J., Schultz, P., & Stone, A. A. (2004). Perceived work overload and chronic worrying predict weekend–weekday differences in the cortisol awakening response. *Psychosomatic Medicine, 66,* 207–214.

Scott, S. B., Bergeman, C. S., Whitehead, B., & Pitzer, L. (2013). Combinations of stressors in midlife: Examining role and domain stressors using regression trees and random forests. *Journals of Gerontology, Series B: Psychological and Social Sciences.* Advance online publication. doi: 10.1093/geronb/gbs166

Scott, S. B., Jackson, B. R., & Bergeman, C. S. (2011). Sorting out stress: A CART analysis. *Psychology and Aging, 26,* 830–843.

Scott-Marshall, H., & Tompa, E. (2011). The health consequences of precarious employment experiences. *Work: A Journal of Prevention Assessment & Rehabilitation, 38,* 369–382.

Seematter-Bagnoud, L., Karmaniola, A., & Santos-Eggimann, B. (2010). Adverse life events among community-dwelling persons aged 65–70 years: Gender differences in occurrence and perceived psychological consequences. *Social Psychiatry and Psychiatric Epidemiology, 45,* 9–16.

Seroczynski, A. D., Cole, D. A., & Maxwell, S. E. (1997). Cumulative and compensatory effects of competence and incompetence on depressive symptoms in children. *Journal of Abnormal Psychology, 106*(4), 586–597.

Sliwinski, M. J., Almeida, D. M., Smyth, J., & Stawski, R. S. (2009). Intraindividual change and variability in daily stress processes: Findings from two measurement-burst diary studies. *Psychology and Aging, 24,* 828–840.

Sroufe, A. L. (1997). Psychopathology as an outcome of development. *Development and Psychopathology, 9,* 251–268.

Stawski, R. S., Sliwinski, M. J., Almeida, D. M., & Smyth, J. (2008). Reported exposure and emotional reactivity to daily stressors: The roles of adult age and global perceived stress. *Psychology and Aging, 23,* 52–61.

Steinberg, D., & Colla, P. (1997). CART: Classification and regression trees. San Diego, CA: Salford Systems.

Steinberg, D., & Golovyna, M. (2006). *CART 6.0 User's Manual.* San Diego, CA: Salford Systems.

Steptoe, A., Owen, N., Kunz-Ebrecht, S. R., & Brydan, L. (2004). Loneliness and neuroendocrine, cardiovascular, and inflammatory stress responses in middle-aged men and women. *Psychoneuroendocrinology, 29,* 593–611.

Strobl, C., Boulesteix, A., Kneib, T., Augustin, T., & Zeileis A. (2008). Conditional variable importance for random forests. *BMC Bioinformatics, 9*(307). http://www.biomedcentral.com/1471-2105/9/307

Strobl, C., Boulesteix, A., Zeileis, A., & Hothorn, T. (2007). Bias in random forest variable importance measures: Illustrations, sources and a solution. *BMC Bioinformatics, 8*(25). http://www.biomedcentral.com/1471-2105/8/25

Strobl, C., Hothorn, T., & Zeileis, A. (2009). Party on! A new, conditional variable-importance measure for Random Forests available in the party package. *R Journal, 1/2*, 14–17.

Strobl, C., Malley, J., & Tutz, G. (2009). An introduction to recursive partitioning: Rationale, application, and characteristics of classification and regression trees, bagging, and random forests, *Psychological Methods, 14*, 323–348.

Tak, S. H., Hong, S. H., & Kenneday, R. (2007). Daily stress in elders with arthritis. *Nursing and Health Sciences, 9*, 29–33.

Thoits, P. (2010). Stress and health: Major findings and policy implications. *Journal of Health and Social Behavior, 51*, S41–S53. doi: 10:1177/0022146510383499

Turner, R., Wheaton, B., & Lloyd, D. A. (1995). The epidemiology of social stress. *American Sociological Review, 60*, 104–125.

van Eck, M., Nicholson, N. A., & Berkhof, J. (1998). Effects of stressful daily events on mood states: Relationship to global perceived stress. *Journal of Personality and Social Psychology, 75*, 1572–1585.

von Bertalanffy, L. (1969) *General systems theory: Foundations, development applications.* George Brazziller: New York.

Wallace, K. A., Bergeman, C. S., & Maxwell, S. E. (2002). Predicting well-being outcomes in later life: An application of classification and regression tree (CART) analysis. In S. P. Shohov (Ed.), *Advances in psychology research* (pp. 71–92). Huntington, NY: Nova Science Publishers.

Wheaton, B. (1983). Stress, personal coping resources, and psychiatric symptoms: An investigation of interactive models. *Journal of Health and Social Behavior, 24*, 208–229.

Wheaton, B. (1994). Sampling the stress universe. In W. Avison and I. Gotlib (Eds.), *Stress and mental health* (pp. 77–114). New York: Plenum.

Wheaton, B. (1997). The nature of chronic stress. In B. H. Gottlieb (Ed.), *Coping with chronic stress* (pp. 48–73). New York: Plenum.

Whitbourne, S. K. (2001). The physical aging process in midlife: Interactions with psychological and sociocultural factors. In M. Lachman (Ed.), *Handbook of midlife development* (pp. 109–155). New York: John Wiley & Sons, Inc.

Widmer, E., & G. Ritschard (2009). The de-standardization of the life course: Are men and women equal? *Advances in Life Course Research, 14*(1–2), 28–39.

Wigers, S. H. (1996). Fibromyalgia outcome: The predictive values of symptom duration, physical activity, disability pension, and critical life events—A 4.5 year prospective study. *Journal of Psychosomatic Research, 41*(3), 235–243.

Williams, D. R., Yu, Y., Jackson, J. S., & Anderson, N. B. (1997). Racial differences in physical and mental health: Socio-economic status, stress and discrimination. *Journal of Health Psychology, 2*, 335–351. doi:10.1177/135910539700200305

Yen, I. H., Michael, Y. L., & Perdue, L. (2009). Neighborhood environment in studies of health in older adults: A systematic review. *American Journal of Preventive Medicine, 37*, 455–463.

15 Recursive Partitioning to Study Terminal Decline in the Berlin Aging Study

Paolo Ghisletta

Introduction

In the past two decades, much research has focused on the terminal decline hypothesis of cognitive performance in very old age. Only in recent years, however, has this hypothesis been assessed with advanced methodologies. One such proposal used the joint longitudinal+survival model to (a) assess cognitive performance and change therein, across a wide set of cognitive tasks, and (b) assess how cognitive performance and particularly change therein could predict survival in a sample of older individuals (Ghisletta, McArdle, & Lindenberger, 2006). In this chapter we expand on that inquiry by making use of recent data mining techniques that more aptly consider the concomitant effects of a large number of predictors and rely on resampling techniques to achieve robust results. Prediction survival trees and random survival forests are used to study the effect of cognition on survival in the Berlin Aging Study.

The hypothesis of terminal decline states that a long-term, moderate decline in cognitive performance may predict mortality (Kleemeier, 1962). This hypothesis has received wide attention in the cognitive aging literature and current evidence converges to a number of conclusions: (a) Cognitive functioning can predict survival, especially in samples of older individuals; (b) This link holds over very long periods of time (e.g., cognitive performance in childhood has been shown to predict survival in old age); (c) This association remains when a number of potentially confounding variables are controlled for (i.e., several indicators of physical and medical conditions, socio-economic status, personality characteristics, etc. are not fully multicollinear with cognition in predicting survival). For reviews, see for instance (Bäckman & MacDonald, 2006b; Bosworth & Siegler, 2002).

A limited number of issues remain however open to discussion (cf. Bäckman & MacDonald, 2006a). The first concerns whether the cognition–mortality association is limited to a few, specific cognitive abilities or is more general across a wide spectrum of cognitive capacities (specificity vs. pervasiveness dilemma). To this date, the empirical evidence in this regard does not converge to a single answer. Second, while most extant studies have focused on the general level of cognitive functioning, only a few have analyzed the relationship between *change* in cognitive performance and mortality. This second type of association, however, seems more appropriate when investigating the terminal decline hypothesis than when focusing on the level of cognitive performance. Third, the amount of empirical evidence that allows researchers to compare a relatively wide set of cognitive tasks and that investigates both level of and change in performance remains to date very scarce (Bosworth & Siegler, 2002). The main substantive purpose of this chapter is to further elucidate these theoretical points by using recent data mining techniques that allow testing of the predictability of the large sets of covariates, even with small sample sizes.

As so often is the case, the divergence in empirical extant results can be partially or largely explained by the differing methodologies and statistical models employed. Most commonly, a single assessment of cognitive functioning is performed. The assessment varies in its generality, from studies with only a single cognitive task to those administering whole psychometric batteries of cognition, where various subdomains of intelligence are compared with respect to their predictability. After some time (usually at least one year to at most about five years) the survival status of the original sample is assessed. Simple logistic regressions or similar classification methods then allow assessment of the degree of relationship between cognition and survival status. If the exact date of death is available, a survival model (most commonly the Cox proportional hazards, the exponential, or the Weibull) is applied to further elucidate the effect of cognitive performance on survival time. Minimal extensions of this cross-sectional design consist in administering twice the cognitive variables, so as to estimate change in cognitive performance. For instance, if variable Y operationalizes cognitive performance, a simple difference score D between times $time = 0$ and $time = 1$ for each individual i is commonly obtained by a simple subtraction ($D_i = Y_{1i} - Y_{0i}$). These difference scores are then used as predictors in the selected survival model. Note however that such difference scores are not adjusted for unreliability in the measure of Y.

Henderson, Diggle, and Dobson (2000) presented the Joint Longitudinal + Survival model to maximize the statistical efficiency in similar designs. In short, this approach allows simultaneous estimation of the parameters describing the longitudinal process, represented by the individual repeated cognitive assessments, and those of the survival component. The model integrates the classical random–effects model for longitudinal data of Laird and Ware (1982) and conditions the survival component upon the individual characteristics of the longitudinal process. Ghisletta et al. (2006) and Ghisletta (2008) have applied this model to study the terminal decline hypothesis in two independent samples of very old individuals.

Following Henderson et al. (2000), the joint distribution of measurement and events for any subject i is modeled through an unobserved, latent bivariate Gaussian process $W_i(time) = \{W_{1i}(time), W_{2i}(time)\}$, realized independently across subjects. This latent process drives two linked sub-models. First, the measurement sub-model determines the sequences of measurements $y_{i1}, y_{i2}, \ldots$ at times $time_{i1}, time_{i2}, \ldots$ as follows:

$$Y_{ti} = \mu_i(time_{ti}) + W_{1i}(time_{ti}) + Z_{ti}, \tag{1}$$

where $\mu_i(time_{ti})$ is the mean response and can be described by a linear model $\mu_i(time) = x_{1i}(time)'\beta_1$ and $Z_{ti} \sim \mathcal{N}(0, \sigma_z^2)$. The second sub-model is the survival, or event intensity, component, modeled by

$$\lambda_i(time) = H_i(time)\alpha_0(time)exp[x_{2i}(time)'\beta_2 + W_{2i}(time)]. \tag{2}$$

The two models are joint if the vectors x_{1i} and β_1 share some elements with the vectors x_{2i} and β_2. Henderson et al. showed that the likelihood of this joint model can be approximated by pseudo maximum likelihood (ML) methods (e.g., Gaussian quadrature). According to the authors, this simultaneous estimation proves more efficient than a two-step procedure, where first the longitudinal model is used to estimate individual random effects, which then are inserted as predictors in the survival model. Guo and Carlin (2004) showed how the joint model can be estimated via pseudo-ML (in SAS NLMIXED; Wolfinger, 1988) and within the Bayesian framework (in WinBUGS; Spiegelhalter, Thomas, Best, & Lunn, 2004). Note that the use of the random-effects model to estimate change allows correcting for unreliability in the measurement of Y.

Previous Analysis of Terminal Decline

Ghisletta et al. (2006) have used this statistical model to address the terminal decline hypothesis in the context of the Berlin Aging Study I (BASE-I; Baltes & Mayer, 1999; Lindenberger, Smith, Mayer, & Baltes, 2010). This study started in 1990 and is still ongoing. At inception, 516 adults (age range 70–103 years, *mean* = 85 years) were assessed. In this application participants were assessed at most 11 times over 13 years on eight cognitive variables measuring four cognitive domains: perceptual speed (assessed by Digit Letter and Identical Pictures), verbal fluency (Categories and Word Beginning), episodic memory (Memory for Text and Paired Associates), and verbal knowledge (Vocabulary and Spot-a-Word); (for details of the cognitive battery see Lindenberger & Baltes, 1997). Participants' survival status was last assessed in December 2004 and by that time 404 of the original participants were deceased.

The joint model was applied by joining the following two components:

$$y_{ti} = (\beta_{11} + u_{1i}) + (\beta_{12} + u_{2i})A_{ti} + \beta_{13}A_{ti}^2$$
$$+ \beta_{14}R1_{ti} + \cdots + \beta_{113}R10_{ti}$$
$$+ \beta_{114}DD_{ti} + \beta_{115}DD_{ti}A_{ti} + \epsilon_{ti}$$

and

$$\lambda_i(A_i) = \sigma A^{\sigma-1} exp[\beta_{21} + \beta_{22}IA_i + \beta_{23}sex_i\beta_{24}SES_i + \beta_{25}SensPerf_i$$
$$+ \beta_{26}MotPerf_i + \beta_{27}PosPers_i + \beta_{28}NegPers_i + \gamma_1u_{1i} + \gamma_2u_{2i}],$$

where A_{ti} is age at time of assessment t of individual i, R1–R10 are 10 binary variables (coding whether the cognitive task has already been presented once, twice, ..., or at most 10 times before any given time point) that estimate retest effects, *DD* is a time-varying dementia diagnostic (indicating whether the individual was likely to be affected by dementia), *IA* is participants' initial age, *sex* is their sex, *SES* their socio-economic status (coded on three levels), *SensPerf* and *MotPerf* indicate their sensory (vision and hearing) and motor (balance and gait) performance, and *PosPers* and *NegPers* are participants' scores on positive (extraversion, openness, positive affect, internal control, optimism, and positive future outlook) and negative (neuroticism, negative affective powerful others, social and emotional loneliness) personality attributes, respectively. Note that both equations contain the terms u_{1i} and u_{2i}, which represent the random effects of the overall level of and linear change in cognitive performance. These random effects, estimated within the longitudinal portion of the overall model, are predictors of mortality in the survival portion of the model and constitute the link between the two components. Of particular interest to the terminal decline hypothesis is the effect of u_{2i}, that is, the effect of the amount of linear change in cognitive performance, on survival. If the hypothesis is tenable, then we expect that individuals with large, negative values on u_{2i}, that is individuals whose cognitive performance deteriorates strongly over the years, are those more likely to die and to do so at earlier ages than individuals with less or no cognitive decline (with shallower or nonnegative u_{2i} values).

When we applied this model (Ghisletta et al., 2006) we faced various computational issues. Because we analyzed the effects of eight different cognitive tasks we could not define a multivariate random-effects model in the longitudinal portion of the joint model. We therefore ran a separate joint model for each cognitive task (however, for a bivariate application of the joint longitudinal+survival model with simultaneous estimation in a similar context see Ghisletta (2008), where the effects of two cognitive tasks on survival were examined concomitantly). This means that we could not compare, within a single joint model, which cognitive tasks appeared to have the strongest survival prediction, given the presence of all other cognitive tasks. To address the specificity vs. pervasiveness dilemma, we therefore also estimated the joint model with the two-step procedure (Guo & Carlin, 2004). First, for each cognitive variable we ran the random-effects model and estimated the random effects u_{1i} and u_{2i} for each individual. In the second step we inserted all estimated random effects (for a total of 16, two per cognitive task) in a Weibull survival model. This two-step

Table 15.1 Estimates of the γ_1 and γ_2 parameters of the joint model for each cognitive task by cognitive domain (numbers in parentheses are standard errors)

Cognitive construct	Cognitive task	γ_1	γ_2
Perceptual speed	Digit Letter	.002 (.002)	.070 (.040)
	Identical Picture	.005 (.003)	−.034 (.042)
Verbal memory	Paired Associates	.001 (.002)	−.005 (.035)
	Memory for Text	.002 (.002)	.019 (.042)
Verbal fluency	Categories	.005 (.002)	.087 (.048)
	Word Beginning	.002 (.002)	.036 (.043)
Verbal knowledge	Vocabulary	−.001(.002)	−.051 (.043)
	Spot-a-Word	.001 (.002)	.015 (.037)

estimation allowed us to compare directly the cognitive tasks with respect to their survival prediction. Although this procedure might have reduced statistical efficiency compared to a simultaneous estimation (Guo & Carlin, 2004), it let us address our theoretical hypotheses of interest and assess which of the cognitive variables was most strongly associated to mortality.

We summarize in Table 15.1 the salient results of the two–step estimation procedure (for complete results see Ghisletta et al., 2006, p. 217). For each cognitive task (column 2) we present the underlying cognitive domain assessed by it (column 1) and the survival predictability of both the level u_{1i} and linear change u_{2i} random effects, represented by the γ_1 and γ_2 parameters in the survival portion of the joint model.

In the two–step multivariate estimation only the level of Categories was predictive of survival. All other cognitive tasks were not associated to survival, neither in their level component (γ_1), nor in their change counterpart (γ_2). That is, when the effects of all cognitive tasks are estimated concomitantly only level in Categories appears to be different from zero. However, when we analyzed each task separately in the joint model with the simultaneous estimation procedure we obtained a significant effect of both level and change of Digit Letter, and of the level of Identical Pictures, Paired Associates, Categories, and Word Beginning. This discrepancy in results could be due to multiple factors. It is likely that the superior efficiency of the simultaneous estimation procedure when compared to the two–step procedure played a crucial role in recovering effects. However, it may also be that the likely correlations among random effects of the different cognitive tasks in the two–step procedure are at the basis of strong statistical control on each single effect. In other words, the random effects of no single cognitive variable could uniquely contribute to the survival prediction, given all other cognitive variables are controlled for (except for level of Categories). Furthermore, note that in all these analyses we controlled for the effects of all covariates mentioned previously, which typically also correlate with cognitive performance and, of course, with age.

To address this issue we show the correlation matrix for initial age and the predictors of the survival model (Matrix 15.1).

Matrix 15.1

```
> library(Hmisc)
> rcorr(sel1.matrix, type="pearson")
        K1ZAGE   sex   ses sensory motor  POSA  NEGA
K1ZAGE    1.00 -0.02 -0.09   -0.67 -0.61 -0.27  0.23
sex      -0.02  1.00  0.17    0.07  0.17  0.09 -0.22
ses      -0.09  0.17  1.00    0.28  0.18  0.21 -0.23
sensory  -0.67  0.07  0.28    1.00  0.58  0.32 -0.32
motor    -0.61  0.17  0.18    0.58  1.00  0.28 -0.35
POSA     -0.27  0.09  0.21    0.32  0.28  1.00 -0.26
NEGA      0.23 -0.22 -0.23   -0.32 -0.35 -0.26  1.00
...
P
        K1ZAGE sex     ses    sensory motor  POSA   NEGA
K1ZAGE         0.6199 0.0450 0.0000  0.0000 0.0000 0.0000
sex     0.6199        0.0000 0.1174  0.0000 0.0353 0.0000
ses     0.0450 0.0000        0.0000  0.0000 0.0000 0.0000
sensory 0.0000 0.1174 0.0000         0.0000 0.0000 0.0000
motor   0.0000 0.0000 0.0000 0.0000         0.0000 0.0000
POSA    0.0000 0.0353 0.0000 0.0000  0.0000        0.0000
NEGA    0.0000 0.0000 0.0000 0.0000  0.0000 0.0000
```

As expected, age correlates quite strongly and negatively with sensory and motor performance, but also slightly negatively with positive personality characteristics and positively with negative personality traits.

Likewise, the estimated random effects of the level components of the cognitive tasks intercorrelate positively (p-values not shown as all were $< .01$) (Matrix 15.2).

Matrix 15.2

```
> rcorr(sel2.matrix, type="pearson")
       lev_dl lev_ip lev_pa lev_mt lev_ca lev_wb lev_be lev_sw
lev_dl   1.00   0.56   0.29   0.25   0.42   0.43   0.33   0.29
lev_ip   0.56   1.00   0.24   0.22   0.40   0.31   0.34   0.33
lev_pa   0.29   0.24   1.00   0.44   0.35   0.42   0.28   0.18
lev_mt   0.25   0.22   0.44   1.00   0.33   0.37   0.30   0.19
lev_ca   0.42   0.40   0.35   0.33   1.00   0.60   0.45   0.28
lev_wb   0.43   0.31   0.42   0.37   0.60   1.00   0.39   0.30
lev_be   0.33   0.34   0.28   0.30   0.45   0.39   1.00   0.56
lev_sw   0.29   0.33   0.18   0.19   0.28   0.30   0.56   1.00
```

The slope components also shared much information (all p-values $< .01$ except for the correlation between the slope components of MT and PA, $p = .03$) (Matrix 15.3).

Matrix 15.3

```
> rcorr(sel3.matrix, type="pearson")
        lslo_dl lslo_ip lslo_pa lslo_mt lslo_ca lslo_wb lslo_be lslo_sw
lslo_dl    1.00    0.55    0.21    0.15    0.39    0.31    0.29    0.27
lslo_ip    0.55    1.00    0.14    0.12    0.27    0.22    0.15    0.32
lslo_pa    0.21    0.14    1.00    0.24    0.20    0.33    0.28    0.13
lslo_mt    0.15    0.12    0.24    1.00    0.31    0.26    0.22    0.10
lslo_ca    0.39    0.27    0.20    0.31    1.00    0.40    0.29    0.14
lslo_wb    0.31    0.22    0.33    0.26    0.40    1.00    0.27    0.22
lslo_be    0.29    0.15    0.28    0.22    0.29    0.27    1.00    0.35
lslo_sw    0.27    0.32    0.13    0.10    0.14    0.22    0.35    1.00
```

Three major conclusions can be drawn from this previous analysis. First, in the joint, univariate analyses using pseudo-ML, level of performance in perceptual speed, verbal memory, and verbal fluency are all related to survival. Hence, the cognition–survival relationship appears to be rather pervasive to multiple cognitive domains, rather than specific to a single one. Second, in joint, univariate analyses based on pseudo-ML only change in perceptual speed performance is related to survival. This lends little support to the terminal decline hypothesis. Lastly, when all cognitive variables are allowed to condition the survival model simultaneously, which is estimable only in the two-stage multivariate analysis, only level in Categories is related to survival. It is not clear at this point whether this last result is due to multicollinearity among the cognitive random effects or simply lack of statistical efficiency. It is however well-established that individual rankings in level of and in change in cognitive performance are associated, especially in older adulthood (e.g., Ghisletta, Rabbitt, Lunn, & Lindenberger, 2012; Lindenberger & Ghisletta, 2009). This appears confirmed by the correlation matrices shown here.

Objectives

A number of questions remain unanswered. At this point the exploration of survival predictability from multiple cognitive tasks is hindered by computational difficulties in estimating a joint longitudinal+survival model multivariately. In the application cited above we would need to estimate in parallel eight joint models, where besides estimating the random effects of each cognitive task (level, linear change, and covariance of the two) all other possible associations among the tasks are also specified and estimated. More precisely, the level and linear change of each cognitive task would correlate with the same terms of all other cognitive tasks. With eight tasks the model would need to estimate 112 additional parameters. At present, this is not feasible with pseudo-ML estimation.

In the present work we extend these analyses by applying a different analytical strategy, namely data mining techniques to further explore the degree of predictability among the cognitive tasks considered, the demographic characteristics, and additional individual psychological features. We rely on recent advances from machine learning and related fields to explore how to predict survival from cognitive performance. In particular, we hope that the multicollinearity issue that was likely present in the multivariate joint model estimated with the two-step procedure will not appear in the context of the data mining techniques we are going to use.

Data Mining

Induction Trees (ITs; also called Classification and Regression Trees, Breiman, Friedman, Olshen, & Stone, 1984) are data mining techniques that recently have gained much popularity in several fields such as genetics, epidemiology, and medicine, where classification is often of chief interest. In the last decade, ITs have also started eliciting interest in the psychology community, where prediction is often the main goal of the analyses. An excellent introduction to this set of techniques, with illustrations, data, computer syntax, and exhaustive references for in-depth reading, is provided by Strobl, Malley, and Tutz (2009).

ITs are built according to recursive partitioning methods, in which, following strict statistical procedures, the overall sample is partitioned into multiple sub-samples that are maximally different, with respect to the outcome, from each other. The observations of any sub-sample share similar, or possibly identical, response values, conditional on a set of classification or prediction variables. Decision trees are typically produced as a graphical result of the partitioning. Such trees usually show at the top the most relevant variable to classify or predict the outcome and that variable's value that best splits the sample into two sub-samples. If the variable is dichotomous the split is naturally made, whereas if it is continuous the most discriminating value is found. Different criteria are available to find the splitting values (e.g., Abellán & Masegosa, 2007; Breiman et al., 1984; O'Brien, 2004; Shih , 1999). In the end, a tree with several leaves (also called nodes), each showing an additional variable with its associated discriminating value, is constituted, at the bottom of which are sub-samples of observations that are no longer possibly split or are stopped by some user-defined criterion. These constitute the final leaves of the tree.

Graphically, the interpretation of such trees is quite intuitive and this major advantage probably increased the popularity of ITs. Moreover, such trees allow identification of complicated interactive effects among the chosen categorization or prediction variables. The estimation of complex interactions is typically difficult or impossible in parametric models such as multiple regressions, because of the high number of parameters needed to do so. Another important advantage of ITs is that they easily adapt to data situations that are arduous for parametric analyses, namely situations in which a large number p of variables ought to be tested in a small sample size n, where $n < p$, the so-called "small n, big p problem" (see Strobl, Malley, & Tutz, 2009 for more detail).

In this application we use ITs to investigate the terminal decline hypothesis in the BASE-I sample. We again predict survival based on indicators of cognitive performance, controlling for demographic information (age, sex, socio-econonomic status), physical functioning (sensory and motor performance), and psychological functioning (positive and negative personality features). The cognitive indicators are the estimated random effects of level and linear change of the longitudinal portion of the joint model estimated with a two-step procedure. We now hope that the correlations among these random effects will not hinder the estimation of their effects on survival.

Single Survival Tree

We start by first examining the single survival tree fit to the whole sample, in which the survival information (the age of death or of the last measurement and the event death) is predicted by all covariates considered previously. We use the package `party` (Hothorn, Hornik, Strobl, & Zeileis, 2012; Hothorn, Hornik, & Zeileis, 2006) of the R system for statistical computing (R Development Core Team, 2011). This package allows, among others, computing trees with a wide variety of advanced options. We use here the default options (e.g., the permutation test framework is used to determine optimal binary split values of covariates). The resulting tree is obtained as follows:

```
> library("party")
> survtree <- ctree(Surv(ADall70,dead) ~ age70 + sexf + ses +
+                   sensory + motor + POSA + NEGA +
+                   lev_dl + lslo_dl + lev_ip + lslo_ip +
+                   lev_pa + lslo_pa + lev_mt + lslo_mt +
+                   lev_ca + lslo_ca + lev_wb + lslo_wb +
+                   lev_be + lslo_be + lev_sw + lslo_sw,
+                   data=sub70)
> print(survtree)

 Conditional inference tree with 6 terminal nodes

Response:  Surv(ADall70, dead)
Inputs:  age70, sexf, ses, sensory, motor, POSA, NEGA, lev_dl,
lslo_dl, lev_ip, lslo_ip, lev_pa, lslo_pa, lev_mt, lslo_mt,
lev_ca, lslo_ca, lev_wb, lslo_wb, lev_be, lslo_be, lev_sw, lslo_sw
Number of observations:  516


1) age70 <= 23.21; criterion = 1, statistic = 170.623
  2) lev_ca <= 57.21666; criterion = 1, statistic = 29.951
    3)* weights = 179
  2) lev_ca > 57.21666
    4) sexf == {1}; criterion = 1, statistic = 19.263
      5)* weights = 119
    4) sexf == {-1}
      6)* weights = 101
```

```
1) age70 > 23.21
  7) age70 <= 29.52; criterion = 1, statistic = 41.526
    8) lev_pa <= 52.57867; criterion = 0.997, statistic = 14.733
      9)*  weights = 66
    8) lev_pa > 52.57867
      10)*  weights = 35
  7) age70 > 29.52
    11)*  weights = 16
```

For simplification we centered both the age of the event (ADall70) and the initial age (age70) around 70 years (hence indicating the number of years since age 70 years). The event variable dead has two values, either 0 for censored data or 1 if the individual deceased. Predictors also include socio-economic status (ses), sensory (sensory) and motor (motor) performance, positive (POSA) and negative (NEGA) personality facets, and, for each cognitive variable considered previously, the overall level (lev_) and the linear change (lslo_) scores, on which individuals vary. The tree is built by applying the permutation test framework of Strasser and Weber (1999), which produces the statistic shown (for details see Hothorn et al., 2012).

We see that the tree distinguishes six situations, in increasing order of survival age:

1   Individuals with age younger than or equal to 93.21 years with a level Categories score lower than or equal to 57.22 (Node 3)
2   Individuals with age younger than or equal to 93.21 years, with a level Categories score superior to 57.22, who are males (Node 5)
3   Individuals with age younger than or equal to 93.21 years, with a level Categories score superior to 57.22, who are females (Node 6)
4   Individuals with age greater than 93.21 but less than or equal to 99.52, with a level Paired Associates score lower than or equal to 52.58 (Node 9)
5   Individuals with age greater than 93.21 but less than or equal to 99.52, with a level Paired Associates score greater than 52.58 (Node 10)
6   Individuals older than 99.52 years (Node 11)

An intuitive graphical representation of the single prediction tree can be obtained with the command

```
> plot(survtree)
```

The resulting tree is depicted in Figure 15.1.

We can obtain the predicted median survival time, that is the age (if we add 70 years) at which the event is predicted to occur for each sub-sample.

The rows correspond to the terminal leaves (or nodes) of the tree, from the one with the smallest (3) to the largest (11) index (going left to right in Figure 15.1).

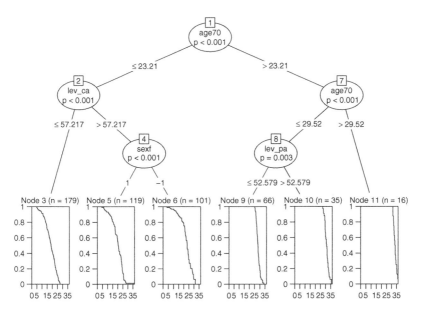

Figure 15.1 Single survival prediction tree with demographic predictors, composite covariates (motor and sensory performance, positive and negative personality facets), and cognitive (intercept and change) predictors.

```
> table(predict(survtree), dead)
             dead
               0   1
 18.736844627 33 146
 20.505092402 29  90
 23.320965092 47  54
 28.242861054  1  65
 31.102621492  2  33
 33.471724846  0  16
```

This single tree is the one that best predicts survival in our sample given the covariates we have chosen. However, it is likely that this tree adapts not only to legitimate characteristics of our sample, but also to the ubiquitous sampling error present in our data. Had we drawn a different sample from the same population, even with the same average characteristics, we would have obtained different combinations of survival information and covariates' values. Thus, a potentially different tree would have resulted from the analysis. Perhaps age would have still resulted as the first node of the tree, but probably the best cutoff value to create two branches would have been different from the one shown above. To avoid the overfit that so easily may occur when analyzing a relatively small data set, *ensemble methods* have been proposed (Breiman, 2001).

Random Forest

The basic principle of *ensemble methods* is to draw a certain number of either bootstap samples (of equal size to the original sample but with replacement) or sub-samples (of smaller size than the original sample but without replacement) and compute a separate single tree on each sub-sample. This usually is achieved in one of two ways: in *bagging* all predictors are considered, as they were in the single tree approach. Given the diversity of the bootstrap samples this approach will produce a set of different trees; in *random forests* different random subsets of predictors are considered in each tree. This will increase even further the heterogeneity of the resulting trees because each tree could potentially be fit to a different bootstrap sample with different variables. In the final step of ensemble methods, the single trees are combined (for instance by means of weighted or unweighted averages) to produce a final unbiased solution that is statistically superior to any single tree (Strobl, Malley, & Tutz, 2009).

The solution of an ensemble method is however not easily depicted because it does not correspond to an average tree with a simple structure. To interpret such a solution, then, we rely on *variable importance* measures, which assess the importance that each predictor has played over all trees of each bootstrap sample or sub-sample. Different criteria can be applied, and some are more desirable than others (for a comparison see Strobl, Boulesteix, Kneib, Augustin, & Zeileis, 2008), but in general they all focus on comparing overall predictions or classification with and without each covariate. Hence, each predictor's importance is assessed as a function of its main effect but also of all of its potentially complicated multivariate interactions with other predictors. The importance value associated to each variable allows comparing, within a single ensemble analysis, of all variables to each other. Note that such measures are not standardized. They do allow comparing variables within a given analysis, but one should not compare importance measures from one ensemble analysis with those of another.

Strobl et al. (Strobl, Boulesteix, Zeileis, & Hothorn, 2007; Strobl et al., 2008) observed that variable importance measures tend to be biased when predictors are of different nature (e.g., a mix of dichotomous and continuous variables), highly correlated, numeric with different values, or even with many missing values. Strobl et al. have therefore developed an unbiased variable selection procedure that is implemented in the `party` package (Hothorn et al., 2012), which also computes ensemble methods (Hothorn, Buehlmann, Dudoit, Molinaro, & Laan, 2006; Strobl et al., 2007, 2008). While in `party` the unbiased selection procedure is available for trees predicting continuous or categorial outcomes, it is not yet implemented for censored outcomes. We hence, for illustration purposes, use the `randomSurvivalForest` package, a different R package that also computes random forests but that allows for variable importance measures for censored outcomes. Note, nevertheless, that in this package the importance measures are not unbiased (Ishwaran & Kogalur, 2007, 2012). Given the different scaling properties of some of our variables (e.g., sex), use of this package is hence not advised (Strobl, Hothorn, & Zeileis, 2009). We used the logrank splitting rule (Ishwaran & Kogalur, 2007), which is different from that used in `party` (a specific conditional statistic).

```
> library("randomSurvivalForest")
> rsf0 <- rsf(Surv(ADall70,dead) ~ age70 + sexf + ses +
+    sensory + motor + POSA + NEGA +
+    lev_dl + lslo_dl + lev_ip + lslo_ip +
+    lev_pa + lslo_pa + lev_mt + lslo_mt +
+    lev_ca + lslo_ca + lev_wb + lslo_wb +
+    lev_be + lslo_be + lev_sw + lslo_sw,
+               data=sub70)
> rsf0

Call:
 rsf(formula = Surv(ADall70, dead) ~ age70 + sexf + ses + sensory +
 motor + POSA + NEGA + lev_dl + lslo_dl + lev_ip + lslo_ip +
 lev_pa + lslo_pa + lev_mt + lslo_mt + lev_ca + lslo_ca + lev_wb +
 lslo_wb + lev_be + lslo_be + lev_sw + lslo_sw, data = sub70)

                       Sample size: 447
                   Number of deaths: 340
                    Number of trees: 1000
          Minimum terminal node size: 3
        Average no. of terminal nodes: 59.931
 No. of variables tried at each split: 4
              Total no. of variables: 23
                     Splitting rule: logrank
              Estimate of error rate: 16.81%
```

The error rate is calculated as $1 - C$, where C is Harrell's concordance index (Harrell, Califf, Pryor, Lee, & Rosati, 1982), a generalization of the area under the ROC curve to measure how well the model discriminates between the two survival predictions. The error rate represents the average proportion of badly classified observations according to their survival function. The closer the error to zero, the higher the prediction accuracy. A value or 50% corresponds to random prediction or classification. This index is based on classifying not observations that were included in a given tree, but so-called out-of-bag observations, which were not part of the sub-sample analyzed by a given tree (Ishwaran & Kogalur, 2007). Here, the error rate is both quite far from the random-prediction value of 50% and in the lower direction, hence indicating a strong predictive value of the covariates considered.

To ascertain which particular variable is predictive of survival we refer to the variable importance index. The basic idea behind this measure consists in comparing the error rate of a tree where out-of-bag observations are randomly assigned to a final node whenever a split on each predictor is encountered with the error rate of the original tree. This procedure is repeated with respect to each predictor, so that in the end a ranking can be established. The predictor with the strongest importance is the one causing the biggest difference in error rate. We obtain the listing and the associated histograms with the following commands:

```
> plot(rsf0)
```

	Importance	Relative Imp
age70	0.0775	1.0000
lslo_sw	0.0068	0.0876
lslo_dl	0.0053	0.0681
lslo_ca	0.0044	0.0571
lslo_wb	0.0044	0.0567
lslo_be	0.0044	0.0562
lslo_ip	0.0043	0.0558
lslo_mt	0.0034	0.0442
sensory	0.0029	0.0370
lev_be	0.0021	0.0277
lev_ip	0.0020	0.0254
lslo_pa	0.0010	0.0135
sexf	0.0010	0.0125
lev_ca	0.0009	0.0121
NEGA	0.0005	0.0059
lev_mt	0.0003	0.0036
lev_sw	0.0002	0.0030
ses	0.0001	0.0015
POSA	0.0000	0.0002
lev_wb	−0.0007	−0.0089
lev_dl	−0.0009	−0.0118
motor	−0.0011	−0.0144
lev_pa	−0.0032	−0.0410

Note that because of random fluctuation the difference in error rate may be negative. Following the advice of Strobl et al. (2008) we can consider the lowest value (here −0.0032 for lev_pa) as the strongest manifestation of this randomness. Besides all negative importance values, we should then also ignore the positive values that are smaller than the absolute value of the lowest negative value. In this application, then, we retain the importance values of the first eight predictors (from age70 to lslo_mt). We can see that age is by far the predictor most strongly associated to survival information. Of interest to the terminal decline hypothesis, we observe that all other important predictors concern change in cognitive performance in all tasks but Paired Associates. The graphical representation of the overall error rate, represented as a function of the number of trees in the forest, and the importance measures of the predictors are depicted in Figure 15.2.

Comparing the random forest results obtained with the randomSurvivalForest package with the single tree of the party package we see that

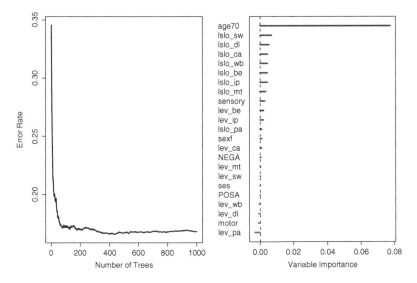

Figure 15.2 Error rate as a function of number of trees (left panel) and out-of-bag importance values for predictors (right panel).

both accord much importance to the age variable. However, the remaining predictors are considered differently by the two approaches. Whereas the random forest approach accords much importance to the changes in cognitive performance, the single-tree approach lists level performance of the Categories and the Paired Associates tasks and sex as more informative. This difference could be due to a number of reasons. First of all, the overall methodological procedure is different in the two packages (Ishwaran, Kogalur, Blackstone, & Lauer, 2008). Second, part of the results of the single tree could be due to overfit, whereas the random forest results should be affected by random sampling to a much lower degree. Third, note that the variable importance measures reported by the random forest analysis basically tell us that participants' age is of chief importance, while all other predictors have minor importance values (relative importance values smaller than 0.1). Apart from age, whose importance appears clearly established, it could hence be that the difference between the important and unimportant predictors is negligible. Finally, it is not clear whether the two packages handle missing data in an alike manner. Comparing the two sets of results would require an in-depth study and clearly goes beyond the scope of this chapter.

As mentioned previously, a major advantage of ITs and, consequently, random forests, is their ability to handle the "small *n*, big *p* problem." Recall that in the simultaneous estimation of the longitudinal+survival model we ran into computational problems when we considered the effects of all covariates singularly. We hence computed four composite scores (sensory and motor performance, positive and negative personality characteristics), to reduce the dimensionality of the computational problem and estimate a solution. This step is not necessary in the IT context, given that likelihood of the complete model is not estimated. The computations of the IT approach are thus much simpler than those of the joint model. In the following analysis we computed a random forest based on all 16 individual covariates originally used to obtain the four composite scores.

```
> rsf02 <- rsf(Surv(ADall70,dead) ~ age70 + sexf + ses +
+    DVEE1 + CVEE1 + HIP1 +
+    I1CROMBG + I1CTURN +
+    ES + OS + PAS + ICS + P1DOPT1 + P1DFUTOR +
+    NS + NAS + POX + LOSS + LOES +
+    lev_dl + lslo_dl + lev_ip + lslo_ip +
+    lev_pa + lslo_pa + lev_mt + lslo_mt +
+    lev_ca + lslo_ca + lev_wb + lslo_wb +
+    lev_be + lslo_be + lev_sw + lslo_sw,
+                data=sub70)
> rsf02

Call:
 rsf(formula = Surv(ADall70, dead) ~ age70 + sexf + ses + DVEE1 +
 CVEE1 + HIP1 + I1CROMBG + I1CTURN + ES + OS + PAS + ICS +
 P1DOPT1 + P1DFUTOR + NS + NAS + POX + LOSS + LOES + lev_dl +
 lslo_dl + lev_ip + lslo_ip + lev_pa + lslo_pa + lev_mt +
 lslo_mt + lev_ca + lslo_ca + lev_wb + lslo_wb + lev_be +
 lslo_be + lev_sw + lslo_sw, data = sub70)

                          Sample size: 438
                     Number of deaths: 331
                      Number of trees: 1000
           Minimum terminal node size: 3
         Average no. of terminal nodes: 58.149
No. of variables tried at each split: 5
                Total no. of variables: 35
                        Splitting rule: logrank
               Estimate of error rate: 17.72%
```

We see that the error rate does not change considerably from the previous random forest. Apparently, then, there is no advantage in disaggregating the four sensory, motor, and positive and negative personality composites. Moreover, when we reconsider the single variables' importance measure within the random forest, we see that the physical and psychological covariates are virtually not related to survival prediction, after the model accounts for the other predictors. Age is again, by far, the most important predictor, followed once more by the indicators of change in cognitive performance. Note, however, that only change in Digit Letter emerges as important (with a value greater than $|-.0047|$). The other cognitive variables do not contribute more importance than that attributable to random sampling fluctuation.

```
> plot(rsf02)
```

	Importance	Relative Imp
age70	0.0682	1.0000
lslo_dl	0.0059	0.0868
lslo_sw	0.0043	0.0635
lslo_ca	0.0031	0.0447
lslo_ip	0.0029	0.0420
lslo_wb	0.0026	0.0386
lslo_mt	0.0025	0.0359
lslo_be	0.0021	0.0303
lev_dl	0.0013	0.0194
lslo_pa	0.0008	0.0117
sexf	0.0007	0.0099
DVEE1	0.0006	0.0090
P1DFUTOR	0.0002	0.0029
P1DOPT1	0.0002	0.0029
I1CTURN	0.0002	0.0023
ES	0.0001	0.0018
PAS	0.0001	0.0009
lev_mt	0.0000	-0.0002
ses	-0.0001	-0.0016
LOES	-0.0001	-0.0018
HIP1	-0.0001	-0.0018
lev_be	-0.0002	-0.0023
I1CROMBG	-0.0002	-0.0032
NS	-0.0006	-0.0095
POX	-0.0008	-0.0113
OS	-0.0008	-0.0115
LOSS	-0.0008	-0.0120
CVEE1	-0.0012	-0.0169
lev_wb	-0.0013	-0.0188
lev_sw	-0.0013	-0.0197
NAS	-0.0015	-0.0215
ICS	-0.0016	-0.0228
lev_pa	-0.0029	-0.0423
lev_ip	-0.0032	-0.0468
lev_ca	-0.0047	-0.0687

Finally, given the major effect of age, we can wonder what happens when this natural covariate is excluded from the analysis. This is a typical procedure in parametric models, where to assess a variable's importance we may want to estimate the model both with and without that variable (e.g., stepwise multiple regression). This is, however, not a typical strategy in the context of ITs, given that multicollinearity is not problematic in such procedures. We nevertheless proceed to examine the overall error rate (which should now increase) and check the robustness of the other variables' importance.

```
> rsf04 <- rsf(Surv(ADall70,dead) ~ sexf + ses +
+    DVEE1 + CVEE1 + HIP1 +
+    I1CROMBG + I1CTURN +
+    ES + OS + PAS + ICS + P1DOPT1 + P1DFUTOR +
+    NS + NAS + POX + LOSS + LOES +
+    lev_dl + lslo_dl + lev_ip + lslo_ip +
+    lev_pa + lslo_pa + lev_mt + lslo_mt +
+    lev_ca + lslo_ca + lev_wb + lslo_wb +
+    lev_be + lslo_be + lev_sw + lslo_sw,
+                data=sub70)
> rsf04

Call:
 rsf(formula = Surv(ADall70, dead) ~ sexf + ses + DVEE1 + CVEE1 +
        HIP1 + I1CROMBG + I1CTURN + ES + OS + PAS + ICS + P1DOPT1 +
        P1DFUTOR + NS + NAS + POX + LOSS + LOES + lev_dl + lslo_dl +
        lev_ip + lslo_ip + lev_pa + lslo_pa + lev_mt + lslo_mt +
        lev_ca + lslo_ca + lev_wb + lslo_wb + lev_be + lslo_be +
        lev_sw + lslo_sw, data = sub70)

                               Sample size: 438
                          Number of deaths: 331
                          Number of trees: 1000
                  Minimum terminal node size: 3
              Average no. of terminal nodes: 58.293
    No. of variables tried at each split: 5
                      Total no. of variables: 34
                           Splitting rule: logrank
                   Estimate of error rate: 22.78%
```

As expected the error rate increases considerably, from about 17% to 23%. This confirms the important contribution of the age variable in predicting survival. The importance measures of the other predictors is displayed next:

```
> plot(rsf04)
```

	Importance	Relative Imp
lev_sw	0.0124	1.0000
lslo_sw	0.0095	0.7659
lev_ca	0.0079	0.6401
lev_mt	0.0069	0.5579
lslo_dl	0.0067	0.5442
lev_be	0.0066	0.5367
lslo_pa	0.0064	0.5168
lslo_ca	0.0061	0.4932
lslo_mt	0.0040	0.3263
lslo_be	0.0038	0.3101
lev_wb	0.0034	0.2727
lslo_wb	0.0033	0.2640
DVEE1	0.0029	0.2304
lslo_ip	0.0026	0.2105
CVEE1	0.0023	0.1868
lev_pa	0.0020	0.1594
lev_ip	0.0013	0.1021

PAS	0.0012	0.0996
NS	0.0009	0.0747
P1DFUTOR	0.0008	0.0623
HIP1	0.0007	0.0535
I1CTURN	0.0004	0.0299
ES	0.0002	0.0199
P1DOPT1	0.0002	0.0125
LOSS	0.0001	0.0062
I1CROMBG	0.0001	0.0062
sexf	0.0000	0.0025
LOES	-0.0001	-0.0100
ses	-0.0001	-0.0112
NAS	-0.0002	-0.0149
ICS	-0.0007	-0.0573
POX	-0.0008	-0.0672
lev_dl	-0.0011	-0.0884
OS	-0.0012	-0.0984

We see that practically all cognitive tasks influence, either by their level or linear slope score, or both, mortality prediction. The physical functioning (except for close and distant visual acuity, CVEE1 and DVEE1, respectively) and personality variables, on the other hand, are not predictive. This again lends some support to the terminal decline hypothesis.

As a comparison we could also try the model-based (mob) recursive partitioning approach, where the same parametric model is estimated in groups of observations (participants) defined by their values of combinations of covariates, determined heuristically. In this application we could

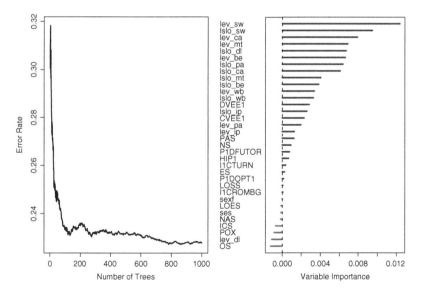

Figure 15.3 Error rate as a function of number of trees (left panel) and out-of-bag importance values for predictors (right panel).

have specified a Weibull survival function and tested the stability of the estimated parameters across different subgroups of participants (see Zeileis, Hothorn, & Hornik, 2008).

Discussion

In this chapter we examined an old hypothesis in the cognitive aging literature that in recent years has become the focus of considerable research. The terminal decline hypothesis necessarily motivates questions that are multivariate in nature: Does a particular cognitive domain singularly predict the timing of death or do several cognitive domains decline prior to death? Moreover, the longitudinal nature of the hypothesis, expressed by the long-term cognitive decline, further complicates matters.

Estimation Issues

We first approached this hypothesis with the joint longitudinal+survival model of Henderson et al. (2000) and estimated its parameters following the advice of Guo and Carlin (2004). We obtained a solution via pseudo-ML estimation but only when considering one (or at most two) cognitive variable(s) at a time. Furthermore, the solution was not very robust, in that at times it depended highly on starting values. Indeed, we had to apply preliminary analyses to obtain educated starting values (by running separately the longitudinal and the survival components, both individually easily estimable, before merging them in the final joint model). In analyses not reported here we also estimated the joint model via Bayesian methods. Estimation matters were not necessarily simpler. The Markov Chain Monte Carlo convergence process was quite slow. Various diagnostics (e.g., trace diagrams, autocorrelations, the coda statistics; Plummer, Best, Cowles, & Vines, 2006; Spiegelhalter et al., 2004) all indicated that a very high number of iterations was necessary to obtain enough independent estimates for a well-behaved posterior distribution. In sum, the parametric model that so nicely matched the substantive questions in its parameters was quite hard to estimate. Recent developments, most notable of which is the inclusion of a similar model in the M*plus* structural equation modeling software (Muthén & Muthén, 1998–2010), will surely facilitate the estimation procedure in the near future.

The recursive partitioning methods, particularly classification and regression trees as well as random forests, discussed here provide a very valuable methodology to further explore hypotheses for which parametric models are not fully satisfactory in applied research settings. Despite the high theoretical relevance of the joint longitudinal+survival model we were faced with estimation problems for which effective solutions can be quite complex and tedious. Conversely, estimating the trees and random forests shown here was a very simple task. Minimal knowledge of the

R environment is necessary, and several published examples of syntax are easily available. R packages such as "party" (Hothorn et al., 2012) and randomSurvivalForest (Ishwaran & Kogalur, 2012), among others, make estimating these data mining techniques straightforward.

Substantive Issues

In our data mining exploration of the ultimate risk of dying we found some support for the terminal decline hypothesis. Age was logically the strongest predictor of death. This result is not surprising to developmental psychologists. As Wohlwill argued decades ago, however, the variable "age" is a recipient for a tremendously large number of effects and at the same time is simply operationalized as time elapsed since birth (Wohlwill, 1970, 1973). One of the tasks of developmentalists is hence to understand which factors, contextual or individual, correlate with age and act on a specific target, such as survival in this application, differentially according to participants' age.

Other well-known survival predictors in old age, such as sex, socio-economic status, and physical functioning (balance and gait) were shadowed by age but also by the indicators of cognitive decline. These results held up also when we disaggregated the physical functioning scores. In other words, change in cognitive performance showed up to be the second most important predictor of mortality, after age. From an epidemiological viewpoint this is an important point. Contextual characteristics that are hardly malleable (socio-economic status) or personal attributes that represent biological effects but also a life-long accumulation of factors, which are crystallized in old age (such as sex), appeared as less determining of survival than change in cognitive performance. Such results motivate researchers to identify those factors, independent of contextual characteristics and personal attributes, that are reflected in shallower cognitive decline. Indeed, recently the field of cognitive epidemiology was born (Deary & Batty, 2007; Lubinski & Humphreys, 1997), to study how cognitive abilities affect health, illness, and ultimately death.

Conclusion

The tools for computing induction trees and ensemble methods are relatively recent, but they have made tremendous progress, both in theoretical terms and with respect to availability and user-friendliness. Of course, broader use by not only statistically oriented researchers but also substantively focused users will inevitably contribute to the fine-tuning of such statistical software. Work remains to be done to further refine data mining software, widen their generalizability, and study their statistical properties. We sincerely thank the authors of such packages and their collaborators for making these valuable tools available to the larger scientific community. Thanks to their efforts, these statistical methods have grown from being interesting exploration tools to powerful confirmatory techniques. We hope that many more researchers will benefit from these and other state-of-the-art recursive partitioning data mining techniques.

Notes

1 Sincere thanks to the Berlin Aging Study I research group, hosted at the Center for Lifespan Psychology of the Max Planck Institute for Human Development of Berlin (Germany), for lending their data and to Ulman Lindenberger for helpful suggestions during the workshop.
2 Parts of these analyses were previously presented at the "Benefits and Limitations of Exploratory Data Mining for Predicting Risk" workshop (July 11–14, 2010, Château la Bretesche, Missilac, France; organized by Prof. John J. McArdle and Prof. Gilbert Ritschard with support provided by the Albert and Elaine Borchard Foundation and the University of Southern California, USA).

References

Abellán, J., & Masegosa, A. R. (2007). Split criterions for variable selection using decision trees. In K. Mellouli (Ed.), *Symbolic and quantitative approaches to reasoning with uncertainty* (Vol. 4724, pp. 489–500). Berlin, Heidelberg: Springer Berlin Heidelberg. Available from http://www.springerlink.com/content/53736706j4168516/

Bäckman, L., & MacDonald, S. W. S. (2006a). Death and cognition. synthesis and outlook. *European Psychologist*, *11*, 224–235.

Bäckman, L., & MacDonald, S. W. S. (2006b). Death and cognition. Viewing a 1962 concept through 2006 spectacles. *European Psychologist*, *11*, 161–163.

Baltes, P. B., & Mayer, K. U. (1999). *The Berlin aging study: Aging from 70 to 100*. New York: Cambridge University Press.

Bosworth, H. B., & Siegler, I. C. (2002). Terminal change in cognitive function: An updated review of longitudinal studies. *Experimental Aging Research*, *28*, 299–315.

Breiman, L. (2001). Random forests. *Machine Learning*, *45*, 5–32.

Breiman, L., Friedman, J. H., Olshen, R. A., & Stone, C. J. (1984). *Classification and regression trees*. New York: Chapman & Hall.

Deary, I. J., & Batty, G. D. (2007). Cognitive epidemiology. *Journal of Epidemiology and Community Health*, *61*(5), 378–384.

Ghisletta, P. (2008). Application of a joint multivariate longitudinal-survival analysis to examine the terminal decline hypothesis in the Swiss interdisciplinary longitudinal study on the oldest old. *Journal of Gerontology: Psychological Sciences*, *63B*, P185–P192.

Ghisletta, P., McArdle, J. J., & Lindenberger, U. (2006). Longitudinal cognition–survival relations in old and very old age: 13-year data from the Berlin aging study. *European Psychologist*, *11*, 204–223.

Ghisletta, P., Rabbitt, P., Lunn, M., & Lindenberger, U. (2012). Two-thirds of the age-based changes in fluid and crystallized intelligence, perceptual speed, and memory in adulthood are shared. *Intelligence*, *40*, 260–268. doi: 10.1016/j.intell.2012.02.008

Guo, X., & Carlin, B. P. (2004). Separate and joint modeling of longitudinal and event time data using standard computer packages. *The American Statistician*, *58*, 16–24.

Harrell, F. E., Califf, R. M., Pryor, D. B., Lee, K. L., & Rosati, R. A. (1982). Evaluating the yield of medical tests. *The Journal of the American Medical Association*, *247*(18), 2543–2546.

Henderson, R., Diggle, P. J., & Dobson, A. (2000). Joint modeling of longitudinal measurements and event time data. *Biostatistics*, *1*, 465–480.

Hothorn, T., Buehlmann, P., Dudoit, S., Molinaro, A., & Laan, M. V. D. (2006). Survival ensembles. *Biostatistics*, 7(3), 355–373.

Hothorn, T., Hornik, K., Strobl, C., & Zeileis, A. (2012). party: A laboratory for recursive partytioning. Available from http://cran.r-project.org/web/packages/party/index.html

Hothorn, T., Hornik, K., & Zeileis, A. (2006). Unbiased recursive partitioning: A conditional inference framework. *Journal of Computational and Graphical Statistics*, *15*(3), 651–674.

Ishwaran, H., & Kogalur, U. B. (2007). Random survival forests for R. *R News*, 7, 25–31.

Ishwaran, H., & Kogalur, U. B. (2012). randomSurvivalForest: Random survival forests. Available from http://cran.r-project.org/web/packages/randomSurvivalForest/index.html

Ishwaran, H., Kogalur, U. B., Blackstone, E. H., & Lauer, M. S. (2008). Random survival forests. *The Annals of Applied Statistics*, *2*, 841–860.

Kleemeier, R. W. (1962). Intellectual changes in the senium. *Proceedings of the American Statistical Association*, *1*, 290–295.

Laird, N. M., & Ware, J. H. (1982). Random-effects models for longitudinal data. *Biometrics*, *38*, 963–974.

Lindenberger, U., & Baltes, P. B. (1997). Intellectual functioning in old and very old age: Cross-sectional results from the Berlin aging study. *Psychology and Aging*, *12*, 410–432.

Lindenberger, U., & Ghisletta, P. (2009). Cognitive and sensory decline in old age: Gauging the evidence for a common cause. *Psychology and Aging*, *24*, 1–16.

Lindenberger, U., Smith, J., Mayer, K., & Baltes, P. (2010). *Die Berliner altersstudie*. Berlin: Akademie Verlag.

Lubinski, D., & Humphreys, L. G. (1997). Incorporating general intelligence into epidemiology and the social sciences. *Intelligence*, *24*(1), 159–201.

Muthén, L. K., & Muthén, B. O. (1998–2010). *M*plus *user's guide* (Vol. 6). Los Angeles: Muthén and Muthén.

O'Brien, S. M. (2004). Cutpoint selection for categorizing a continuous predictor. *Biometrics*, *60*(2), 504–509. Available from http://www.jstor.org/stable/3695779

Plummer, M., Best, N., Cowles, M. K., & Vines, K. (2006). Coda: Convergence diagnosis and output analysis for mcmc. *R News*, *6*, 7–11.

R Development Core Team (2011). R: A language and environment for statistical computing [Computer software manual]. Vienna, Austria. Available from http://www.R-project.org (ISBN 3-900051-07-0)

Shih, Y. (1999). Families of splitting criteria for classification trees. *Statistics and Computing*, *9*, 309–315.

Spiegelhalter, D., Thomas, A., Best, N., & Lunn, D. (2004). Winbugs user manual (version 2.0) [online]. Available from http://mathstat.helsinki.fi/openbugs/

Strasser, H., & Weber, C. (1999). On the asymptotic theory of permutation statistics. *Mathematical Methods of Statistics*, *8*, 220–250.

Strobl, C., Boulesteix, A.-L., Kneib, T., Augustin, T., & Zeileis, A. (2008). Conditional variable importance for random forests. *BMC Bioinformatics*, *9*(307). Available from http://www.biomedcentral.com/1471-2105/9/307

Strobl, C., Boulesteix, A.-L., Zeileis, A., & Hothorn, T. (2007). Bias in random forest variable importance measures: Illustrations, sources and a solution. *BMC Bioinformatics*, *8*(25). Available from http://www.biomedcentral.com/1471-2105/8/25

Strobl, C., Hothorn, T., & Zeileis, A. (2009). Party on! A new, conditional variable-importance measure for random forests available in the party package. *The R Journal*, *1*, 14–17.

Strobl, C., Malley, J., & Tutz, G. (2009). An introduction to recursive partitioning: Rationale, application and characteristics of regression trees, bagging, and random forests. *Psychological Methods*, *14*, 323–348.

Wohlwill, J. F. (1970). The age variable in psychological research. *Psychological Review*, *77*, 49–64.

Wohlwill, J. F. (1973). *The study of behavioral development*. New York: Academic Press.

Wolfinger, R. D. (1988). Fitting nonlinear mixed models with the new NLMIXED procedure. *Analysis*, *872*, 1–10.

Zeileis, A., Hothorn, T., & Hornik, K. (2008). Model-based recursive partitioning. *Journal of Computational and Graphical Statistics*, *17*, 492–514.

16 Predicting Mortality from Demographics and Specific Cognitive Abilities in the Hawaii Family Study of Cognition

Yan Zhou, Kelly M. Kadlec, and John J. McArdle

Introduction

This study investigates the relation between specific cognitive abilities and mortality using data from over 1,800 pairs of parents in the Hawaii Family Study of Cognition (HFSC; DeFries et al., 1979). Three methods of survival analysis are used: (1) Cox regression, (2) Conditional Inference Survival Trees, and (3) Random Survival Forests. The results are consistent with other studies showing that cognitive abilities contribute to the prediction of mortality above and beyond gender and education, with approximately 5% increase in prediction accuracy based on Harrell's concordance index (Harrell et al., 1982). Specific cognitive abilities are very close in their prediction strength, and provide little to the prediction of mortality above and beyond general intelligence. Sex-specific effects may exist; survival trees suggest that numerical ability may be more important in predicting women's mortality and that visual-spatial ability may be more important in predicting men's mortality. Due to the high percentage of censoring (85.1%), conclusions that can be drawn from the data are limited, and the study requires additional follow up. We also demonstrate that statistical learning techniques provide new opportunities for studying survival problems, but they are still in the development stage and interpretations should be made with caution—some explicit recommendations for their practical use are discussed. We also relate these findings to other recent findings in the field of cognitive epidemiology.

Evidence is accumulating that intelligence in early life is inversely associated with later mortality, and this association seems to be present irrespective of study population, cohort, and the cognitive tests used (see a review by Batty et al., 2007). With the association being established, the mechanism underlying the association is as yet unclear. This gives rise to an emerging field of study—labeled "cognitive epidemiology" by some researchers (Deary & Batty, 2007; Deary, 2009). One of the key aspects to understanding the phenomenon is to examine how different domains of intelligence are related to mortality. The current study uses a variety of standard survival analysis and novel exploratory data mining techniques to investigate the roles of specific cognitive abilities in the prediction of mortality. Data from the parental generation of the HFSC (DeFries et al., 1979) are used in all analyses.

The empirical studies on IQ-mortality association can be placed into one of two categories depending on when cognitive function was assessed— either in childhood to early adulthood (i.e., premorbid IQ) or in later life. Interpretation of the association varies greatly between the categories of research, partly because cognitive abilities change with age (Horn & Cattell, 1967). In older adults, cognitive function may decline as a consequence of poor heath, and the research is often focused on what has been termed "terminal decline" (Kleemeier, 1962; Riegel & Riegel, 1972). The parental generation participants of the HFSC were typically measured in their midlife, a stage when cognitive abilities are relatively stable.

In addition to the standard method for survival analysis, we also used exploratory data mining techniques to achieve optimal prediction. These methods originated from the Classification and Regression Trees (CART; Breiman et al., 1984) and the ensemble method that followed—Random Forests (RF; Breiman, 2001). The tree algorithm is a recursive binary partitioning of a predefined covariate space into smaller and smaller regions, containing observations of homogeneous response (i.e., dependent variable) values. Thus, we are trying to group people who have similar survival curves. The main function of trees is classification and prediction. Single trees have a major problem of instability (Berk, 2008; Strobl et al., 2009), and ensemble methods provide a solution by aggregating a large number of trees. Random Forests (RF) is one of the ensemble methods that has the strongest prediction power (Berk, 2008) because it can respond to highly local features of the data. For example, it can detect complex interaction patterns (particularly the higher-order interactions) that are usually hidden when a few strong predictors dominate, and are not examined in standard survival analysis (Strobl et al., 2009). Additional merits include its ability to handle a large number of predictors, even when the sample size is not as large, and a variable importance measure that ranks the predictors. Data mining, or statistical learning, as these methods are termed, does not make any assumptions about the data. Therefore, they are useful in the absence of a priori hypotheses.

In the survival context, the censoring problem poses difficulty in the application of the standard tree algorithms, and some kind of adaption is needed for the treatment of censored response. Many survival tree methods have been proposed (Gordon & Olshen, 1985; Ciampi et al., 1986; Segal, 1988; LeBlanc & Crowley, 1992, 1993; Zhang & Singer, 1999; Molinaro et al., 2004), but the conditional inference tree by Hothorn et al. (2006b) seems to be the most stable and least likely to overfit the data.

Survival ensembles are more advanced and still in the development stage. Hothorn et al. (2006a) proposed a random forest algorithm for censored data using a weighting scheme, but it is not suitable in circumstances where the percentage of censoring is high. The random survival forest by Ishwaran et al. (2008) sticks to Breiman's original prescription of random forests (2001), and seems to have worked well with a few empirical datasets. It has mainly been applied to studies in the health sciences (for example, Weichselbaum et al., 2008; Rizk et al., 2010) but also in economics (Fantazzini & Figini, 2009).

Aided by these novel techniques, we compared the predictability of mortality when using (1) the first principal component of all test scores, which is considered to represent general intelligence, (2) specific cognitive factor scores, and (3) raw test scores. The first question of interest is whether the specifics of cognitive abilities contribute to the prediction of mortality above and beyond general intelligence, and if so, which domains are most relevant. The second question is whether there is a sex difference with respect to different domains of cognition in relation to mortality.

Method

Participants

The Hawaii Family Study of Cognition (HFSC) included 1,816 nuclear families (a total of 6,581 individuals) living on the island of Oahu, Hawaii in 1972–1976. These families consisted of two parents and one or more children aged 13 years or older. The principal objective of the HFSC was to assess genetic and environmental bases of cognitive performance on various tests of cognitive abilities.

The current analyses are based on the 3,631 parents (1,815 men and 1,816 women) in these families. The offspring generation is younger and most likely still alive, so they were not included in these analyses.

The parents received an average of 14.3 years of education ($SD = 2.8$ years), with 91.4% of the sample achieving at or above the high school level. At the time when the cognitive tests were administered, they were between 28 and 71 years of age, but most (90%) were between 35 and 55. The average age was 44.2 years ($SD = 6.1$ years). Americans of European ancestry (870 couples) and Americans of Japanese ancestry (311 couples) were the two largest ethnic groups in the sample.

Measures

Mortality

A National Death Index (NDI) search was initially conducted in May 1987 and repeated in August 2006, which searched death records on file with the National Center for Health Statistics between 1975 and 2004 (the latest records available at the time the search was conducted). The accuracy of each NDI match was evaluated using the NDI suggested approach. Based on a match score returned for each record by NDI, we used the cut-off score recommended by Horm (i.e., Horm's 1996 Assignment of Probabilistic Scores to the National Death Index Record Match; see Appendix A of the National Death Index Plus: Coded Causes of Death Supplement to the NDI User's Manual) for determining the likelihood of a true match. According to Horm (1996), records with scores greater than the cut-off are considered true matches while records with scores lower than the cut-off are considered false matches. For the HFSC participants, their Social Security numbers were unknown and there were fewer than eight matches on either first name, middle initial, last name, birth day, birth month, birth year, sex, race, marital status, or State of birth, so in the NDI system they fell in Class 4, for whom the cut-off score is set at 32.5.

NDI assigned a final search status ("no match," "possible match," or "probable match") to each person, and only the "probable match" category indicated a probability cut-off score greater than 32.5. We used this criterion to create our death status variable. Participants in the "no match" or "possible match" categories were assumed to be alive and participants in the "probable match" category were assumed to be dead. A total of 84 people (53 men and 31 women) were matched in the 1987 search, and the death date we use for these people is December 31, 1986. A total of 454 people (311 men and 143 women) were matched in the 2006 search, and we use the NDI death date for these people. For those assumed to be alive, their ages at the end of the year 2004 were calculated as the last age in the survival analysis. Most of the children were assumed to be alive based on their ages in 2004, so most (91.49%) were not included in the National Death Index (NDI) search (see Table 16.1).

Table 16.1 NDI search status from the 2006 search

	Frequency	*Percent*
1st generation (missing = 4)		
Not included in NDI search	13	.36
Included in NDI search, but no match	1454	40.09
Included in NDI search, possible match	1706	47.04
Included in NDI search, probable match	454	12.52
2nd generation (missing = 10)		
Not included in NDI search	2688	91.49
Included in NDI search, but no match	127	4.32
Included in NDI search, possible match	119	4.05
Included in NDI search, probable match	4	.14

Cognitive Measures

Fifteen tests of specific cognitive abilities were administered: *Vocabulary* (VOC), *Visual Memory Immediate* (VMI), *Things Category* (TH), *Mental Rotation* (MR), *Subtraction and Multiplication* (SAM), *Lines and Dots* (LAD), *Word Beginnings and Endings* (WBE), *Card Rotations* (CR), *Visual Memory Delayed* (VMD), *Pedigrees* (P), *Hidden Patterns* (HP), *Paper Form Board* (PFB), *Number Comparisons* (NC), *Social Perception* (which includes two parts— *Identification*, SPN, and *Description*, SPV), and *Progressive Matrices* (PMS). Details about the test procedure and psychometric properties of these tests have been published earlier (Wilson et al., 1975; DeFries et al., 1979).

Because of the presence of marked age effects (Wilson et al., 1975), test scores were standardized within age groups in the previous reports (DeFries et al., 1976, 1979). This problem is less of a concern because we only used data from the parents. In fact the 35–55 age range, which includes 90% of the parents, is treated in their correction procedure as one age group. We used percentage scores calculated from the uncorrected raw scores, which were divided by the highest possible score on each test, and multiplied by 100. However, we still examined how test age effects could possibly affect the results in survival analysis. Summary statistics of these scores and their inter-correlations are listed in Table 16.2.

Earlier analysis showed that the first principal component of these scores correlates .73 with the Wechsler Adult Intelligence Scale (WAIS; Wechsler, 1955) full-scale IQ (Kuse, 1977), and four composite scores representing spatial, verbal, perceptual speed, and visual memory were identified based on principal component analysis (DeFries et al., 1974).

Analyses

Factor Analysis

Instead of using principal component analysis, we conducted a factor analysis with maximum likelihood estimation to find a simpler structure among the 16 cognitive test scores. We increased the number of factors consecutively until a satisfactory model fit (based on RMSEA) was obtained. Promax rotation was then applied to find an interpretable factor pattern. We calculated Bartlett factor scores (Bartlett, 1937) based on the final solution, which produced by using maximum likelihood estimates—a statistical procedure considered to produce the unbiased estimates of the true factor scores (Hershberger, 2005).

Cox Regression

In the second step, we examined the relationship between the specific cognitive abilities and mortality using standard survival analysis. The Cox

Table 16.2 Summary statistics and inter-correlations of Cognitive Test Scores

	VOC	VMI	TH	MR	SAM	LAD	WBE	CR	VMD	P	HP	PFB	NC	SPN	SPV	PMS
VOC	1															
VMI	.27	1														
TH	.53	.19	1													
MR	.21	.13	.27	1												
SAM	.41	.14	.23	.16	1											
LAD	.22	.11	.24	.36	.24	1										
WBE	.58	.22	.44	.24	.36	.22	1									
CR	.34	.18	.35	.57	.29	.39	.33	1								
VMD	.23	.50	.18	.09	.12	.10	.20	.16	1							
P	.62	.30	.44	.36	.42	.36	.51	.46	.26	1						
HP	.44	.21	.38	.42	.36	.37	.38	.49	.16	.52	1					
PFB	.37	.19	.40	.44	.24	.34	.34	.52	.16	.42	.51	1				
NC	.38	.23	.20	.11	.53	.19	.33	.31	.20	.48	.37	.28	1			
SPN	.54	.23	.45	.32	.28	.27	.40	.38	.21	.56	.43	.36	.32	1		
SPV	.58	.26	.45	.33	.27	.26	.43	.38	.22	.57	.43	.38	.31	.84	1	
PMS	.51	.27	.45	.48	.33	.37	.45	.50	.20	.63	.56	.54	.36	.53	.57	1
Age	.02	-.08	-.04	-.10	-.02	-.11	-.02	-.15	-.11	-.17	-.10	-.07	-.20	-.19	-.17	-.14
Education	.38	.11	.33	.27	.29	.17	.29	.22	.07	.34	.32	.32	.18	.31	.34	.40
N	3629	3583	3615	3602	3585	3602	3626	3619	3610	3625	3626	3584	3627	3592	3554	3600
M	76.3	80.8	37.8	47.2	53.4	46.9	27.0	43.5	63.4	57.8	36.8	39.7	40.8	75.7	66.6	60.3
SD	20.5	11.2	11.1	20.3	18.3	17.8	10.7	15.2	15.9	17.5	12.7	14.9	10.3	14.6	17.2	17.8

proportional hazards (PH) model is the most popular method for studying survival problems. Unlike alternative parametric survival models (e.g., exponential, Weibull, log-logistic, etc.), it does not make any restriction about the shape of the baseline hazard function but instead assumes proportional hazards (Cox, 1972). It is useful when we don't have any knowledge about what the baseline hazard is likely to be. It will closely approximate a correct parametric model and is thus considered robust. The effects of covariates are modeled in a regression form, with the assumption that the effects on the hazard are multiplicative (or proportional) over time, and the hazard ratio is a constant independent of time.

Gender and education are two demographic variables to be adjusted for in the model, and this is our baseline model. We added cognitive factors to the baseline model to examine any gains in the goodness of fit. We compared models using the specific factor scores versus the model using the first principal component score. We also considered a gender by cognitive test interaction.

Survival Trees

In the third step, we used the R package "party" (Hothorn et al., 2010) to construct conditional inference trees developed by Hothorn et al. (2006b).

The basic idea of the survival tree method is to group people with homogeneous survival functions by recursive binary partitioning. Starting with the whole set of people, it searches for the best initial split of a certain cut-off value in a certain covariate, resulting in two groups of people. The resulting groups are called "nodes." It then searches for the next best split to further divide the nodes and so on. A group that cannot be partitioned any more is called a "terminal node" or a "leaf." The terminal nodes represent the final grouping of the original sample.

Among the proposed tree algorithms for censored outcomes, Hothorn's algorithm is based on a theory of permutation tests. In splitting a node, it selects the split with the minimum p-value. According to the authors, this procedure attempts to overcome the problem of selection bias toward predictors with many possible splits or missing values, which is a major drawback in other recursive partitioning procedures. Instead of a pruning procedure which is adopted in most tree algorithms to solve the problem of overfitting, conditional inference trees stop growing when no p-value is below a pre-specified α-level, so α may be seen as a parameter determining the size of the tree. For censored responses, it seems that the log-rank statistic is embedded in its calculation in the conditional inference framework.

We applied this algorithm to the entire sample using gender, education, and the cognitive factor scores as predictors. Because single trees are unstable, we then drew 10 bootstrap samples and repeated the procedure for each random sample. We simply compared the tree structures by eye to see if any common pattern existed.

Random Survival Forests

In the fourth step, we applied the random survival forest procedure of Ishwaran et al. (2008) implemented in the R package "randomSurvivalForest" (Ishwaran & Kogalur, 2007, 2010). This algorithm adapts the standard random forests (Breiman, 2003) to the censored responses. Random forests is a procedure that aggregates over a number of (unpruned) single trees, each from a bootstrap sample of the data. In addition, in the construction of each tree, a pre-specified smaller number of predictors are randomly selected before each node is split, and the splitting variable is searched within the reduced set of predictors.

The program "ransomSurvivalForest" is different from the program "party" in tree splitting criteria. Instead of a permutation-based conditional inference framework, it is closer to the classic CART approach in building single trees. Four alternative splitting rules are available for single survival trees: log-rank splitting, conversation-of-events principle, log-rank score (standardized log-rank statistic) splitting, and random log-rank splitting (Ishwaran et al., 2008). The predicted value for a terminal node is the cumulative hazard function (CHF) estimated by the Nelson–Aalen method. The CHF for a tree is defined by conditioning the terminal node CHFs on the covariates. The forest is constructed in three steps:

(1)   Draw B bootstrap samples from the original data. On average each bootstrap sample excludes 37% of the data, which are called out-of-bag (OOB) observations.

(2)   Grow a survival tree for each bootstrap sample. At each node, select a split from a random reduced set of p predictors using one of the four aforementioned splitting criteria. Trees are grown to full size under the constraint that a terminal node should have at least one death.

(3)   Calculate a CHF for each tree, and average the CHF over B trees to obtain the ensemble CHF.

Harrell's concordance index (C-index; Harrell et al., 1982) is used as the measure of prediction error. It estimates the probability that, in a randomly selected pair of cases, the sequence of events is correctly predicted. The prediction error is therefore $1 - $ C-index. OOB observations are used to calculate the predictor error for the ensemble CHF.

Like the standard random forests, a variable importance measure is calculated for each predictor. It is defined as the original prediction error subtracted from the prediction error obtained by randomizing the values in that predictor, given that the forest is unchanged. As always, OOB observations are used in calculating the prediction errors.

We applied this procedure to the data using the default number of trees, 1000, and the default setting for number of predictors sampled at each node—the square root of the total number of predictors. We used the random log-rank splitting rule. This splitting rule selects a random split value for each of the p candidate predictors in a node, and then uses the log-rank criterion to select the split predictor with its random split value. It is the fastest in computation speed while maintaining fairly good performance (Ishwaran et al., 2008). Given the size of our data we chose this splitting rule to improve the computation speed.

Results

Results from Factor Analysis

Results from the factor analysis show that five factors are sufficient to summarize the 16 cognitive measures (see Table 16.3), yielding an RMSEA of .043 (90% confidence interval is .039 to .047). Rotated factor pattern, inter-factor correlations, and factor correlations with test age and education are presented in Table 16.4. This seems to be a clean simple structure. The first factor loads on Mental Rotation, Lines and Dots, Card Rotation, Hidden Patterns, Paper Form Board, and Progressive Matrices, and represents a visual-spatial domain of cognition. The second factor loads on Vocabulary, Things, Word Beginnings and Endings, and Pedigrees, representing verbal abilities or crystallized intelligence. The third factor loads on Social Perception Identification and Social Perception Description, representing social perception. The fourth factor loads on Subtraction and Multiplication, and Number Comparison, representing basic numerical abilities. The fifth factor loads on Visual Memory Immediate and Visual Memory Delayed, and thus can be labeled visual memory.

There are moderate to high correlations among the factors ranging from .33 to .72. These factors have small negative correlations with age ($r = -.10$ to $-.17$) except the verbal factor ($r = -.01$). They all have positive correlations with education. Visual memory is least associated with education ($r = .14$) while visual-spatial, verbal, and social perception seem to have stronger associations with education.

Table 16.3 Results from factor analysis with maximum likelihood estimation

	$F = 0$	$F = 1$	$F = 2$	$F = 3$	$F = 4$	$F = 5$
χ^2	26403	6531	3848	1999	1172	387
df	120	104	89	75	62	50
RMSEA	—	.130	.108	.084	.070	.043
90% RMSEA	—	.128 −.133	.105 −.111	.081 −.087	.067 −.074	.039 −.047
P(close)	—	.000	.000	.000	.000	.998

Table 16.4 PROMAX rotated factor pattern and inter-factor correlations

	Visual-spatial	Verbal	Social perception	Numerical	Visual memory
VOC	−12	82	9	4	0
VMI	0	−1	−2	−3	89
TH	18	52	7	−17	0
MR	83	−8	3	−15	−2
SAM	3	21	−8	56	−7
LAD	50	−2	0	5	−3
WBE	3	66	−6	7	1
CR	77	−6	−3	8	0
VMD	−3	3	2	2	56
P	20	36	13	23	6
HP	50	16	1	15	0
PFB	60	14	−4	1	1
NC	−1	−13	5	88	4
SPN	3	1	86	2	−2
SPV	−1	7	91	−2	1
PMS	47	22	15	4	3
Visual-spatial	1				
Verbal	.66	1			
Social perception	.59	.72	1		
Numerical	.49	.62	.43	1	
Visual memory	.33	.42	.37	.36	1
Age	−.13	−.01	−.17	−.13	−.10
Education	.38	.48	.37	.28	.14

Note: The factor loadings are multiplied by 100 and rounded.

Results from Cox Regression

Results from the Cox regression are presented in Table 16.5. The gender variable is coded 1 for females and 0 for males, so we labeled it female in the table. The baseline model (Model 0) includes only gender and education, and both reach statistical significance. The estimate for female is −.59, which translates to a hazard ratio of .555, meaning that the hazard of death is 44.5% lower for the females than for the males. The effect of education is weak but significant; the hazard is 3.3% lower for each additional year of education. The addition of the interaction term of gender by education does not improve the model fit (LR = 44.7, df = 3).

Then we add the cognitive factors to the model. Each significantly improves the fit compared to the baseline model when the other four factors are not included in the model (Models 1, 2, 3, 4, and 5). All the specific cognitive abilities are negatively associated with hazard (or positively associated with life time). In terms of hazard ratios, the hazard is 15.0% lower for each additional standard deviation increase in the visual-spatial factor, 14.2% for verbal, 11.4% for social perception, 17.4% for numerical, and 13.6% for visual memory. The effect of education becomes insignificant

Table 16.5 Parameter estimates and fit indices from the Cox regression for mortality

	Model 0	Model 1	Model 2	Model 3	Model 4	Model 5	Model 6	Model 7
Estimate (standard error)								
Female	−.59	−.70	−.61	−.63	−.57	−.60	−.61	−.65
	(.10)	(.10)	(.10)	(.10)	(.10)	(.10)	(.11)	(.10)
Education	−.03	−.02	−.01	−.02	−.02	−.03	−.01	−.005
	(.01)	(.02)	(.02)	(.02)	(.02)	(.02)	(.02)	
Visual-spatial		−.16					−.06	
		(.05)					(.07)	
Verbal			−.15				.03	
			(.05)				(.09)	
Social perception				−.12			−.03	
				(.05)			(.07)	
Numerical					−.19		−.13	
					(.05)		(.07)	
Visual memory						−.15	−.08	
						(.05)	(.06)	
1st PC								−.18
								(.05)
Hazard Ratio								
Female	.555	.495	.542	.534	.563	.549	.544	.520
Education	.967	.984	.989	.981	.982	.972	.988	.995
Visual-spatial		.850					.939	
Verbal			.858				1.029	
Social perception				.886			.971	
Numerical					.826		.876	
Visual memory						.864	.920	
1st PC								.835
Goodness-of-fit								
Likelihood ratio	44.7	53.0	51.9	50.1	56.8	52.9	61.2	57.1
df	2	3	3	3	3	3	7	3
AIC	7954	7107	7108	7109	7103	7107	7106	7102
BIC	7963	7119	7120	7122	7115	7119	7136	7115

when cognitive factors are added in. According to the AIC and BIC, these five models are close in fit but Model 4 is slightly better than the other four models, suggesting that numerical ability may be the strongest predictor. Model 7 uses the first principal component as a predictor, and the fit seems to be better than using one of the specific factors, but very close to that of Model 4.

When two or more cognitive factors are added in simultaneously, their effects cancel each other out. Model 6 includes all five specific factors, and the fit (LR = 61.2, df = 7) is not significantly better than any of the models with one single cognitive factor. In order to find an optimal model, an exploratory variable selection procedure is used. Forward selection, backward deletion, and the stepwise procedure all yield the same final set of predictors—gender and numerical ability.

When sex difference is considered and the interaction term between gender and each cognitive factor is added to the model, no interaction is found to be significant.

To examine the possible influence of age effects in the cognitive scores, we extracted a subgroup of 3,264 individuals whose ages at the time of testing were between 35 and 55, and refit the model with gender, education, and a factor score. The estimates of the coefficients change to $-.21$, $-.17$, $-.14$, $-.21$, $-.17$, respectively, for visual-spatial, verbal, social perception, numerical, and visual memory. All become more salient compared to the estimates from using the entire age range (the coefficients in Models 1, 2, 3, 4, and 5 in Table 16.5), suggesting that age effects may have attenuated the cognition-mortality association in the survival analysis.

In sum, the results from the standard survival analysis show that cognitive variables provide extra contribution in predicting mortality above and beyond gender and education. The effects of the five specific cognitive abilities largely overlap, but the effect of numerical ability seems relatively stronger. There is no gender by cognition interaction. Age effects in the test scores may have an influence in the survival analysis—they tend to attenuate the strength of the cognition–mortality association.

Survival Tree Results

Figure 16.1 is the survival tree plot from the R program "party" using seven predictors—gender, education, and the five cognitive factor scores, and the tuning parameters were left unchanged at their default values. At the top of the tree, the original sample (denoted by Node 1) is first split by gender, separating the males (Node 2) and the females (Node 5). Node 2 is further split by the visual-spatial factor with the cut-off value of .249, resulting in one group of males scoring lower on the visual-spatial factor (Node 3) and the other group of males scoring higher on the visual-spatial factor (Node 4). There is no more splitting of Nodes 3 and 4, so they are terminal nodes. Similarly, the females in Node 5 are split by the numerical factor at the cut-off value of -1.735, into terminal Nodes 6 and 7. The final partition of the original sample results in four groups, each indicated by a Kaplan–Meier curve in their respective terminal nodes. Figure 16.2 shows the Kaplan–Meier curves for the four groups in plot (readers can find the colored version of this figure on the book's web page).

The summary statistics for each terminal node are listed in Table 16.6. The predicted value in the conditional inference survival tree is the Kaplan–Meier survival function. In the table we list the median life time and its 95% confidence bound. Overall, the male group with better visual-spatial abilities has a lower mortality (16.0%) than the other male group (25.6%), and the female group with better numerical ability has a lower

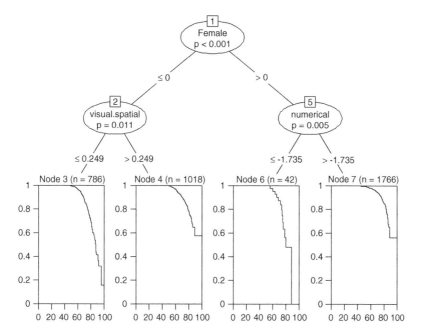

Figure 16.1 Survival tree for mortality.

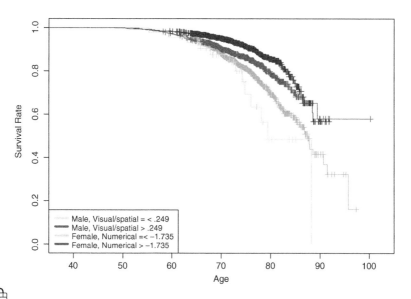

Figure 16.2 Kaplan–Meier curves by survival tree terminal node.

Table 16.6 Summary statistics for survival tree terminal nodes

	N	# Death (%)	Median	95% C.I.
Node 3: Male, visual-spatial ≤ .249	786	201 (25.6%)	87.5	(85.8, 91.4)
Node 4: Male, visual-spatial > .249	1018	163 (16.0%)	—	(89.4, ∞)
Node 6: Female, numerical ≤ −1.735	42	13 (31.0%)	79.4	(76.0, ∞)
Node 7: Female, numerical > −1.735	1766	161 (9.1%)	—	(88.4, ∞)

mortality (9.1%) than the other female group (31.0%). However, due to the high percentage of censoring, the information conveyed by the survival functions is very limited. For Node 4 (84.0% censoring rate) and Node 7 (90.9% censoring rate), the median life time and the upper confidence bound cannot be estimated. The upper confidence bound also cannot be estimated for Node 6 because of small group size ($n = 42$).

The structure of the tree suggests a pattern of interaction—that the males and the females are split by different cognitive abilities. Also note that the cut-off value categorizing men in terms of visual-spatial ability is a little above the mean, so the resultant two groups are close in size; however, the cut-off value categorizing females in terms of numerical ability is 1.7 standard deviations below the mean, resulting in two rather unbalanced groups.

As noted above, the results of this particular tree structure should not be generalized given the instability of single trees. In fact, we obtained 10 different trees by bootstrapping the original sample 10 times. These trees vary greatly in terms of selection of predictors and their split values. The number of terminal nodes ranges from 3 to 8 and the average is 4.8. Still, we observe some common characteristics in these trees which may suggest some systematic pattern in the data. The first split is by gender 9 out of 10 times (by numerical at the other 1 time). The males are most frequently split by visual-spatial (6 out of 10 times), and the females are most frequently split by numerical ability (5 out of 10 times).

Results from Random Survival Forests

The first survival forest is constructed by entering the seven predictors of mortality—gender, education, and the five cognitive factors. Two candidate predictors are sampled for each node, and the average number of terminal nodes is 177. This forest obtains an error rate of 40.42%, an approximate 10% gain by using these predictors compared to random guess. The left plot in Figure 16.3 shows the stabilization of the error rate as the number of trees increases. The right plot presents the variable importance rank and the exact values are listed in Table 16.7. A negative variable importance value (red lines in the plot, see the book's website) is considered as a result of random variation and within the range of zero, suggesting the predictor is irrelevant (Strobl et al., 2009).

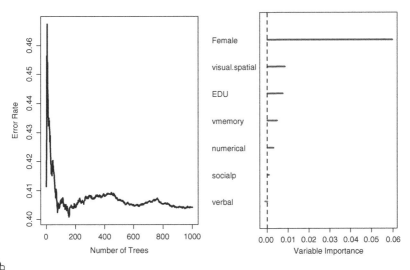

Figure 16.3 Error rate stabilization and variance importance rank from the random survival forest.

Table 16.7 Variable importance and prediction error for Nested forests

	Importance	Error rate		Error drop
Female	.0598	Female	.5867	—
Visual-spatial	.0085	Female + Visual-spatial	.4185	.1683
Education	.0074	Female + Visual-spatial + Education	.4087	.0098
Visual memory	.0048	Female + Visual-spatial + Education + Visual memory	.4048	.0039
Numerical	.0029	Female + Visual-spatial + Education + Visual memory + Numerical	.4080	−.0033
Social perception	.0008	Female + Visual-spatial + Education + Visual memory + Numerical + Social perception	.4107	−.0027
Verbal	−.0011	Female + Visual-spatial + Education + Visual memory + Numerical + Social perception + Verbal	.4109	−.0002

There is no clear criterion for how big the importance value should be to be a meaningful predictor. Ishwaran et al. (2008) add noise variables to the dataset as reference variables. Strobl et al. (2009) suggest a conservative strategy to only include predictors whose importance scores exceed the amplitude of the largest negative scores. We follow the approach used in Ishwaran and Kogalur (2007) to examine the incremental effect of each predictor using a nested analysis. Based on the importance rank in Table 16.7, we construct a nested sequence of forests starting from the top predictor—gender, followed by the forest with the top two predictors—gender and visual-spatial, then the forest with the top three variables, and so on. The error rate of these nested forests is listed, and the difference in the error rate between two consecutive forests is calculated. Combining all the information, it seems that using the top four predictors (gender, visual-spatial, education, and visual memory) is enough to reach the optimal prediction accuracy that would be obtained by including all seven predictors.

Variable importance is a relative measure and its value depends on the other predictors used in growing the forest (see Strobl et al., 2009). The nested comparison in this particular sequence may not lead to the conclusion that the bottom variables are not "important" predictors. Next we grow six forests using two predictors—gender and each of the other six predictors. The error rate is 41.81% with gender and visual-spatial (slightly different from the previous 41.85% because it is a different run), 45.4% with gender and education, 42.04% with gender and visual memory, 44.29% with gender and numerical, 42.95% with gender and social perception, and 42.43% with gender and verbal. So it seems that the variable importance rank does not necessarily reflect reliable difference in the prediction power, probably due to the fact that the difference in the importance values is small for the six predictors. Visual-spatial and visual memory are slightly more important, but the five cognitive factors are generally close in predicting mortality. Cognitive factors improve the prediction of mortality compared to using gender and education alone—consistent with the conclusion from the Cox regression, and the extra reduction in prediction error is about 5%.

We also compare the prediction error resulting from using the five factor scores with the prediction errors resulting from using the first principal component and the 16 test scores. The forest with the first principal component, along with gender and education, yields a prediction error of 41.41%, slightly higher than the first forest (40.42%). The error rate is 39.78% when using 16 test scores plus gender and education.

Conclusion

Substantive Findings

The conclusions from the Cox regression and the random survival forests are largely consistent and in agreement with findings reported by others, that cognitive abilities contribute to the prediction of mortality above and beyond gender and education. This is about a 5% increment in prediction accuracy based on Harrell's concordance index using all five factors of cognition. Educational attainment is considered to be strongly associated with intelligence; however, the extra part predicted by intelligence but not education, suggests that intelligence may be more internally or biologically determined, whereas educational attainment may be more socially driven.

Specific cognitive abilities are similar in strength in terms of predicting mortality. Using all of them does not improve prediction accuracy by much compared to using any one of them, and is similar to using a more general form of intelligence score (i.e., first principal component, see Table 16.5). Of course, we then would be relying on a single factor model of measurement which does not actually fit these five factor HFSC data (e.g., Horn & McArdle, 2007). Possibly of more importance may be that these five factors measured in the HFSC may not be the best cognitive indicators of mortality – i.e., there is no direct measure of "fluid reasoning" in the HFSC archive (e.g., Horn & Cattell, 1967). In addition, our results from the Cox regression suggest that numerical ability may be the strongest predictor, whereas results from the random survival forest suggest that visual-spatial and visual memory are more important predictors. However, the difference is generally small among the five cognitive abilities based on either method, and the results may not be generalized unless they can be replicated in future research.

Although no interaction effect is detected in the Cox regression, there are some clues from the survival tree analyses that sex-specific effects might exist with respect to different domains of intelligence in the cognition–mortality link. Specifically, we found that numerical ability might be a stronger predictor for women while visual-spatial might be a strong predictor for men. But again, the evidence is not strong, since no gender by cognition interaction is significant in the Cox regression, meaning that these sex-specific effects are not supported on a probability basis via the formal hypothesis testing approach.

Limitations

The major limitation of our study is the high percentage of censoring in the data (85.1% on the whole; 79.8% for men and 90.4% for women). This substantially affects the statistical power of the Cox regression analysis, and may bias the estimate if the predictors are related to censoring. Such a high percentage of censoring also constrains the strength of survival trees and survival forests. Some results or patterns may become more salient, and new facts which are hidden now may be revealed when participants' vital status can be updated in the future.

Secondly, the results of the Cox regression analysis conducted with a narrower age range (35–55) compared to the full sample, showed that the effect of age seems to weaken the cognitive–mortality association. But exclusion of people who were tested at older than 55 years would further increase the censoring, and we are reluctant to lose any more of the already limited information. Again, when more death records are available in future, we can afford to examine a narrower age ranged group.

The third limitation is that the HFSC data we used here are from married couples, so they are not strictly independent observations. DeFries et al. (1979) pointed out that the spouse correlations in the cognitive scores were not as high as they appeared when the scores were corrected for age. In addition, they argued that if the scores obtained in this study were restricted in range, lower spouse correlations would occur. Although there is no clear evidence for the restriction of range in the cognitive scores, some homogenous characteristics of these people may restrict the range of longevity in the data. For instance, marital status is believed to be associated with longevity, so these married couples may benefit in this respect. Such a restricted range of data could decrease the estimate for the cognition–mortality association.

Methodological Issues

Random survival forests analysis, as promising as it is, is still being developed and needs more testing; interpretations should be made with caution. One potential flaw we found through the analyses was that when only gender was used in the forest, the prediction error (i.e., $1 - C$-index) obtained was 58.67% with 1,000 trees, higher than the chance level, 50%. When the number of trees is varied, the error rate does not stabilize and monotonically increases with the number of trees. This suggests that gender is used to point to the wrong direction in the prediction. But when a second predictor is put in, gender is used effectively and makes the most contribution to the prediction. We think that the failure of the forest when gender is used as the only predictor is because of two reasons. First, when only gender is available in growing the trees, there is only one possible split of the node, so the forest is 1,000 trees of two terminal nodes. Second, in aggregating these trees, the fitted value this algorithm employs is the cumulative hazard function, which may give the wrong prediction due to the confusion in the data—more men are dead than women (364 deaths vs. 174 deaths), but on average husbands are older than wives. This may be a flaw of using CHF in this particular situation, while the Cox regression still obtained the correct answer using gender only. This problem might be corrected when more death records are available and censoring is less serious.

On the other hand, with the inclusion of the other predictors, the forests start to work properly. The fact that the gender information is only utilized correctly in combination with the other predictors, suggests interactions between gender and the other variables. But what really happens inside the forest is unknown, and in this sense the random survival forest is like a black box—there is no way to depict how the input variables are related to the response variable. Their contributions are only reflected in the reduction of the prediction error.

Thus, we can recommend that survival trees and survival ensemble methods be used only in combination with standard survival analysis. Statistical learning is at its best when the goal is forecasting, and is particularly useful in exploring large datasets without specific theories about the phenomenon under investigation. When the research aim is to test specific hypotheses, these methods are no substitute for long established, testable models.

Note

This work was supported by a pilot grant awarded to the third author from the USC/UCLA Center on Biodemography and Population Health (P30 AG017265-07).

The technical appendix will appear on the website http://kiptron.usc.edu/

References

Bartlett, M. S. (1937). The statistical conception of mental factors. *British Journal of Psychology, 28*, 97–104.

Batty, G. D., Deary, I. J., & Gottfredson, L. S. (2007). Premorbid (early life) IQ and later mortality risk: Systematic review. *Annals of Epidemiology, 17*, 278–288.

Berk, R. A. (2008). *Statistical Learning from a Regression Perspective.* New York: Springer.

Breiman, L. (2001). Random forests. *Machine Learning, 45*, 5–32.

Breiman, L. (2003). Manual—setting up, using and understanding random forests V4.0. Retrieved from ftp://ftp.stat.berkeley.edu/pub/users/breiman/Using_random_forests_v4.0.pdf

Breiman, L., Friedman, J. H., Olshen, R., & Stone, C. J. (1984). *Classification and Regression Trees.* New York: Chapman & Hall.

Ciampi, A., Thiffault, J., Nakache, J. P., & Asselain, B. (1986). Stratification by stepwise regression, correspondence analysis and recursive partition: A comparison of three methods of analysis for survival data with covariates. *Computational Statistics and Data Analysis, 4*, 185–204.

Cox, D. R. (1972). Regression models and life tables. *Journal of the Royal Statistical Society Series B, 34*(2), 187–220.

Deary, I. J. (2009). Introduction to the special issue on cognitive epidemiology. *Intelligence, 37*, 517–519.

Deary, I. J., & Batty, G. D. (2007). Cognitive epidemiology. *Journal of Epidemiology and Community Health, 61*, 378–384.

DeFries, J. C., Ashton, G. C., Johnson, R. C., Kuse, A. R., McClearn, G. E., Mi, M. P., Rashad, M. N., Vandenberg, S. G., & Wilson, J. R. (1976). Parent–offspring resemblance for specific cognitive abilities in two ethnic groups. *Nature, 261*, 131–133.

DeFries, J. C., Johnson, R. C., Kuse, A. R., McClearn, G. E., Polovina, J., Vandenberg, S. G., & Wilson, J. R. (1979). Familial resemblance for specific cognitive abilities. *Behavior Genetics, 9*(1), 23–43.

DeFries, J. C., Vandenberg, S. G., McClearn, G. E., Kuse, A. R., Wilson, J. R., Ashton, G. C., & Johnson, R. C. (1974). Near identity of cognitive structure in two ethnic groups. *Science, 183*, 338–339.

Fantazzini, D., & Figini, S. (2009). Random survival forests models for SME credit risk measurement. *Methodology and Computing in Applied Probability, 11*, 29–45.

Gordon, L., & Olshen, R. A. (1985). Tree-structured survival analysis. *Cancer Treatment Reports, 69*, 1065–1069.

Harrell, F., Califf, R., Pryor, D., Lee, K., & Rosati, R. (1982). Evaluating the yield of medical tests. *Journal of the American Medical Association, 247*, 2543–2546.

Hershberger, S. L. (2005). Factor scores. In B. S. Everitt and D. C. Howell (Eds.), *Encyclopedia of Statistics in Behavioral Science* (pp. 636–644). New York: John Wiley.

Horm, J. (1996). *Assignment of Probabilistic Scores to National Death Index Record Matches*. Hyattsville, MD: National Center for Health Statistics.

Horn, J. L., & Cattell, R. B. (1967). Age differences in fluid and crystallized intelligence. *Acta Psychologica, 26*, 107–129.

Horn, J. L., & McArdle, J.J. (2007). Understanding human intelligence since Spearman. In R. Cudeck & R. MacCallum (Eds.), *Factor Analysis at 100 Years* (pp. 205–247). Mahwah, NJ: Lawrence Erlbaum Associates.

Hothorn, T., Bühlmann, P., Dudoit, S., Molinaro, A., & van der Laan, M. J. (2006a). Survival ensembles. *Biostatistics, 7*(3), 355–373.

Hothorn, T., Hornik, K., Strobl, C., & Zeileis, A. (2010). Package 'party': A laboratory for recursive part(y)itioning (R package Version 0.9-9997) [Computer software]. Retrieved from http://cran.r-project.org/web/packages/party/index.html

Hothorn, T., Hornik, K., & Zeileis, A. (2006b). Unbiased recursive partitioning: A conditional inference framework. *Journal of Computational and Graphical Statistics, 15*, 651–674.

Ishwaran, H., & Kogalur, U. B. (2007). Random survival forests for R. *R News, 7*(2), 25–31.

Ishwaran, H., & Kogalur, U. B. (2010). Package 'randomSurvivalForest': Random Survival Forest (R package Version 3.6.3) [Computer software]. Retrieved from http://cran.r-project.org/web/packages/randomSurvivalForest/index.html

Ishwaran, H., Kogalur, U. B., Blackstone, E. H., & Lauer, M. S. (2008). Random survival forests. *The Annals of Applied Statistics, 2*(3), 841–860.

Kleemeier, R. (1962). Intellectual changes in the senium. In *Proceedings of the Social Statistics Section of the American Statistical Association* (pp. 290–295). Washington, DC: Statistical Association.

Kuse, A. R. (1977). Familial resemblances for cognitive abilities estimated from two test batteries in Hawaii. Unpublished doctoral dissertation, University of Colorado.

LeBlanc, M., & Crowley, J. (1992). Relative risk trees for censored survival data. *Biometrics, 48*, 411–425.

LeBlanc, M., & Crowley, J. (1993). Survival trees by goodness of split. *Journal of the American Statistical Association, 88*, 457–467.

Molinaro, A. M., Dudoit, S., & van der Laan, M. J. (2004). Tree-based multivariate regression and density estimation with right-censored data. *Journal of Multivariate Analysis, 90*, 154–177.

Riegel, K. F., & Riegel, R. M. (1972). Development, drop, and death. *Developmental Psychology, 6*, 306–319.

Rizk, N. P., Ishwaran, H., Rice, T. W., Chen, L. Q., Schipper, P. H., Kesler, K. A., et al. (2010). Optimum lymphadenectomy for esophageal cancer. *Annals of Surgery, 251*(1), 46–50.

Segal, M. R. (1988). Regression trees for censored data. *Biometrics, 44*, 35–47.

Strobl, C., Malley, J., & Tutz, G. (2009). An introduction to recursive partitioning: Rational, application, and characteristics of classification and regression trees, bagging, and random forests. *Psychological Methods, 14*(4), 323–348.

Wechsler, D. (1955). *Manual for the Wechsler Adult Intelligence Scale*. New York: The Psychological Corporation.

Weichselbaum, R. R., Ishwaran, H., Yoon, T., Nuyten, D. S. A., Baker, S. W., Khodarev, N., et al. (2008). An interferon-related gene signature for DNA damage resistance is a predictive marker for chemotherapy and radiation for breast cancer. *PNAS, 105*(47), 18490–18495.

Wilson, J. R., Defries, J. C., McClearn, G. E., Vandenberg, S. G., Johnson, R. C., Mi, M. P., & Rashad, M. N. (1975). Cognitive abilities: Use of family data as a control to assess sex and age differences in two ethnic groups. *International Journal of Aging and Human Development, 6*, 261–276.

Zhang, H. P., & Singer, B. (1999). *Recursive Partitioning in the Health Sciences*. New York: Springer.

17 Exploratory Analysis of Effects of Prenatal Risk Factors on Intelligence in Children of Mothers with Phenylketonuria

Keith F. Widaman and Kevin J. Grimm

Introduction

Phenylketonuria is an inborn error of metabolism that can have devastating effects on child intelligence. Children of mothers with phenylketonuria can suffer damage in utero, even if they do not have the full phenylketonuria genetic mutation. One crucial question is the relation between prenatal phenylalanine exposure by the fetus in utero and intelligence in childhood and later. Prior research and theory predicted a threshold effect, with prenatal phenylalanine exposure up to a threshold having no effect, but exposure above the threshold having a teratogenic effect. To investigate this issue, we estimated the relation between average prenatal phenylalanine exposure and child intelligence assessed at ages 4 years and 7 years. Two analytic models were used: two-piece linear splines, and multivariate adaptive regression splines (MARS), the latter a new data-mining approach. For intelligence at 4 years of age, the two-piece linear spline model identified a threshold around 4 mg/dL of exposure, whereas the MARS model with multiple predictors identified multiple thresholds, the first around 8.5 mg/dL. For intelligence at 7 years of age, the two-piece linear spline model identified a threshold around 6.8 mg/dL of exposure, whereas the MARS model with multiple predictors identified multiple thresholds, the first around 4.5 mg/dL and the second around 9 mg/dL. Discussion centered on the trade-off between simple models and simple policy implications of the two-piece linear spline results versus the more complex representations and greater explained variance provided by MARS analyses.

Phenylketonuria, or PKU, is an in-born error of metabolism that disrupts the metabolism of phenylalanine (PHE) into tyrosine. Appearing normal at birth, an infant with PKU, if fed a standard diet with normal levels of PHE, will suffer brain damage due to the high levels of PHE in his/her blood. By 2 years of age, children with PKU on a normal diet have an average IQ around 50, and most fall at the moderate level of intellectual disability (ID; formerly called mental retardation, or MR). This brain damage and its associated low level of intelligence are permanent and therefore not remediable.

As discussed below, infants placed early and consistently on a low-PHE diet develop in ways that appear normal in most or all respects. Indeed, the mean IQ of groups of children and adolescents with good dietary practice is around 100, the population mean IQ. Because of this, research on and treatment of PKU are, in general, a story of scientific success. That is, a genetic defect that can cause very serious developmental deficiencies can be blocked from having its effect through the provision of an optimal environment—specifically, a diet that is low in phenylalanine. Researchers and practitioners in the field of PKU research can rightly feel proud with regard to advances in the identification and treatment of PKU.

But, even if individuals with PKU are saved from developmental defects by eating a low-PHE diet, these individuals are carriers of the PKU genetic defect, so the genetic defect is lurking, waiting to have an effect. As we discuss below, this issue of a lurking genetic defect is exemplified most powerfully in the study of mothers with PKU, women with the PKU genetic trait who inadvertently may damage their developing fetuses by consuming a diet that is high in PHE. Children born to mothers with PKU are at risk for a number of negative developmental outcomes including lowered intelligence, and we will provide empirical results that illustrate these problems.

Before providing additional details regarding PKU and maternal PKU, we would like to highlight the fact that results discussed in this chapter are consistent with two major new trends in research on environmental and genetic contributions to behavioral traits and their developmental emergence. First, results discussed below on effects of prenatal exposure to phenylalanine are consistent with the work by Barker (1998), who reviewed and documented effects of prenatal environmental factors on a wide variety of developmental outcomes, even behavioral outcomes that manifest relatively late in life. In the current chapter, we concentrate on effects of prenatal exposure to phenylalanine on offspring intelligence during childhood, a more restricted range of the life span. Still, our results are in line with those by Barker and extend the reach of prenatal environmental effects to as important a behavioral outcome as intelligence.

Second, our results underscore the fact that genetic inheritance is not destiny and that environments can alter the effects of genetic risk factors on psychological traits in important ways. Behavioral genetic studies often conclude with estimates of the heritability of traits, where heritability is the proportion of population variance on a trait that is due to genetic factors. Experts on behavior genetics continually stress that high heritability does not mean that a trait is immutable by the environment. However, many practicing developmental researchers presume that high heritability implies, in general, that development of a trait will unfold in a relatively routine fashion controlled by genetic factors.

But, research published during the past decade defies this "immutability" conclusion and demonstrates the interactive effects of genes and environments even on highly heritable traits such as depression. In a series of studies, Caspi and colleagues (e.g., Caspi et al., 2002) investigated the effects of single nucleotide polymorphisms (SNPs) on behavioral outcomes. The most important findings concerned interactions of environmental factors with the SNPs. Persons with one version of a particular SNP may be relatively unaffected by environment stress, whereas persons with a different version of the same SNP may be relatively strongly affected, with increased environmental stress leading to increased psychopathology.

Research on maternal PKU extends this research on interactions of genetic and environmental factors in fascinating ways. In particular, the genetic defect in one organism (i.e., the mother) in interaction with her environment (i.e., her diet) can create a teratogenic prenatal environment that damages another organism (i.e., the developing fetus). Because the developing fetus does not have PKU, this offspring would never manifest the negative developmental outcomes associated with the PKU syndrome. Moreover, current behavior genetic methods are incapable of evaluating the heritability of the effects of this genetic defect. But, research on maternal PKU provides strong evidence for effects of both genes and environments in the development of intelligence in children of mothers with PKU. To understand these statements better, we turn next to an overview of research on PKU.

Brief History of PKU Research and Practice

To provide context for the data used in this chapter, we first discuss general findings concerning PKU and then discuss the additional problems associated with maternal PKU. A brief, yet very informative history of milestones in research on PKU was presented by Koch and de la Cruz (1999a), the first article in a special journal issue devoted to documenting state-of-the-art information on PKU (Koch & de la Cruz, 1999b). PKU was first identified by Følling in 1934. Følling noted the presence of phenylketone in the urine of several patients who had severe mental retardation. Phenylketone is a byproduct of the metabolism of phenylalanine, and incomplete metabolism of phenylalanine led to high levels of phenylketone expelled in urine.

Later work established that PKU is caused by identifiable genetic defects and is associated with high levels of phenylalanine in the blood. Phenylalanine is an essential amino acid. Essential amino acids, including phenylalanine, are necessary for protein synthesis and must be derived from the diet. Normally, phenylalanine is metabolized (i.e., broken down and converted) into tyrosine, a nonessential amino acid. In turn, tyrosine is a precursor of several key neurotransmitters and hormones, including epinephrine, norepinephrine, and dopamine, that are necessary for normal brain development and functioning. In PKU, metabolism of phenylalanine into tyrosine is disrupted, leading to high levels of phenylalanine and low levels of tyrosine in the blood. The high levels of PHE and low levels of tyrosine are both potential factors leading to brain damage. But, research has shown that the high levels of blood PHE is the teratogenic factor (i.e., the factor causing damage), not the low levels of tyrosine, because adequate levels of tyrosine can be obtained from a normal diet. High levels of blood PHE during infancy and childhood lead to severe, permanent brain damage, as noted above. Interestingly, the exact nature of the biological impact on neuronal function in the brain is not yet understood. Two mechanisms by which PKU disrupts and damages brain function have been proposed. One mechanism emphasizes the deficiency of dopamine in the brain, which results from the lowered levels of tyrosine in PKU-affected individuals. The second mechanism centers on a generalized slowing of the development of the myelin sheath of neuronal fibers in the brain (Dyer, 1999). Current data support both hypotheses, and researchers have not yet identified either as the more crucial underlying mechanism.

In the early 1950s, a German researcher, Bickel, and his colleagues demonstrated that a low-phenylalanine diet had clear beneficial effects in treating a young child with PKU. A low-phenylalanine diet typically consists primarily of a phenylalanine-free medical formula drink. However, the low-PHE diet can also include carefully selected amounts of fruits, vegetables, and low-protein breads and pastas. The most important characteristic of a low-PHE diet is the avoidance of high-protein foods.

In 1961, Guthrie devised a simple test to identify PKU within the first week after birth. The Guthrie test involves a heel stick and blotting of the infant's blood on filter paper. These specimens can be sent to state laboratories, and infants with PKU can be identified very early in life. The ease, accuracy, and inexpensiveness of the Guthrie test led to its rapid adoption across the United States, and screening tests are now routinely used on births throughout the world. With an early-identification test and low-PHE diet for affected individuals, the negative, teratogenic effects of PKU could largely be avoided.

Incidence and Severity of PKU Genetic Mutation

PKU is one of the most common inborn errors of metabolism, occurring in 1 in 10,000 to 15,000 live births, although the rate varies across populations. PKU is caused by mutations on the gene coding for the enzyme phenylalanine hydroxylase (PAH). PKU is an autosomal recessive trait, so a person will exhibit symptoms of PKU only if she or he receives a defective PAH gene from both parents. If a person receives a defective PAH gene from only one parent, the person will show no symptoms of PKU but would be a carrier of the defect. The PAH Locus Knowledgebase Web site (http://www.pahdb.mcgill.ca/) lists state-of-the-art information about PKU, including the over 500 different mutations on the PAH gene identified to date.

PAH gene mutations vary in their severity, with more severe mutations leading to lower levels of PHE metabolism than milder mutations. If an infant is on a normal diet, levels of blood PHE between 0 and 3 milligrams per deciliter (mg/dL) are considered normal. If blood PHE levels fall between 3 and 10 mg/dL, the infant is diagnosed with mild hyperphenylalaninemia, or MHP, which indicates some disruption in the metabolism of PHE into tyrosine. However, infants with MHP are often considered so little affected that they are not placed on the low-PHE diet, under the common assumption that blood PHE levels under 10 mg/dL are not teratogenic and thus do not lead to brain damage. Infants with blood PHE levels over 10 mg/dL are identified as having PKU. The severity of mutations within the PKU class are often listed as: (a) mild PKU, with blood PHE levels between 10 and 15 mg/dL; (b) moderate PKU, with blood PHE levels between 15 and 20 mg/dL; and (c) classic PKU, with blood PHE levels over 20 mg/dL. By placing infants with PKU on a low-PHE diet by the third week after birth, physicians hope to keep blood PHE levels between 3 and 10 milligrams per deciliter (mg/dL), because they assume that levels below 10 mg/dL do not lead to brain damage.

Debate on the level of phenylalanine required to ensure normal brain development continues at present. Some experts argue that blood PHE levels should be kept below 6 mg/dL, and others argue that PHE levels of 10 mg/dL or below are good enough and are not teratogenic. The bright line of 10 mg/dL that separates the presumably safe vs. teratogenic levels of PHE is the basis for standard treatment of infants with MHP, who are not subjected to the difficult-to-maintain low-PHE diet because they have levels of blood PHE that are not considered dangerous. But, surprising variability in standards is seen in the developed countries. Indeed, professional recommendations regarding the level of blood phenylalanine to maintain and the age at which it is safe to discontinue the restricted diet vary widely across European countries and the United States (Burgard, Link, & Schweitzer-Krantz, 2000; Schweitzer-Krantz & Burgard, 2000). In the United States, experts on PKU often debate the boundary of the safe levels of PHE exposure, with values between 6 and 10 mg/dL being the most commonly considered. But, some European countries consider levels of PHE exposure up to 15 mg/dl to be unproblematic. Furthermore, across organizations, the recommended age for low-PHE diet discontinuance varies considerably, with some recommending the diet until 10–12 years of age, others recommending the diet until 20 years of age (i.e., the end of the developmental period), and still others recommending that the low-PHE diet be continued throughout life and thus never discontinued.

In addition, many researchers design studies with groups of PKU patients divided into groups, such as groups for 3–10 mg/dL, 10–15 mg/dL, 15–20 mg/dl, and greater than 20 mg/dL. The results of research designed in this fashion tend to reinforce the idea that pre-defined levels of PHE exposure are the key levels to be investigated. For example, assume that the 10–15 mg/dL group performs only a bit lower than the 3–10 mg/dL group, but the remaining two groups show much lower performance. This might lead researchers to argue that levels below 15 mg/dL are not problematic. But, effects of phenylalanine exposure are likely to be threshold effects, with essentially no effect of increasing exposure up to a threshold of PHE exposure, and a negative, teratogenic effect of exposure beyond the threshold. Thus, more informative results are likely to come from treating phenylalanine level as a continuous variable and searching for the crucial threshold of exposure, than from continuing to categorize persons coarsely into PHE-level groups and basing policy recommendations on resulting group comparisons. Useful information on many aspects of diagnosing and managing PKU is contained in the statement by an NIH Consensus Panel, *Phenylketonuria: Screening and Management* (NIH, 2000).

Behavioral Effects of High PHE

The potentially devastating effects of PKU on development are well known. If the diet of an infant with PKU is unrestricted, its intellectual functioning declines precipitously. Apparently normal at birth, such an infant will fall to the level of severe mental retardation (mean IQ of 50) by age 2—a decline that cannot be reversed. But if an infant with PKU is placed on a low-phenylalanine diet early in life and continues strictly on such a diet until age 20 years or later, he or she will exhibit normal or near-normal development.

Two recent studies that included cognitive measures deserve special mention. In the first, Koch et al. (2002) retested adults who had participated in a large study of diet adherence and discontinuance during the developmental period. Interestingly, individuals were randomly assigned to diet discontinuance groups so that firmer causal conclusions could be supported. Results revealed that the mean IQ of a group of persons who had never discontinued their low-phenylalanine diets was 17 points higher than that of a group of persons who had discontinued their diets at mean age 8 years. Further, of 16 individuals who later resumed the low-phenylalanine diet, 9 persons who maintained the diet into adulthood showed a significant rise in mean IQ from childhood to adulthood, whereas 7 individuals who later discontinued the diet had a significant drop in mean IQ from childhood to adulthood.

In the second study, Diamond, Prevor, Callender, and Druin (1997) tested infants and children with PKU or MHP and compared the performance of a low-phenylalanine group (levels between 2 and 6 mg/dL) and a high-phenylalanine group (levels between 6 and 10 mg/dL) against the performance of multiple comparison groups on cognitive tasks relying on prefrontal cortex functions. The low-phenylalanine group performed similarly to comparison groups, whereas the high-phenylalanine group exhibited significant deficits in performance. Thus, the Koch et al. study supports recommendations that a low-phenylalanine diet should be continued at least through the developmental period, and the Diamond et al. study suggests that phenylalanine levels should be maintained below 6 mg/dL.

In many ways, research on PKU represents a scientific success story (see Koch & de la Cruz, 1999b, for a review of PKU research). Until the early 1960s, biology was destiny for persons with PKU. Diagnosis of PKU usually happened so late during development—even if during infancy or early childhood—that severe brain damage had already occurred and could not be remedied. If behavior-genetic studies had been conducted, the heritability (which estimates the proportion of variance in a phenotypic trait that is attributable to genetic variation among individuals) for the PKU syndrome would have been very high. However, after development of the Guthrie screening test, infants could be placed early and continuously on low-phenylalanine diets, and brain damage could be largely or completely prevented. Thus, heritability of the PKU syndrome would have fallen to low levels in a single generation because of a key manipulation of the environment—the use of a low-phenylalanine diet.

Maternal PKU

The success of medical science in treating PKU is well known, but less widely known are the potentially devastating effects of maternal PKU. Lenke and Levy (1980) reported results of over 500 pregnancies of women with PKU who were not on low-phenylalanine diets during pregnancy. Infants born to mothers with PKU had a high rate of birth defects and developmental disabilities. These infants never would have exhibited symptoms of PKU, because they received a PAH gene defect only from their mothers, so the non-defective PAH gene inherited from the father would have allowed normal metabolism of PHE. But, phenylalanine in the mother's blood during pregnancy passed the placental barrier and exposed the fetus to high levels of phenylalanine prenatally. These prenatal-exposure effects showed a dose-response relation, with higher levels of exposure leading to higher levels of disability.

The striking results that Lenke and Levy (1980) presented are worthy of note. Of children born to mothers with classic PKU (PHE $\geq$ 20 mg/dL), over 90% had ID, 73% had microcephaly, 12% had congenital heart disease, and 40% had low birthweight. A general dose-response effect held, as children of women with moderate PKU (20 mg/dL > PHE $\geq$ 15 mg/dL) had lower, but still significantly elevated, levels of ID (73%), microcephaly (68%), and congenital heart disease (15%). Mothers with mild PKU (15 mg/dL > PHE $\geq$ 10 mg/dL) had less compromised offspring, of whom 22% had ID, 35% had microcephaly, and 6% had congenital heart disease. Interestingly, even the women with MHP (10 mg/dL > PHE $\geq$ 3 mg/dL), with PHE levels thought to be in a non-problematic range, had offspring who exhibited some deficits. For example, 21% of children born to mothers with MHP had ID and 24% had microcephaly.

The Lenke and Levy (1980) study suggested that research on maternal PKU might productively center on levels of maternal blood PHE that are "safe," meaning that these levels lead to no notable negative effects on the developing fetus, and levels of maternal blood PHE that can lead to damage. The results of the Lenke and Levy study, particularly the results for the women with MHP, suggest that even some PHE levels that are in a range that is often considered non-problematic (i.e., PHE $\leq$ 10 mg/dL) may lead to negative developmental outcomes in offspring.

Nonlinear Modeling of PHE—Outcome Relations

Two-piece Linear Spline Models

One statistical approach that can identify one or more points at which a regression function changes slope is the use of linear spline regression models. Two-piece linear spline models divide the regression space into two parts: one regression slope prior to the knot point, a second regression slope after the knot point, and an estimated knot point. The two-piece linear spline model can be written as:

$$y_n = b_0 + b_1 \cdot \min(x_n - c, 0) + b_2 \cdot \max(x_n - c, 0) \tag{1}$$

where b_0 is the intercept and predicted value of y_n when $x_n = c$, b_1 is the pre-knot slope indicating the expected amount of change in y_n for a one-unit change in x_n when $x_n < c$, b_2 is the post-knot slope indicating the expected amount of change in y_n for a one-unit change in x_n when $x_n > c$, and c is the knot point.

The estimation of this model can be approached from multiple angles. At times, the knot point is fixed to test specific hypotheses regarding the location of the change point and such models can be estimated using linear regression models. Additionally, the series of knot points can be tested and compared utilizing a profile likelihood method. That is, values of -2 times the log likelihood of each fitted model can be plotted against the location of the knot point, and the model with the smallest -2(log likelihood) value is selected as the optimal model. The knot point for this selected model is used as the optimal value of the knot point. A second approach is to estimate the model of Equation 1 directly. In this model, the optimal knot point (along with appropriate standard error) can be estimated directly; however, estimation can only take place within nonlinear regression programs, such as PROC NLIN in SAS.

In terms of specification, the model of Equation 1 can be specified in various ways to tests specific hypotheses regarding the need for certain model parameters. For example, the model can be simplified by removing a pre-knot (or post-knot) slope (i.e., by fixing the relevant slope parameter to zero). If this approach were taken, no effect of x_n before (or after) the knot point would be estimated, but an effect of x_n after (before) the knot point would be estimated. This type of model is thought of as a threshold model, such that x_n is unrelated to y_n up to a certain level (the threshold) of x_n, after which x_n begins to have an impact on y_n. On the other hand, additional terms can be added in the form of more than one knot point (e.g., c_1 and c_2) or nonlinear effects (e.g., x_n^2) to model higher degrees of nonlinearity in the association between x_n and y_n within defined intervals.

The spline regression model carries several benefits for modeling nonlinear associations, including interpretation of the fit of the model to data and elegance of the resulting predicted function. However, the model has three noticeable limitations. First, the number of knot points is chosen by the researcher, and many researchers do not examine more than a single knot point. Second, spline regression models are often fit with a single input variable (x_n) due to the complexity of model fitting and researchers lacking theories regarding knot points for multiple input variables. Third, higher-order effects and interaction terms are rarely considered because of the confirmatory nature of model fitting.

Multivariate Adaptive Regression Splines

An alternative to the confirmatory nature of fitting the spline regression is to use Multivariate Adaptive Regression Splines (MARS; Friedman, 1991). MARS takes the model of Equation 1 and places it within an exploratory framework. That is, in the univariate case, MARS searches for the optimal number and location of knot points. Additionally, higher-order terms of the input variable (x_n^2) can be included within the search by simply changing the order of the search (order = 1 = linear; order = 2 = quadratic). In the multivariate case, a collection of potential input variables can be entered. MARS searches all variables (potentially including all interactions and higher-order terms) and potential knot points to derive an optimal solution. Both the input variables to use and the knot points for each variable are found via a brute force, exhaustive search procedure, using very fast updating algorithms and efficient program coding.

Input variables, knot points, and interactions are optimized simultaneously by evaluating a "loss of fit" (LOF) criterion—choosing the LOF that most improves the model at each step. In addition to searching variables one by one, MARS also searches for interactions between variables, allowing any degree of interaction to be considered. An optimal model is selected in a two-phase process. In the first phase, a model is grown by adding spline functions (i.e., new main effects, knots, or interactions) until an overly large model is found. In the second phase, spline functions are deleted in order of least contribution to the model until an optimal balance of bias and variance is found. By allowing for any arbitrary shape for the response function as well as for interactions, MARS is capable of reliably tracking very complex data structures that often hide in high-dimensional data.

MARS is a more flexible approach to regression compared with standard approaches to multiple linear regression, even spline regression modeling. It yields simple to understand findings, handles automatic variable selection, and tends to have good bias–variance tradeoff. Thus, using an admittedly exploratory procedure, MARS answers the several limitations of standard fitting of spline regression models in a confirmatory manner.

The Present Study

The goals of the present study are to compare and contrast the results of applying different forms of spline models to IQ scores of children born to mothers with PKU. In particular, we will compare results from use of traditional two-piece linear splines, MARS analyses with a single input variable, and MARS analyses with multiple input variables. Our principal goal is to understand better the form of the function relating prenatal PHE exposure to child intelligence during childhood, and we therefore used different analytic approaches that might elucidate this relation more completely.

Method

Participants

The Maternal PKU Collaborative (MPKUC) Study (Koch, de la Cruz, & Azen, 2003) was initiated in 1984 to monitor pregnancies of women with PKU, maintain mothers on a low-phenylalanine diet throughout their pregnancies, and study relations between levels of phenylalanine in the mothers' blood during pregnancy and birth and developmental outcomes of their offspring. All 413 children in the MPKUC Study received a gene defect only from their mothers. Consequently, they never would have exhibited symptoms of PKU as they would have metabolized phenylalanine normally. However, these children were exposed prenatally to differing levels of phenylalanine, with varying teratogenic effects.

Due to less-than-complete adherence to low-phenylalanine diets during their own development, mothers in the MPKUC Study tended to have lower-than-average IQ scores (mean WAIS Full Scale IQ = 86). Because pregnancy alters food preferences and makes it more difficult to adhere to a diet, the phenylalanine level in a mother's blood was monitored at each prenatal visit (if possible) throughout her pregnancy. Prenatal visits occurred weekly for some mothers, biweekly for others, and less regularly for the rest. Despite attempts by medical personnel to maintain MPKUC Study participants on the diet, mothers exhibited wide individual differences in mean blood phenylalanine level across the pregnancy (ranging from 1.3 to 28.3 mg/dL).

Measures

Input variables

A number of background or demographic variables were available for the mothers in the MPKUC Study. These included (Table 17.1):

- Socioeconomic status, using the Hollingshead index, a five-level SES index scored so that the highest score (5) indicated the highest level of socioeconomic status.
- Mother's education, in years
- Mother's Verbal IQ, from the Wechsler Adult Intelligence Scale
- Mother's Performance IQ, from the Wechsler Adult Intelligence Scale
- Mother's age (in years) at conception of the target child
- Child sex, coded 1 = male, 0 = female
- Mother's assigned PHE level, or PHE level in the blood when on a regular diet
- Mother's Guttler score, an index of severity of her PKU mutation (range 0–16)
- Average PHE exposure for the child in utero, assessed as the mean PHE level in the mother's blood during pregnancy
- Weeks in gestation when mother's PHE level consistently fell below 6 mg/dL
- Child birth head circumference

Table 17.1 Descriptive statistics for the manifest variables from the MPKUC study

Variable	N	Mean	SD	Min	Max
SES (Hollingshead)	389	2.03	0.95	1	5
Mother's education	397	4.09	1.08	1	7
Mothers Verbal IQ	376	85.57	12.74	22	129
Mother's Performance IQ	378	88.50	14.43	47	132
Child sex (Male = 1)	413	0.49	0.50	0	1
Mother's assigned PHE level	413	22.03	9.18	3.3	51.1
Mother's Guttler score	247	4.00	2.55	2	12
Average PHE Exposure	412	8.23	4.49	1.3	28.3
Weeks gestation PHE < 6 mg/dL	413	25.24	14.39	0	42.3
Child birth head circumference	403	32.84	1.95	26.0	38.0
McCarthy GCI	276	85.18	21.24	45	132
WISC Full Scale IQ	284	91.35	23.21	35	139

Outcome Variables

With regard to mental ability, offspring in the MPKUC Study were assessed using the Bayley Scales of Infant Development at 1 year and 2 years, the McCarthy Scales of Children's Abilities at 4 years, and the Wechsler Intelligence Scale for Children–Revised (WISC-R) at 7 years. For brevity, we will concentrate on McCarthy General Cognitive Index (GCI) scores at 4 years of age and WISC-R Full Scale IQ scores at 7 years of age, although similar results were found for outcomes at earlier ages.

Analyses

Data for the Maternal PKU study contained records for 413 participants. The data were reduced separately due to missing outcome data (McCarthy General Cognitive Index at 4 years or WISC Full Scale IQ at 7 years). A total of 276 participants had McCarthy GCI scores at 4 years, and 284 participants had WISC Full Scale IQ scores at 7 years. The remaining data were largely complete as input variables had less than 4% incomplete values. Data missing from these reduced datasets were then imputed using the "mi" (Su, Gelman, Hill, & Yajima, 2011) library in R. Squared and cubed terms were added to the imputation to help account for any nonlinear relationships between variables.

These data were then subjected to spline modeling, including the use of Multivariate Adaptive Regression Spline (MARS) (Friedman, 1991) models using the "earth" (Milborrow, 2011) library in R. First, we present two-piece linear spline regression analysis results reported by Widaman and Azen (2003) for the McCarthy GCI scores assessed when participants were 4 years of age and WISC IQ scores when participants were 7 years of age. Then, MARS models were fit separately using McCarthy GCI scores and WISC Full Scale IQ as outcome variables, with the child's average PHE exposure during gestation as the only input variable. Subsequently, the following variables were added as input variables to the MARS analyses: socioeconomic status, maternal education, maternal verbal IQ, maternal performance IQ, mother's age at conception of target child, child gender, mother's assigned PHE level on a regular diet, months of gestation with no PHE levels over 6 mg/dL, and the child's head circumference at birth. Stepwise inclusion of the additional predictors was performed, presuming the presence of correlational overlap among these additional predictors.

Results

McCarthy General Cognitive Index at 4 years

Our first set of analyses used McCarthy General Cognitive Index (GCI) scores obtained when offspring were 4 years of age as the outcome variable. We report results in three sections, first describing the fitting of two-piece linear spline models to the data using the SAS PROC NLIN program, then the fitting of adaptive spline models with a single independent variable using the MARS package in R, and finally the fitting of adaptive spline models using MARS in which additional predictors were included.

Two-piece Linear Spline Modeling

In prior analyses of data, Widaman and Azen (2003) reported results of the fitting of two-piece linear spline models to data from the MPKUC Study. The two-piece linear spline model for McCarthy GCI scores led to the following model:

$$GCI = 98.5 - 3.08 \, [\max(PHE - 3.91, 0)] \tag{2}$$

which is interpreted in the following way: The intercept estimate is 98.5, and the slope for PHE level is -3.08, which means that predicted IQ drops by 3.08 points for every 1-per-mg/dL increase in PHE levels beyond the estimated knot point of 3.91 mg/dL. The max( ) function evaluates the expression (PHE $-$ 3.91, 0). At PHE levels below the knot point of 3.91, the value of (PHE $-$ 3.91) would be a negative value, so the max( ) function would return the value 0. As a result, for any PHE level below 3.91, the predicted GCI equals 98.5. For PHE values above 3.91, the predicted GCI would be 98.5 minus 3.08 times the number of mg/dL of PHE exposure above 3.91 mg/dL.

The two-piece linear spline model provided a strong account of the GCI data, with a squared multiple correlation of .381. That is, prenatal exposure to PHE explained over 38% of the variance in GCI scores at 4 years of age.

MARS Model with a Single Independent Variable

Next, we used the "earth" package in R, with offspring McCarthy GCI scores at 4 years of age as outcome variable. When the child's average PHE exposure during gestation was the only input variable, only one knot point was identified. The resultant predicted spline regression model was:

$$GCI = 53.4 + 3.40 \, [\max(17.6 - PHE, 0)] \tag{3}$$

The knot point was located at a PHE exposure level equal to 17.6 with a regression coefficient equal to 3.40. Thus, based on this MARS model, PHE exposure was negatively related to GCI scores until the level of exposure reached 17.6 mg/dL. Prior to 17.6 units, GCI scores were expected to decline 3.40 units for each one-unit increase in PHE exposure. Subsequent to 17.6 units, GCI scores were unrelated to PHE exposure. This spline regression model accounted for 42% of the variance in GCI scores. Figure 17.1 is a display of the predicted association between PHE Exposure and McCarthy GCI scores at age 4.

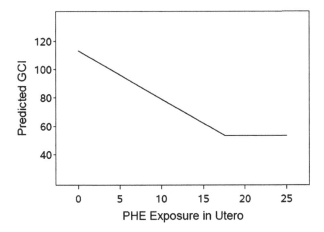

Figure 17.1 Plot of predicted offspring McCarthy GCI scores at 4 years of age as a function of PHE exposure in utero based on MARS analysis.

MARS Model with Additional Predictors

When the additional input variables were allowed to enter into the model, the explained variance increased to 50% and five of the ten input variables were selected for inclusion. Moreover, a total of eight knot points were identified across the five input variables. The following variables were selected and appear in order of importance: PHE exposure, maternal verbal IQ, months of gestation with no PHE levels of 6 mg/dL, mother's age at conception, and gender.

We note that, when the additional variables were included in the equation, the location of the knot point for PHE exposure changed. In this model, two knot points for PHE exposure were found—the first knot point at 8.7 units, and the second at 17.4 units. This MARS model predicted no relationship between PHE levels and GCI scores until PHE exposure hit 8.7 units. Subsequent to 8.7 units, GCI scores were negatively related to PHE exposure with an effect of 4.24 points per year for each one-unit increase in PHE exposure. Subsequent to PHE exposure of 17.4, a predicted positive relation between PHE exposure and intelligence was estimated: a 2.4 IQ point increase per one-unit change in PHE exposure.

WISC Full Scale IQ at 7 years

Our final set of analyses used WISC Full Scale IQ scores obtained when offspring were 7 years of age as the outcome variable. We report results for WISC IQ scores in three sections, first describing the fitting of two-piece linear spline models to the data using the SAS PROC NLIN program, then the fitting of adaptive spline models with a single independent variable using the "earth" package in R, and finally the fitting of adaptive spline models using MARS in which additional predictors were included.

Two-piece Linear Spline Model

Widaman and Azen (2003) reported results of the fitting of two-piece linear spline models to data from the MPKUC Study. However, Widaman and Azen reported spline models only for the Verbal IQ and the Performance IQ from the WISC, not the Full Scale IQ. The two-piece linear spline model for WISC Verbal IQ (VIQ) scores led to the following model:

$$VIQ = 98.3 - 4.09 \, [\max(PHE - 6.80, 0)] \tag{4}$$

and the spline model for WISC Performance IQ (PIQ) was:

$$PIQ = 99.7 - 4.25 \, [\max(PHE - 6.79, 0)] \tag{5}$$

Thus, for both VIQ and PIQ, the intercept fell close to the population mean of 100. At PHE levels below the knot point around 6.8 mg/dL, the value of (PHE − knot) would be a negative value, so the max( ) function would return the value 0. As a result, for any PHE level below around 6.8 mg/dL, the predicted VIQ and PIQ values would be 98.3 and 99.7, respectively. For PHE values above 6.8, predicted VIQ and PIQ values decline by 4.09 and 4.25 points, respectively, for every 1 mg/dL increase in PHE exposure.

The two-piece linear spline model led to strong prediction of WISC VIQ and PIQ scores, with squared multiple correlations of .388 and .425, respectively. That is, prenatal exposure to PHE explained over 38% of the variance in WISC VIQ and PIQ scores at 7 years of age.

MARS Model with Single Independent Variable

Next, we used offspring WISC Full Scale IQ score at 7 years of age as the outcome variable, but implemented our spline models in the "earth" (Milborrow, 2011) library in R. When the child's average PHE exposure during gestation was the only input variable, three knot points were identified.

The resultant adaptive spline regression model was:

$$WISC \; IQ = 103.5 - 1.98 \, [\max(PHE - 4.6, 0)]$$
$$- 5.15 \, [\max(PHE - 9.1, 0)] + 6.22 \, [\max(PHE - 14.3, 0)] \tag{6}$$

The three knot points were located at PHE exposure levels equal to 4.6, 9.1, and 14.3, with regression coefficients equal to −1.98, −5.15, and 6.22. Thus, based on this MARS model, PHE exposure was unrelated to Full

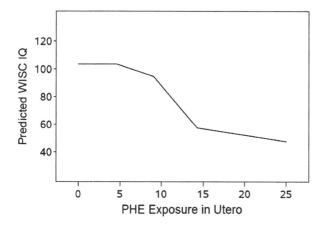

Figure 17.2 Plot of predicted offspring WISC full scale IQ scores at 7 years of age as a function of PHE exposure in utero based on MARS analysis.

Scale IQ scores until the level of exposure reached the first knot point, 4.6 mg/dL. Subsequent to this level of exposure, IQ scores were predicted to decline at the rate of 1.98 IQ points for each one-unit increase in PHE exposure until PHE exposure hit the second knot point, 9.1 mg/dL. At PHE exposure levels above the second knot point, the predicted decline in IQ scores changed to a decline of 7.13 (or −1.98 + (−5.15)) IQ points for each one-unit increase in PHE exposure. After the third knot point at 14.3 mg/dL of exposure, predicted IQ scores decreased at a rate of 0.91 (or −7.13 + 6.22) IQ points for each one-unit increase in PHE exposure. This spline regression model accounted for 50% of the variance in Full Scale IQ scores. Figure 17.2 is a display of the predicted association between PHE Exposure and Full Scale IQ at age 7.

MARS Model with Additional Predictors

When the additional input variables were entered into the model, the explained variance increased to 61%. Five of the ten input variables were selected, and a total of eight knot points were estimated for the five selected variables. The following variables were selected and appear in order of importance: PHE exposure, child's head circumference at birth, mother's age at conception, maternal verbal IQ, and social economic status.

With the inclusion of the additional predictor variables, only a single knot point for PHE exposure was identified, and its location was at a PHE exposure level of 4.6 mg/dL. Subsequent to this level of exposure, predicted IQ declined by 2.50 points for each one-unit increase in PHE exposure.

Conclusion

In prior published work, Widaman and Azen (2003) used two-piece linear spline regression to model the relation between prenatal PHE exposure and intelligence in children born to mothers with PKU. These linear spline models always had adequate fit to data, whether the outcome variables were general ability scores or more specific scores (e.g., language scores) at 1, 2, 4, or 7 years of child age. In the present chapter, we have concentrated on general intelligence outcomes at only the 4-year and 7-year assessments. For these outcome variables, the two-piece linear spline models fit well, had interpretable parameter estimates, and explained the relation between prenatal PHE exposure and ability test scores better than did linear regression models (cf. Widaman & Azen, 2003). Thus, the two-piece linear spline models were more adequate representations of the data and provided a more solid basis for policy recommendations regarding safe levels of prenatal PHE exposure than did linear regression models. In general, across a wide array of outcome variables at different age levels, the knot point for two-piece linear spline regression models ranged between 5 and 7 mg/dL. These consistent findings imply that practitioners whose patients are pregnant women with PKU should strive to have their patients keep blood levels of PHE below approximately 6 mg/dL throughout pregnancy to avoid teratogenic effects of fetal exposure to PHE.

One potentially biasing aspect of the Widaman and Azen (2003) results is associated with the form of two-piece linear spline model specification they used. In particular, based on theory regarding the expected threshold effect of PHE exposure, Widaman and Azen forced the initial slope of the PHE-intelligence relation (i.e., below the knot point) to be zero, assuming that low levels of PHE exposure would have minimal effects. The resulting estimates from the two-piece spline models were consistent across outcome variables, but did not allow non-zero slopes for PHE exposure below the knot point. When implementing models in this fashion, non-zero slopes below the knot point—which might lead to a better fitting two-piece spline regression model—were never considered.

The results obtained from analyses using the MARS program call some of the Widaman and Azen (2003) conclusions into question. For example, adaptive spline models can easily identify more than a single knot point, and the resulting predicted values for WISC Full Scale IQ scores under the adaptive spline models suggested that the rate of decline in predicted IQ per 1 mg/dL increase in PHE exposure varied as a function of overall level of PHE exposure. That is, although no decrease in predicted IQ occurred until the first knot point, the predicted decline between the first and second knot points (a decline of about 2 IQ points per unit increase in PHE exposure) was less extreme than that between the second and third knot points (decline of about 7 IQ points per unit increase in PHE exposure) for the WISC IQ at 7 years and was less than the decline (decline of about

4 IQ points per unit increase in PHE exposure) predicted by the two-piece linear splines estimated using PROC NLIN. That is, the predicted decline between the second and third knot points in the MARS analysis was considerably stronger than that estimated using the two-piece linear spline model in PROC NLIN. Thus, the MARS analysis suggests a more nuanced understanding of the relation between prenatal PHE exposure and offspring IQ at 4 and 7 years of age, with multiple knot points and differential slopes for predicted decline between knot points.

It is tempting to argue in favor of one or another of the present approaches—PROC NLIN vs. MARS—on the basis of fit of the model to data. For example, for McCarthy GCI scores, the PROC NLIN analyses explained 38% of the variance, the MARS analysis with one predictor explained 42% of the variance, and the MARS analysis with multiple predictors explained 50% of GCI variance. A similar pattern of explained variance was seen in the WISC IQ scores, where the PROC NLIN analyses explained 40% of the variance, the MARS single-predictor model explained 50% of the variance, and the MARS multiple predictor model explained 61% of the variance. Thus, MARS adaptive spline modeling with a single predictor appears to explain more variance than the PROC NLIN analyses, albeit often with the need to estimate additional knot points and additional slope values. One might wonder if the increased explained variance is worth the price of additional predictors, additional estimates, and more difficult, nuanced interpretation. We offer no firm conclusion on this matter, but leave it as an interesting rhetorical question.

Moreover, the MARS results were altered in substantial ways with the addition of additional predictors in the equation. One very striking result was the rather different pattern of predicted results for the McCarthy GCI. In the single predictor case, the MARS analysis found a single knot point. In this model, predicted GCI values declined rapidly from 0 mg/dL PHE exposure to the knot point located at over 17 mg/dL of PHE exposure; after the knot point, predicted values were unrelated to additional PHE exposure. This result is drastically different from the two-piece linear spline model using PROC NLIN, which had found no relation between PHE exposure and GCI scores up to the knot point around 6 mg/dL, and a strong negative relation after the knot point. One might wonder how two alternative two-piece linear spline models could arrive at such different results.

Before pondering this question in any detail, it is useful to consider the MARS analysis when additional predictors were added to the equation. In this model, the relation between prenatal PHE exposure and GCI was much more similar to the two-piece linear spline model from PROC NLIN. In particular, this MARS model predicted no relationship between PHE levels and GCI scores until PHE exposure hit the first knot point of 8.7 mg/dL. Subsequent to the knot point, predicted GCI scores were negatively related to PHE exposure, with an effect of a decline of 4.24 IQ points for every 1 mg/dL increase in PHE exposure. This knot point of 8.7 mg/dL was somewhat higher than the corresponding knot point from the PROC NLIN analyses, which fell at 3.9 mg/dL, and the slope coefficient from the MARS analysis (−4.24) was larger than that from the PROC NLIN analysis (−3.08). But, these MARS results were also tempered by a second knot point at 17.4 mg/dL, after which predicted IQ increased by 2.4 points for every 1 mg/dL increase in PHE exposure. Thus, by including additional predictors, the MARS analysis resulted in a substantially higher estimate of the point at which damage from prenatal PHE exposure is likely to occur and a steeper decline in damage after this point than estimated by the two-piece linear spline model fit with PROC NLIN.

Similar alterations in model predictions occurred when MARS was used to fit the adaptive spline model to the WISC IQ scores from 7 years of child age. In the single-predictor analyses, the MARS analysis identified three knot points, as shown in Figure 17.2. Then, in the multiple-predictor analysis, the MARS analysis suggested only a single knot point. Although a single knot point is more in line with results from the PROC NLIN modeling, the MARS analysis suggested a lower knot point (at 4.6 mg/dL) than did the PROC NLIN analysis (around 6.8 mg/dL), and the MARS analysis led to a less extreme predicted decline after the knot point (about a 2.5 point decline per unit increase in PHE exposure) than did the PROC NLIN analysis (about a 4.2 point decline per unit increase in PHE exposure).

Working with real empirical data is very exciting, particularly when the empirical data have as much import as do the current data. That is, the data from the MPKUC Study has been the basis of policy recommendations regarding safe levels of blood PHE in pregnant women with PKU, and prior policy recommendations were made on the basis of two-piece linear spline analyses. Moreover, the spline models we fit with PROC NLIN and MARS can provide very useful estimates of the likely damage to the developing fetus that might occur with various levels of prenatal exposure to PHE. Analyzing simulated Monte Carlo data—simulated to follow a particular generating mechanism or mechanisms—is usually not as interesting to the analyst on an analysis-by-analysis basis, because the processes generating the data are known. Admittedly, analyses of simulated data are of interest to practicing scientists because these analyses can offer important insights into the likelihood with which researchers will "find" a statistical model that matches the true generating mechanism.

The downside of analyzing empirical data is the lack of precise knowledge of the processes generating the data. Research and clinical practice suggest that prenatal PHE exposure can play a potent role in affecting growth and development in a number of behavioral domains, including the domain of intellectual abilities. But, whether prenatal PHE exposure is the only important variable or whether other variables—including maternal background variables—also play a role is crucial in understanding trends in the data. One of the most intriguing results presented in this chapter was the alterations in MARS results when additional predictors were added to the single-predictor equation. When additional predictors were added to the equation, the MARS analyses showed a different number of knot points and substantially altered slope coefficients than had been estimated in the single-predictor equation.

On the basis of the results presented in this chapter, we feel that it would be premature to designate one form of analysis the "generally correct" or "generally useful" model. We applied three general approaches to our data: two-piece linear spline models implemented with PROC NLIN, single-predictor adaptive spline models with MARS, and multiple-predictor adaptive spline models with MARS. Each approach appears to have its strengths, and each has its weaknesses. We look forward to continued research on data-mining methods that will offer additional insights into the special ways in which these methods are applicable and the special insights they bring to substantive research applications. We trust that the present chapter will serve to illuminate the differing understandings of data that accompany the use of these methods. And, we hope that the interactive effects of genetic risk factors and environmental factors has been more fully illustrated in our results.

Note

Work by Keith F. Widaman on this research was supported in part by grant HD 064687 from the National Institute of Child Health and Human Development (Rand D. Conger, PI) and grant DA 017902 from the National Institute of Drug Abuse and the National Institute of Alcohol Abuse and Alcoholism (Rand D. Conger, Richard W. Robins, and Keith F. Widaman, Joint PIs). Kevin J. Grimm was supported in part by National Science Foundation Reece Program Grant DRL-0815787.

References

Barker, D. J. P. (1998). *Mothers, babies, and health in later life* (2nd ed.). New York: Churchill Livingstone.

Bickel, H., Gerard, J., & Hickmans, E. M. (1953). Influence of phenylalanine intake on phenylketonurics. *Lancet, 262*, 812–813.

Burgard, P., Link, R., & Schweitzer-Krantz, S. (Eds.), (2000). Phenylketonuria: Evidence-based clinical practice. *European Journal of Pediatrics, 159*(Suppl. 2), S69–S168.

Caspi, A., McClay, J., Moffitt, T. E., Mill, J., Martin, J., Craig, I. W., et al. (2002). Role of genotype in the cycle of violence in maltreated children. *Science, 297*, 851–854.

Diamond, A., Prevor, M. B., Callender, G., & Druin, D. P. (1997). Prefrontal cortex cognitive deficits in children treated early and continuously for PKU. *Monographs of the Society for Research in Child Development, 62*(No. 4, Serial No. 252).

Dyer, C. A. (1999). Pathophysiology of phenylketonuria. *Mental Retardation and Developmental Disabilities Research Reviews, 5*, 104–112.

Følling, A. (1934). Über Ausscheidung von Phenylbrenztraubensäure in den Harn als Stoffwechselanomalie in Verbindung mit Imbezillität. *Zeitschrift für Physiologische Chemie, 227*, 169–181.

Friedman, J. H. (1991). Multivariate adaptive regression splines. *Annals of Statistics, 19*, 1–67.

Guthrie, R. (1961). Blood screening for phenylketonuria. *Journal of the American Medical Association, 178*, 863.

Koch, R., & de la Cruz, F. (1999a). Historical aspects and overview of research on phenylketonuria. *Mental Retardation and Developmental Disabilities Research Reviews, 5*, 101–103.

Koch, R., & de la Cruz, F. (Eds.), (1999b). Phenylketonuria. *Mental Retardation and Developmental Disabilities Research Reviews, 5*, 101–161.

Koch, R., Burton, B., Hoganson, G., Peterson, R., Rhead, W., Rouse, B., et al. (2002). Phenylketonuria in adulthood: A collaborative study. *Journal of Inherited Metabolic Diseases, 25*, 333–346.

Koch, R., de la Cruz, F., & Azen, C. G. (Eds.), (2003). The Maternal Phenylketonuria Collaborative Study: New developments and the need for new strategies. *Pediatrics, 112*, 1513–1587.

Lenke, R. R., & Levy, H. L. (1980). Maternal phenylketonuria and hyperphenylalaninemia. *New England Journal of Medicine, 202*, 1202–1208.

Milborrow, S. (2011). Derived from mda:mars by Trevor Hastie and Rob Tibshirani. earth: Multivariate Adaptive Regression Spline Models. R package version 3.2-1. http://CRAN.R-project.org/package=earth

NIH (2000). *Phenylketonuria: Screening and Management.* NIH Consensus Statement (October 16–18, 2000), 17(3), 1–27. http://www.ncbi.nlm.nihttp://consensus.nih.gov/2000/2000Phenylketonuria113html.htmh.gov/books/bv.fcgi?rid=hstat4.chapter.20932

Schweitzer-Krantz, S., & Burgard, P. (2000). Survey of national guidelines for the treatment of phenylketonuria. *European Journal of Pediatrics, 159*(Suppl. 2), S70–S73.

Su, Y-S., Gelman, A., Hill, J., & Yajima, M. (2011). Multiple imputation with diagnostics (mi) in R: Opening windows into the black box. *Journal of Statistical Software, 45*, 1–31.

Widaman, K. F., & Azen, C. (2003). Relation of prenatal phenylalanine exposure to infant and childhood cognitive outcomes: Results from the International Maternal PKU Collaborative Study. *Pediatrics, 112*, 1537–1543.

Index

For Product Safety Concerns and Information please contact our EU
representative GPSR@taylorandfrancis.com Taylor & Francis Verlag GmbH,
Kaufingerstraße 24, 80331 München, Germany

Printed and bound by CPI Group (UK) Ltd, Croydon, CR0 4YY
01/05/2025
01858337-0001